D1689159

Edited by
Yang Gao

Contemporary Planetary Robotics

Edited by Yang Gao

Contemporary Planetary Robotics

An Approach Toward Autonomous Systems

WILEY-VCH
Verlag GmbH & Co. KGaA

Editor

Professor Yang Gao
University of Surrey
Surrey Space Centre
STAR Lab, Stag Hill
GU2 7XH, Guildford
United Kingdom

Cover
Courtesy NASA/JPL-Caltech

All books published by **Wiley-VCH** are carefully produced. Nevertheless, authors, editors, and publisher do not warrant the information contained in these books, including this book, to be free of errors. Readers are advised to keep in mind that statements, data, illustrations, procedural details or other items may inadvertently be inaccurate.

Library of Congress Card No.: applied for

British Library Cataloguing-in-Publication Data
A catalogue record for this book is available from the British Library.

Bibliographic information published by the Deutsche Nationalbibliothek
The Deutsche Nationalbibliothek lists this publication in the Deutsche Nationalbibliografie; detailed bibliographic data are available on the Internet at <http://dnb.d-nb.de>.

© 2016 Wiley-VCH Verlag GmbH & Co. KGaA, Boschstr. 12, 69469 Weinheim, Germany

All rights reserved (including those of translation into other languages). No part of this book may be reproduced in any form – by photoprinting, microfilm, or any other means – nor transmitted or translated into a machine language without written permission from the publishers. Registered names, trademarks, etc. used in this book, even when not specifically marked as such, are not to be considered unprotected by law.

Print ISBN: 978-3-527-41325-6
ePDF ISBN: 978-3-527-68494-6
ePub ISBN: 978-3-527-68495-3
Mobi ISBN: 978-3-527-68496-0
oBook ISBN: 978-3-527-68497-7

Cover Design Grafik-Design Schulz, Fußgönheim, Germany
Typesetting SPi Global, Chennai, India
Printing and Binding Markono Print Media Pte Ltd, Singapore

Printed on acid-free paper

Contents

List of Contributors *XIII*

1   **Introduction** *1*
Yang Gao, Elie Allouis, Peter Iles, Gerhard Paar, and
José de Gea Fernández
1.1   Evolution of Extraterrestrial Exploration and Robotics *1*
1.2   Planetary Robotics Overview *4*
1.3   Scope and Organization of the Book *6*
1.4   Acknowledgments *9*

2   **Planetary Robotic System Design** *11*
Elie Allouis and Yang Gao
2.1   Introduction *11*
2.2   A System Design Approach: From Mission Concept to Baseline Design *12*
2.2.1   Mission Scenario Definition *14*
2.2.2   Functional Analysis *14*
2.2.3   Requirements Definition and Review *15*
2.2.4   Design Drivers Identification *17*
2.2.5   Concept Evaluation and Trade-Off *17*
2.3   Mission Scenarios: Past, Current, and Future *19*
2.3.1   Lander Missions *19*
2.3.1.1   Luna Sample-Return Landers *20*
2.3.1.2   Viking Landers *21*
2.3.1.3   Mars Surveyor Lander Family and Successors *23*
2.3.1.4   Huygens Lander *25*
2.3.1.5   Beagle 2 Lander *28*
2.3.1.6   Philae Lander *29*
2.3.2   Rover Missions *30*
2.3.2.1   Lunokhod 1 and 2 Rovers *31*
2.3.2.2   Prop-M Rover *33*
2.3.2.3   Sojourner Rover *34*
2.3.2.4   Spirit and Opportunity Rovers *36*

2.3.2.5	Curiosity Rover	*39*
2.3.2.6	Chang'E 3 Rover	*42*
2.3.2.7	ExoMars Rover	*43*
2.3.2.8	Mars 2020 Rover	*46*
2.3.3	Future Mission Concepts	*46*
2.3.3.1	Toward New Business Models	*46*
2.3.3.2	Medium-Term Mission Concepts	*47*
2.3.3.3	Long-Term Mission Ideas	*47*
2.4	Environment-Driven Design Considerations	*47*
2.4.1	Gravity	*49*
2.4.2	Temperature	*49*
2.4.3	Atmosphere and Vacuum	*51*
2.4.4	Orbital Characteristics	*52*
2.4.4.1	Distance to the Sun	*53*
2.4.4.2	Length of Days	*53*
2.4.5	Surface Conditions	*54*
2.4.5.1	Rocks	*54*
2.4.5.2	Dusts	*55*
2.4.5.3	Liquid	*56*
2.4.6	Properties of Planetary Bodies and Moons	*56*
2.5	Systems Design Drivers and Trade-Offs	*56*
2.5.1	Mission-Driven System Design Drivers	*58*
2.5.1.1	Mass	*59*
2.5.1.2	Target Environment	*59*
2.5.1.3	Launch Environment	*59*
2.5.1.4	Surface Deployment	*60*
2.5.1.5	Surface Operations	*63*
2.5.2	System Design Trade-Offs: A Case Study	*63*
2.5.2.1	Mission Scenario Definition: MSR/SFR	*63*
2.5.2.2	SFR System Design Drivers	*64*
2.5.2.3	SFR Subsystem Design Drivers	*65*
2.5.2.4	SFR Design Evaluation	*68*
2.6	System Operation Options	*72*
2.6.1	Operation Sequence	*72*
2.6.2	Operational Autonomy	*75*
2.6.2.1	Autonomous Functions	*76*
2.6.2.2	Autonomy Levels: Teleoperation versus Onboard Autonomy	*77*
2.7	Subsystem Design Options	*79*
2.7.1	Power Subsystem	*79*
2.7.1.1	Power Generation	*80*
2.7.1.2	Power Storage	*84*
2.7.2	Thermal Subsystem	*90*
2.7.2.1	Sizing Warm/Cold Cases	*92*
2.7.2.2	Heat Provision	*92*

2.7.2.3	Heat Management (Transport and Dissipation)	94
2.7.2.4	Trade-Off Options	96
	References	96

3 Vision and Image Processing 105

Gerhard Paar, Robert G. Deen, Jan-Peter Muller, Nuno Silva, Peter Iles, Affan Shaukat, and Yang Gao

3.1	Introduction	105
3.2	Scope of Vision Processing	108
3.2.1	Onboard Requirements	110
3.2.2	Mapping by Vision Sensors: Stereo as Core	112
3.2.3	Physical Environment	113
3.3	Vision Sensors and Sensing	114
3.3.1	Passive Optical Vision Sensors	116
3.3.2	Active Vision Sensing Strategies	117
3.3.3	Dedicated Navigation Vision Sensors: Example Exomars	118
3.3.3.1	Navigation (Perception/Stereo Vision)	119
3.3.3.2	Visual Localization and Slippage Estimation	119
3.3.3.3	Absolute Localization	120
3.4	Vision Sensors Calibration	120
3.4.1	Geometric Calibration	121
3.4.2	Radiometric Calibration	124
3.4.3	The Influence of Errors	125
3.5	Ground-Based Vision Processing	128
3.5.1	Compression and Decompression	129
3.5.2	3D Mapping	129
3.5.3	Offline Localization	133
3.5.4	Visualization and Simulation	135
3.6	Onboard Vision Processing	138
3.6.1	Preprocessing	138
3.6.2	Compression Modes	138
3.6.3	Stereo Perception Software Chain	139
3.6.4	Visual Odometry	140
3.6.5	Autonomous Navigation	141
3.7	Past and Existing Mission Approaches	142
3.7.1	Lunar Vision: Landers and Rovers	142
3.7.2	Viking Vision System	143
3.7.3	Pathfinder Vision Processing	144
3.7.4	MER and MSL Ground Vision Processing Chain	145
3.7.5	ExoMars Onboard Vision-Based Control Chain	150
3.7.6	ExoMars Onboard Vision Testing and Verification	151
3.7.7	ExoMars PanCam Ground Processing	152
3.7.8	Additional Robotic Vision Systems	155
3.8	Advanced Concepts	157
3.8.1	Planetary Saliency Models	157

3.8.2	Vision-Based Rover Sinkage Detection for Soil Characterization *162*	
3.8.3	Science Autonomy *166*	
3.8.4	Sensor Fusion *166*	
3.8.5	Artificial Intelligence and Cybernetics Vision *169*	
	References *170*	

4 Surface Navigation *181*
Peter Iles, Matthias Winter, Nuno Silva, Abhinav Bajpai, Yang Gao, Jan-Peter Muller, and Frank Kirchner

4.1	Introduction *181*
4.2	Context *181*
4.2.1	Definitions *182*
4.2.2	Navigating on Extraterrestrial Worlds *182*
4.2.3	Navigation Systems on Current and Past Flight Rovers *183*
4.2.3.1	Lunokhod I and II *183*
4.2.3.2	Apollo Lunar Roving Vehicle *184*
4.2.3.3	Sojourner Microrover *184*
4.2.3.4	Mars Exploration Rovers *185*
4.2.3.5	Mars Science Laboratory/Curiosity *188*
4.2.3.6	Yutu/Jade Rabbit *190*
4.3	Designing a Navigation System *190*
4.3.1	Requirements *191*
4.3.1.1	Performance Requirements *191*
4.3.1.2	Environmental Requirements *192*
4.3.1.3	Resource Requirements *193*
4.3.1.4	Other Requirements *193*
4.3.2	Design Considerations *194*
4.3.2.1	Functional Components *194*
4.3.2.2	Sensors *194*
4.3.2.3	Software *196*
4.3.2.4	Computational Resources *198*
4.3.2.5	Rover Control Strategy *199*
4.4	Localization Technologies and Systems *201*
4.4.1	Orientation Estimation *201*
4.4.1.1	Sun Finding *202*
4.4.1.2	Star Trackers *203*
4.4.1.3	Inertial Measurement Units *203*
4.4.1.4	Vision Techniques *204*
4.4.1.5	Antenna Null-Signal Technique *204*
4.4.2	Relative Localization *205*
4.4.2.1	Wheel Odometry *205*
4.4.2.2	Visual Odometry *205*

4.4.2.3	Other Vision-based Techniques *208*
4.4.2.4	3D Visual Odometry *208*
4.4.2.5	Speed Sensing *208*
4.4.3	Absolute Localization *209*
4.4.3.1	Rover-to-Orbiter Imagery Matching *209*
4.4.3.2	Rover-to-Orbiter Horizon Matching *211*
4.4.3.3	Rover-to-Orbiter Digital Elevation Model Matching *211*
4.4.3.4	Orbiting Asset- or Earth-based Localization *212*
4.4.3.5	Fixed Assets/Beacons Localization *213*
4.4.3.6	Celestial Localization *214*
4.4.4	Combining Localization Sources *214*
4.4.4.1	Gaussian Filters *215*
4.4.4.2	Particle Filters *216*
4.4.4.3	Simultaneous Localization and Mapping *216*
4.4.5	Example Systems *218*
4.5	Autonomous Navigation *219*
4.5.1	Sensing *219*
4.5.2	Mapping *220*
4.5.3	Terrain Assessment *222*
4.5.4	Path Planning *223*
4.5.4.1	Local Path Planning and Obstacle Avoidance *224*
4.5.4.2	Global Path Planning *225*
4.5.5	Control *227*
4.6	Future of Planetary Surface Navigation *228*
4.6.1	Planned Flight Rovers *228*
4.6.1.1	ExoMars Rover *229*
4.6.1.2	Mars 2020 Rover *233*
4.6.2	Future Rover Missions *234*
4.6.2.1	Mars Precision Lander *234*
4.6.2.2	Resource Prospector Mission *234*
4.6.3	Field Trials as Proving Grounds for Future Navigation Technologies *235*
4.6.3.1	RESOLVE/Resource Prospector Mission (NASA/CSA) *236*
4.6.3.2	SEEKER and SAFER (ESA) *239*
4.6.3.3	Teleoperation Robotic Testbed (CSA) *240*
4.6.3.4	Other Field Trials *241*
4.6.4	Future Capabilities *242*
4.6.4.1	SLAM Systems *242*
4.6.4.2	Cooperative Robotics and New Mobility Concepts *242*
4.6.4.3	Enhanced Processing Capabilities *244*
4.6.4.4	New Sensors *244*
4.6.4.5	New Applications of Orbital Imagery *245*
	References *247*

5	**Manipulation and Control** *255*
	José de Gea Fernández, Elie Allouis, Karol Seweryn, Frank Kirchner, and Yang Gao
5.1	Introduction *255*
5.1.1	Review of Planetary Robotic Arms *255*
5.1.1.1	Mars Surveyor '98/'01 *256*
5.1.1.2	Phoenix *258*
5.1.1.3	MARS Exploration Rovers (MERs) *259*
5.1.1.4	Beagle 2 *260*
5.1.1.5	Mars Science Laboratory *261*
5.2	Robotic Arm System Design *263*
5.2.1	Specifications and Requirements *263*
5.2.1.1	Performance Requirements *264*
5.2.1.2	Design Specifications *264*
5.2.1.3	Environmental Design Considerations *267*
5.2.2	Design Trade-Offs *268*
5.2.2.1	Arm Kinematics *268*
5.2.2.2	Structure and Material *269*
5.2.2.3	Sensors *269*
5.3	Robotic Arm Control *270*
5.3.1	Low-Level Control Strategies *271*
5.3.1.1	Position Control *271*
5.3.1.2	Force Control *273*
5.3.1.3	Dynamic Control *278*
5.3.1.4	Visual Servoing *283*
5.3.2	Manipulator Trajectory Generation *284*
5.3.2.1	Trajectory Interpolation *285*
5.3.2.2	On-Line Trajectory Generation *286*
5.3.3	Collision Avoidance *287*
5.3.3.1	Self-Collision Avoidance *288*
5.3.4	High-Level Control Strategies *289*
5.3.4.1	Path Planning *289*
5.3.4.2	Telemanipulation *290*
5.3.4.3	Higher Autonomy (E2–E4) *293*
5.4	Testing and Validation *294*
5.4.1	Testing Strategies *295*
5.4.2	Scope of Testing Activities *297*
5.4.2.1	Kinematic Calibration *298*
5.4.2.2	Beyond Calibration *299*
5.4.3	Validation Methods *299*
5.4.3.1	Use of ABTs *301*
5.5	Future Trends *304*
5.5.1	Dual-Arm Manipulation *304*
5.5.2	Whole-Body Motion Control *306*
5.5.3	Mobile Manipulation *309*

5.5.3.1	Mobile Manipulators as Research Platforms	*309*
5.5.3.2	DARPA Robotics Challenge (DRC)	*310*
5.5.3.3	Mobile Manipulators for Space	*311*
	References	*313*

6 Mission Operations and Autonomy *321*

Yang Gao, Guy Burroughes, Jorge Ocón, Simone Fratini, Nicola Policella, and Alessandro Donati

6.1	Introduction	*321*
6.2	Context	*323*
6.2.1	Mission Operation Concepts	*323*
6.2.2	Mission Operation Procedures	*326*
6.2.3	Onboard Segment Operation Modes	*329*
6.3	Mission Operation Software	*330*
6.3.1	Design Considerations	*331*
6.3.2	Ground Operation Software	*332*
6.3.3	Onboard Operation Software	*336*
6.3.4	Performance Measures	*341*
6.4	Planning and Scheduling (P&S)	*343*
6.4.1	P&S Software Design Considerations	*343*
6.4.2	Basic Principles & Techniques	*343*
6.4.2.1	Classical Methods	*344*
6.4.2.2	Neoclassical Methods	*344*
6.4.2.3	Solving Strategies	*346*
6.4.2.4	Temporal Planning	*347*
6.4.2.5	Scheduling	*350*
6.4.2.6	Handling Uncertainties	*350*
6.4.2.7	Planning Languages	*351*
6.4.3	P&S Software Systems	*354*
6.4.4	P&S Software Development Frameworks	*357*
6.5	Reconfigurable Autonomy	*362*
6.5.1	Rationale	*362*
6.5.2	State-of-the-Art Methods	*363*
6.5.3	Taxonomy	*365*
6.5.4	Design Examples: Reconfigurable Rover GNC	*367*
6.5.4.1	Application Layer	*369*
6.5.4.2	Reconfiguration Layer	*370*
6.5.4.3	Housekeeping Layer	*372*
6.5.4.4	Ontology Design	*372*
6.5.4.5	Rational Agent Design	*374*
6.5.4.6	Impact on Mission Operations	*376*
6.6	Validation and Verification	*377*
6.6.1	Simulation Tools	*379*
6.6.2	Model Checking	*381*
6.6.3	Ontology-based System Models	*383*

6.7	Case Study: Mars Rovers' Goal-Oriented Autonomous Operation *384*	
6.7.1	Design Objectives *385*	
6.7.2	Onboard Software Architecture *385*	
6.7.3	Implementation and Validation *388*	
6.7.4	Integration with Ground Operation *392*	
6.7.5	Design Remarks *394*	
6.8	Future Trends *394*	
6.8.1	Autonomic Robotics *395*	
6.8.2	Common Robot Operating System *395*	
6.8.3	MultiAgent Systems *396*	
	References *396*	

Index *403*

List of Contributors

Elie Allouis
Airbus Defence and Space Ltd.
Future Programmes
Gunnels Wood Road
SG1 2AS Stevenage
United Kingdom

Abhinav Bajpai
University of Surrey
Surrey Space Centre
STAR Lab, Stag Hill
GU2 7XH Guildford
Surrey
United Kingdom

Guy Burroughes
University of Surrey
Surrey Space Centre
STAR Lab, Stag Hill
GU2 7XH Guildford
Surrey
United Kingdom

Robert G. Deen
California Institute of Technology
Instrument Software and Science Data Systems Section
Jet Propulsion Laboratory
4800 Oak Grove Drive
Pasadena
CA 91109
USA

Alessandro Donati
European Space Agency
ESOC, OPS-OSA
Robert-Bosch-Strasse 5
64293 Darmstadt
Germany

José de Gea Fernández
DFKI GmbH
Robotics Innovation Center
Robert-Hooke-Str. 1
D-28359 Bremen
Germany

Simone Fratini
European Space Agency
ESOC, OPS-OSA
Robert-Bosch-Strasse 5
64293 Darmstadt
Germany

Yang Gao
University of Surrey
Surrey Space Centre
STAR Lab, Stag Hill
GU2 7XH Guildford
Surrey
United Kingdom

Peter Iles
Neptec Design Group
Space Exploration
302 Legget Drive
Ottawa
ON K2K 1Y5
Canada

Frank Kirchner
DFKI GmbH
Robotics Innovation Center
Robert-Hooke-Str. 1
D-28359 Bremen
Germany

and

University of Bremen
Faculty of Mathematics and
Computer Science
Robert-Hooke-Str. 1
D-28359 Bremen
Germany

Jan-Peter Muller
Imaging Group
Mullard Space Science
Laboratory
UCL Department of Space &
Climate Physics
Holmbury St Mary
RH5 6NT Surrey
United Kingdom

Jorge Ocón
GMV Aerospace and Defense
Avionics On-board Software
Division
Space Segment Business Unit
Isaac Newton, 11 (PTM)
Tres Cantos
28760 Madrid
Spain

Gerhard Paar
JOANNEUM RESEARCH
Institute for Information and
Communication Technologies
Machine Vision Applications
Group
Steyrergasse 17
8010 Graz
Austria

Nicola Policella
European Space Agency
ESOC, OPS-OSA
Robert-Bosch-Strasse 5
64293 Darmstadt
Germany

Karol Seweryn
Space Research Centre of the
Polish Academy of Sciences
(CBK PAN)
18a Bartycka str.
00–716 Warsaw
Poland

Affan Shaukat
University of Surrey
Surrey Space Centre
Department of Electrical &
Electronic Engineering
GU2 7XH Guildford
United Kingdom

Nuno Silva
Airbus Defence and Space Ltd.
Department of AOCS/GNC and
Flight Dynamics
Gunnels Wood Road
Stevenage
SG1 2AS Hertfordshire
United Kingdom

Matthias Winter
Airbus Defence and Space Ltd.
Department of AOCS/GNC and
Flight Dynamics
Gunnels Wood Road
Stevenage
SG1 2AS Hertfordshire
United Kingdom

1
Introduction

Yang Gao, Elie Allouis, Peter Iles, Gerhard Paar, and José de Gea Fernández

Planetary robotics is an emerging multidisciplinary field that builds on knowledge of astronautics, terrestrial robotics, computer science, and engineering. This book offers a comprehensive introduction to major research and development efforts for planetary robotics, with a particular focus on autonomous space systems, which will enable cost-effective, high-performing, planetary missions. Topics covered in this book include techniques and technologies enabling planetary robotic vision processing, surface navigation, manipulation, mission operations, and autonomy. Each topic or technological area is explained in a dedicated chapter using a typical space system design approach whereby design considerations and requirements are first discussed and followed by descriptions of relevant techniques and principles. Most chapters contain design examples or use cases that help demonstrate how techniques or theoretical principles can be implemented in real missions. Since any space engineering design or development is a system engineering process, this book also dedicates one chapter to planetary robotic system design – from mission concepts to baseline designs. As a result, this book can be used as a text or reference book for relevant engineering or science courses at the undergraduate and postgraduate level, or a handbook for industrial professionals in the space sector.

This chapter introduces the book by offering a chronicle on how planetary exploration and robotics have evolved to date, a systematic overview of planetary robotics, as well as an explanation on the organization and scope of the book.

1.1
Evolution of Extraterrestrial Exploration and Robotics

The need for humans to explore beyond the realm of Earth is driven by our inherent curiosity. Throughout our history, new worlds have been discovered by daring explorers who set out to discover new lands, find riches, or better understand these little-known territories. These journeys were fueled by the technological advances of the times such as the compass, maritime maps, or plane, and in return contributed tremendously to the scientific knowledge of humankind. For all the good provided by these exploratory endeavors, history also reveals that *exploration* is difficult, perilous, and can be fraught with unforeseeable consequences. For examples, within early maritime exploration, only a fraction of all the ships that aimed for the new worlds eventually achieved their goals. There have also been countless instances where the discovery of the new lands was detrimental to the indigenous populations. The past and lessons learned serve as a stark reminder to all new exploration endeavors.

Outer space has provided real, new exploration frontiers for mankind since the 1950s. With the capability and the irresistible attraction to go beyond our planet Earth, minimizing the impact of mankind on other extraterrestrial bodies (be it a planet, a moon, a comet, or an asteroid) is paramount. Strong with the hindsight and knowledge provided by humans' own history, we are continuously learning about these new space frontiers and taking precautions to avoid repeating mistakes learned from the past exploration activities.

The onset of space exploration in the late 1950s to early 1960s focused on sending humans into space and the Moon, a key priority for the two main adversaries of the Cold War. However, it was true then as it is now, in parallel to the expensive development of manned space programs, the use of cheaper robotic proxies was deemed important for understanding the space environment where the astronauts will be operating. The USSR had the first set of robotics missions, successfully launching a series of Luna probes starting from 1959. Within a year, the Luna 1 managed a flyby of the Moon, Luna 2 crash landed on the Moon, and Luna 3 took pictures of the Moon's far side. It took another 7 years before both the USSR and the United States, within a few months from each other, performed soft landing on the Moon with their respective probes, Luna 9 and Surveyor 1. These missions paved the way for the first human landing on the Moon in 1969 by the United States. Building on these earlier successes, robotic exploration missions have extended their reach to Mercury, Venus, Mars (known as the *inner solar system*), and subsequently the *outer solar system* where tantalizing glimpses of the volcanic Io, the frozen Europa, or the methane rains of Titan have been obtained.

Planetary missions can use various ways to explore an extraterrestrial body, often starting with reconnaissance or remote sensing using orbiting satellites. More advanced approaches (such as landing, surface operation, and sample return) enabled by sophisticated robotic systems represent a giant leap in terms of mission complexity and risk, but more importantly scientific return. Not surprisingly, advanced extraterrestrial exploration is littered with unsuccessful missions bearing witness to serious technical challenges of such endeavors. Table 1.1 presents statistics of successful surface missions aimed for the solar system (excluding manned missions). The relatively low success rate is a clear reflection on the technical difficulties involved in designing, building, and operating the required robotic spacecraft. It is worth noting that space engineers and scientists have created the landscape of what we know today. With sheer determination, they continue to address countless challenges, failing often, but regrouping until they succeed.

Within the existing successful unmanned missions, various types of robotic systems have played significant roles, including **robotic platforms** (such as the

1.1 Evolution of Extraterrestrial Exploration and Robotics

Table 1.1 Statistics on planetary unmanned landing missions as of 2015.

	Venus	Moon	Mars	Titan	Asteroids/comets
Total landing missions launched	19	35	16	1	6
Successful surface operation	9	13	9	1	1
Successful sample return	0	3	0	0	1

Table 1.2 Successfully flown robots on Mars, the Moon, and small bodies as of 2015.

Mission	Country	Target	Rover	Arm	Sampler	Drill
Surveyor 3	United States	Moon			×	
Luna 16/20/24	USSR	Moon		×	×	×
Luna 17/21	USSR	Moon	×			
Viking	United States	Mars		×	×	
Mars Path Finder	United States	Mars	×			
Hayabusa (or Muses-C)	Japan	Asteroid			×	
Mars Exploration Rovers	United States	Mars	×	×	×	
Phoenix	United States	Mars		×	×	
Mars Science Laboratory	United States	Mars	×	×	×	
Chang'E 3	China	Moon	×			
Rosetta	Europe	Comet		×	×	×

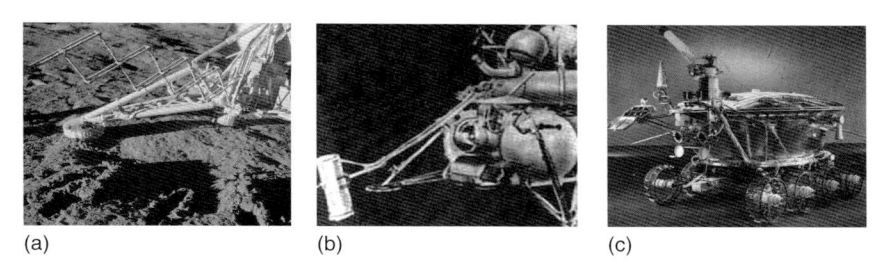

Figure 1.1 First successfully flown planetary robotic systems. (a) Surveyor 3 scoop, (b) Luna 16 arm-mounted drill, (c) Luna 17 rover (Lunokhod 1). (Credits NASA, Lavochkin Association).

surface rovers) or **robotic payloads** (such as the manipulators or robotic arms, subsurface samplers, and drills). Table 1.2 summarizes those successfully flown robots found on the Moon, Mars, and small bodies. The first genuine robotic payload successfully operated on an extraterrestrial body was a scoop (i.e., a manipulation cum sampling device) onboard the Surveyor 3 lander launched in 1967 to the Moon (as shown in Figure 1.1a). Following that, Luna 16 succeeded with the first planetary robotic arm-mounted drill in 1970 (as shown in Figure 1.1b), and Luna 17 succeeded with the first planetary rover called Lunokhod 1 in 1970 (as shown in Figure 1.1c).

There is no denying that these "firsts" led to incredible mission successes and science discoveries as a result of unabated and relentless launch attempts during the *space race* between the superpowers. Building on these foundations, the new generation of planetary exploration has since the 1990s not only traveled further into the solar system but also studied deeper fundamental scientific questions. The desire to go and explore is as strong as ever. Past space powers have been gradually joined by a flurry of new nations eager to test and demonstrate their technologies and contribute to an increasing body of knowledge. Commercial endeavors also have eyes on space and actively promote the Moon and Mars as possible destinations for long-term human presence or habitation. Shall the future exploration missions be manned or unmanned, planetary robots are always desired to deliver the robotic "avatars" and perform *in situ* tasks to proxy or assist through their "eyes," "ears," and "hands."

1.2 Planetary Robotics Overview

A typical robot on Earth is an unmanned electromechanical machine controlled by a set of automatic or semi-autonomous functions. Industrial standard robots are typically used to address the "3D" activities: Dirty, Dull, and Dangerous. This notion was created in reference to the Japanese concept of "3K" (kitanai, kiken, and kitsui) describing the major areas where the robots should be used to effectively relieve human workers from working environments such as with the construction industry. Therefore, robotic systems are envisaged to work on repetitive, long-duration or high-precision operations in the environment where humans are expected to perform poorly or where it is impractical for human presence.

Robotics, as an engineering or scientific subject, emerges from a number of traditional disciplines such as electronics, mechanics, control, and software, as illustrated in Figure 1.2. Designing a robotic system, therefore, involves the design of hardware subsystems (e.g., sensors, electronics, mechanisms, and materials) and software subsystems (such as perception, control, and autonomy). A planetary robotic system is functionally similar to a terrestrial robotic system, with different performance characteristics to cope with things such as stringent space mission requirements (often in aspects such as radiation hardness to survive the space environment), scarce power and computational resources, and high autonomy demanded due to communication latency.

A robotic system is not required to possess a fixed level of automation or autonomy. In fact, it can employ a wide range of control modes from remote or teleoperation, semi-autonomous to fully autonomous operation as appropriate to its mission goal, location, and operational constraints. Fundamental differences between the *automatic* and the *autonomous* control mode relates to the level of judgment or self-direction in the action performed. An autonomic response or control is associated to a reflex, an involuntary behavior that is "hard-wired" into the robot with no decision making involved. An autonomous behavior on the

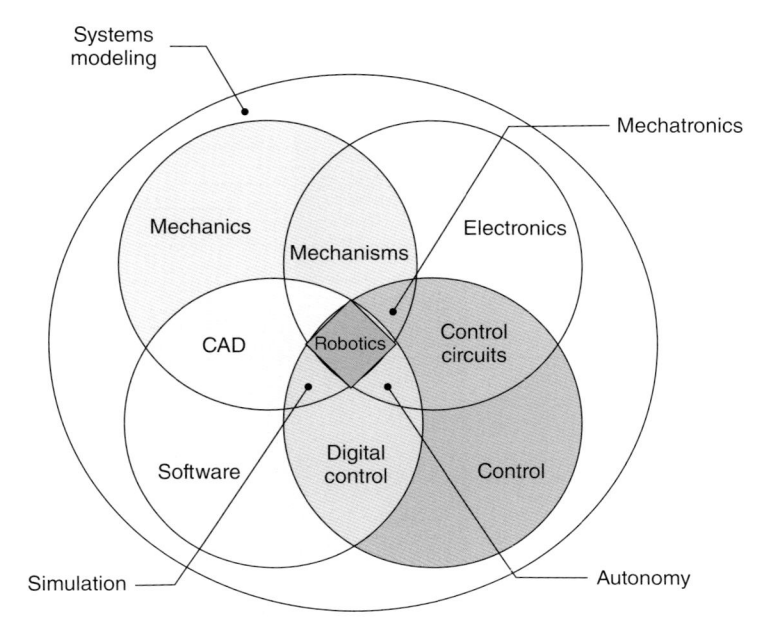

Figure 1.2 Robotics: a multidisciplinary subject.

other hand represents a complex independent response not controlled by others (meaning there is decision making involved). An analogy can be drawn from the nature that evolved from the monocellular organisms behaving similar to an automatic system reacting to external stimuli to complex organisms such as mammals and birds that exhibit significantly more advanced independent logical behaviors.

According to the European Cooperation for Space Standardization (ECSS), a spacecraft (or a planetary robotic system in this case) is standardized to work on four different levels of autonomy or control modes as follows:

- Level E1: execution mainly under real-time ground control, that is, remote or teleoperation;
- Level E2: execution of preplanned mission operations onboard, that is, automatic operation;
- Level E3: execution of adaptive mission operations onboard, that is, semi-autonomous operation;
- Level E4: execution of goal-oriented mission operations onboard, that is, fully autonomous operation.

Planetary robots can also be classified into three groups depending on their capabilities of achieving different ECSS levels of autonomy:

- **Robotic agents** that act as human proxies in space to perform exploration, assembly, maintenance, and production tasks in the level E1–E3 operations.

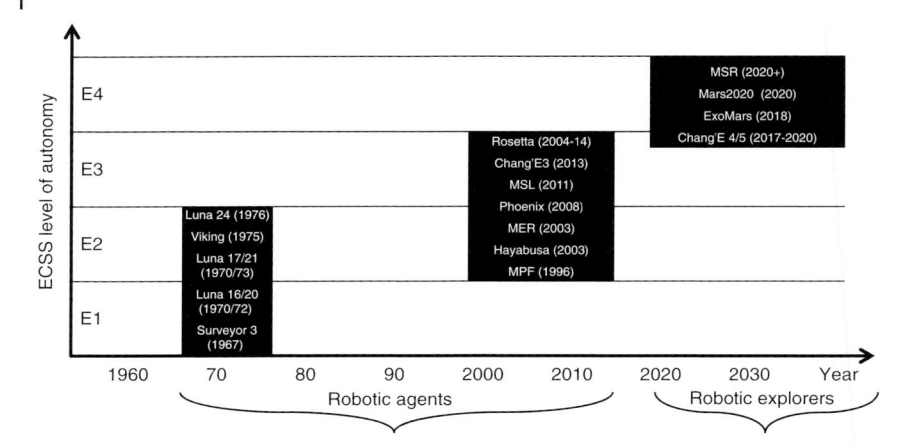

Figure 1.3 ECSS defined Level of autonomy for existing and planned planetary robotic systems.

- **Robotic assistants** that help human astronauts to perform tasks quickly and safely, with higher quality and cost efficiency using the level E3 or potentially E4 operation.
- **Robotic explorers** that explore the extraterrestrial targets using the level E4 operation.

Figure 1.3 presents the timeline of existing and foreseen planetary robotic systems with respect to the ECSS levels E1 – E4. Existing, successfully flown planetary robots are all within the robotic agent category. It is evident that as time proceeds, modern planetary missions with increasingly challenging goals require increased level of autonomy within the robotics systems, hence a shift from robotic agents to robotic explorers.

1.3
Scope and Organization of the Book

The book focuses on R&D topics that directly influence the onboard software and control capabilities of a planetary robot in achieving greater level of autonomy. It does not aim to cover design issues of any hardware subsystems such as sensors, mechanisms, electronics, or materials. However, discussions on hardware-related issues can be provided whereby they influence the design of required functions, or to provide a wider context of the robotic system design. The rest of the book is organized in such a way that each chapter focuses on a specific technical topic so that it can be read or used without much dependency on other chapters. At the same time, the technical chapters are cross-referenced among each other when there are crossovers between their subjects to help the readers establish an understanding of the system engineering philosophy that is fundamental to any space system design and development.

The design of a planetary robotic mission is complex. Past and current missions have shown how the endeavor can be treacherous and fraught with design, implementation and operational challenges. From the impact of the environment, the management of resources, to the operational concept, the system design and development must be approached as a whole rather than the sum of discrete elements. Chapter 2 conveys that a top-level view of mission-driven considerations should be established at the early stage of the robotic system design assessment, and hence introduces the space system design methodology and tools. Following the introduction Section 2.1, Section 2.2 presents a system engineering approach required to design the planetary robotic systems. Section 2.3 introduces a range of planetary robotic systems used as part of past and present exploration missions, as well as how they can address specific mission challenges and contribute to the return of the valuable science data. This section also looks ahead at future robotic systems that are currently being investigated to implement more adventurous mission concepts and operational scenarios. Section 2.4 reviews a range of planetary environmental factors that are driven by mission targets and at the same time drive designs of various robotic systems and subsystems. Section 2.5 demonstrates using a case study how to define system-level design drivers and perform subsystem design trade-offs. Finally in this chapter, Sections 2.6 and 2.7 provides insightful design options for key system operations and subsystems that have major influence to the overall system design.

A planetary robot is expected to interact with the environment and other assets of a mission, and perceive information. Similar to humans, visual sensing to the robot is the most effective and powerful way for collecting information in an unknown environment and situation. Chapter 3 addresses the vision aspects as being a prerequisite for navigation, autonomy, manipulation, and scientific decision-making. The introduction Sections 3.1 and 3.2 presents the scope, aims, terms, and most important requirements as well as constraints for robotic vision in the planetary context. Vision sensors and sensing are addressed in Section 3.3 including representative examples. Section 3.4 describes the radiometric and geometric sensor calibration that is the key to objective and meaningful sensor data interpretation and exploitation with important error influences listed as well. The following sections cover the complementary approaches of ground-based vision processing (Section 3.5) and onboard vision processing (Section 3.6) offering complementary material to the surface navigation and localization in Chapter 4. Section 3.7 presents past and present mission approaches exploiting robotic vision techniques and highlights the vision processing mechanisms used in Mars missions MER, MSL, and ExoMars. The chapter closes with a set of advanced concepts in Section 3.8.

Planetary surface navigation is among the key technologies in any robotic exploration missions, particularly involving mobile robotic platforms such as the rovers. Navigation technologies allow the rover (and hence the ground operators) to know where the robot is, where the robot should go next, and to guide the robot along a selected path. In the presence of obstacles, the navigation system enables safe and efficient exploration of its environment. Chapter 4 investigates all aspects of the rover navigation system. Following the introduction Section 4.1, Section 4.2 presents challenges of navigating on different extraterrestrial bodies and describes relevant flight rover systems including the Apollo LRV, the Russian Lunokhods, the Mars Exploration Rovers, and Curiosity. Section 4.3 presents the navigation system design process through a discussion of requirements and major design concepts. A thorough description of localization technologies is given in Section 4.4, including orientation estimation, relative localization, absolute localization, and fusion of localization sources. This is followed by Section 4.5 with a discussion of all the steps necessary to achieve autonomous navigation, from sensing to control. Finally, in Section 4.6 of the chapter, the prospect of planetary robotic navigation is presented, with a review of planned flight rovers, missions, and as enabling future technologies.

As evident from existing planetary robotic missions, robotic manipulators have played an important role, such as serving scientific experiments by grabbing samples or delivering the drills to access rocks or soil. The first part of Chapter 5 reviews past manipulators and their technical characteristics (see Section 5.1). Section 5.2 provides an overview of design criteria, specifications, and requirements for constructing a planetary manipulator, a lot of which have synergies to constructing a rover. Section 5.3 discusses control algorithms, from the low-level control of an actuator to high-level motion planning for the arm including trajectory generation, teleoperation, and possible autonomous mode. Section 5.4 further discusses testing and validation procedures for a planetary robotic arm system. Future planetary robots are envisaged to possess not only sophisticated manipulation skills and the ability to reuse these skills for different tasks but also a high level of autonomy to cope with complex mission scenarios (such as building lunar outpost). Hence, the last section (Section 5.5) of the chapter investigates various novel capabilities of planetary manipulators in the long term, for example, the use of two arms, the use of whole-body control algorithms, which considers the mobile platform as part of the manipulation system, or the ability to act in dynamically changing environments.

There is no doubt that future planetary robotic missions aim for high operational autonomy and improved onboard software capabilities. Chapter 6 offers a systematic, thorough discussion on mission operations and autonomy. Section 6.1 introduces the background and Section 6.2 sets the context of the topic by introducing the basic concepts of mission operations, processes and procedures, and typical operation modes of planetary robotic systems. Section 6.3 discusses the first step in developing the mission operation software, that is, how to establish the software architecture (both onboard and on ground) for a given mission operation. The following three sections investigate the main design aspects or core technologies in mission operations: Section 6.4 discusses the planning and scheduling (P&S) techniques and representative design solutions that can enable high level of autonomy; Section 6.5 presents the technology that allows reconfiguration of autonomous software within mission operation; and Section 6.6 covers various tools and techniques for validation and verification of autonomous software. To demonstrate the practicality of the theoretical principles, Section 6.7 presents a design example of mission operation software for Mars rovers. The last Section 6.8 of the chapter outlines some over-the-horizon R&D ideas in achieving autonomous operations and systems for future planetary robotic missions.

1.4
Acknowledgments

The editor and coauthors of the book thank Wiley-VCH for publishing this work and the editorial team for their support.

Some parts of the book are based on results, experience, and knowledge gained from past funded R&D activities. The following funded projects should be acknowledged:

- FASTER funded by European Community's Seventh Framework Programme (EU FP7/2007–2013) under grant agreement no. 284419.
- PRoViDE funded by European Community's Seventh Framework Programme (EU FP7/2007–2013) under grant agreement no. 312377.
- ExoMars mission development funded by European Space Agency (ESA) and European national space agencies.
- GOAC funded by European Space Agency Technology Research Programme (ESA-TRP) under ESTEC contract no. 22361/09/NL/RA.
- RoboSat funded by UK Royal Academy of Engineering (RAEng) Newton Research Collaboration Programme under grant agreement NRCP/1415/89.
- Reconfigurable Autonomy funded by UK Engineering & Physical Sciences Research Council (EPSRC) under grant agreement EP/J011916/1.
- Consolidated grant to MSSL funded by UK Science and Technology Facilities Council (STFC) under grant agreement ST/K000977/1.
- BesMan funded by German Federal Ministry of Economics and Technology under grant agreement FKZ 50 RA 1216 and FKZ 50 RA 1217.

The editor and coauthors also thank the following organizations or colleagues for their support to this book project: Neptec Design Group, Airbus Defence and Space Future Programme and ExoMars GNC Team, Surrey Space Centre STAR Lab, JR 3D Vision Team led by Arnold Bauer, GOAC team, including Antonio Ceballos, Michel Van Winnendael, Kanna Rajan, Amedeo Cesta, Saddek Bensalem, and Konstantinos Kapellos. The book is also in memory of Dave Barnes, a longtime contributor to the planetary robotics R&D community in Europe.

Last but surely not the least, many loved ones of the coauthors are greatly appreciated for their unreserved support when this book project was carried out mostly during everyone's spare time, to name a few Yang's mom and dad, Peter's wife Tracey and mom, Elie's wife Anneso, and others.

2
Planetary Robotic System Design

Elie Allouis and Yang Gao

2.1
Introduction

At the inception of a new mission concept, the system design of a spacecraft (planetary robots included) is a critical phase that should not be overlooked. Galvanized by the prospects of an exciting concept, it is often tempting to delve too quickly into subsystem design, at the peril of the mission and its development team. Identification of key links, interactions, and ramifications of design decisions are crucial to the feasibility of the concept and success of the mission. Only then can the optimized system design be found, which may not be the sum of the optimized subsystems.

Building on a comprehensive review of existing and future planetary robotic missions, this chapter uses the system engineering design philosophy and discusses the mission-driven design considerations, the system-level design drivers as well as the subsystem design trade-offs for a robotic system, whether the mission is set to explore the lunar craters, the Martian landscape, or beyond. It demonstrates the design thought process starting from the mission concept up to the baseline design at the system and subsystem levels. As a result, the chapter offers a number of system design tools as the foundation or integrator for technologies discussed in subsequent chapters.

The chapter is structured systematically as follows:

- **Section** 2.2: This section describes in detail the system design approach and implementation steps applicable to planetary robotic missions and required robotic systems. The process starts from defining the mission scenario, which provides inputs for system-level functional analysis and determination of functional objectives for the robot(s). This then allows the progression to the next phase of the system definition by specifying and reviewing design requirements (e.g., using the S.M.A.R.T. method). Design drivers are subsequently identified and used to evaluate and trade-off different design choices, which results in the baseline design.

Contemporary Planetary Robotics: An Approach Toward Autonomous Systems, First Edition.
Edited by Yang Gao.
© 2016 Wiley-VCH Verlag GmbH & Co. KGaA. Published 2016 by Wiley-VCH Verlag GmbH & Co. KGaA.

- **Section 2.3:** Having introduced the system design philosophy and general tools, this section presents the state-of-the-art design and development examples of primary robotic platforms demonstrated through existing and future exploration missions. This provides a comprehensive overview of key robotic systems, subsystems, and their performance. The section provides extensive real-world examples of relevant robotic concepts, designs, and technologies.
- **Section** 2.4: This section identifies the space environmental factors that should be considered by any planetary robotic system design for a given mission, namely gravity, temperature, atmosphere/vacuum, orbital characteristics, and surface conditions, and so on. Each factor is investigated thoroughly in terms of its impact on different system- or subsystem-level designs for the planetary robot(s). It also presents the properties of various popular extraterrestrial targets showing rationales behind the resulting differences in robotic system design for different targets.
- **Section** 2.5: This section describes the design drivers based on the mission targets and objectives. A case study on Sample-Fetching Rover (SFR) for a Mars sample-return (MSR) mission concept is used to demonstrate how to implement system and subsystem design drivers, and how to perform trade-off analysis and design evaluation given the design drivers and options. It also presents the design tools that systematically capture the requirements (i.e., ripple graph) and evaluate design options (i.e., H.E.A.R.D.).
- **Section** 2.6: This section reviews the robotic system operation sequence and design options that allow the robot to vary its level of autonomy between teleoperation and full autonomy. It identifies that the autonomy functions and levels of the robot are interrelated and/or to be defined driven by design requirements and trade-off constraints.
- **Section** 2.7: This section presents design options for a couple of selected subsystems for planetary robots (i.e., power and thermal). These two are chosen because they are crucial subsystems, which also drive the entire system design of planetary robots. In addition, they cover complementary design aspects of planetary robotic systems to those addressed by the subsequent chapters in the book. The section also presents a wide range of design options when each subsystem is discussed, covering both the state-of-the-art and future technologies.

2.2
A System Design Approach: From Mission Concept to Baseline Design

Similar to any space system, the design of a planetary robot is an iterative process starting from a mission concept to a consolidated design. This chapter discusses the system engineering approach where the readers are guided through the typical system definition activities where background information is provided to understand the state of the art and to critically review and understand the design decisions that have shaped past and current planetary missions. This process is composed of a number of important steps that should be investigated sequentially in order to avoid being drawn too quickly from a concept to a technological solution, otherwise such pitfalls can significantly constrain the implementation of the target system and lead to costly changes at a later stage in the mission development. The system design approach, as shown in Figure 2.1, therefore concentrates on the preliminary specification and definition of the target system, and the identification of a wide range of possible implementations options. Then, these options can be critically compared against their merits (e.g., performance, cost, and complexity) before being downselected as a baseline design. Hence, the definition of the initial problem is critical to ensure the final system fulfills its original purpose. In the remainder of this section, an example mission concept is used to illustrate the key steps shown in Figure 2.1 and how the various stages of the design process provide critical data or inputs required by the following step of the process.

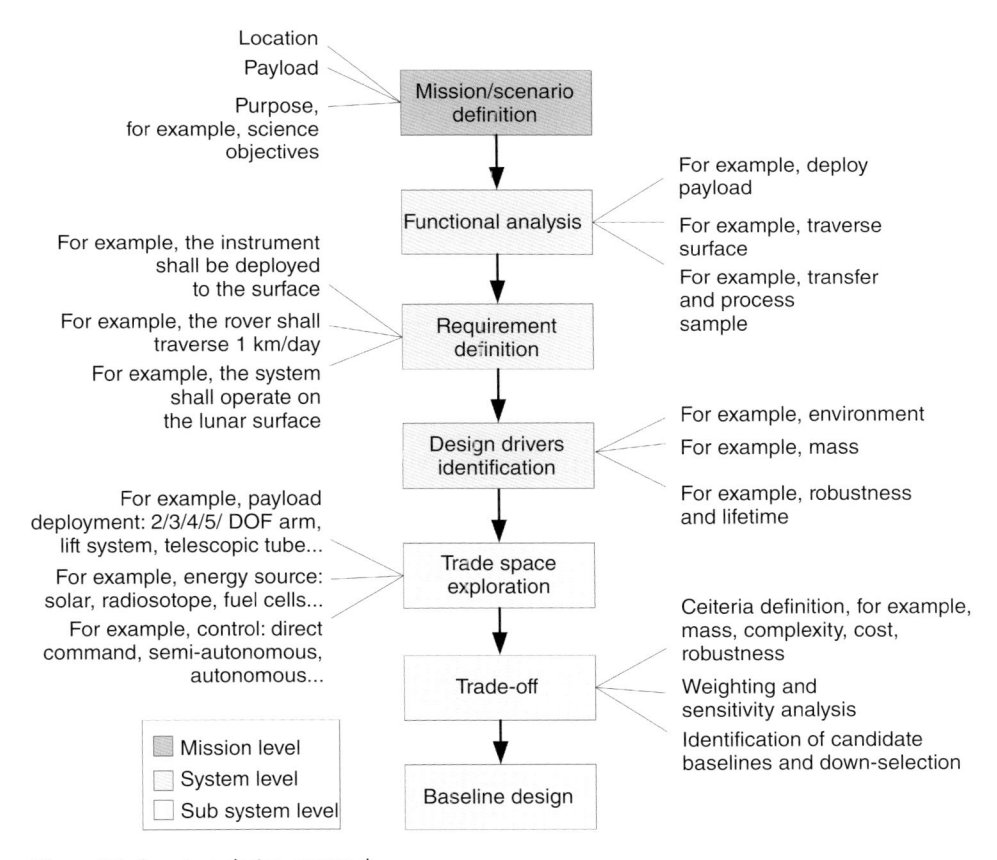

Figure 2.1 A system design approach.

2.2.1
Mission Scenario Definition

A new mission concept starts with a number of key criteria that the mission is expected to achieve. Mission concepts are more often than not either led by a science team pitching the concept to a space agency, or reciprocally initiated by the space agency itself involving a mission definition team to flesh out a mission concept. To date, most planetary exploration missions have been primarily science driven to answer some fundamental questions about the target bodies that the robotic systems are used to explore (e.g., find traces of life on Mars, or understand and characterize the presence of water in the shadowed craters on the Moon). In the future, the robotic systems may be used to support human settlements on the Moon and beyond, in which case the mission scenario is to concentrate on addressing needs beyond pure science such as the need to build the necessary infrastructures prior to the arrival of Humans (e.g., habitat, local production of oxygen or water from *in situ* resources utilization).

Outputs – Mission objective, science objectives.
Example – A challenging mission to Mars is proposed to explore caves on the red planet. It is anticipated that a number of candidate locations have been identified and that a mobile platform is required to investigate whether traces of water ice can be found as well as possible signs of past or present life. The platform will carry a suite of instruments to characterize the environment through optical systems, contact and noncontact sensing (e.g., drill and spectrometers).

2.2.2
Functional Analysis

A functional analysis is required at a number of design stages. At the system level, it captures and reformulates the mission objectives in terms of functions that the overall system must perform to help with the requirement definition. It is important to capture the functions, not the implementation of the function at this stage. These can be expressed diagrammatically or through short statements.

Outputs – Identification of the system-level functionalities and operations.
Example – A mobile platform will be deployed on the surface of Mars. The mobile platform will carry a suite of instruments. The mobile platform will access the Martian caves. The platform will deploy a suite of instruments in the cave to characterize soils and rocks. The platform will communicate directly to an orbiter. The platform will not rely on radioisotopic power systems. The platform will be compliant with planetary Protection Level X. In a typical industrial setup, the Mission Definition and the System-Level Functional Analysis tend to be both performed by the customer (e.g., a space agency) at the inception of the project. These can also be originated by a working group fleshing out mission concepts to address a specific science need.

2.2.3
Requirements Definition and Review

Building upon the functional analysis, this stage formalizes the functions into more formal requirements that define the system or represent the expected outcome of the project. These requirements capture functional and nonfunctional aspects to frame the definition of the system. Functional requirements include the features of the system, its behaviors, its capabilities, and the conditions that must exist for its operation. Nonfunctional requirements capture the environmental conditions under which the system must remain effective or describe the performance or quality of service of the solution. This process is critical as the set of requirements bound the design of the system and flow down to every subsystem and operation. Given a great number of people are often involved in a typical space mission project from inception to operation, it is, therefore, important to define these requirements in such a way that they do not initially overconstrain the design or do not prescribe a specific technological solution too early. In addition, they must provide an unambiguous way of framing the design of the system despite the different perspectives of the stakeholders. As such, a requirement for a mobile platform could be written as "The platform shall have six wheels." This particular wording is not advised as it constrains already the type of locomotion method (i.e., wheels and the number of them). It is possible that a specific design heritage is anticipated as part of the project, leading to the need to reuse a specific solution. Nevertheless, the requirement can be made more general such as "The platform shall provide the necessary mobility to access and return from the cave environment." This wording keeps options open and allows a subsequent trade-off study to investigate how the mobility could be achieved in the target environment, for example, four or six wheels, or maybe legs? To help define good and useful requirements, Doran [1] proposed the S.M.A.R.T. method, a mnemonic acronym that consists of five criteria to set objectives and requirements as detailed in Table 2.1. Pending on the context and the application, a number of alternative descriptions for each of the S.M.A.R.T. criteria can be found. For the purpose of defining technical requirements, definitions for the criteria draw upon further work in Refs [2, 3] are also summarized in Table 2.1.

Outputs – Set of functional and nonfunctional requirements.
Example – For the rover concept discussed earlier, the set of requirements could include:
- FR-10 – The rover shall traverse 1 km/sol.
- FR-20 – The rover shall be capable of accessing caves with terrain compatible with ENV-20.
- ENV-10 – The rover shall be capable of operating between −50 and 40 °C.
- POW-10 – The rover shall only use photovoltaic for energy generation.
- ROV-40 – The mass of the deployed rover shall be less than 250 kg.

Table 2.1 Defining suitable requirements through S.M.A.R.T. criteria.

ID	Meaning	Definition
S	Specific	*Concise and Complete* – A requirement should contain the minimum amount of detail necessary and be consistent with the terminology that has been used throughout the specification to describe the same system or concept. Nonsingular requirements should be avoided, for example, "the rover shall traverse 1 km and perform its science operations." These are two activities that should have their own requirement
M	Measurable	*Testable and unambiguous* – A requirement must ultimately be verified to make sure the system fulfills the requirement. It is, therefore, necessary to quantify as much as possible what the success criteria is. For example, the requirement "The rover shall traverse several kilometers" is not quantifiable and, therefore, ambiguous. It could be reworded as "The rover shall traverse at least 2 km." This measurable requirement can be tested and validated at a later stage in the project and will be used to size the various subsystems during the definition phase
A	Achievable	*Realistic and attainable* – An achievable requirement must be physically possible for the system to fulfill this goal. For example, the requirement "The rover shall have a reliability of 100%" is, despite the best efforts of the design team, unlikely to be achievable. Trying to achieve such a goal would lead to a prohibitive development cost and would ultimately be unlikely to succeed. The requirement could be captured as "The rover shall be single-point failure tolerant" to captures the need of being able to perform its mission despite a failure in the system
R	Relevant	*Necessary and Feasible* – To keep the number of necessary requirements to a minimum, the requirement must demonstrate that it has an effect on the desired goal. In other words, is this requirement needed in order for the solution to function properly? Alternatively, is this requirement achievable given what is known of the project (e.g., resources, budget, and timeframe)? For example, "The rover shall implement a warp drive to explore the Martian surface." Despite being set and written as a requirement, it is nevertheless impossible to implement within the scope of the project
T	Traceable	*Identifiable and Linked* – Requirements traceability provides the ability to trace a requirement (forward and backward) through all levels of specification, design, implementation, and test. It is critical to understand how top-level requirements are implemented at low level, to verify that each requirement has been implemented and to ensure that any modification has been implemented consistently and completely. Traceability is achieved through a numbering hierarchy to provide a unique identifier for each requirements. For example, a top-level functional requirement "FR-10 – The rover shall traverse 1 km/ sol" could subsequently lead to a locomotion subsystem (LSS) requirement "LOC-30 – The locomotion subsystem shall be capable of traversing 1 km/sol in the terrain defined in ENV-10," where ENV-10 would capture the key parameters of the terrain conditions

2.2.4
Design Drivers Identification

With the set of requirements at hand, identification of the key design drivers is critical to obtain a complete overview of the system and identify the major aspects of the functionalities, operation, and environment that fundamentally drive the design of the robotic system. Through this exercise, it is possible to establish the required understanding of the interdependencies among the various systems and subsystems. While mass and energy are typical drivers for planetary robotic systems, each mission normally presents specific challenges that require a dedicated assessment to deliver specific solutions appropriate to the mission objectives and the systems involved. By building this intimate system understanding, optimization can emerge based on the relations between the subsystems, some of which are presented in Section 2.5.

Outputs – Identification of the critical system drivers and the relations between subsystems.

Example – A typical rover design is limited by mass and energy. The more solar arrays that the rover embarks to generate more energy, the more mass the rover needs to carry, requiring the locomotion subsystem to work harder and leading to an additional energy draw.

2.2.5
Concept Evaluation and Trade-Off

Once the key drivers have been identified, this stage provides the first opportunity to assess the design workspace, looking both at architectural or system-level options and technological solutions for each of the subsystems. It starts with the identification of candidate solutions that fulfill a requirement. These solutions are then thoroughly evaluated and analyzed to mature the concepts to a level consistent across the solutions identified. Once a set of options has been consolidated, a trade-off can be performed. It provides a rigorous assessment of the pros and cons of each solution based on a set of predetermined and measurable criteria (e.g., mass, complexity, and maturity) and a scoring system that enables designers to select the best option based on its specific merits. Note that an optimized system may not be the sum of the optimized subsystems. A top-level system understanding is required to trade efficiently some of the key aspects of the design.

Output – Set of candidate options (e.g., for subsystems), trade-off criteria, downselected option leading to a baseline design.

Example – The evaluation of locomotion subsystem options for the proposed rover has identified four options: four wheels system, six wheels system, six legs, or an aerobot. To compare these options, three criteria are identified, namely mass, maturity, and power requirement. Each of these systems is assessed against these criteria as summarized in Table 2.2. A scoring system

Table 2.2 Trade-offs example: assessment of the various system characteristics.

Candidate	Mass	Maturity/TRL	Power requirement
Four wheels	70	5	50
Six wheels	100	9	90
Six Legs	110	2	100
Aerobot	120	2	50

Table 2.3 Trade-offs example: setting up defined weighing factors.

Criteria	Score		
	1	2	3
Mass	>120 kg	90 kg<x<120 kg	<90 kg
Maturity/TRL	1–3	4–6	>7
Power requirement	>100 W	50 W<x<100 W	<50 W

can then be proposed to provide a traceable and consistent assessment across the proposed concepts as shown in Table 2.3. Here, in line with the proposed mission concept, a higher score is given to low mass, high technology readiness level (TRL), and low power solutions. A number of sublevels are provided for each criterion to provide adequate granularity in the scoring system. A higher score, therefore, corresponds to the more favored solutions closer to the trade objectives.

The data can then be generated based on the predefined scoring and weighting system. The final score for each solution result from the sum of each score with its weighting as shown in Table 2.4. Furthermore, to confirm the suitability of the selected solution, a sensitivity analysis can be performed to check how the overall scoring would evolve with different weighting. Based on the results, the four-wheel option is concluded as the most suitable solution in this example based on the criteria selected.

Table 2.4 Trade-offs example: weighting assessment of the various system characteristics.

	Weighting	Four wheels		Six wheels		Six legs		Aerobot	
		Score	Total	Score	Total	Score	Total	Score	Total
Mass	3	5	15	3	9	3	9	1	3
Maturity/TRL	3	3	9	5	15	1	3	1	3
Power	1	5	5	3	3	1	1	5	5
Total			29		27		13		11

2.3 Mission Scenarios: Past, Current, and Future

The definition of a new mission concept is a daunting and exciting prospect as the mission designers explore ways to address the new mission challenges. Whether they are addressing the need for new science data (e.g., finding life on Mars) or preparing future human settlements on the Moon, the design of the robotic systems starts with the assessment of the minimal number of platforms that need to be placed on the planetary surface and/or in the environment where the robotic system needs to survive and operate in. The notion of "platform" is a generic term to define a planetary asset that carries a range of payloads on the planetary surface and executes a range of actions and operations with a range of automatic and autonomous functions, for example, the mostly common mobile platforms are planetary rovers. A variety of platforms can be considered across different mission concepts and the best suited one needs to be chosen for a specific mission concept.

Given the introduction in Chapter 1, a planetary robot can take many guises, from large to small, from a robotic platform to a robotic payload. To help consolidate a mission concept and anticipate robotic system design challenges, this section presents major past, current, and future planetary missions involving platforms such as the lander and the rover (onboard which the robotic payloads can reside), as well as future mission concepts yet to be materialized.

2.3.1 Lander Missions

A planetary lander does not generally fit in the typical profile of a planetary robot. Looking closely at its functions and operations, however, there is more than meets the eye. A lander is a complex system that needs to perform a number of functions, such as providing the payload with mechanical structure, energy, data processing, and data relay to and from the ground control. Their robotic credentials are gained even before touchdown. During the descent and landing phase, the timely operation of actuators is essential to a safe landing. Whether it involves the deployment of parachutes (for an atmospheric planet, e.g., Mars Pathfinder (MPF) [4]) or the use of an active landing system with thrusters (e.g., Moon landings or Viking mission on Mars), the lander relies on a set of autonomic and autonomous functions to ensure the survival of the spacecraft. These functions are crucial to the success of the mission, and their failure is a painful reminder of the complexity of landing safely on another celestial body. As shown previously in Table 1.1, historical statistics suffer from many failed landings that contribute largely to the low mission success rate, for example, 1/3 to Venus and Mars and 1/2 to the Moon. As new missions are being prepared, landers are becoming endowed with increasingly complex autonomous functions to maximize their robustness in dealing with unknown or dynamic environments hence to maximize the landing success. Autonomous hazard avoidance for the terminal descent phase is one of those powerful functions that rely on inputs from landing cameras [5, 6], laser-illuminated detection and ranging (LiDAR) [7], and/or radar sensors [8] to guide the spacecraft away from hazards (such as steep slopes or large rocks) that can potentially put the lander safety in jeopardy.

A lander fulfills a range of purposes. In its simplest form, it can house and support a range of instruments to perform *in situ* analysis of the planetary environment. For example, the Venera landers (that were deployed to the surface of Venus) performed purely automatic and timed operations, and survived only briefly on the planetary surface before being crushed by the extreme heat and temperature. Some landers only act as a delivery system for other robotic platforms, such as the Luna 17 and 21 landers that were used to deliver the Soviet Lunokhod rovers. Advancing from the simple platforms, landers can perform more advanced robotic activities with additional physical appendages that enable them to interact with the environment after landing. Beyond the use of complex robotic arms or manipulators, landers can also implement deployment mechanism of cameras, extendable booms, or soil-sensing devices, which provide the lander with the attributes of a fully fledged robotic system (Figure 2.2).

2.3.1.1 Luna Sample-Return Landers

The USSR's Luna program spanned from 1959 to 1976 with a series of impactor, flyby, rover (Lunokhod), and ultimately sample-return missions. The Luna 16 launched in September 1970 was the first unmanned spacecraft to return samples from another celestial body through fully automated processes. With two additional successful sample-return missions, Luna 22 (February 1972) and Luna 24

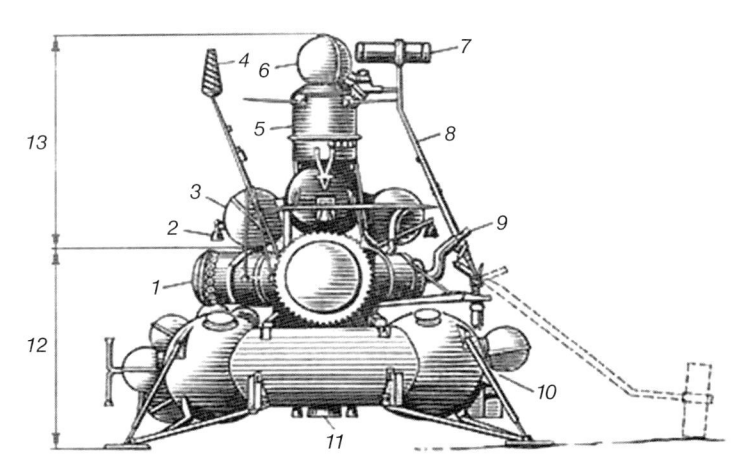

Figure 2.2 Luna 20 lander (sample-return configuration [9]). 1 - instrument module of descent stage; 2 - attitude control thrusters; 3 - propellant tanks; 4 - antenna; 5 - instrument section of ascent vehicle; 6 - return capsule; 7 - drilling mechanism; 8 - rod of drill mechanism; 9 - telephotometer; 10 - propellant tank; 11 - propulsion system of descent stage; 12 - descent stage; 13 - ascent vehicle for Moon–Earth transfer.

(August 1976), the Luna program achieved to return about 326 g of samples. Though this is almost 3 orders of magnitude less than the 382 kg recovered by the US manned program to the Moon, the success of the Luna missions defines a critical moment for the robotic exploration where complex operations were performed by machines in an extremely remote and hostile environment.

With some minor differences, the configuration of the Luna landers for sample return was fairly consistent among missions Luna 16, 18, 20, 23, and 24. With a launch mass of 5727 kg and a landed mass of 1880 kg on the Moon, the surface platform consists of a landing system, a return vehicle (512 kg), and a reentry capsule (50 cm diameter, 34 kg) for returning the sample down to the Earth [9, 10]. The platform carried a stereo imaging system, a radiation sensor, and a sampling arm/drill. Later, Luna sample-return landers also included a radio altimeter. The one degree of freedom arm delivered a drill to the surface before bringing it back up to the sample capsule. The lander was approximately 4 m tall with a diameter of 4 m including the legs. It was entirely powered by silver–zinc non-rechargeable batteries.

2.3.1.2 Viking Landers

The two Viking landers were successful missions to Mars and were the first US missions to land safely on another planet and return images from the surface. The two identical spacecrafts were launched and operated in parallel, a few months apart. The primary mission objectives were to obtain high-resolution images of the Martian surface, characterize the structure and composition of the atmosphere and surface, and search for evidence of life. Viking 1 was launched on August 20, 1975 and reached Mars on June 19, 1976. After a few months of surface mapping from orbit, Viking 1 landed on July 20, 1976 at the Chryse Planitia (22.48°N, 49.97°W). Similarly, Viking 2 was launched on September 9, 1975 and landed at Utopia Planitia (47.97°N, 225.74°W) on September 3, 1976. The landers were deployed to the surface through a combination of parachutes and powered descent and landing system. The payload suite comprised a wide range of instruments including a biological experiment, gas chromatograph/mass spectrometer, X-ray fluorescence spectrometer, seismometer, meteorology instrument, and stereo color cameras. The payloads were tasked to analyze and characterize the physical and magnetic properties of the soil, as well as the aerodynamic properties and composition of Martian atmosphere over altitudes during the descent.

The Viking lander platform as shown in Figure 2.3 had a launch mass (including aeroshell) of 1060 kg and a landed mass of 603 kg. Its power system comprised two 13.5 kg radioisotope thermoelectric generators (RTGs) each producing around 30–35 W [11]. The 3 m reeled sampling boom was made of two sheets of metallic foil fused at their edges and furled during stowage, akin to a metallic tape measure. At deployment, the sheets bounced back to form a tube that provided the necessary mechanical support to deploy the sample collection system [12, 13].

Figure 2.3 Viking lander. (a) Photo [14]. (b) Configuration [12] (Courtesy NASA/JPL/University of Arizona.)

2.3.1.3 Mars Surveyor Lander Family and Successors

The Mars Surveyor Program (MSP) envisaged the simultaneous delivery of an orbiter and a lander to study Mars by using the two assets at the same time. The program anticipated to launch every two years to benefit from the cyclical favorable transfer conditions to Mars. The program has, however, seen a number of challenges and setbacks that led to the loss of a lander (i.e., Mars Polar Lander, MPL) and the cancellation of its successor (i.e., Mars Surveyor '01 Lander). Nevertheless, the legacy of this program can be seen in more recent missions such as the successful Phoenix and the upcoming InSight lander. The lander platforms within this program should not be considered as four separate developments, but an evolution of the same design throughout the program. Building upon the Viking development, these platforms feature a similar three-leg configuration with retrothrusters to decelerate the lander during terminal descent and landing. The body of the platform is somewhat simplified from the Viking era and only consisting of a small service module about 1.5 m diameter covered by the science deck where most of the payloads are located.

The MPL, also called the Mars Surveyor '98 Lander (see Figure 2.4), was designed and built to land and explore the *Planum Australe*, a region 76° south of the equator near the south pole on Mars [15]. The lander scientific payload aimed to gather climate data as part of a dual-asset concept involving the simultaneous acquisition of data from the lander and the orbiter. Instruments on the lander (including an arm) were to analyze the surface materials, frost, weather patterns, and interactions between the surface and atmosphere to better understand how the climate of Mars changes over time, and infer where the water resides now on

Figure 2.4 Mars Polar Lander configuration [16]. (Courtesy NASA/JPL.)

Figure 2.5 Mars Surveyor '01 lander configuration [17]. (Courtesy NASA/JPL.)

the planet. The lander, delivered with two microprobes on December 3, 1999, was unfortunately unsuccessful and failed during the landing phase.

As part of the 2-yearly Surveyor Program, the Mars Surveyor '01 Lander (Figure 2.5) was meant to deliver a similar platform to Mars, including the deployment of a small rover based on the MPF Sojourner. Unfortunately, due to the successive failures of MPL and the Mars Climate Orbiter in late 1999, the lander was canceled in May 2000. The lander was meant to be equipped to study soil, atmospheric chemistry, and radiation on the surface, and to deploy the small 13.8 kg rover platform with a robotic arm to study the vicinity of the landing site on the Martian surface [17]. Later, the orbiter of the Mars Surveyor '01 was launched successfully and renamed Mars Odyssey in 2001.

The Phoenix mission (Figure 2.6) was rebuilt from the "ashes" of the two previous Surveyor landers by reusing extensively various pieces of their hardware. As part of the Scout Program, the Phoenix lander was designed to study the surface and near-surface environment of a landing site in the high northern area of Mars. Its science aimed to understand the history of the water in all its phases, whether the Martian arctic soil could support life and study the Martian weather from a polar perspective [18]. It also featured a robotic arm to perform local trenching around the lander. The platform landed successfully on May 25, 2008, in the Green Valley of Vastitas Borealisat at 68.22°N 125.7°W. The lander continued operations until November 2, 2008, 2 months beyond its 3-month nominal mission until it lost communication with the ground control due to low energy level. On May 12, 2010, after a number of unsuccessful attempts to communicate with the lander,

Figure 2.6 Phoenix lander configuration. (Courtesy NASA/JPL.)

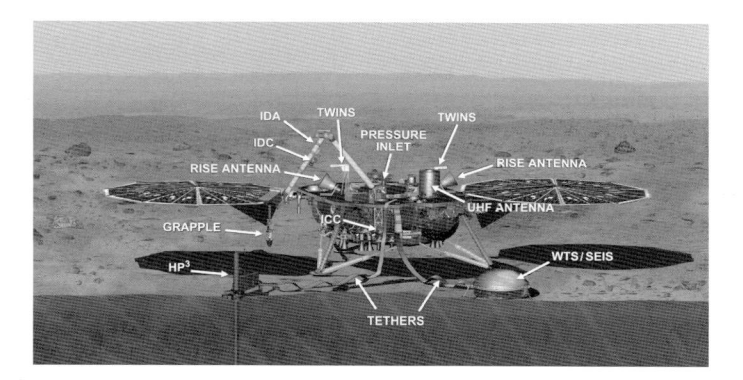

Figure 2.7 InSight lander and payload configuration. (Courtesy NASA/JPL.)

the mission officially ended, concluded all its experiments, and achieved its prime science objectives.

Building on the success of Phoenix, the InSight lander (see Figure 2.7) is expected to land on Mars in 2016, as part of the NASA Discovery Program. The name stands for *Interior Exploration using Seismic Investigations, Geodesy and Heat Transport* and highlights the key mission objectives. The lander will deploy a suite of payload on the surface with a robotic arm.

From MPL to InSight, the design of the lander and its subsystems has evolved to support various incarnations of the mission. Table 2.5 provides a summary of these missions and specific characteristics of the landers.

2.3.1.4 Huygens Lander

Part of the NASA/ESA Cassini/Huygens mission to explore Saturn and its moons, the Huygens lander was deployed to Saturn's largest moon Titan. Shrouded in

Table 2.5 Characteristics of the Surveyor landers and successors.

Lander	Status	Landing date	Landing site	Mass (kg)	Max deployed width (m)	Height (m)	Power system
Mars Polar Lander (Surveyor '98)	Failed	03/12/1999	76S 195W	290	3.6	1.06	200 W – gallium arsenide solar array, NiH_2 battery
Mars Surveyor '01 Lander	Canceled	22/01/2002	12S 315W	328	5.5 (Ø1.5 m deck)	1.2	450 W – 2× Circular UltraFlex solar arrays (2.15 m diameter each) and NiH_2 battery
Phoenix (Scout Program)	Success	25/05/2008	68.22N 125.7W	350	5.5 (Ø1.5 m deck)	2.2	450 W – 2× Circular UltraFlex solar arrays (2.15 m diameter each) and NiH_2 battery
InSight (Discovery Program)	Pending	20/09/2016	3N 154.7E	350	5.5 (Ø1.5 m deck)	1.5	450 W – 2× Circular UltraFlex solar arrays (2.15 m diameter each) and NiH_2 battery

a thick cloud layer, little was known of Titan. The lander was hence devised to characterize the atmosphere during the descent of the probe and provide more data about the surface. The short-lived probe landed successfully on January 14, 2005 after 2 h and 30 min of the atmospheric descent and operated a further 90 min on the ground, capturing details of hydrocarbon lakes and icy surfaces.

According to Ref. [19], the primary objectives of the Huygens lander were to

- investigate the upper atmosphere, its ionization, and its role as a source of neutral and ionized material for the magnetosphere of Saturn;
- determine the abundance of atmospheric constituents (including noble gases); establish isotope ratios for abundant elements; and constrain scenarios of formation and evolution of Titan and its atmosphere;
- observe vertical and horizontal distributions of trace gases; search for more complex organic molecules; investigate energy sources for atmospheric chemistry; model the photochemistry of the stratosphere; and study the formation and composition of aerosols;

- Measure winds and global temperatures; investigate cloud physics, general circulation and seasonal effects in Titan's atmosphere; and search for lightning discharges;
- determine the physical state, topography, and the composition of the surface, and infer the internal structure of the moon.

To maximize science return of the mission, control of the descent was paramount to gather the necessary science data. Onboard autonomy was critical to ensure that the required rate of descent was met in a completely unknown environment. Table 2.6 summarizes the control strategies implemented as part of the parachute descent system for the lander.

As illustrated in Figure 2.8, the lander consisted of a 1.3 m diameter platform designed to house the six main instrument packages. The mass of the probe was

Table 2.6 Autonomy characteristics of the Huygens lander [20].

Phase	Atmospheric descent
Objectives	To achieve atmospheric descent profile suitable for scientific measurement
Main environmental uncertainty	Atmospheric density profile
Controlled states	Descent profile
Available actuators	Timing of main parachute deployment
Available sensors	Accelerometers
Control criterion	Acceleration/velocity profile
Constraints	Radio relay link geometry, power consumption, deployment velocity, and altitude

Figure 2.8 Huygens lander on Titan surface (artist impression). (Courtesy ESA.)

319 kg [21] and was specifically designed to cope with any surface conditions (including soft and hard surfaces, and liquids), which particularly drove the evaluation of the buoyancy of the lander, a rare engineering challenge so far for planetary landers.

The thermal system used 35 one-watt radioisotope heater units (RHUs) to keep the lander warm. Due to the short mission lifetime and the lack of solar energy at Titan surface, five primary batteries were implemented with a total capacity of at least 1800 W h. The battery capacity was sized for a mission duration of 153 min, corresponding to a maximum descent time of 2 h and 30 min plus at least 3 min on the surface [19].

2.3.1.5 Beagle 2 Lander

The Beagle 2 lander was launched as a payload of ESA's Mars Express mission. The UK-led probe was designed to investigate traces of past or present life within the Martian regolith. The lander separated from the Mars Express mother spacecraft on December 19, 2003 for landing taking place on December 25, 2003. Unfortunately, the probe did not make contact after landing, which abruptly stopped the mission. Based on more recent data from the Mars Climate Orbiter (the HiRise camera) in January 2015, it seemed that the lander did land and start its nominal operations on the surface, but the deployment of the solar panels was interrupted covering the antenna and preventing any communication with the lander [22].

The lander targeted Isidis Planitia, a large flat sedimentary basin between the ancient highlands and the northern plains of Mars, and a landing site centered on $11.53°N$ $90.50°E$ [23]. The science objectives of the lander were to [24, 25]

- characterize the landing site geology, mineralogy and geochemistry;
- provide a chemical and physical analysis of the atmosphere;
- measure the dynamic environmental processes;
- provide astronomical observations of the Sun, bright stars, and Phobos and Deimos.

The 72.7 kg probe (including the lander and heat shield) deployed the 33.2 kg lander on the Martian surface carrying ~9kg of payload. The lander (as shown in Figure 2.9) featured a clamshell design in two parts that protected its content during its airbag landing and subsequently opened up on the Martian surface to deploy its solar panels and instrumented arm. The 0.75 m arm carried the position adjustable workbench (PAW) that allowed the platform to deploy the sensor heads and cameras to specific locations on the surface. From the PAW, the PLanetary Undersurface TOol (PLUTO, also known as the Mole) was to be deployed to further characterize the local subsurface environment. The lander was expected to operate for about 180 days. However, based on the projected degradation of the solar panel due to dust deposition, an extended mission of up to one Martian year (687 Earth days) was thought possible. The power system relied on four solar panels providing ~1m^2 of solar cells designed to generate ~320 W h at the beginning of the mission. The power subsystem used a 2.63 kg battery composed of 54 cells

Figure 2.9 Beagle 2 lander: artist impression on Mars with the instrumented PAW upright. (Courtesy ESA.)

providing 13.5 A h of energy to the lander to supplement the solar power during the day and to ensure the survival of the lander at night.

2.3.1.6 Philae Lander

The Philae lander was launched aboard ESA's Rosetta spacecraft on March 2, 2004 for heading to a comet. Upon landing on the nucleus of 67P/Churyumov-Gerasimenko on November 12, 2014, the lander became the first successful soft landing on a comet unlike the past Deep Impact mission that purposefully impacted with comet Tempel-1 on July 4, 2005. However, the harpoon system designed to anchor the lander onto the surface at touchdown malfunctioned, causing the lander to rebound twice on the surface causing it to land significantly further than its intended landing location. The platform ultimately rested canted at an angle of about 30° in the shadow of a cliff, making the power generation (hence communication with Earth) difficult after landing. After a total of 60 h during which the lander was able to perform a large proportion of its intended science, it fell silent. Owing to the seasonal illumination on the comet, the resting location of the lander became increasingly illuminated in mid-June 2015, providing the necessary power for the lander to resume intermittent contact with the Earth ground control.

The lander had a mass of 97.9 kg and consists of a central body on three legs (Figure 2.10). It carried no less than 10 instruments whose key science objectives were to [26]

- characterize, by *in situ* measurements and observations, the composition of the cometary material down to its microscopic scale;
- investigate the physical properties of the nucleus, its environment, its large-scale structure, and its interior;
- contribute to the monitoring of the long-term evolution (activity) of the comet.

To support the mission goals, the lander power subsystem consisted of solar panels, primary batteries (nonrechargeable), and secondary batteries

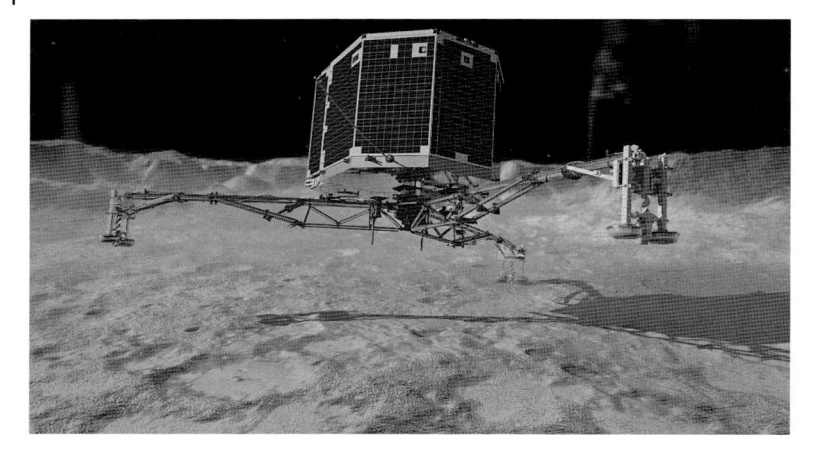

Figure 2.10 Philae lander: artist impression on touchdown. (Courtesy ESA.)

(rechargeable). As discussed in Ref. [26], the primary batteries were used to support operations following release and for the next 5 days, to secure a first scientific sequence including the operation of each instrument at least once. It consisted of Li/SOCl cells with a capacity of about 1200 W h at beginning of life (BOL). The secondary batteries with a total capacity of about 150 W h consisted of two blocks of 14 Li-ion cells each and were meant to be the main energy source for the long-term operations on the comet. The secondary batteries were recharged by the solar array or during cruise via current lines from the orbiter. The solar array was implemented with low-intensity low-temperature (LILT) silicon solar cells sized to provide a daily average of 10 W at 3AU distance to the Sun. The thermal system did not implement any exotic elements such as the RHUs and instead focused on efficient thermal insulation and subsystems with a wide operating temperature range ($-55-70\,°C$) [26].

The early operations presented a challenge, owing to the distance between the comet and the Earth ground control. The mission profile required precise landing on the unknown cometary body without human operators in the loop. A number of autonomous functions were, therefore, necessary to deal with the landing as well as the planned initial soil sampling activities. Details are summarized in Table 2.7.

2.3.2
Rover Missions

Although landers are particularly suited to carry a complex suite of payloads to investigate the landing site, the need to explore beyond the immediate vicinity of the platform requires some sort of mobility to explore further. The mobile platforms then need to be sized with enough payload capability to carry out the key objectives of the mission. The way that these platforms achieve mobility can

Table 2.7 Autonomy characteristics of the Philae lander [20].

	Descent and landing	Soil sampling
Objectives	Safe landing near specified location	Safe acquisition of subsurface samples
Main environmental uncertainty	Dynamic/kinematic properties, topography of landing site	Mechanical soil properties for anchoring and drilling
Controlled states	Spacecraft attitude orbit parameters	Attitude, drill pushing force, rotation rate
Available actuators	Hydrazine/cold gas thrusters	Drill motors, cold gas thrusters
Available sensors	Laser range finder, radar altimeter, gyros, accelerometers	Force sensors, gyros, accelerometer
Control criterion	Velocity/attitude at touch-down, distance to target	Adaptive control law to minimize forces acting at the legs
Constraints	Antenna pointing to Earth, obstacle avoidance	Sample heating limits, anchor sustainable forces

be extremely diverse and provide different exploration capabilities, from using wheels to skis, from free-raging to tethered.

2.3.2.1 Lunokhod 1 and 2 Rovers

As the race to the planets raged on between the USSR and the United States, the Luna missions were the workhorse of the Soviet Lunar exploration, growing in capability and scope. After a string of unsuccessful launches in 1969, resulting in the loss of the country's first rover (dubbed Lunokhod 0) and a number of sample-return missions, the Luna 16 became the first to successfully land and return samples from the Moon as discussed in Section 2.3.1. Building on this success, the Luna 17 mission landed in the vicinity of Mare Imbrium on November 17, 1970 and deployed the Lunokhod 1 rover (see Figure 2.11). The robot, literally meaning *Moon walker*, was a significant milestone and became the first rover being freely teleoperated on the surface of another extraterrestrial body. It was deployed from the lander to the lunar surface through a system of ramps providing forward and backward egress paths. Lunokhod 1 was initially designed to operate through 3 lunar days (about 3 Earth months). In reality, it successfully operated for at least 7 lunar days (212–220 days) [27] and up to even 11 lunar days (302 days) [28]. Over its lifetime, the rover traversed 10.54 km. Two years later in January 1973, the Luna 21 landed in Le Monnier crater and delivered Lunokhod 2. Building upon Lunokhod 1, the second rover provided an upgraded science and control package including higher resolution cameras and an improved scientific payload. Similar to its predecessor, Lunokhod 2 was teleoperated from the Earth during the lunar days and put in hibernation mode at night. It explored the Moon for approximately 4 lunar days (125 Earth days, ~4 months), a duration significantly shorter than Lunokhod 1. It is believed that the rover suffered from overheating, due to

Figure 2.11 Lunokhod 1 rover with stowed solar panel. (Credit Lavochkin Association.)

a failure of the solar array deployment system after a lunar night, that prevented the radiator from dissipating the heat generated by the polonium-210 RHU. However, over its shorter lifetime, the rover covered 3 times the surface distance of its predecessor (i.e., about 37.5 km across the lunar landscape [28]).

The Lunokhod 1 rover had a mass of 756 kg, while Lunokhod 2 was 840 kg [28] owing to an enhanced payload suite. The payload of both platforms mainly consisted of four television cameras, an X-ray spectrometer, an X-ray telescope, cosmic ray detectors, and an extendable device to test the lunar soil for soil density and mechanical properties.

The dimensions of both Lunokhod rovers were similar at about 1.35 m tall, 1.7 m long and 1.6 m wide [28]. To deal with the complexity of building space and vacuum compatible hardware, the rovers bodies were built using a large hermetically sealed bathtub filled with nitrogen. This method conveniently sidestepped a number of issues related to the survival of electronics in vacuum but also facilitated somewhat the thermal control inside the warm box by using convection as well as conduction to dissipate the heat to a top-mounted radiator. To support the survival of the platforms at night, a polonium-210 isotopic heat source was used to provide a continuous source of heat. To further limit the heat dissipation at night, a single solar panel formed a curved lid that could be closed on top of the radiator during the 14-day long lunar nights. During the day, it provided enough power to support long teleoperated traverses.

The locomotion subsystem consisted of eight wheels with a diameter of 51 cm mounted on four bogies. The mesh and spoke wheels provided a lightweight yet efficient traction system and were actuated by DC brushed motors in a pressurized hub. Yet again, this provided a simple solution to make use of DC brushed motors and mitigate their notoriously short life in vacuum due to the fast erosion of the brushes caused by sparking. A gearing system in the hub provided the Lunokhod rover with two driving speeds of 0.93 km h^{-1} and 2 km h^{-1} [28, 29]. An additional small ninth free wheel was located behind the rover and used as an odometer to provide an accurate record of the distance traversed, while being unaffected by wheel slip in the fine lunar dust.

2.3.2.2 Prop-M Rover

The Prop-M rovers were small Soviet rover platforms launched with the twin mission Mars-2 and Mars-3, 9 days apart. Though the orbiter segments were successful, both the Mars-2 lander (with a planned landing on November 27, 1971) and Mars-3 lander (that landed on December 2, 1971) were unsuccessful. The Mars-2 lander failed during entry, descent, and landing (EDL), and Mars-3 only provided 14.5 s of data after landing.

The 4.5 kg platforms were about $215 \times 160 \times 60$ mm (as illustrated in Figure 2.12) and tethered to the lander to provide power and data links. It would have allowed the rover to traverse about 15 m from the lander and gather science data every 1.5 m. The payload suite was comprised of a dynamic penetrometer and a radiation densitometer to measure soil density.

The rover was deployed from the lander to the surface by means of an articulated deployment mechanism as illustrated in Figure 2.13-top. The Prop-M platform provided a new paradigm for surface mobility, implementing the first nonwheeled vehicle for planetary applications. Built with two rotating skis, the rover provided a forward motion by pivoting its skis through rotating lever arms connected to

Figure 2.12 Prop-M rover. (Credit Lavochkin.)

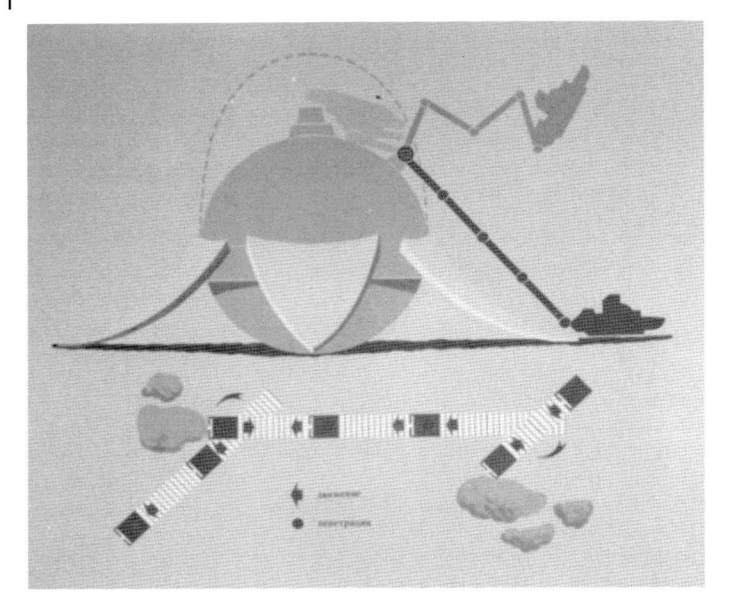

Figure 2.13 Prop-M rover deployment concept (top) and obstacle avoidance scheme (bottom). (Credit Lavochkin Association.)

its body. It placed on the surface alternatively its skis and its body to provide a forward motion. Rotation was performed through skid steering by moving one ski forward and one ski backward [30]. A simple autonomous behavior of the rover was implemented to allow the platform to avoid obstacles during its traverse. If the tactile collision sensor in front of the vehicle detected an obstacle, it initiated a sequence where it moved one step backward, turned slightly, and moved forward again as illustrated in Figure 2.13.

2.3.2.3 Sojourner Rover

The MPF was one of the early missions in the NASA Discovery Program, proposed to foster rapid developing, low-cost spacecraft with highly focused science objectives [15]. MPF's purpose was also to demonstrate an innovative way of deploying an instrumented lander on the planetary surface through the use of airbag technology [31]. Despite the new landing technique, the MPF landed successfully on July 4, 1997 in Ares Vallis on Chryse Planitia. The lander carried a number of meteorological and atmospheric instruments as well as a small free-ranging rover called Sojourner, as illustrated in Figure 2.14a. This was the first mobile platform on Mars that carried a number of instruments to characterize the soil and rocks in the vicinity of the lander including an alpha-proton X-ray spectrometer (APXS), one color and two black and white cameras [15] (see Figure 2.14b).

Compared regularly with a microwave oven, the Sojourner rover was 65 cm long, 48 cm wide, 30 cm tall (18 cm stowed), and 10.5 kg in mass [32, 33]. The platform comprised a warm electronic box (WEB) housing all the electronics

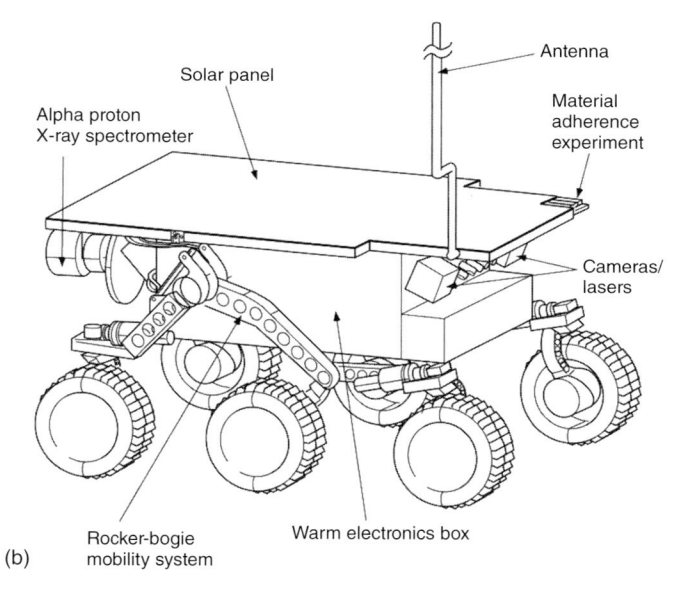

Figure 2.14 Sojourner rover. (a) Photo and (b) Configuration. (Courtesy NASA/JPL.)

that could not survive the cold temperature of the Martian night (−110 °C), the payloads, and locomotion subsystem. The locomotion subsystem consisted of a six-wheel rocker-bogie suspension system with four steerable corner wheels providing a point-turn capability. With the 13 cm diameter wheels, the rover was capable of overcoming obstacles in the order of a wheel size (up to 20 cm) and traversing at a maximum speed of 0.4 m min^{-1} [32] and possibly up to 0.6 m min^{-1} [15]. The power subsystem of the platform consisted of a solar panel, the first demonstration of working solar arrays on Mars at the time. At 0.22 m^2, it provided a peak power of 16 W and was complemented by a primary (i.e., nonrechargeable) battery providing up to 150 W h of energy. The decision to select a primary battery was driven by the original duration of the mission (i.e., a week) and the lower energy-to-mass ratio of rechargeable batteries. During traverses, the platform required around 10 W to power the locomotion subsystem [32]. The communication system relies on the use of a UHF link between the rover and the lander. The lander then acted as a relay station to forward the telecommands to the rover and return the telemetry and science data back to the ground control. Over the course of the 83 sols of operation, Sojourner covered a distance of approximately 100 m, at a maximum of 12 m from the lander. The mission represents a significant milestone where a large amount of planetary data was being returned autonomously. The rover returned about 550 photos and captured the data of 16 sites around the lander. The lander itself returned some 16 500 images from the surface and made 8.5 million measurements with its atmospheric structure instrument and meteorology (ASI/MET) package.

The anticipated lifetime of the lander and the rover were up to a month and a week, respectively. The mission, however, far exceeded the initial expectation by having operated for about 3 months for both the platforms, that is, lasting until September 27, 1997, the day of the last successful transmission. In honor of the late astronomer and writer, Carl Sagan, the lander has been later renamed the Carl Sagan Memorial Station. One of the legacies of the MPF mission was the unconventional way to build the platforms. As part of the "faster, better, cheaper" motto of the program, the mission engineers investigated commercial off-the-shelf (COTS) components and turned them to space-qualified ones, including UHF modems, motors, and actuators [32]. In the approach used, only specific system had built-in redundancy, accepting the risk of failures in other parts of the system. This approach, although riskier than other developments, was pivotal to the success of this mission. However, later missions such as MPL may have suffered from this design philosophy as it failed during landing.

2.3.2.4 Spirit and Opportunity Rovers

Building on the success of MPF, the Mars Exploration Rovers (MER) illustrated in Figure 2.15a were launched in 2003 and targeted two separate locations on Mars. The two rovers MER-A and B, later dubbed *Spirit* and *Opportunity*, were deployed in January 2004. Spirit landed in Gusev Crater on January 4, 2004 and Opportunity landed after 3 weeks on 25th January in Meridiani Planum. The science objectives of the missions revolved around the concept of "following the water," that is, to identify clues of past and present traces of water on Mars conducive to life. As such, the rovers were tasked mainly to search for and characterize a diversity of rock sand soils that hold clues to past water activities (i.e., presence of water-bearing minerals deposited through various processes) [34]. This involved a thorough investigation of the selected landing sites and determination of the

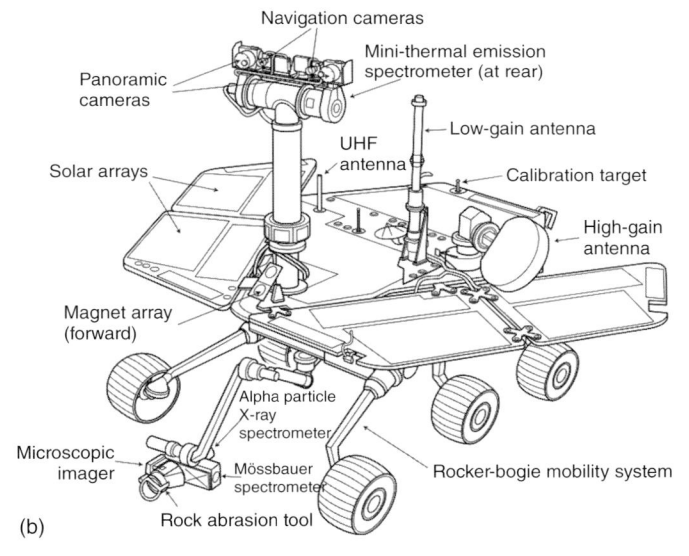

Figure 2.15 Mars Exploration Rover [34]. (a) Artist impression and (b) Configuration. (Courtesy NASA/JPL-Caltech.)

spatial distribution and composition of minerals, rocks, and soils surrounding the landing sites.

The rovers carried a suite of instruments both body-mounted and deployable ones. On the rover body, there were Panoramic Cameras (PanCam) and magnet arrays. To deploy the surface payload, the rovers featured a robotic arm approximately 1-m long. As illustrated in Figure 2.15b, the instrument deployment device (IDD) functioned as a dexterous appendage for the rover [35], delivering a cluster of four interchangeable scientific instruments to the areas of interest. These included the rock abrasion tool (RAT), which grinds away a consistently sized, circular area on a rock specimen; the microscopic imager (MI), which takes extreme close-up images of rocks and soil; the Mössbauer spectrometer (MB), which is used for close-up investigations of the mineralogy of iron-bearing rocks and soils; and the Alpha Particle X-ray Spectrometer (APXS), which performed close-up analyses to study the abundance of elements that make up rocks and soils.

The rovers were an order of magnitude larger and heavier than Sojourner. At 180 kg, the rover platform is 1.5 m tall, 2.3 m wide, and 1.6-m long. Similar to Sojourner, it implemented a similar six-wheeled rocker-bogie locomotion subsystem with steering for the four corner wheels. The wheel diameters was set to 26 cm and designed to traverse the Martian landscape at a maximum of 5 cm s^{-1}, although in practice the average distance was closer to 1 cm s^{-1}. The rover EDL system implemented the same landing scheme as MPF in 1997 (i.e., parachutes, retrorockets, and airbags), pushing the technology to its limits and identifying in the process the upper boundary for the applicability of this method. Unlike MPF, the MER lander platform was not used beyond the landing nor did it provide a communication relay with the Earth. This setup provided the rover the necessary freedom to explore beyond the vicinity of the landing site without constraints. The rover featured a range of communication options [35]: two X-band antennas (high and low gains) were used to support the EDL phase as well as provide a valuable direct-to-Earth (DTE) link for controlling the rovers when necessary; however, bandwidth of this link drastically fluctuates during a typical year with the respective distance of the Earth and Mars (max 2 kbps telecommand and 28.8 kbps telemetry); a UHF link was also used to download engineering data at a much higher rate through a number of orbiters (up to 128 kbps) including the Mars Global Surveyor, Mars Odyssey, and Mars Reconnaissance Orbiter.

The rover power system consisted of an arrangement of solar panels, two of which were folded during cruise. They provided a total area of 1.3 m^2 of triple-junction photovoltaic cells, generating up to 140 W power and up to 900 W h/sol at BOL. Two lithium-ion batteries, weighing 7.15 kg each, were housed in the warm box and provided up to 600 Wh (BOL) at 30 V [36]. Both platforms suffered from dust deposition on the solar arrays that degraded the energy generation capability of the panels. Over the course of lifetime, the rovers saw a number of "cleaning events" attributed to local dust devils that passed by them, effectively cleaning the dust off the solar arrays and improving the power generation of the platform. To survive the cold Martian nights with temperature reaching as low as −105 °C, the thermal control of the rover was much reliant on the heat dissipated by the electronics based on eight RHUs [34]. These small devices each provided 1 W of heat based on the nuclear decay of an isotope, in this case plutonium dioxide.

The original lifetime of the rovers were around 3 months and aimed at a total traverse distance of 600 m. Impressively, the missions for both Spirit and Opportunity have been extended numerous times, thanks to the good health of the rovers. In late 2009, Spirit got stuck in a patch of soft sand that immobilized the rover resulting in its demise on March 22, 2010 after 7 years of service and 7.73 km of traverse. Opportunity, on the other hand, has exceeded all expectations by continuously operating on Mars for more than 12 years and by the end of 2015 covered in excess of 42 km, beating the record of the longest traverse established by Lunokhod 2 (i.e., 39 km while being teleoperated) back in 1973.

2.3.2.5 Curiosity Rover

As part of the Mars Science Laboratory (MSL) mission, the Curiosity rover (see Figure 2.16a) landed in Aeolis Paulus inside Gale Crater on August 6, 2012. In stark contrast with its predecessor, the landing phase did not employ a lander platform nor did it use airbag technologies. Instead, it employed a daring "sky crane" concept to deploy the platform to the surface, which was a separate vehicle that provided controlled powered descent until complete stop some 8 m above ground where the rover was reeled down onto the surface.

The science objectives of the mission built on the MER findings and the "following the water" strategy, and also aims at identifying other basic ingredients for life. In addition, the mission aimed to gather valuable information to assess whether or not Mars is a potential habitat for future manned mission. As such, it focused on four main classes of objectives [37]:

- **Biological objectives**: The rover aims to determine the nature and inventory of organic carbon compounds, perform an inventory of the chemical building blocks of life (carbon, hydrogen, nitrogen, oxygen, phosphorous, and sulfur) and identify features that may represent the effects of biological processes.
- **Geological and geochemical objectives**: During the traverses, the rover investigates the chemical, isotopic, and mineralogical composition of the Martian surface and near-surface geological materials to interpret the processes that have formed and modified rocks and soils.
- **Planetary process objectives**: Based on the data acquired, the mission aims to assess the long-timescale (i.e., 4 billion-year) atmospheric evolution processes and determine the present state, distribution, and cycling of water and carbon dioxide.
- **Surface radiation objectives**: The mission aims to complete the analysis by characterizing the broad spectrum of surface radiation, including galactic cosmic radiation, solar proton events, and secondary neutrons.

To support the science and mission objectives, the rover (see Figure 2.16b) carries a suite of no less than 10 instruments with a combined mass of 75 kg, compared with the 5 kg onboard the MERs. They include a range of rover-mounted payloads and, similar to MER, a number of instruments fitted at the tip of a deployable arm. On the rover mast, MastCam provides visual imagery through a number of medium- and narrow-angle cameras, resolving up to 7.4 cm/pixel

Figure 2.16 Curiosity rover. (a) Full-scale model and (b) configuration. (Courtesy NASA/JPL-Caltech.)

at 1 km. On top, ChemCam, a Chemistry and Camera complex, combines a laser-induced breakdown spectrometer (LIBS) and a remote microimager (RMI) to shine a laser to a target (e.g., rock) 7 m away to derive its composition. RMI can also resolve 1 mm target at 10 m. The Rover Environmental Monitoring Station (REMS) can be found at half-mast and provides meteorological and environmental data such as humidity, wind, temperature, pressure, and UV levels. Inside the body of the rover, a suite of two instruments provide the rover with the capability to perform advanced analyses of surface, rock, and atmospheric samples. The Chemistry and Mineralogy (CheMin) focuses on X-ray powder diffraction and fluorescence, and the Sample Analysis at Mars (SAM) consists of various separate instruments (spectrometers and chromatographs) to provide further data on the isotopic composition of the samples. On the rover itself, the radiation assessment detector (RAD) characterizes the radiation environment while the Dynamic Albedo of Neutron (DAN) measures hydrogen found in water and ice.

The rover features a 2.1 m long robotic arm to place a suite of instruments on the surface [38]. The 30 kg instrument turret comprises four instruments: the powder acquisition drilling system (PADS) and the dust removal tool (DRT) are responsible to acquire samples for CheMin and SAM, while the other two contact instruments consist of the alpha particle X-ray spectrometer (APXS) and the Mars hand lens imager (MAHLI), a microscope.

Though Curiosity draws heavily upon the technologies used since Sojourner, the rover is significantly larger than past Martian rovers. With a mass of 899 kg [37], it is the heaviest mobile platform ever sent to another planetary surface, beating the 840 kg Lunokhod 2 rover. With a length of 3.0 m (not counting the arm), a width of 2.8 m, and a height at the top of mast of 2.1 m, Curiosity is typically compared with a small car. The locomotion subsystem is similar to the MER, albeit scaled up, featuring the same rocker-bogie concept with six-wheel drive and four-wheel steer. Unlike the previous MER and Sojourner platforms, a mechanical differential is implemented with a system of levers around Curiosity's body, rather than the geared implementation that passed through the body of previous rovers. The 50-cm-diameter wheels are made of aluminum with titanium spokes to provide some level of suspension. During navigation and traverse, no less than 12 "engineering cameras" can be used [37], such as the navigation cameras (NavCam) located on the mast that provide the necessary stereo data for navigation, and the hazard cameras (HazCam) located quite low on the platform to keep a watchful eye on local obstacles during traverses.

One obvious departure from the past rover designs is the lack of solar array on the top of the platform. Instead, an RTG is used to provide the platform with the required power to operate and heat and survive the cold Martian nights. The multi-mission radioisotope thermoelectric generator (MMRTG) generates power through the decaying of 4.8 kg of plutonium oxide [39]. In the device, the heat generated by the Pu-238 is converted to electricity through the use of thermocouples. A significant by-product of the use of an RTG is the constant generation of heat ($\sim$2000 W) that can be used to keep the rover within operating temperature [40].

Thermal control is critical to prevent overheating as well. To provide a controlled dissipation of the heat, the RTG is located at the back of the rover surrounded by cooling fins. At beginning of life, the 43 kg device was generating some 110 W [41] of power and during the surface phase is also supported by two batteries providing 42 A h of energy storage [37]. To date, the rover has traversed in excess of 10 km. Unlike rovers that are limited by solar energy, Curiosity can potentially operate more productively for a lot more years to come if no major issue arises.

The rover communication architecture is similar to that of the MER (i.e., medium- and high-gain X-band antenna), providing DTE via the Deep Space Network (DSN). A separate UHF link is used to relay the engineering data via a number of orbiters around Mars including Mars Reconnaissance Orbiter, Mars Odyssey, as well as the European Mars Express [11].

2.3.2.6 Chang'E 3 Rover

The Chang'E 3 mission followed the successful Chang'E 1 (2007) and Chang'E 2 (2010) orbiter missions around the Moon. It landed on December 14, 2013 northwest of Mare Imbrium (19.51W, 44.12N) and consisted of a lander platform carrying a rover. The mission represented the first lunar landing since the Soviet Luna 24 and demonstrated China's first soft landing and rover deployment on the lunar surface. The key science objectives on the mission were summarized as [42]

1) the investigation of the morphological features and geological structures in the vicinity of the landing area;
2) the integrated *in situ* analysis of mineral and chemical composition of the exploration sites;
3) the exploration of the terrestrial-lunar space environment and lunar-based astronomical observations.

Accordingly the rover supported the lander instrumentation by carrying four instruments [43]:

- a panoramic camera (PCAM),
- a VIS–NIR imaging spectrometer (VNIS),
- an active particle-induced X-ray spectrometer (APXS),
- a lunar-penetrating radar (LPR) with two antennas.

The rover (see Figure 2.17), dubbed Yutu (Jade Rabbit in Chinese), had a mass of 140 kg and a payload mass of around 20 kg. It was mainly teleoperated but carried a number of sensor to provide real-time hazard detection. The locomotion subsystem featured a six-wheel drive and four-wheel steer configuration with a rocker-bogie suspension. The use of an internal differential system is inferred, as per the MER configuration. The platform was designed to accommodate 20° slopes and 20 cm-tall obstacles. The power subsystems relied on two top-mounted panels and a lithium-ion battery. The thermal subsystem included a number of RHUs (Chinese made) to keep the rover warm at night.

Figure 2.17 Chang'E 3 rover on the lunar surface. (Credit CNSA.)

The platform aimed to explore an area of about 3 km^2. Unfortunately, it traversed only a total of 114 m over the course of two lunar days before a mechanical control issue prevented the rover from making further progress. Over its traverse, the ground-penetrating radar (GPR) was used successfully and all the instruments onboard had been exercised.

2.3.2.7 ExoMars Rover

ExoMars is a two-mission program aims to deliver (i) in 2016 the ExoMars Trace Gas Orbiter carrying a lander demonstrator, and (ii) two years later in 2018 the ExoMars rover, the first European planetary mobile platform, which will be delivered by a Russian lander. Past and current missions have concentrated so far on the investigation of conditions favorable to the emergence of life such as the presence of water or remnant of by-products of life. The planned 218-sol mission, however, is very much focused on exobiology and geochemistry to detect directly possible biosignatures of past and present Martian life. In addition, it aims to characterize the water and geochemical environment as a function of depth in the shallow subsurface.

To achieve the key science objectives, the ExoMars payload features a comprehensive suite called Pasteur Instrument Suite, comprising nine remote, contact, and analytical instruments [44, 45] supported by two robotic payloads (i.e., the drill and the sample preparation drawer system (SPDS)). The drill is capable of sampling at a depth of 2 m [46] hence is critical to enable the sampling below the anticipated 1-m thick oxidation layer anticipated to be unfavorable to life. Once acquired, the samples are deposed in a small drawer in the SPDS and prepared through a complex chain of preparation that crushes and grinds the samples before being distributed to various analytical instruments of the Pasteur suite [47] (Figure 2.18).

Figure 2.18 ExoMars rover: artist impression. (Courtesy ESA.)

The rover body and its external appendages accommodate a number of instruments including [44]:

- a Panoramic Camera (PanCam) on the top of the mast, consisting of a wide-angle cameras (WAC) with multiple filters and a high-resolution camera (HRC) to help characterize the geology of the rover surroundings and identify potential science targets [48],
- the "infrared spectrometer for ExoMars" (ISEM) at the top of the mast that provides bulk mineralogy characterization, remote identification of water-related minerals and support the PanCam with the target selection process,
- a GPR dubbed WISDOM (i.e., Water Ice and Subsurface Deposit Information On Mars) [49, 50] located at the back of the rover,
- a Mars multispectral imager for subsurface studies (Ma-MISS), an infrared spectrometer located inside the drill bit,
- a close-up imager (CLUPI) that provides microscopic imaging of targets and the sampling locations [51],
- a neutron spectrometer (ADRON) to quantify the amount of subsurface hydration and the possible presence of water ice.

Inside the rover, the analytical laboratory drawer (ALD) is kept at a temperature below freezing and comprises the SPDS and three main instruments [45, 52]:

- Mars Organic Molecule Analyzer (MOMA) that looks for organic molecules extracted from samples through thermal volatilization and laser desorption,
- the Infrared imaging spectrometer (MicrOmega-IR), that provides mineralogy analyses to infer past and present geological processes,

- a Raman spectrometer that helps to further identify organic compounds and biosignatures [53, 54].

The total mass of the rover reaches 310 kg including a science payload mass of ~26 kg. However, the combined mass of science payloads and robotic payloads (drill and SPDS) rises to 86 kg [55], providing a payload to platform mass fraction of about 27%. The rover platform is 1.2-m long, 1.1-m wide, and 2-m tall including the mast. To provide the necessary mobility to the payloads, the rover vehicle consists of a capable locomotion subsystem providing both six-wheel drive and six-wheel steer [56]. Unlike any previous rovers, ExoMars rover allows the platform to "crab" along the surface if appropriate in order to precisely control the heading while traversing and placing the drill. The passive suspension system consists of a three-bogie configuration with two lateral bogies at the front and a transverse bogie at the back. This configuration provides the same six-wheel contact on uneven ground and similar mobility performance as the NASA-JPL's rocker-bogie suspension. However, it removes the need to implement a differential linkage as implemented internally within Sojourner and MER through an internal geared shaft or externally on Curiosity through a mechanical linkage around the rover body. The 30 cm diameter wheels implement a flexible-wheel concept, optimized to provide more compliance with the surface and increase the contact patch with the terrain [57, 58]. As a result, it improves the traction performance over typical solid wheels, enabling the platform to overcome a greater range of hard and soft obstacles.

The rover power subsystem comprises a set of fixed and foldable solar panels capable of producing 1200 W h working in combination with a Li-ion battery of 1142 W h (nominal) [59]. The body of the rover consists of two separate compartments, each with its specific thermal requirement and thermal design [60]. The service module (SVM) is a warm enclosure that houses the rover electronics and batteries, and a cold enclosure houses the ALD and maintains the samples in a frozen state. The thermal control hence must accommodate both temperature zones effectively throughout the mission lifetime. The thermal control of the warm enclosure relies on two Russian RHUs (8.5 W each) and a set of loop heat-pipes (LHP) to transfer the excessive heat to side-mounted radiators. Similarly, the heat generated by the instruments in the ALD is managed with LHP and in places thermoelectric coolers (TEC) to keep specific elements of the instruments at low temperatures [60, 61].

The rover concept of operation relies on the introduction of more onboard autonomy compared with past missions. With a traverse target of 70 m/sol and a range of science activities to perform daily, the rover is being designed to operate without ground loop for two sols including autonomous traverses if required [62]. To achieve this level of autonomy, the platform implements a complex navigation system where 3D digital elevation maps (DEM) of the terrain is generated [63], obstacles are identified, and an optimized path is planned. The rover is then dispatched along that path with accurate trajectory control [64, 65]. Finally, a number of supervisory fault detection, isolation, and recovery (FDIR) functions constantly monitor the state of the rover during the traverse to guarantee its safety.

2.3.2.8 Mars 2020 Rover

The NASA's Mars 2020 mission is slated to be launched in 2020 and features a rover that builds heavily on the Curiosity heritage. It is currently set to advance the astrobiological exploration of Mars, spearheaded by Curiosity and ExoMars [66, 67]. The mission objectives, therefore, focus on exploring locations susceptible of harboring past and present traces of life. In this instance, the rover will not only accommodate a wide range of onboard payload but also carry the necessary hardware to acquire a number of samples and cache them into sealed containers that could be retrieved by a future mission and returned to the Earth for further analysis [68]. In addition, the mission aims to demonstrate a number of valuable technologies for future robotic and manned missions, such as extraction of oxygen from the Martian atmosphere [69]. Similar to Curiosity, the rover platform will be powered by an RTG, initially put aside as a spare unit for the Curiosity mission.

2.3.3 Future Mission Concepts

2.3.3.1 Toward New Business Models

Beyond the original space race in the 1960s to 1970s when rockets and spacecraft were launched at a breathtaking pace, one could think the inertia and drive to explore extraterrestrial landscapes had since waned. However, while space exploration is a more calculated affair, the drive to explore today is as strong as ever. The world is seeing a resurgence of enthusiasm for space exploration, thanks to two key aspects challenging the establishment and the *status quo*: the growing maturity of new players such as China, Japan, and India among others, as well as the rise of new commercial endeavors from private companies such as SpaceX, Orbital Sciences, or Reaction Engines Ltd, aiming to provide cheaper access to space.

The Ansari XPRIZE, started in 1996, has been a pivotal moment where 26 teams from 7 nations competed to *build and launch a spacecraft capable of carrying three people to 100 km above the Earth's surface, twice within 2 weeks* [70]. On October 4, 2004, the Mojave Aerospace Ventures's Scaled Composites team won the $10 million prize after developing a new vehicle and launcher unlike any other built to date. Beyond the success of the winning team, the activity generated more than $100 million of separate investment [70], significantly pushing the field of cheaper suborbital access. Since the completion of the Ansari XPRIZE, the same model has been adopted to encourage technological developments that could benefit mankind, including the Google Lunar XPRIZE. The $30 million competition, announced on September 13, 2007, requires its competitors to successfully land and operate a mobile platform on the Moon across a 500 m traverse while transmitting back high-definition pictures and videos. The activity similar to the original XPRIZE aims to inspire and foster new private investments to develop cost-effective technologies for exploration of other planets.

Table 2.8 Planned lunar missions.

Country	Mission	Launch	Comments
Russia	Luna-Glob 1 (Luna 25) [71]	2016	Lunar pole lander
India	Chandrayaan-2	2017	Landing, roving
China	Chang'E 5	2017	Sample return
Japan	SLIM	2019	Pinpoint landing
Russia/ESA	Luna Resurs (Luna 27) [71]	2019	South pole exploration
China	Chang'E 4	2018/2019	Landing and roving
China	Chang'E 6	2020	Sample return
NASA	Resource Prospector Mission	2020	ISRU rover, prospection
Russia	Luna-Grunt rover (Luna 28)	2025	Rover
Russia	Luna-Grunt sample return (Luna 29)	2025	Sample return

2.3.3.2 Medium-Term Mission Concepts

Building on the past robotics missions and new knowledge gained about Mars and the Moon, new mission concepts are being investigated by major space agencies and commercial entities. Table 2.8 summarizes the timeframe of various planned future lunar exploration missions as announced by 2015. This is in addition to the planned Mars missions detailed in Section 2.3.2.

2.3.3.3 Long-Term Mission Ideas

Despite the successful exploration performed to date of the solar system, planetary robotic systems have literally only scratched the surface. To further advance our knowledge of these destinations, a cornucopia of missions and platforms have been proposed to explore the vast swathes of unexplored landscapes. It is impractical to try to provide an exhaustive list containing every concept that has ever been proposed in the literature. Therefore, the following three tables give an organized view and summary of these future mission ideas in terms of robotic platforms (Table 2.9), mission concepts for different extraterrestrial targets (Table 2.10), and operational concepts for different mission scenarios (Table 2.11).

2.4 Environment-Driven Design Considerations

As the robotics systems are deployed to a range of planetary surfaces, they have to contend with a wide range of environmental conditions and deployment options. For example, a planetary body with atmosphere that allows the use of parachutes will present fundamentally different design challenges compared with airless moons where a hard landing (such as using a penetrator) or powered landing is necessary. Although there are some commonalities between various mission scenarios, a robotic system needs to be designed for the specific planetary

Table 2.9 Future missions: robotic platforms.

Platform	Implementation
Rover	– Wheels
	– Tracks
	– Legs
	– Rolling, for example, ball or sphere
	– Small hopper, for local exploration
	– Overcraft
Lander	– Large hopper, for regional or global exploration
Airborne	– Quadcopters, helicopters, or ornithopters
	– Planes or gliders
	– Balloons
Subsurface	– Submarine
	– Subsurface, for example, corer, ice drilling or melting
Ship	– Robotic buoys
	– Ship

Table 2.10 Future missions: mission destinations and concepts.

Destination	Mission concepts
Mercury	Landers [72, 73], rovers [74]
Venus	Balloons [75, 76], landers
Mars	Airborne [77–79], swarms [80, 81], hopper [82–84], sample return, ISRU [85–87]
Moon	Sample return, ISRU, exploration of permanently shaded craters, prepare for manned base
Titan	Balloon [88], lander [89], co-operative platforms, submarine [90–92], ship [93]
Europa	Subsurface [94], submarine [95], hopper
Enceladus	Subsurface [96], submarine, hopper
Gas giants	Balloon [97]
Asteroid	Hoppers [98, 99], sample return

Table 2.11 Future missions: operational mission concepts.

Operational concepts	Mission scenarios
Teleoperation	– Earth to Moon or Moon orbit
	– Mars/Phobos orbit to surface
Homogeneous Systems	– Multiplatforms
	– Swarms $\gg$10 platforms
	– Self-replicating
Heterogeneous Systems	– Collaborative rover platforms, for example, cliff climbers
	– Aerobots, lander, rovers

target and mission hence take into considerations of specific nature of relevant planetary environment.

This section presents environmental conditions of some popular targets for robotic exploration, and discusses their implication and impact on the system design of robotic systems.

2.4.1
Gravity

- **Descent and landing design:** An airless and low gravity body such as an asteroid or a small moon may not require a powerful descent system and can rely instead on controlled free fall to the surface. On the other hand, a larger planet such as Mars would require some form of decelerator or powered landing to reach the ground at a speed compatible or safe to the landing system. These factors drive the mechanical and structural design of the robotic platform in order to survive the landing loads and vibration of the landing phase.
- **Actuators design and sizing:** As the gravity increases, the weight of a system increases. This increases the torque required from the actuators to perform their tasks (e.g., rover wheel steering and manipulator joint), which affects the size and mass of the actuator design as well as the supporting structure.
- **Structure and actuator dynamics:** During the operation of a robotic system such as a manipulator, the local gravity will affect the dynamic behavior of the joint. On larger planetary bodies, gravity provides a damping effect on the structure of the arm and a constant loading on the joint. In a free fall or low-gravity scenario, this damping effect is minimal and changes the dynamic characteristics of the robotic system hence takes more time to settle. These situations, therefore, require different control approaches.

2.4.2
Temperature

As an exploration tool, the robotic system can be exposed to a wide spectrum of temperatures from extremely low to high. Consider the Moon as an example, temperature at the equator can vary between 100 K in the dark or in deep shadows to 390 K in full sun with a thermal gradient ~300°; at the poles, permanently shaded craters can feature temperatures down to 80 K, making these location a prime target for scientists and engineers for detecting water ice. Even local boulders and terrain depressions can affect the direct solar and infrared fluxes received by the system causing extreme regional gradients. According to transient thermal analyses [100], the total infrared heat flux experienced by a rover platform can increase by up to 331% when it transits in front of sunlit boulders.

- **Thermal control:** The thermal design and control of the robotic system is critical to ensure the survivability in both hot and cold cases. Not only the system must take into account the thermal energy coming into the system (e.g., from the Sun and the surroundings), but it must also deal with the inherent internal power dissipation from the onboard subsystems such as the computer, communication electronics, power conditioning unit, and actuators including the motor and gearboxes. To control the temperature of the system, a range of heaters, thermistors (i.e., temperature sensors), and radiators are used to keep the internal temperature to within the safe operating range of the units. However, a daily operation cycle of the system in both lit and dark environment inevitably imposes contradictory requirements [101]. During operation in a lit environment, the system needs to dissipate the surplus of thermal power through adequately sized radiators that must be kept from facing the Sun. The more powerful or the hotter the system, the larger the radiators need to be. Conversely, hibernation of the platform or operation in a dark location aims to keep the internal heat within the body of the robot, hence the system needs to minimize the radiator size to minimize heat loss.

- **Actuator design:** Temperature drives the design and implementation of actuators in a number of ways. As various materials and metals are heated up and cooled down to very low temperature, they expand and contract based on their own coefficient of thermal expansion (CTE). When parts made out of different materials interact (e.g., within a mechanism), the matching of CTE between different parts can be of importance to make sure the actuator does not seize due to an expanding or contracting piece. Similarly, lubrication of such mechanism across a range of temperature is critical to ensure necessary longevity of the actuator. The field of *tribology* typically addresses the challenges of "interacting surfaces in relative motion," and focuses on the principles of friction, lubrication, and wear. The space system design concentrates on limiting both friction and wear through either solid lubricants such as coatings or liquid lubricants such as greases and oils [102]. At very low temperatures, for instance, below $-75\,°C$, even greases and oil can become solid. Local heaters are, therefore, used to facilitate the flow of lubricant through the mechanism. Martian surface systems such as the Sojourner and MER rovers have used perfluoropolyethers (PFPE) grease as lubricant [102]. Similarly, Beagle 2 and MSL mobility mechanisms/arm have used sputter-deposited molybdenum disulfide (MoS2) coating [38, 103]. Additional discussions on lubrication of space mechanisms and the specificity of different lubrication options can be found in Refs [104, 105].

- **Electronics design:** Similar to the actuators, the design and operation of electronics over a wide range of thermal conditions need to be carefully thought through. The location of the electronics, inside or outside a thermally controlled housing, will significantly drive the design and implementation of the system. As discussed earlier, the CTE plays a significant role as each of the components on a printed circuit board (PCB) can deform differently from the board and the solder, potentially creating fatigue through thermal cycling or cracks potentially leading to loss of functionalities or overall failure of the board. The NASA Extreme Temperature Electronics Program provides a valuable account of the effect of temperature on a range of circuits and devices [106]. Over the last decade, new developments to implement "cold electronics" have made significant steps to allow electronics operate outside a warm box in temperatures ranging from −120 to +220 °C. To cope with these extreme temperatures, implementations such as the ESA Motion-Control Chip developments [107] forego with the packaging of the electronic chips itself and uses the bare die mounted on a highly thermally conductive silicon substrate. This method also has the benefit of reducing the footprint of the chips, hence the overall board, which can also benefit the boards inside warm enclosures. Unlike cold electronics, typical boards need to be thermally controlled within a narrower operating temperature range (−20 and +20 °C) to keep them running in optimum conditions. These designs, therefore, concentrate on heat rejection and local thermal control should the temperature of the board drop too low.

2.4.3 Atmosphere and Vacuum

During different mission phases, the robotic system needs to deal with a range of environmental conditions in both hibernation and operating states. The system can reach the deep vacuum of space within minutes from the launch pad on Earth with atmosphere. Depending on the target destination, it either stays in vacuum for months/years (e.g., operation on an airless body) or encounters pressure if the destination supports atmosphere. The presence or absence of the atmosphere and its composition drive a number of design aspects.

- **Thermal subsystem:** As discussed earlier, adequate thermal management of the platform is crucial to the reliability and survival of the robotic system. Atmospheric or vacuum environment will affect the ability of the system to dissipate efficiently the heat generated. Thermal heat dissipation relies on conduction, radiation, and convection. Without air, convection is not possible; therefore, the system needs to rely on the other modes of heat transfer. A typical probe to Mars will need to survive long months of cruise in vacuum and perform equally well on Mars, which only has a thin atmosphere. Therefore, during cruise, the body of the rover is thermally connected to an external radiator located on the outside of the transfer vehicle (e.g., aeroshell) to radiate the extra thermal energy into the space. The problem is, however, exemplified if the system uses RHU or RTG both relying on the heat dissipation of radioisotopic material, which by definition cannot be "turned off." The design of the thermal control during the cruise phase is especially challenging to make sure the system does not overheat. Once on the surface, the thermal subsystem is sized to meet the capability of the environment to dissipate the heat from the platform. In vacuum, a radiator needs to avoid facing the Sun or any illuminated face that radiates heat toward the platform (e.g., lunar surface). A range of design options are available to limit the exposition of the radiator to the Sun or the surface, either by providing physical shielding or by designing reflectors such as the paraboloid radiator reflectors used during the early days of lunar exploration.

In a thin atmosphere like that of Mars, the convection is significantly reduced compared with a typical terrestrial atmosphere but still allows the surface systems to dissipate their heat more efficiently. Rovers such as the MERs and the planned ExoMars use RHUs to support the thermal control at night. This heat needs to be managed during the day when all the onboard systems are running concurrently. Similarly, the Curiosity rover uses an RTG as the sole power generation system, which requires a rather large setup of radiator surfaces on the RTG. The efficiency of these radiators is important to maximize the efficiency of the thermal control that relies on the temperature difference between the radioisotope and the environment to generate energy through Peltier devices.

- **Material and structure:** Despite the relatively short history of space exploration, robotic systems have already explored the whole spectrum of pressure they are likely to encounter, from the vacuum of the Moon and the thin atmosphere on Mars, to the denser atmosphere of Titan, and the high pressure inferno of Venus with 92 bar on the surface at 464 °C. For a structure left open to the vacuum, the difference in pressure between the moment the material was manufactured and the vacuum means that the gases trapped during manufacturing tend to leak out of the part in a process called *outgassing*. While it may not be critical for a wide range of parts, it could adversely affect devices such as optical systems where the gases could condense or solidify on valuable parts such as the CCDs. Multilayer insulation blankets are especially perforated to allow the controlled escape of these gases. For enclosed systems, they can similarly be made to be "breathable" like a Martian rover body where a suitable valve allows the inside of the body to depressurize when going into vacuum and repressurize with the local atmosphere on Mars. Alternatively, a fully airtight system with its own pressurized vessel can present some benefits. For the Venera landers on Venus, the platforms offered a structurally efficient way to survive, albeit briefly, at high pressure and temperature, akin to the design of a deep sea submarine. They were also protected from the corrosive atmosphere containing sulfuric acid. A number of past Russian landers and more recently the Russian Fregat upper stage implemented a number of pressurized vessels housing the critical electronic systems, effectively side-stepping the typical issues of design, implementation, and operation in vacuum. This is a way to flight rugged COTS hardware with minimal modification. However, this comes at a price as the pressurized vessels require significantly extra structural mass that affect the overall launch mass of the spacecraft, hence the overall cost.

2.4.4
Orbital Characteristics

The orbital characteristics of the target body such as the distance to the Sun over a local year (a whole revolution around the Sun), the length of the local days, and the tilt of the body axis drastically affect the design and operation of the robotic systems. These parameters particularly drive the design of the power generation system, power storage, and more broadly the energy management of the system over the course of the mission.

2.4.4.1 Distance to the Sun

- **Power subsystem:** Distance to the Sun dictates the amount of energy the system can transform through photoelectric solar panels. As the energy reduces by the cube of the distance to the Sun, the amount of energy within the Earth orbit is ample and decreases significantly beyond Mars. The efficiency and benefits of solar panels, therefore, significantly decrease, favoring solutions such as the RTGs or solar concentrators that can deal with the lower solar energy away from the Sun. Past missions to the outer solar system (such as Saturn/Titan, Jupiter, and beyond) relied heavily on RTGs and RHUs. However, the only robotic platform that landed beyond Mars (i.e., Huygens) used only a primary, nonrechargeable battery to fulfill its short mission on Titan. To address the power generation challenges of future exploration missions to the Saturnian and Jovian systems, one can anticipate that a combination of these design options will be used to survive and explore the surface of these moons.
- **Communication subsystem:** The distance from Earth to the target body dictates the amount of *RF* power required to sustain communication between the ground control and the robotic system. In addition, it drives the selection of the type of antenna and whether there is a need for fine pointing accuracy that requires a dedicated pointing mechanism, adding extra mass. As a rule of thumb, the electrical power needed by a communication system is around 4 times the RF power. As such, a 5 W UHF transponder requires 20 W of electrical power. Direct-to-Earth communication may not always be possible for practical reasons, for example, limited power generation capabilities or limited direct view of the ground station. The use of a relay satellite can alleviate some of these issues by acting as an intermediary between the robotic platforms and the ground station. This then reduces the power requirements of the communication subsystem of the robot and can save significantly on mass. However, these gains would still need to be offset against the cost of the orbiter as part of the overall mission. Nevertheless, it is expected that the orbiter could also be used to fulfill other mission objectives such as providing science and surface mapping from orbit.

2.4.4.2 Length of Days

Power generation and energy management: The length of the local day and especially the night is critical to the sizing of the entire energy chain from power generation, conditioning to storage. In practice, for robots relying primarily on solar power, their daily operation cycle is centered around the local noon when the solar flux is at its strongest. It also needs to complete its day in a state that allows it to survive the night and power a range of heaters to keep the subsystems within their nominal temperature range even when the environment temperature plummets at night. Typical operation of a Martian rover cannot be performed over a typical 8 h/day. The platform needs to warm up in the morning, preferably through solar illumination, up to about 10:00. Then, nominal operation can proceed over the local noon until 14:00, maximizing the use of the solar energy supplemented with battery power. The rover then puts itself in a charging state to replenish the batteries until sundown so that it can support the necessary subsystems overnight. A Martian day (or a sol) is close to an Earth day at 24 h 45 min, and a day on the Moon is a succession of 15 Earth days of sunlight and 15 days of darkness, making the sizing of the power subsystem very challenging. The use of RTG can alleviate some of these operational constraints, however, the daily management of power is still required to ensure the total energy required by the platform is within the production capabilities of the RTG.

The inclination of the axis of the planet or the moon affects the seasons and the variability of the Sun's energy reaching the surface over a local year. High latitudes will see greater variability in solar fluxes compared with the equator, leading to greater variability in temperature and energy throughout the year. To ensure the survivability of the system, the sizing of the power system must be performed for the worst energy conditions, that is, the local "winter" that accounts for lower environmental temperature, reduced energy on the surface, and shorter days.

Beyond Mars that has a day and an inclination similar to Earth, the other planets and moons have significantly different characteristics that require a careful case-by-case evaluation on the design implications.

2.4.5
Surface Conditions

The surface conditions vary depending on the target destination. In most cases, the initial knowledge of the surface will be limited and based on orbital imagery and scarce *in situ* data from past probes. Dust, rocks, fluids, or ices are potential surfaces where the robotic platforms are required to operate on or in, which will affect the system design in many different ways.

2.4.5.1 Rocks
The size and distribution of rocks drive the design of a number of subsystems, which include the following:

- **Lander structure, descent and landing subsystem:** The descent and landing system needs to either provide the capability to avoid large obstacles during the descent phase (i.e., hazard avoidance) or ensure safe landing even if an obstacle is encountered (e.g., airbags). In addition, once landed, its structure and its configuration need to be able to deal with the possible obstacles without making the lander unstable, for example, landing on a rock or a slope.
- **Mobility subsystem:** For mobile platforms, the rock size and distribution drive the design of the locomotion subsystem and the guidance, navigation, and control (GNC) functionalities. The design trades therefore concentrate on either sizing the locomotion subsystem to overcome obstacles ultimately impacting

the system mass, or developing a capable GNC subsystem to avoid the obstacles, affecting the required onboard processing capabilities and the cost/risk involved in developing and testing advanced software.

2.4.5.2 Dusts

The dust drives the design of a range of subsystems including mechanisms, power, and thermal as well as the GNC to some extent. Most of the moons and planets have little or no weathering through water or aeolian processes unlike Earth. The resulting dust tends to have sharp edges making it particularly abrasive.

On Mars, winds and static charging are the main *vectors* of dust transportation. The yearly cycle of Mars also includes a "dust storm season" typically between the Solar Longitude (Ls) 180 (autumn equinox) and 330 (post-winter solstice at Ls 270). During this period, local and regional-scale storms can lift the dust up into the atmosphere, making it more opaque and affecting the amount of light reaching the surface. To quantify the state of the atmosphere, the optical depth parameter (OD) is defined and used as a measure of the attenuation of the radiant power coming from the Sun through the atmosphere, that is, how clear the atmosphere is on any given day. Starting at 0 for an extremely clear day, it rises to level 5 for a full dust storm with minimal visibility.

On the Moon and possibly other airless bodies, dust can also lift from the surface through static charging just before dawn, creating a very loose cloud of particulates that can then fall onto unprotected surfaces [108–110].

- **Power subsystem:** The OD can help size the power subsystem (e.g., to maintain nominal operation for up to OD 2) and provide a way to identify non-nominal situations that require the robot to hibernate due to lack of energy.
- **Thermal control subsystem:** Periods of high OD tend to be warmer due to the dust in the atmosphere, marginally reducing the need for sustained heating during the day of hibernation. However, by warming up the atmosphere it also reduces in density, hence the ability to provide effective braking for aerodynamic decelerators (i.e., aeroshell and parachutes). Finally, dust deposition or "sand blasting" of radiator surfaces will reduce their emissivity and, therefore, decrease the heat rejection capability of the system.
- **Mechanisms:** The abrasive nature of the dust allows it to seep into mechanisms through seals and reduce the efficiency of the actuators, leading to jamming and ultimately failure. Some dusts such as the lunar or Martian can be loaded with iron, making them especially prone to static charging through friction with the surroundings and attracted to metallic structures of the robot (such as the actuators), hence make the removal of the dust particularly difficult.
- **Solar panels:** Dust deposition on the solar panels can affect generation of solar power, as it slowly degrades the efficiency of the panel and limits the capability of the panel to produce energy.
- **GNC:** The above-mentioned issues also apply to all the optics on the robotic system, usually at the core of any GNC subsystem. For example, the robot would be sensitive to dust deposition if it uses a dedicated sun tracker to compute its heading.

2.4.5.3 Liquid

To date, all the surface missions have landed on firm grounds. However, Huygens was designed to float in case it landed on lakes or seas of methane on Titan. From buoyancy analysis to the design of a liquid proof system, implementation of such a probe is still fairly unconventional but will become a vanguard for things to come. As concepts are being drafted to target Titan or the ice-covered oceans on Europa, floating or submersible robots will be needed in the future. The development of these robots will certainly draw experience from relevant terrestrial systems (e.g., robotic sea exploration), but still needs to address design challenges specific to the space environments such as the local power generation or effective communication with Earth.

2.4.6
Properties of Planetary Bodies and Moons

Tables 2.12 and 2.13 summarize some key parameters of the rocky planets and a few popular moons in the solar system. These data can be used to infer and derive how the characteristics of these extraterrestrial bodies will influence the design of robotic exploration missions and systems.

2.5
Systems Design Drivers and Trade-Offs

The design and operation of a robotic system to perform planetary surface exploration is a multidisciplinary activity. Due to the interleaved nature of the mission objectives, platform design and operations, it is necessary to identify as early as possible the key design drivers to build an understanding of the system as a whole. This is performed by identifying the impact of the system operation in the anticipated environment over the projected length of the mission, as well as identifying and mapping the design and operation relationship among each subsystem comprising the robotic platform.

Though the top-level assessment can appear straightforward, delving into a detailed investigation can be deceptively complex. It is, therefore, valuable to spend enough efforts to evaluate these aspects to prevent jumping too quickly into subsystem design activities. This can significantly narrow down the design space and implementation options too early, potentially leading to costly repercussions if the selected option needs to be changed or redesigned. Spending adequate time in system-level design is, therefore, important to achieving a robust design that fulfills the key mission requirements. It can also provide valuable data and tools to perform an optimization of the whole system later in the design phase, keeping in mind that an optimized system may not be a system composed of optimized subsystems.

This section introduces a systematic investigation of system design drivers, starting with relatively generic mission-driven factors followed by subsystem

Table 2.12 Rocky planets' characteristics [111].

Property	Mercury	Venus	Earth	Mars	Pluto
Mass (10^{24} kg)	0.33	4.87	5.97	0.642	0.0131
Diameter (km)	4879	12 104	12 756	6792	2390
Density (kg m^{-3})	5427	5243	5514	3933	1830
Gravity (m s^{-2})	3.7	8.9	9.8	3.7	0.6
Escape velocity (km s^{-1})	4.3	10.4	11.2	5	1.1
Rotation period (h)	1407.6	−5832.5	23.9	24.6	−153.3
Length of day (h)	4222.6	2802	24	24.7	153.3
Distance from Sun (10^6 km)	57.9	108.2	149.6	227.9	5870
Perihelion (10^6 km)	46	107.5	147.1	206.6	4435
Aphelion (10^6 km)	69.8	108.9	152.1	249.2	7304.3
Orbital period (days)	88	224.7	365.2	687	90 588
Orbital inclination (°)	7	3.4	0	1.9	17.2
Orbital eccentricity	0.205	0.007	0.017	0.094	0.244
Axial tilt (°)	0.01	177.4	23.4	25.2	122.5
Mean Temperature (°C)	167	464	15	−65	−225
Surface pressure (bars)	0	92	1014 mbar	6.36 mbar at mean radius (4.0–8.7 mbar depending on season)	0
Number of moons	0	0	1	2	5
Global magnetic field	Yes	No	Yes	No	TBC
Atmospheric composition	–	CO_2 96.5%; N_2 3.5%	N_2 78.08%; O_2 20.95%	CO_2 95.32%; N_2 2.7%; Ar 1.6%; O_2 0.13%; CO 0.08%	–
Temperature range	–	Average 737 K/ 464 °C; Diurnal range: ~0°	Average 288 K/15 °C; Diurnal range: 283–293 K (10–20 °C)	Average 210 K/ − 63 °C; Diurnal range: 184–242 K (−89 to −31 °C) (Viking 1 site)	–

Table 2.13 Characteristics of popular moon targets [111].

Planet	Earth	Jupiter				Saturn
Moons	Moon	Callisto	Ganymede	Europa	Io	Titan
Diameter (km)	3475	4821	5262	3122	3643	5150
Mass (10^{21} kg)	73.5	107.6	148.2	48	89.3	134.6
Density (kg m^{-3})	3340	1830	1940	3010	3530	1881
Gravity (m s^{-2})	1.6	1.24	1.43	1.31	1.8	1.35
Escape velocity (km s^{-1})	2.4	2.4	2.7	2	2.6	2.6
Rotation period (h)	655.7	400.5	171.7	85.2	42.5	382.7
Orbital distance (10^3 km)	384	1883	1070	671	422	1222
Periapse (10^3 km)	363	1870	1068	664	420	1186
Apoapse (10^3 km)	406	1896	1072	678	424	1258
Orbital period (days)	27.3	16.7	7.2	3.6	1.8	15.9
Orbital inclination (°)	5.1	0.51	0.21	0.47	0.04	0.33
Orbital eccentricity	0.055	0.007	0.0015	0.0101	0.004	0.029
Axial tilt		0	–	–	–	0
Mean temperature (K)	**Equator** min 100 K mean 220 K max 390 K 85°**N** min 70 K mean 130 K max 230 K [112]	min 80 K mean 134 K max 152 K	min 70 K mean 110 K max 152 K	min 50 K mean 102 K max 125 K	min 90 K mean 110 K max 130 K	−93 K
Global magnetic field	No	No	Yes	No	Possible	Unknown

design factors that are subjective to mission and platform specifics. A case study is used here to illustrate how the various subsystems relate to each other and ultimately drive the overall design.

2.5.1
Mission-Driven System Design Drivers

The discussion of mission-level constraints requires identification of the flow of requirements that are inherited from the mission and flown down to various system platforms (e.g., lander or rover) and their respective subsystems. These top-down mission requirements are, as far as the robotic system is concerned, inherent to the mission, hence impossible to change, for example, the need to operate in vacuum or the requirement for surface deployment through hard landing. As a result, these constraints are not opened for trade-offs at the subsystem level and must be taken for granted.

Alternatively, some constraints emerge during the design process that cannot be fully consolidated at the beginning of the assessment, and are fed from the subsystem level to the system/mission level. For example, the launch mass of the mission is only known after a first iteration of the design, including cruise stage, propellant, delivery vehicle, and the landed element. This then drives the selection of the launcher (if not already a constraint) and only then confirms the launch environment that the system needs to survive. Similarly, if the robotic systems require a complex communication scheme, this may constrain the deployment or operation of other mission elements (such as an orbiter) significantly beyond the design boundaries of the robotic system itself.

2.5.1.1 Mass

For any mission, mass is one parameter that needs to be minimized at all levels since it will single-handedly drive the sizing of the entire mission. At the component, unit, subsystem, and system level, every effort must be made to minimize the expense of mass. At the mission level, mass of the delivery system (such as the cruise stage, propellant, structure, or surface deployment) is driven by the mass of the landed element. This means that a small increase in the landed system mass can lead to a major increase in the landing system mass (including structure, propulsion, and parachute size), leading ultimately to much bigger launch mass, such as the snowball effect. In the context of a small mission to Mars according to Ref. [113], one additional kilogram on the Martian surface requires an additional 1.5 kg for the EDL system. Once the impact on fuel and carrier structure is taken into consideration, every kilogram on Mars requires the launch mass of 5 times this amount. At the market rate for launch in 2010 [114], a kilogram in geostationary transfer orbit (GTO) costs around $20k to 30k and on Mars costs around $100k to 150k, which can lead to the selection of a larger, more expensive launcher and, therefore, a more expensive mission.

2.5.1.2 Target Environment

As discussed previously in Section 2.4, the target environment provides a range of bounding requirements that constrains either the design or operation of the robotic system and subsystems. Among others, gravity, temperature, atmospheric condition or vacuum, orbital characteristics, and surface conditions will drive significantly the sizing and design of many subsystems and ultimately the mass of the robotic system.

2.5.1.3 Launch Environment

The launch phase provides a harsh and mechanically demanding environment. Within minutes, the system goes from the launch pad to space during which it is subjective to a wide range of mechanical vibrations, acoustic noise and vibrations (through air pressure), sudden shocks during the separation of various stages, and finally relatively quick depressurization as the launcher transitions from the Earth atmosphere to the vacuum of space.

2.5.1.4 Surface Deployment

As mentioned previously in Section 2.4, the method used to deploy a robotic system from orbit to the surface will affect the design of the surface element including any robotic system onboard. If the target does not have an atmosphere, the use of a powered landing system based on thrusters is likely to be necessary to achieve a landing velocity compatible with the mechanical design of the landing system. During descent, shocks and vibrations from the thrusters and the final touchdown are transmitted to the robotic elements. These need to be sized appropriately to survive this critical phase, particularly when the landing happens to be non-nominal as experienced by the Philae lander (as described in Section 2.3.1 the lander could not secure touchdown in the first attempt and rebounded another 2 times).

If the target destination possesses an atmosphere, the lander needs to be protected by an aeroshell covered with a thermal protection system (TPS). As it enters and descends through the atmosphere, the aeroshell can provide passive and progressive deceleration. Additional instantaneous shocks or deceleration events consist of the aeroshell separation, parachute deployment, and the final touchdown. All these events and their magnitudes depend upon the selected entry and descent strategy as well as the selected landing system. In 1996, the MFP mission deployed a 360 kg lander carrying the 11.5 kg Sojourner rover platform. Figure 2.19a shows that the mission used a combination of aeroshell, parachute, and solid retrorockets. Figure 2.20a further illustrates that significant loading was produced during entry/descent, that is, loads up to 15.9 g during entry and 6.2 g during parachute opening. The EDL subsystem was designed to almost stop the lander at an altitude of 20 m above ground where the bridle was cut, leaving the spacecraft to freefall the remaining height. The impact was cushioned by an airbag-based landing system. The lander and rover systems were designed to sustain these repetitive loading over a number of bounces on the Martian surface until it finally rested. As shown in Figure 2.20b, 10–15 g were measured both in the normal and axial directions, adding significant stress to the system during the landing phase with an impact velocity of 10 m s^{-1} [4].

For comparison, Figure 2.19b shows the deployment of the Curiosity rover, which shared some similarities in terms of using the aeroshell and parachute deployment (but with 12 g peak deceleration and 6 g parachute opening). However, MSL mission implemented a novel landing system relying on an innovative sky crane concept to provide the final descent and deployment of the rover without the use of a supporting lander. The EDL subsystem provided a controlled touchdown of the robotic platform at 0.77 m s^{-1} [115].

2.5 Systems Design Drivers and Trade-Offs | 61

- HRS Freon Venting: Landing – 1:35 min
- Cruise stage separation: 8500 km, 6.1 km s^{-1}, Landing – 35 min
- Entry: 130 km, 7470 m s^{-1}, Landing – 304 s
- Peak heating/deceleration: 40–80 W cm^{-2}, 15.9 g, Landing – 228 s
- Parachute deploymet: 9.4 km, 370 m s^{-1}, Landing – 134 s
- Heatshield separation: Landing – 114 s
- Lander separation: Landing – 94 s
- Bridle deployed: Landing – 84 s
- Radar ground aquisition: 1.59 km, 68 m s^{-1}, Landing – 114 s
- Airbag inflation: 355 m, Landing – 10.1 s
- Rocket firing: 98 m, –61. m s^{-1}, 3.4 g (6 g pk), Landing – 6.1 s
- Bridle cut: 21.5 m, +1 m s^{-1} residual, Landing – 3.8 s
- Landing: 14 m s^{-1}, 18 a. @2:58 am Mars LST. @16:56:55.3 UTC
- Bounces: >15, 1st was 15.7 m high, distance 1 km?
- Deflation/petal latch firing: Landing + 20 min
- Airbags retracted: Landing + 74 min
- Petals opened: Landing + 87 min

(a)

(b)

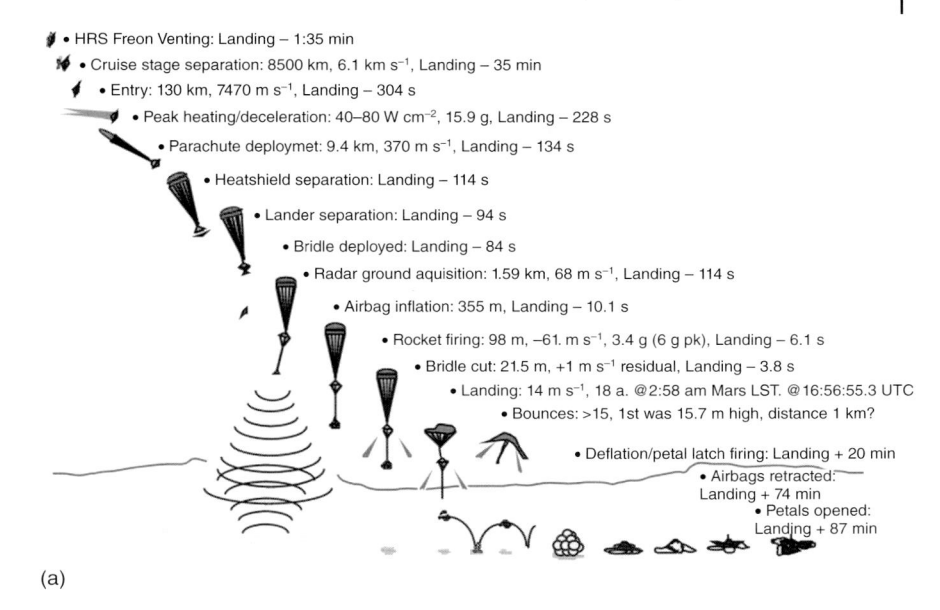

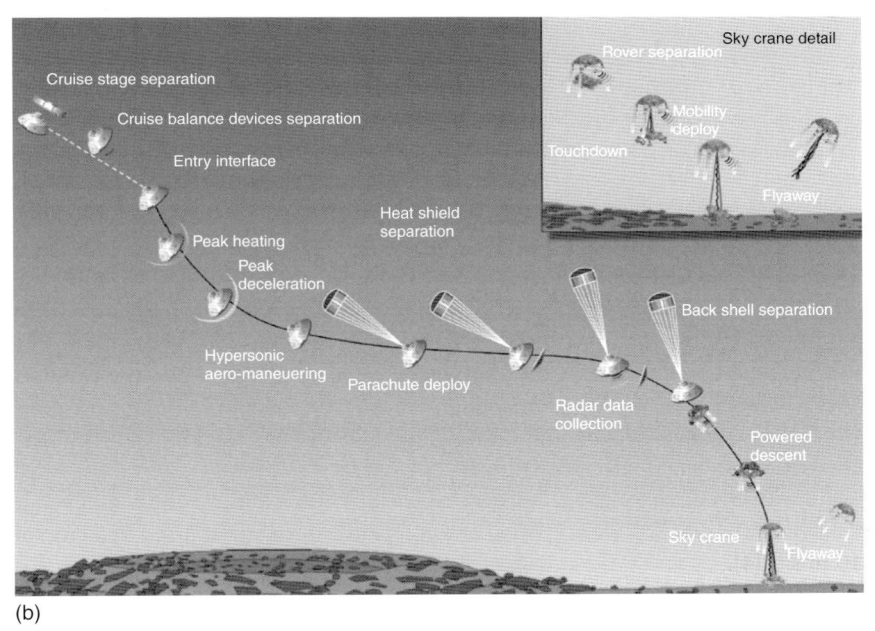

Figure 2.19 EDL sequence. (a) MPF [4] and (b) MSL [115]. (Courtesy NASA/JPL-Caltech.)

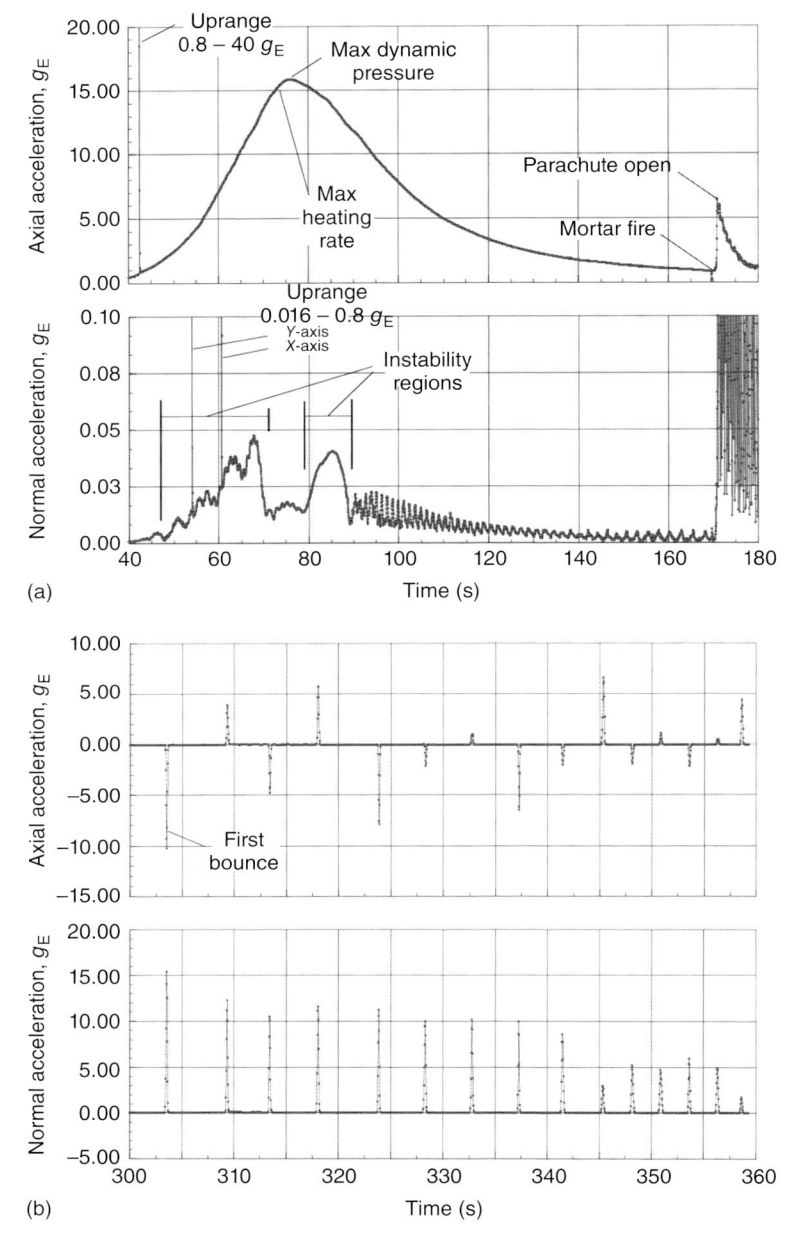

Figure 2.20 MPF landing [4]. (a) Deceleration profile and (b) Loads profile.

2.5.1.5 Surface Operations

Mission operational scenarios (also called Concept of Operations or ConOps) capture various phases of the mission operation, the elements involved and the functionalities expected of them over time. The operational scenario of the mission or the surface operations (in the case of planetary robots) drives the sizing of various subsystems of the robotic platform. For example, a mission required to last only days will be designed, built, and operated very differently from a mission required to work over several years, due to difference in the required longevity of all the subsystems such as the mechanisms and the power subsystem particularly exposed to the environment. Similarly, a teleoperated lunar mission with a delay of a few seconds will not have the same operational needs of a Mars mission where teleoperation is not feasible due to much longer communication latencies.

2.5.2
System Design Trade-Offs: A Case Study

To discuss some of the design challenges at the system level, a future planetary mission scenario is used here to demonstrate key design drivers and trade-off options to evaluate different robotic system design options.

2.5.2.1 Mission Scenario Definition: MSR/SFR

Since the Viking missions, the return of samples from Mars has been seen as the next critical step to advance our understanding of Mars and more generally the formation of the solar system. The goal of an MSR campaign would be to return a number of samples of Martian soils to Earth for analysis in detail. For a number of years, NASA and ESA have drafted mission architectures and concepts for an MSR Program [116], including a joint venture (circa 2006) planned for the launch timeframe 2018–2023 but subsequently canceled due to budgetary issues. Nevertheless, the scientific community remains committed to support and pursue such a mission in the decades to come, potentially starting with the launch of a caching rover in 2020. In the meantime, other old and new space players (such as Russia or China) have also thrown down the gauntlet with new plans to return fresh materials from the Martian surface, stemming potentially the start of a new space race to Mars.

The MSR mission concepts have evolved since its initial conception. The current architecture as drafted by ESA and NASA assumes a program composed of three missions [116]:

- **Caching rover deployment and operation:** The rover will acquire and store a range of surface samples into a container (i.e., a cache) that will be left on Mars.
- **MSR orbiter mission deployment:** An orbiter will act as a telecommunications relay to provide critical event coverage during the future MSR lander descent and ascent, as well as to capture the orbiting sample and return it to Earth.
- **MSR lander mission deployment:** A single-purpose fetching rover will pick up the sample cache and return it to the Mars ascent vehicle (MAV) located on the MSR lander.

In the remainder of the section, the sample-fetching rover (SFR) concept is chosen to discuss the impact of mission goals and requirements on the design of various rover subsystems. To date, exploration rovers have been supporting science missions dedicated to the *in situ* characterization of the environment, carrying a suite of instruments to analyze the area accessible by the rover. Therefore, these rovers traverse until they reach a prescribed target, stop, perform the necessary science data acquisition, and move on. The SFR is unlike any past and current rovers due to the unique challenges defined by the MSR mission, including the following:

- The rover must traverse 15 km over the course of a 180 sol nominal mission (120 sols of actual traverse), leading to a challenging, minimum average traveling speed of 120 m/sol on a variety of terrains while locating, navigating, and retrieving the sample cache.
- The rover must be as light as possible while providing the necessary robustness expected of the mission.
- The primary objective is not to acquire science from various locations but to retrieve a sample cache from a know location.

The nominal SFR surface operation is limited to 180 sols including post-landing checkout, egress, and rover commissioning. Over the course of the mission, additional time allocations must be considered to identify the sample cache location, approach, and pick up the sample. Return operations to the MAV also need to be factored in. As a result, the total time available for the 15 km traverse is at maximum 125 sols. This includes an operational contingency of 25 sols allocated to consider effects such as adverse weather conditions (e.g., dust storm), reduced power generation/mobility (e.g., due to dust or sand trap), and also the operational impact of evolving from a cautious direct-drive mode in the first sols onto the more autonomous drive. The net traveling speed for the rover should achieve a minimum average rate of 120 m/sol. It is worth noting that the NASA SFR mission baseline aims to complete the cache return operations within 3 Earth months or 90 sol. This would lead to a speed of about 166 m/sol. Given the current assumption for the landing is approximately 3 months before the global dust season, maximizing the traverse speed would be beneficial to the mission concept as high optical depth would impair the rover mobility and its return to the MAV hence jeopardize the timeline of the overall mission.

2.5.2.2 SFR System Design Drivers

To account for the cache position knowledge error and inherent landing dispersion errors, the mission-driven requirements require the SFR to traverse 15 km in less than 6 months to meet the optimum window of opportunity for the orbiter journey back to Earth. Given the fact that existing rovers are only capable of traveling a fraction of the required distance in the required time (e.g., MER covered 15 km in 7 years), this requires drastic improvements in the rover's mobility driven by the platform design.

The term **mobility** is used here referring to both the locomotion and the GNC subsystems, which are interleaved, for example, a more capable locomotion subsystem can negotiate larger rocks and decrease the need to avoid them or to the complexity of the GNC subsystem. Similarly, a highly capable GNC subsystem can avoid rocks efficiently and minimize the need for a capable locomotion subsystem. z, environmental conditions play an important part in driving the design by considering the season, optical depth, surface and terrain conditions, temperatures, and so on.

In the MSR/SFR mission scenario, relevant requirements and constraints can be synthesized into three overarching operational factors that govern the rover design, namely [117]:

- **factors relating to path length** that tend to increase the distance the rover has to travel to reach the target location,
- **factors relating to speed along the path** that tend to decrease the average speed of the rover while reaching to the target location,
- **factors relating to the time available** that tend to decrease the time available to the rover to reach the target location.

These factors directly or indirectly drive the design of the rover's mobility and power subsystems, which are primary contributors to the overall mass and in turn affect the design of the other subsystems of the rover. To capture the system dependencies, a range of visual aids or tools can be used among which is the *ripple graph*. This tool can help visualize how a specific requirement on one subsystem ripples through the system, similar to throwing a rock in a pond that creates propagating waves across the surface. This is especially useful to identify the impact of a requirement on a system with no direct or obvious link to the requirement. To create the graph, the first step is to identify all the subsystems and drivers involved in the design and capture how they are linked with each other. For the SFR, based on the three operational drivers identified earlier, and the knowledge of typical rover subsystems, the main interdependencies within the rover are identifiable that ultimately drive the system mass. The results are illustrated in Figure 2.21, starting from the environmental constraints (left in the figure) down to the rover mass (right in the figure).

Identification of the interactions and their importance requires a number of iterations. In the case of SFR, the design drivers have been ranked in term of importance of their sensibility to the operational drivers. This provides a convenient approach to identify quickly where the optimization can be performed as part of the trade-space at a system level and materialized by arrows going both ways between the elements concerned. In this instance, it can be concluded that the mobility and the power subsystems are critical to the rover platform design, and hence are further discussed in the following section.

2.5.2.3 SFR Subsystem Design Drivers

Mobility of the surface rover is a critical capability to provide local, regional, and possibly global reach of the explored location. Here, an overview of the key trades

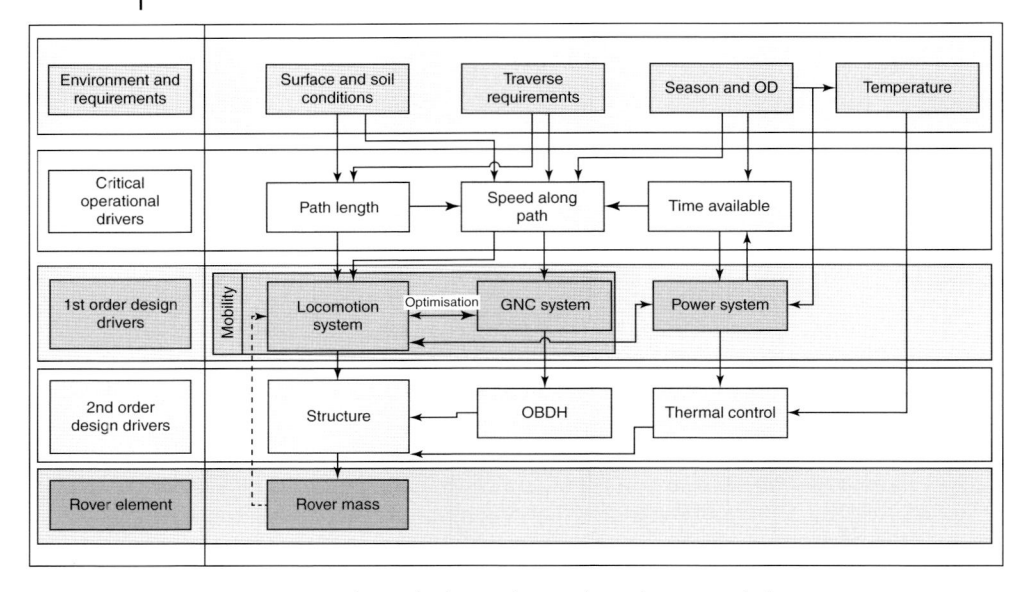

Figure 2.21 SFR Ripple graph: design drivers, dependencies, and ultimate impact on rover mass [117].

and design considerations is provided to address the challenges of surface mobility. As discussed earlier, the SFR mobility design is driven by three operational drivers: the factors affecting path length, speed along path, and time available. By investigating closely each of these factors, one can identify some of the fundamental (and nonexhaustive) performance and design criteria that can improve or reciprocally impede the performance of the mobility subsystem. The factors relating to path length are centered around three design aspects including the locomotion performance, GNC performance and terrain topography as illustrated in Figure 2.22a. The topography plays a major role as it heavily influences both the locomotion and GNC subsystems.

Similarly, the factors relating to the speed along the path can be summarized as illustrated in Figure 2.22b. It shows again how both the locomotion and GNC performance are affected. Although at a first glance it would seem that they are only affected in the way they are intrinsically designed (e.g., wheel size and drive speed), the overall system optimization may not rely on having optimized subsystem on both sides. For example, having a more robust locomotion subsystem capable of traversing rough terrain may relieve the GNC subsystem from having to avoid rocks. However, there will be a mass penalty for selecting this solution. If mass budget is critical, one could argue that a smaller locomotion subsystem with a highly capable GNC subsystem is the way to go because the added complexity is offloaded to the software (considered "mass-free" by default). However, this approach would require more frequent actuation of the steering drive, resulting in more power consumption and possibly more mass. A desirable design for the

2.5 *Systems Design Drivers and Trade-Offs* | **67**

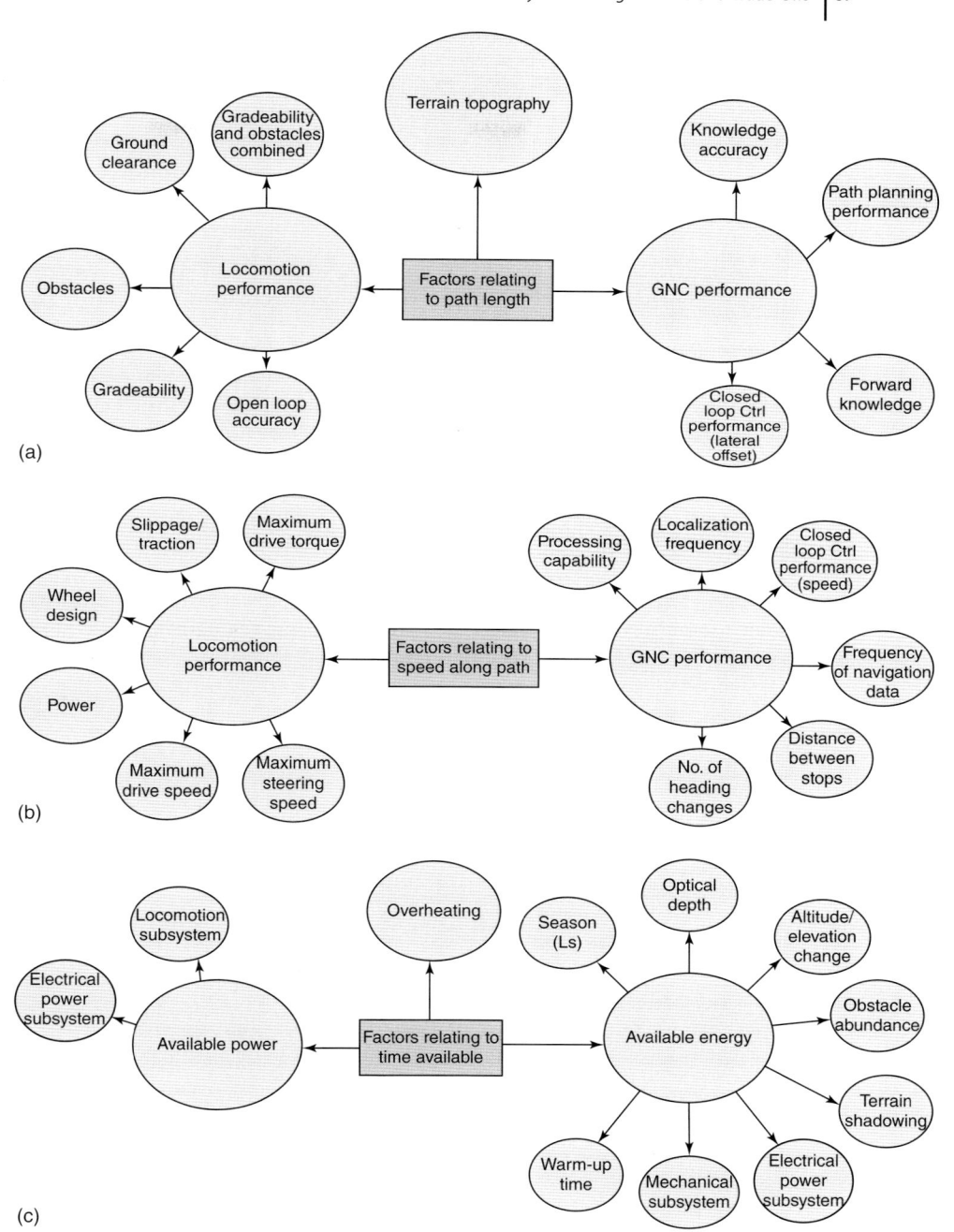

Figure 2.22 Mobile platform subsystem design [117]. (a) Driven by factors relating to path length, (b) Driven by factors relating to speed along the path, and (c) Driven by factors relating to the time available.

rover should, therefore, target the optimal balance between the two with reference to the overall system requirements.

Thirdly, the factors influencing the time available to reach the target are summarized in Figure 2.22c. Here, the available energy and power dominate the situation and are intimately linked to a number of environmental factors, subsystem design, and operation. The landing in September 2025 at Ls133 will only provide about 40 sols before the global dust season. The increase in the atmosphere optical depth can drastically affect the power generation for the rover and its ability to perform a traverse at a sustainable regime. Some additional interactions between the subsystems may not be obvious but worth noting as follows:

- Warm-up time relates to the energy required at dawn to warm up the rover up to its operating temperature. In this case, energy is coming from both the battery and the solar panels. This means that there are times where power might be generated but not available to perform other operations.
- Obstacle abundance will require more steering, hence affect the energy budget over the traverse.
- Electrical Power Subsystem (EPS) design relates to the specific subsystem design decision, such as the battery sizing or the inherent power consumption of the power control and distribution unit (PCDU).
- The main element of the mechanical subsystem that affects the energy available for a transverse is the size of the solar array. Physical size of the solar array (hence the energy that can be generated by the rover) is constrained by the available stowage volume in the lander or descent stage, the acceptable level of deployment mechanism complexity and mass.

The SFR shares many specifications with existing rover designs, but its operation concept is rather unique; for example, it is required to cover vast distances quickly with additional tasks to perform close to the start of the global dust storm season. The dust storm has implications on the mobility performance, including the ability to maintain a stable thermal environment and the ability to generate enough power to maintain sufficient operational functions. To avoid having to deal with harsher environmental conditions beyond the SFR mission baseline, the locomotion and power subsystem should be optimized to ensure that the cache acquisition and delivery can be completed within the shortest amount of time. Minimizing the time it takes to complete the traverse is the best way to ensure that the rover baseline design does not heavily depend on the environmental challenges predicted during the global dust storm season.

As demonstrated by Figure 2.21, the main system design drivers have impact on the rover mass, which then drives the rest of the mission. The next natural step is to investigate and evaluate various rover design concepts that address issues of mobility and energy in view to minimize the rover mass.

2.5.2.4 SFR Design Evaluation

As with any typical design challenges, evaluation of potential design concepts is critical to explore the trade space, that is, the range of possible implementation options. At this stage, it is necessary to capture a full spectrum of options without yet worrying about their implementation feasibilities and leave the downselection of specific concepts to a later stage. Brainstorming sessions involving experts in various subsystems can be useful to gather different perspectives. Evaluating the problem from different subsystem perspectives (e.g., power, mechanical, thermal) can help capture various aspects of the problem under question. Once a range of options are selected, they can be assessed from qualitative viewpoints and subsequently quantitative viewpoints.

During the evaluation, it is of interest to identify where the concepts sit in terms of TRL, hence the risk incurred to the mission development in both operation and cost. This does not necessarily mean that a low TRL solution should not be selected, but it highlights on where the solution departs from existing heritage and where additional development is required. For example, the sky crane design for the Curiosity EDL represented a radical departure from the typical soft and hard landing technologies with flight heritage (e.g., the Viking/Phoenix powered landing and the airbag solutions found in the MPF and MER missions) [115]. The proposed design concepts should consider a full scale of solutions from high-heritage ones to more radical ones. The H.E.A.R.D. scale as shown in Table 2.14 provides a convenient tool to investigate concepts across the scale allowing the

Table 2.14 H.E.A.R.D. scale: assessment of concepts for a mobile robotic platform to explore hazardous areas.

Solution type	Approach	Comments/issues	Examples
Heritage	Optimize existing solutions	Optimization of heritage; low risk/high confidence; good retention of TRL	Rover platform reusing a previously designed wheeled mobility subsystem
Evolution	Evolution of heritage solution	Based on heritage item; medium uncertainty/risk	Optimized locomotion subsystem, for example, high capability to access difficult terrains
Alternative	New development, same functions	Design needs validation and qualification	Cliff rover with tethered deployment
Radical	Clear departure from heritage, thinking outside the box	Novel architecture, technology innovative and ground braking; full validation and qualification required – usually high risk	Legged rover
Displacement	Remove need for a solution	Remove the need for the solution from the design; loss of functionality needs to be assessed	The mobile platform is an aerobot

investigation of a wider trade space. After brainstorms, the scale can also be used to classify the concepts generated for further analysis. The table illustrates within each category an example of mobility concepts to explore hazardous areas, for example, boulder field, cliffs, or caves.

In the case study of SFR, this process can be used to explore a variety of concepts addressing the issue of mobility and energy while minimizing the overall mass [117]. As shown in Figure 2.23, the mass distribution of a typical rover platform is heavily concentrated on the locomotion subsystem, power subsystem, and structure. Therefore, these three subsystems are prime targets for the mass reduction exercises aiming to minimize the rover's overall mass.

In order to achieve both the mass and performance goals established in the mission and system requirements, five design approaches are identified, which are not specifically rover design options but design philosophies for delivering a low mass rover and maintaining performance. They are neither exhaustive nor the only possible approach for designing the SFR. Nevertheless, these approaches are representative and significantly different from each other, which can be further investigated or used at the subsystem level.

1) **Stripped-out approach:** This approach targets any mass that is not essential for achieving the major goals of the mission; therefore, the approach is centered around identifying the key goals of the rover. Each subsystem of the rover is pared down to its minimum required functionality and performance while still ensuring that the original mission goals can be met. As mentioned earlier, the SFR must not become the weak link in the MSR architecture. As a result, subsystem functionality must be distinguished from subsystem redundancy, that is, subsystems can be de-scoped and performance can be traded to achieve a lower mass solution but component redundancy must be maintained unless a less robust system is mandatory for achieving the mass target.

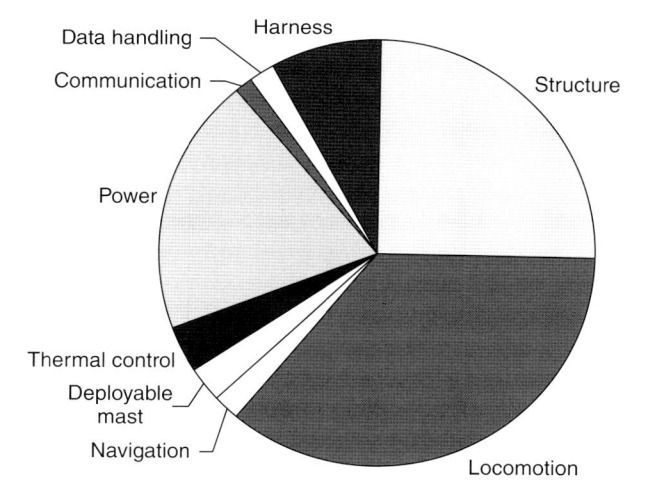

Figure 2.23 Mobile platform design: representative rover mass distribution [117].

2) **Ready-to-go-approach:** This approach targets the need for deployments before operating the rover at egress. It seeks to remove any deployments or mechanisms that are not essential (i.e., utilization of fixed solar arrays, fixed mast, and no wheel deployment, etc.), hence reduces risk during the commissioning phase. The gain by removing mechanisms must be carefully traded against the impact on performance and the reciprocal effects on the system design. For example, a rover with a fixed solar panel may have a smaller total array size available due to the constraints imposed by the lander configuration. A lower power generation then leads to the rover traveling a shorter distance each day since the power cannot support longer drives. A reduction in the distance traveled per sol requires a rover design with a longer operating life, driving the design of the subsystems too. The design of subsystems with longer life is likely to affect their mass, potentially delivering a higher overall system mass than if hinge mechanisms had been used to deploy solar panels in the first place.

3) **Locomotion subsystem optimized approach:** This approach is driven by the optimization of the locomotion subsystem. The rover is designed around the LSS ensuring that mobility is not compromised by any design decision. To achieve the low mass and long transverse requirement set for SFR, the rover may need a rethink of the basic locomotion principle, reevaluating thoroughly the configuration built for a previous rover mission with different locomotion requirements. The implementation strategies optimized for the rover will need to be investigated to ensure that the new rover design meets the mission requirements.

4) **24/7 drive approach:** A perpetual power supply can allow locomotion at any time during the Martian day. This is achieved if the power subsystem is not constrained by solar flux availability as seen in many rovers using solar arrays. To satisfy the perpetual power supply, an RTG-powered Stirling cycle generator can be used. An RTG provides a constant source of power and allows the SFR to traverse with a high duty cycle, that is, constant roving as required. Though the RTG can be a large system in itself, applying it for the power subsystem can lead to a more compact rover by avoiding the solar array structure and deployable mechanisms. However, the RTG must be robust and reliable to ensure safety of the planetary environment and the rover. Curiosity has successfully implemented an RTG providing a constant power level of about 100 W. However, the RTG also dissipates 2 kW of thermal power that requires a thermal control subsystem capable of radiating the constant heat. Curiosity's power subsystem implemented in the MMRTG has a mass of 45 kg already, which makes this power option less attractive for SFR.

5) **Day rover approach:** Conceived as the opposite of the 24/7 approach, this approach relies exclusively on solar power and panels. In a simplest form, it eliminates the battery. As a result, there is no night-time operation possible and the solar flux availability defines the operation period of the rover. The major mass savings are derived from the removal of the battery and the need to thermally support the battery at night, that is, without a battery the body

can be made smaller resulting in a small warm box and less harnessing. On the other hand, the lack of a battery limits the operation of the rover during the day impacting its performance, especially if communication windows can only be established during the night time.

Once a range of design approaches have been identified, it becomes clear that not all of them can result in possible, practical implementations, either due to the risks involved or the complexity or inherent mission constraints. At the mission level, it may be beneficial to constrain the power subsystem using photovoltaic panels to ease launch procedures (i.e., no radioisotopes) and to require communication support during night time. If so, the "24/7 drive approach" that relies on a constant power source such as an RTG and the "day rover approach" that does not support energy storage cannot be considered here. Therefore, the design should be focused on implementing other strategies and incorporating them into subsystem designs. In addition, the strategies introduced are not always mutually exclusive and can be combined to deliver the most efficient design. For example, a rover design may implement a ready-to-go approach by removing deployment mechanisms and also implement a day rover approach by utilizing cold electronics and minimizing the chassis size.

Going through a range of conventional and controversial solutions provides valuable insights of key system elements that are critical to its operation and performance. While this design example assesses the system-level design options, similar investigations can and need to happen at subsystem level to converge to a suitable solution.

2.6
System Operation Options

2.6.1
Operation Sequence

It is important to understand the operational sequences of the robot to be able to produce a system that fulfills all the mission requirements and is practical to operate. Unlike a Hollywood blockbuster, the deployment and operation of a robotic system on an extraterrestrial body is a calculated operation, taking days rather than minutes and requiring a number of steps to guarantee the safety of the robot.

A high-level sequence is provided here, which outlines the operational activities and approximate timing of these events. The times are important in establishing certain resource capacities (e.g., battery level), deployment logic robustness, and to help identify areas of the design that need strengthening. The "days" used refer to Earth days or operational days rather than the local days, which may be shorter or longer depending on the target planet. The exact time duration depends on manpower and duty cycles of the ground operation team.

1) *Landing System Commissioning, time allocation 3–4 days*
   After the EDL, a checkout of the landed system (consisting of the surface robots) is important for two reasons: first, the lander and its payload must be in good health to proceed with the deployment of the robotic platform (e.g., rover); second, it is important to characterize the environment where the robotic platform is to be deployed and ensure a safe delivery to the surface. The rover deployment sequence aims to ensure its safety and sustainability at all times. Activities during this phase consist of two main sequences as follows:
   - lander post-landing checkout and characterization;
   - initial checkout and preparation for rover deployment.

2) *Robotic Platform Deployment, time allocation 1 day*
   The deployment of the rover involves a carefully timed sequence that put the rover in a state ready for egress from the lander. The sequence will have been uploaded to the rover via the lander prior to the physical separation of the two. Once triggered, the sequence initiates a chain of activities that must be completed for the rover to remain safe and in a sustainable condition.
   Another consideration relates to the deployment of the rover's communication antenna(s). If an antenna does not need to be deployed to communicate, the risk of deployment failure is averted. A signal channel is then opened and available to support any necessary recovery operations. The landed system commissioning and rover deployment activities of the MER Opportunity lasted from sol 1 to sol 6 and comprised the following [118]:
   - deployment of the PanCam mast;
   - deployment of the high-gain antenna;
   - characterization of the lander;
   - characterization of the landing site and the surrounding terrains;
   - rover stand up, from its stowed and locked down position;
   - calibration of science instruments;
   - selection of a suitable egress path.

3) *Robotic Platform Egress, time allocation 1 day*
   Egress must be a carefully preplanned operation. Though initially triggered by the ground command, the Egress sequence is reinitiated by timer in the event of loss of communications. The reason for this is that the rover has a limited battery life and is vulnerable at this time of the mission. If the egress sequence is interrupted, the rover health is under threat as it may not be in a safe and sustainable state at the time of failure. Conditions must be carefully checked prior to triggering the egress sequence to ensure that it can be fulfilled successfully.
   The rover must move to a safe distance from the lander to provide room for deployment of critical subsystems such as solar panels and other instruments. Once the rover is at a safe distance, the mechanisms that contribute to achieving sustainability (e.g., thermal) can be executed if not done already and observed by the lander if it carries a camera.

4) *Robotic Platform Consolidation, time allocation 2–3 days*
   If the landing system only consists of a small relatively passive robotic platform, the initial checkout would have included this phase. If the rover is not yet in its final operational configuration, this consolidation phase is required. Following egress, it is important to consolidate the condition of the rover, particularly in terms of communications, electrical power generation, and thermal stability. During this phase, not only the instantaneous state of the rover is monitored but also the underlying trends are important in planning anticipatory actions and operations. In addition, communications must be confirmed reliable under all conditions, and optimizing adjustments must be made to the thermal, electrical, and communications to support efficient operations through the on-surface life of the rover. During the phase, the Earth control must devise ongoing optimal strategies for the operation of the rover, which includes both commissioning and science operations.

5) *Robotic Platform Commissioning, time allocation 4–6 days*
   Commissioning the rover involves the following steps that bring the rover to a full operational readiness:
   - Complete checkout of all subsystems mechanisms.
   - Release remaining hold-downs (if any).
   - Deploy and exercise all deployables and characterize their performance.
   - Check all equipments for function.
   - Monitor thermal and electrical situation during commissioning.
   - Remote sense and characterize the environment (survey).
   - Explore local terrain and characterize surface properties near to the lander.
   - Relay environmental data to the ground control and enable environment model to be refined, for example, locomotion (trenching, turning, or skidding), solar isolation, and surface thermal condition.

   The first part of this sequence is a check of the rover itself while the second seeks to understand and characterize the environment where the rover does the science. With such knowledge, the operations team is then able to plan subsequent activities, anticipate possible issues, plan optimal routing, and manage the rover resources to keep it safe within margins of full operations. For example, in the case of the MERs, the clear egress path and favorable egress conditions allowed the rover to exercise traverse, deployment of its instrument arm, and operation of its tools between sol 8 and 11, in the direct vicinity of the lander [118].

6) *Nominal Operation*
   Once the rover elements have been fully commissioned, the ground control operators can concentrate on achieving the mission objectives. As the mobile platform progresses on the surface across the local landscape, the need for accurate remote sensing (imaging of the surrounding) is crucial to the operators for planning the rover nominal operation from one day to the next.
   A typical day of operation requires careful planning by the operators, taking into account constrains from the rover capabilities, the communication

windows, and the anticipated mission timeline. As such, the rover starts the day by waiting for the temperature of its subsystem to rise to the operating range. For a Mars mission, a 10 am local time target can be used to maximize the inputs from the warmth of the rising Sun rather than relying exclusively on the use of electrical heat from the batteries. Depending on the communication windows with Earth, the ground control and the rover may have communication opportunities at different times of the day across the mission. These windows are used to upload new commands and download science or telemetry data from the rover. Those power-demanding operations such as traverses should, as much as practical, take place around the local midday (e.g., 10.00–14.00) to maximize the use of the solar power supplemented by batteries if required. In the second part of the afternoon, the battery should be left to recharge, remote sensing can be performed, and the day's telemetry can be downloaded. The rover then should be operated in a way to complete the day's operation in a state that enables it to survive the night, that is, with a suitable battery charge. Chapter 6 provides further, in-depth discussions on the nominal operation of planetary robots and involved techniques/technologies.

2.6.2 Operational Autonomy

The use of autonomy in the robotic mission primarily aims at reducing the ground intervention to a minimum and allowing effective and efficient pursuit of the operational goals. Low latency reactivity achieved through autonomy could help keep the rover safe and in good health in case unexpected events and situations arise. Chapter 1 has previously introduced the ECSS defined level of autonomy (i.e., E1–E4). Taking the SFR as an example, the rover would require a minimum autonomy level at E3. This requirement implies that the GNC subsystem should collaborate with the rover's onboard executive control software and ensure that operations involving locomotion and navigation are carried out efficiently. This includes rover position adjustments associated with sample cache retrieval as well as traverses from one sample site to another. More rapid progress of the rover's operation schedule can be achieved by removing many approval loops via the ground control. For example, when the rover is instructed to pick up the sample cache, it should be able to move to the target location and refine its position to enable the acquisition device to reach the cache. This involves collaboration between the sample cache acquisition control process and the GNC. Similarly, when the rover is directed to a remote target site it should be able to reach the target without the ground intervention while protecting the rover from any hazards or unsafe situations.

As presented in Chapter 1 and Figure 1.3, future robotic missions are envisaged to focus on robotic explorers that are fully autonomous (i.e., autonomy level E4) without the need of ground intervention. However, there are typical concerns over fully autonomous systems due to the lack of basic trust that machines can or will make the same decisions as the trained human operators when dealing with non-nominal situations. To facilitate the future implementation of autonomous functions, current research in machine intelligence focuses on solutions that can be implemented in a transparent, robust way where decision making via the robot is validated and verified against its human counterpart. The adoption of these autonomous functions is currently facing two main obstacles: (i) psychological issues from human in general, which demand sufficient reassurance that the autonomous system will perform as planned given the lack of human-like accountability and (ii) political issues from the space sector where historically more reserved approaches tend to be used to limit any perceived risks that could jeopardize an entire mission. Chapter 6 presents detailed descriptions of what operational autonomy entitles and how it can be designed for planetary robotic missions and systems. In this section, only system-level design considerations are presented.

2.6.2.1 Autonomous Functions

Autonomy can be implemented or achieved in different ways for different purposes or functions. The lower level function is often used for ensuring safety of the robot during normal operations, for example, using FDIR, which is a monitoring executive onboard the robot to cope with anomalies, failures, and major errors. Additional autonomy functions are also envisaged to detect hazards and take the necessary action to keep the robot safe and stable, which will involve building perceptions using NavCam/HazCam, tilt sensors, and inertial measurement units (IMUs), and so on. Independence from other processes running on the robot is vital to ensure that anomalies or errors do not propagate and cause the functions to halt or fail. This requires autonomy functions to be designed and implemented thoroughly and to ensure effective protection of the robot. Additional safeguards are possible by adding ground monitoring loop with dedicated operation breaking points.

Once confidence has been established in operating the robot on planetary surfaces, higher level autonomy functions can be considered to reduce the human operators' load. In case of rovers, examples include functions allowing rover to tackle precise tasks such as accurate positioning under tightly controlled conditions, or autonomous navigation and hazard avoidance. The instruments onboard the rover used for navigation must provide necessary performance for both teleoperated and autonomous operations, with various design issues to note: (i) sensor inputs from the IMUs to the cameras must provide reliable and accurate data (meaning free from drift and error as possible); (ii) mounting the NavCam low on the rover body can lead to more stable images as the rover gets closer to hazards; and (iii) mast-mounted cameras can have a better vantage point and horizon reach, but suffer from swaying from side to side as the vehicle rolls over the surface features that can potentially result in blurring images and need to be taken into account when designing vision/navigation processing loops. As successfully demonstrated by NASA's MER and MSL rovers, the body-mounted cameras can be used as the primary visual navigation aid, and the mast-mounted cameras are free to observe the planetary surface with a pan and tilt mechanism either for science or as an additional aid to navigation.

Future planetary missions (such as the SFR) will require more advanced autonomy functions (in addition to FDIR and safety-driven navigation) to allow consideration of unexpected events and replanning of operations. For example, if a rover encounters an insurmountable obstacle on the path to a target, it should be able to seek alternative routes and pursue them while keeping its overall goal of reaching that target unaided. This represents E4 level of autonomy or goal-oriented operation.

2.6.2.2 Autonomy Levels: Teleoperation versus Onboard Autonomy

Planetary robotic systems are generally deemed to perform semi or fully autonomous systems (i.e., E2–E4) in the long term, but undoubtedly teleoperation of the robotic systems (i.e., E1) can remain viable and relevant in certain future mission scenarios, for example, the lunar missions as demonstrated by the former Lunokhod missions. It is also anticipated that humans may operate rovers on the Martian surface from the Mars orbit or a settlement on Phobos. For both the Moon and Mars, teleoperated robots could be used to prepare the surface infrastructure prior to the arrival of humans or to explore hazardous sites where the robot GNC could benefit from the human-in-the-loop operation. Sites such as the shaded craters and caves are prime science targets, but require rapid and precise exploration due to limited power storage (if no RTG is available) and harsh illumination conditions. Human operators are inherently better at perceiving features and controlling robots operating in those challenging environments, and making intelligent decisions rapidly to help keep the robot safe and healthy while on its way to the targets. However, teleoperations are subject to vulnerabilities that can seriously compromise the performance of human operators primarily related to link latency and accuracy, as detailed in the following:

- **Anticipation:** If the human operator stands beside the robot for real-time "joystick" control, the only significant delay in the control loop is the operator's reaction time, provided the operator can clearly see and immediately modify the robot's behavior in the environment. Given the inherent communication latency in space mission (even the shortest Moon–Earth link takes about 1.5 s for one-way communication), the operator is required to anticipate the robot situation in advance. If any additional communication latency is introduced into the remote control loop, the operator's performance is affected due to the need to guess the next change in robot's movement.
- **Link stability:** Stability of communication link and latency cannot be guaranteed due to possible link dropouts, burst errors, or changes in propagation conditions; therefore, timing of the remote control response becomes unpredictable. Reliable links guarantee that data arrive in order without error or duplication, but this is provided by retransmission techniques at the expense of link latency stability. In the case when the link drops out and requires reestablishment of the link, communication delays could be considerable. If a rover has received a command by the ground control to continue the current traverse before the link drop-off, by the time the link has been reestablished the rover may

be in a different or dangerous location. Loss of control in this case can result in loss of mission.
- **Feedback quality:** Human feedback control in terms of quality and resolution is limited, which can introduce errors and blind spots or lack of resolution. Anticipating robot behaviors by image and telemetry data is not always reliable due to bandwidth/resolution constraints on the data used by human operators. The analogy is like comparing a teleconference with a face-to-face meeting. The teleconference is adequate for providing majority of the basic functions of a meeting but at a much reduced quality compared with a physical meeting. The compromise in quality in this case for controlling an expensive and sophisticated space hardware can be an issue. For driving on flat and open ground, the consequence of a driver error or control inaccuracy may not be severe. However, if the operator is required to place the rover precisely during the placement of an instrument arm, lack in control precision can cause damage to the robot (e.g., collisions with the target or the surrounding environment). As a result, some form of precise control is required.
- **External or environmental factors:** Some environmental factors may affect the timing of the control response. For example, rover's locomotion performance varies depending on soil characteristics, traction conditions, and slope. This adds uncertainties to real-time operations. An onboard control layer (meaning autonomy) can be used to achieve immediate response to measured and perceived danger so as to prevent potential disaster propagation.

To address the above-mentioned issues and make teleoperations potentially reliable and safe, a number of design options can be implemented:

- **Link persistence:** The use of a persistent remote control loop revolves around the principle of *dead-man switch*. This design requires that all control input signals must be continually submitted to the robot so that it continues the control action. The control action ceases when the control signal ceases. Thus, the robot stops and waits if the communications link drops, preventing a situation whereby it could continue to move over the edge of a cliff when the operator is no longer in control.
- **Link strength margin monitoring:** For a teleoperated robot, the loss of communication with the robot means a lack of both control and feedback putting it at risk. This can happen when the robot goes out of range or passes into an RF shadow (e.g., behind a large rock). This situation can be avoided by monitoring the link strength and ensuring that the signal margins never fall below a predefined threshold. If the signal margins drop below the threshold, the robot stops, reports an alarm, and awaits further instruction. This requires an override by the operator who then takes cautious steps to restore the signal to appropriate levels.
- **Limited life validity of control signals:** The fundamental principle of teleoperation relies on the validity and timely sending of control commands and the reception of valid feedback. If the feedback data are delayed on the return link, it should be labeled as invalid and the control link to the rover as inhibited.

By time-stamping each monitoring or control signal at the time and place of its origin, it is possible to determine the age of the signal when it is read and just before it is to be used. If a time limit is placed on control signals and used as a validity criteria, the use of out-of-date and potential harmful information can be eliminated. This can prevent the operator from relying on out-of-date information for controlling the robot. Similarly, if a control signal to the rover is delayed in transit and passes the programmed time limit threshold, the signal will be treated as invalid by the robot, which then stops.

- **Detection and management of errors, failures and hazards:** Teleoperators rely on data feedback to control the robot, which is limited by the communication channel bandwidth, variety and accuracy of sensors, and timeliness of information. Response times by operators to potential hazards may be too slow, erroneous, inappropriate, or imprecise. It is, therefore, necessary to have onboard safeguards (e.g., watchdogs, FDIR, limits protections) that monitor the robot situation and intervene in event of a perceived problem or anomaly (e.g., stop, or switch to safe mode). For example, an onboard hazard avoidance function can detect potential danger to the robot and react in a timely manner before the situation worsens, or disable the operator's remote control inputs until it is deemed safe to continue.
- **Achieving right balance between teleoperation and onboard autonomy:** As already shown earlier, the use of teleoperation would benefit from some onboard autonomy functions to enhance the operation or ensure safety of the robot. In reality, varying level of autonomy (from E1 to E4) or a balance of both teleoperation and full onboard autonomy are often implemented to use the inherent abilities of human operators as well as some autonomy functions at critical situations to alleviate the pressure from the operators (such as recovering operation after a lost signal, or precise instrument placement).

2.7 Subsystem Design Options

This section uses two examples to demonstrate design options for subsystem-level design for a planetary robotic system, including the power subsystem that provides required energy for the robot to function and the thermal subsystem, a critical aspect for the overall system design.

2.7.1 Power Subsystem

The power subsystem must provide the robotic system with sufficient energy to support its needs and meet peak power demand to fulfill mission objectives. The subsystem must be sized to maximize the operation and mobility time; furthermore, it must be able to cope with non-nominal situation (e.g., high optical depth

80 | *2 Planetary Robotic System Design*

on Mars) and be easy to stow, deploy, and mass-efficient to limit its impact on the overall configuration.

Fundamentally, the power subsystem size and mass are intrinsically linked to the environment and the surface operation requirements as shown in Figure 2.21. On average, the power subsystem represents 20–30% of a typical spacecraft in-orbit mass and around 20% of its cost [119]. The subsystem consists of three main components: the power generation/conversion, the energy storage, and the power management and distribution. Design of the power subsystem has a range of options and can be tailored to fulfill the specifications of a robotic system (see Figure 2.24).

2.7.1.1 Power Generation

The power generation functions of a robotic platform can be implemented in a number a ways including solar power, nuclear, or chemical. Since the fundamental principles of the chemical approach is also used for power storage, hence discussed in the power storage section later.

- **Photovoltaic power**: Photovoltaic or solar cells use solar energy to generate electricity. Through an arrangement of semiconductor layers, photons from the light source are converted into electrons. A photovoltaic power subsystem is typically composed of solar cells, a substrate, a panel, or an array structure, and possibly a deployment mechanism. An energy storage is then used to store the excess energy generated for future use. This technology is used extensively in various terrestrial applications and is by far the typical power generation method for most spacecraft. The range of applicability of solar panels is driven by the local solar energy that can be harnessed at a given location. The solar

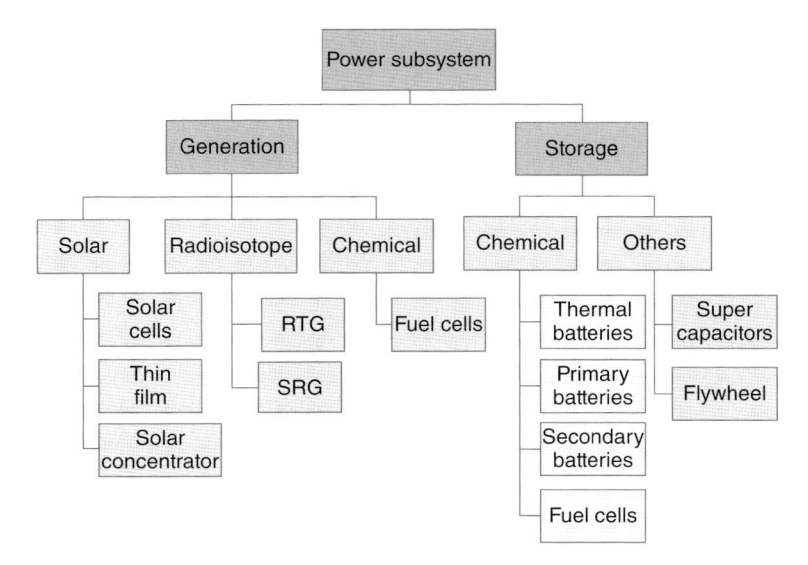

Figure 2.24 Power subsystem design options.

Table 2.15 Solar irradiation across the solar system.

Planet	Distance (10^9 m)	Mean solar irradiance (W m^{-2})
Mercury	57	9116.4
Venus	108	2611.0
Earth	150	1366.1
Mars	227	588.6
Jupiter	778	50.5
Saturn	1426	15.04
Uranus	2868	3.72
Neptune	4497	1.51
Pluto	5806	0.878

radiation at Earth (called the solar constant) is around 1366 W m^{-2}, and it decreases as the distance to the Sun increases (as shown in Table 2.15). This tends to limit the use of solar panels for the inner planets. Beyond Mars, solar panels need to be excessively large, which can constrain their implementation on spacecraft, making them unlikely to be deployed to the surface.

Once on the surface, atmospheric effects (if present) may affect the power generation capability or efficiency of the solar array. To maximize their efficiencies, solar cells may need to be "tuned" to the wavelength of the location where they will be operating in. For example, solar cells on Mars are slightly different from the terrestrial solar cells to account for the different wavelengths blocked by the Martian atmosphere. Over the course of the recent successful Mars rover missions such as MPF/MER/MSL, greater experience has been gained about impact of atmospheric effects and dust deposition on solar arrays over a few Martian years [120, 121]. The regular accumulation of airborne dust is unavoidable and on average led to the degradation of the solar power output by 0.028% /sol for Sojourner. However, "cleaning events" have been shown to occur a number of times for both MERs where Martian dust devils had blown the dust away from the solar arrays hence cleaned them. Solar panel cleaning techniques have also been investigated for some time such as using mechanical brushes, vibration of the array, or by blowing a compressed gas [122]. Another technique involving an electrodynamic screen (EDS) can provide a convenient cleaning option. The technique uses a thin transparent film through which an electrodynamic wave is propagated, lifting and shifting the dust away from target areas [123–125].

Solar panels represent a mature technology that is easy to implement and relatively cost-effective. Table 2.16 summarizes the major pros and cons of solar technology. The main design requirements of this technology are based on the power level to generate, the specific power (W kg^{-1}), and the packaging/stowage options for the mission. The solar panels for planetary robots can be configured in various ways depending on the mission and destination (such as body mounted, panels, thin films, or rely on solar concentrators). The current state of the art for space solar technology can provide around 150 W kg^{-1} at

Table 2.16 Solar power technology: pros and cons.

Advantage	Disadvantages
Lightweight and economic source of electrical power (compared with radioisotopes)	Highly dependent on environmental conditions (e.g., dust, temperature, radiation) and no power generation in shaded areas
Mature heritage in space for in orbit and surface spacecraft	May require a mechanism for deployment and correct orientation (especially for landed assets)
Simple manufacturing, integration, and assembly processes	Discontinuous production of electrical power due to solar cycle and site topology/morphology (e.g., shadows)

1 AU [119]. The triple-junction cells provide an efficiency around 29%, while current developments are projecting 33–36% efficiency cells around 2017 [119]. New developments focus on low-light, low-temperature (LILT) cells that can be used for the exploration of the outer planets [126, 127].

- **Radioisotopic power:** In its simplest form, an RTG is a nuclear electrical power source that generates power from radioactive decay. In such a device, the heat released by the decay of a suitable radioactive material is converted into electricity by the Seebeck effect using an array of thermocouples. These devices provide both power and heat to the system. Typically, the RTGs are used for destinations to the outer planets where the amount of solar energy is significantly limited or to the inner planets where extreme conditions or operations are expected. By working around the constraints imposed by solar fluxes on the surface, operations in harsh environment (e.g., low temperature, low light intensity) can be enhanced with the availability of additional power, providing extended daily operation, increased mobility, or increased communication bandwidth.

To date, a number of nuclear fuels have been used in space applications and their respective properties are summarized in Table 2.17. The ideal option for an RTG should provide a good power density, a manageable working temperature (to prevent overheating the robot), as well as low radiation (otherwise it requires more shielding mass to prevent a significant acceleration of the degradation of electronics, other subsystems and materials). To date, plutonium-based RTGs provide a good power-to-weight ratio and require less shielding than other options. However, the supply of plutonium globally is extremely limited [128] and the United States has only restarted production of Pu^{238} at a rate of 1.5 kg/year [129].

Currently, the United States is developing and operating mainly plutonium-based RTG systems such as the MMRTG to power the MSL rover as mentioned previously [39, 41]. The MSL MMRTG is built around a Pu^{238} core comprising eight individually shielded heat source modules that can generate 125 W at BOL (typically ∼100 W) for a mass of 43 kg [39]. An alternative concept is the advanced Stirling radioisotope generator (ASRG) [133] development (a NASA program canceled as of November 2013 [134]), which is a mechanical

Table 2.17 Radioisotope fuel options.

Fuel	Properties
Plutonium (Pu^{238}) [130]	—High specific power (0.5 W g^{-1}); Lifetime (1/2 life 87.7 years) —Mainly from United States with dwindling supplies, limited availability in Europe —Lower radiation than polonium and americium —High TRL and strong heritage
Polonium (Po^{210})	—High specific power (140 W g^{-1}) but limited lifetime (1/2 life 138 days) —Used mainly as radioisotope heat source and not for power generation —Used in Lunokhod 1 and 2 as RHUs [28]
Americium (Am^{241}) [131]	—Long lifetime (1/2 life 432 years), stable, long-lasting power output —Lower specific power (less W kg^{-1}) and higher gamma radiation than plutonium —ESA is keen to develop the technology for future missions [132]

implementation of the simpler RTG that uses a working fluid to operate on a Stirling thermodynamic cycle. The closed-cycle system converts the heat from a radioisotopic heater module into reciprocating motion converted into electricity through an alternator resulting in a AC or DC electrical power output. The Stirling process provides a higher efficiency system compared with the thermoelectric conversion of typical RTG, offering a fourfold reduction of plutonium fuel for the same output. Compared with the specific power of 2.8 W kg^{-1} provided by the MMRTG, the ASRG is expected to provide 140 W for a 32 kg system (i.e., specific power of 4.5 W kg^{-1}) [134]. Future RTGs are expected to reach 6.4 W kg^{-1} providing a significant improvement over the current generation. However, such improvements are pending upon appropriate funding support being provided to advance the development of these systems. While the radioisotopic power source have obvious advantages, they also possess some disadvantages that can make the implementation difficult as shown in Table 2.18.

Since 1961, the United States has successfully launched 42 nuclear power sources (i.e., 41 RTGs and 1 nuclear reactor) on 24 space missions along with hundreds of RHUs [135]. Table 2.19 summarizes some of these developments. Over the past few years, European americium-based RTG concepts have been evaluated as an alternative to Pu^{238}-based systems [136, 137]. Am^{241} being only a 1/4 of the power density of Pu^{238}, RTG and RHU systems based on americium will be invariably heavier and bulkier. However, americium is easier to get access to as a by-product or waste from the decay of plutonium [131].

Table 2.18 Radioisotopic power: pros and cons.

Advantage	Disadvantages
Stability – Continuous provision of electrical (and heat) power	Mass – The mass of such system is substantial, making it prohibitive for smaller robotic platforms
Stand alone – Independent from environmental conditions (suitable for extended missions in dark areas, and with low/variable illumination levels)	Thermal – Continuous production of large amount of heat power, which could be difficult to manage during illuminated periods
Longevity – Suitable for extended missions, long-range rovers, missions affected by solar panel aging/degradation	TRL – Low TRL for non-Pu-based system, European developments are ongoing
TRL – Significant heritage in space even if mainly for targeted missions (only a few on the surface)	Availability – Pu^{238} provides the most efficient radioisotope but is extremely limited in resource
	Security – Human handling during integration and testing, fuel containment, requiring late accommodation into the spacecraft, transport, and launcher failure scenarios have to be investigated to mitigate associated risks
	Political – US launchers for US RHU/RTG, Russian launchers for Russian system

- **Other power solutions:** Apart from the use of photovoltaic and radioisotopic power generation, a range of methods can provide alternative options for future implementation. The chemical options such as the Stored Chemical Energy Power Systems (SCEPS) can provide a short life (~days) and high power output option. The concept based on the exothermic reaction of lithium-metal rods with sulfur hexaflouride is used as part of a closed Rankine cycle engine to generate electrical power [139]. The engine used extensively in torpedoes (e.g., MK 50) [140] is being investigated to power the spacecraft for a Venus mission proposal [141, 142]. Similarly, other concepts based on energy harvesting are being investigated, such as electrodynamic tethers using the strong radiations in the Jovian orbit or the gravity of Jupiter itself [143].

2.7.1.2 Power Storage

Unless the power generation function and the environment provide a constant supply of power, the robotic platform needs some form of power storage to deal with high energy demands (e.g., fast traverse) or time of low/no power (e.g., crater exploration, or survival at night). Factors (such as irregular illumination patterns due to the environment, operation plan, or day/night cycle inherent to specific location) make sizing of power storage critical to the mission operations. Similar to power generation, the power storage can be implemented in a number of ways.

Table 2.19 NASA's RTG implementations [135, 138].

Model	Mission heritage (no. used)	Max power output (W)	Max heat output (W)	Isotope	Mass fuel (kg)	Mass (kg)	Specific power (W kg^{-1})
ASRG	Discovery Program	140 (2 × 70)	500	238Pu	1	34	4.1
MMRTG	MSL/Curiosity rover	110	2000	238Pu	4	<45	2.4
GPHS-RTG	Cassini (3), New Horizons (1), Galileo (2), Ulysses (1)	300	4400	238Pu	7.8	55.9–57.8	5.2–5.4
MHW-RTG	LES-8/9, Voyager 1 (3), Voyager 2 (3)	160	2400	238Pu	4.5	37.7	4.2
SNAP-3B	Transit-4A (1)	2.7	52.5	238Pu	2.1	1.3	
SNAP-9A	Transit 5BN1/2 (1)	25	525	238Pu	1	12.3	2.0
SNAP-19	Nimbus-3 (2), Pioneer 10 (4), Pioneer 11 (4)	40.3	525	238Pu	1	13.6	2.9
Modified SNAP-19	Viking 1 (2), Viking 2 (2)	42.7	525	238Pu	1	15.2	2.8
SNAP-27	Apollo 12-17, ALSEP (1)	73	1480	238Pu	3.8	20	3.65

The typical methods rely on a chemical process such as batteries and fuel cells. Selection and sizing of the power storage function are driven by specific applications and mission requirements. These will vary based on the mission duration, the amount of power instantaneously provided to support the power generation, and the space environment. Figure 2.25 summarizes the operation range and domain of applicability for different technologies, which indicate that (i) batteries are suitable choices when high specific power is required for a limited duration as characterized by their lower specific energy and (ii) fuel cells are usually suitable when high specific energy is required for long periods of time, which makes them particularly efficient for applications requiring low power consumption for extended periods.

- **Batteries:** Batteries have been used extensively for various space and exploration missions. To support short- and long-lived applications, different kind of batteries have been designed to answer specific needs. Typical surface missions implement two to three types of batteries such as thermal batteries, primary batteries, and rechargeable batteries. Using the MER mission as an example, these battery types are described as follows [36, 144]:

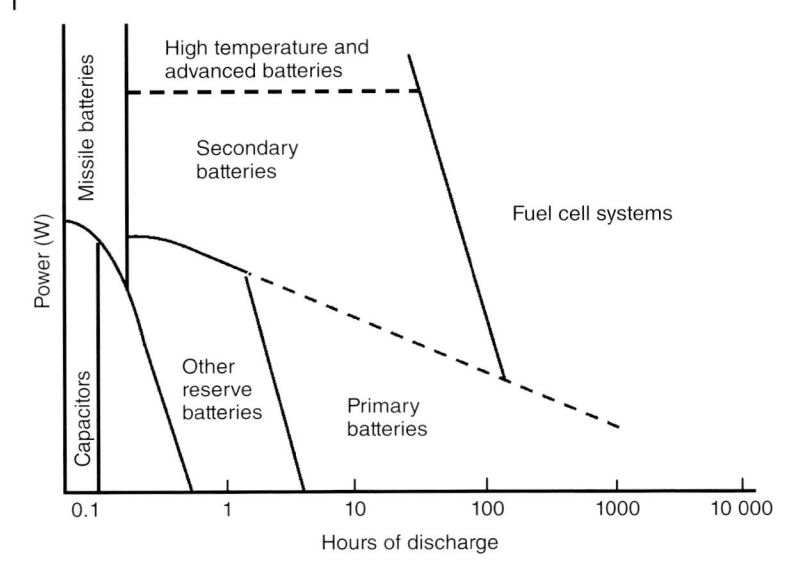

Figure 2.25 Energy storage options [126].

— *Thermal batteries* are aimed at energetic, short power release such as the actuation of pyros. When energy is required, the battery triggers a unique and irreversible chemical reaction causing a rapid generation of power, which provides only minutes of discharge. These batteries are ideal for transient operations such as EDL where bursts of power are required to actuate various system components. For example, MER used two Li–FeS_2 (prime and redundant) to actuate six pyros (6A each) to separate the cruise vehicle from the lander element.

— If long-duration operation is required, *primary batteries* can provide hours of discharge but are also designed to be used only once. Their chemical makeup provides higher specific energy compared with secondary batteries. Lithium–sulfur is among the most widely used types for primary batteries in space. Recent developments have shown that significant improvements can be expected from a type called lithium carbon monofluoride (Li-CF_x). Desirable characteristics of these batteries are high rate of discharge, low heat generation to remove the need for active cooling, high specific energy and energy density, and good life time. The MER mission implemented a primary battery to provide 650 W h over 1.25 h with a peak power of 250 W, which supported the critical EDL sequence up to the rover standup on the surface prior to the rover and solar arrays' full deployment. The battery assembly consisted of five parallel strings each of 12 D-size cells, being a 36 V and 34 A h (−1200 W h) power source. Table 2.20 summarizes typical characteristics of primary batteries and Table 2.21 on pros and cons.

Table 2.20 Typical primary batteries characteristics [144].

Type	Form factor	Mission	Specific energy (W h kg^{-1})	Energy density (W h l^{-1})	Operating temperature range (°C)	Mission life (years)	Issues
Li–SO$_2$	Battery	Galileo Probe, Genesis SRC, MER Lander, Stardust SRC	238	375	−40 to 70	<10	Voltage delay
Li-SOCL$_2$	Battery	Sojourner, Deep Impact, DS-2, Centaur Launch	200–250	380–500	−20 to 30	<5	Severe voltage delay
Li–CF$_x$	Cell	–	614	1051	−20 to 60		Poor power capability

Table 2.21 Primary batteries: pros and cons [143, 145].

Advantage	Disadvantages
Low cost	Not ideally suited for heavy load/high discharge rate performance
High specific power (W kg^{-1}) and low specific volume (l kg^{-1})	Not ideally suited for load-leveling, emergency backup, hybrid battery
Low dependency from environmental conditions (except from temperature but less so than secondary batteries)	Moderate specific energy (100–250 W h kg^{-1})
Operating temperature range (−40 to 70 °C) tends to be wider than for current rechargeable technology	Radiation tolerance poorly understood
	Voltage delay

—Complementary to the primary batteries, *secondary batteries* have gained a significant heritage in space because of their high adaptability to the required loads and type of applications (in-orbit or on planetary surface), their scalability, and modularity. They are typically used to provide hours of discharge time between recharge cycles. Despite their broad use, secondary batteries suffer from the harsh conditions of the space environment, which can dramatically reduce their performances if no mitigation strategy is implemented. In particular, the inherently tight operation and survival temperature ranges can represent a challenge for the thermal control design when missions require long-term exposure to extreme hot and cold conditions. Among the implementation options, Li-based technologies are so far the most promising options. Li-ion rechargeable batteries typically provide the best performances in terms of specific energy (up to 190 W h kg^{-1}). Table 2.22 summarizes

Table 2.22 Space secondary batteries' characteristics [119, 144].

Type	Mission	Specific energy (W h kg^{-1})	Energy density (W h l^{-1})	Operating temperature range (°C)	Design life (years)	Cycle life	Issues
Ag–Zn	Pathfinder Lander	100	191	−20 to 25	2	100	Electrolyte leakage limited life
Ni–Cd	Landsat, TOPEX	34	53	−10 to 25	3	25–40 K	Heavy, poor low temp Perf.
Super Ni–Cd	Sampex	28–33	70	−10 to 30	5	58 K	Heavy, poor low temp Perf.
IPV Ni-H$_2$	ISS, HST, Landsat 7	8–24	10	−10 to 30	6.5	>60 K	Heavy, Poor low temp Perf.
CPV Ni-H$_2$	Odyssey, Mars 98 MGS, Stardust, MRO	30–35	20–40	−5 to 10	10–14	50 K	Heavy, poor low temp Perf.
SPV Ni-H$_2$	Clementine, Iridium	53–54	70–78	−10 to 30	10	<30 K	Heavy, poor low temp Perf.
Li-ion	MER Rover, Curiosity [119]	90–110 [119]	250	−20 to 30	1	>500	Limited life – TRL9
Li-Ion	InSight [119]	140 [119]	TBC	−30 to 35	TBC	>500	TRL6
Li-Ion	Europa Flyby [119]	220 [119]	TBC	−10 to 40	TBC	TBC	TRL5

typical characteristics of secondary batteries used in past and current space missions. Growing energy needs in particular to support advanced exploration missions require extra improvements to the current Li-ion technology. This drives the need to investigate new technologies to provide higher capacity rechargeable batteries, among which the lithium metal batteries have shown promises although issues regarding safety and life cycle performance still persist, and lithium–air is also of interest. The boom in renewable energies and electric vehicles drive rapid terrestrial development in rechargeable battery technologies to provide increased energy density (see the roadmap for the coming decade in Table 2.23. Space-qualification of these technologies is anticipated in parallel in order to support future mission concepts.

- **Fuel cells:** Fuel cells (FCs) are electrochemical devices capable of converting the energy of a chemical reaction directly into electrical energy. Unlike the battery, a fuel cell is an energy conversion system that can generate electricity as long as the electrodes receive the fuel and oxidant. The main factor that limits the lifetime of FCs is the deterioration or corrosion of the constituting material. There are two main categories of FCs: the nonregenerative ones where the product of the chemical reaction cannot be converted back to their reagents, and the regenerative ones that can use an additional reaction to recover the reagents from their products. FCs have been already extensively used in space missions, in particular for human spaceflights such as the nonregenerative

Table 2.23 Development roadmap of terrestrial rechargeable batteries.

Type	Energy density (W h kg^{-1})	Timeframe
LiFePO$_4$	120	2012
LiCoO$_2$	180	2012
NCM/Gr	230	2012
NCM/Alloy	260	2013
TMO/TiO$_x$	310	2014
Adv TMO/TiO$_x$	410	2015
Zn–air	1100 theoretical	2015+
Li–S	2600	2015+
	>350 [119]	2015
Li–air	5200 theoretical	2020+

Source: (Updated from Refs [146] and [119])

hydrogen–oxygen FCs onboard the Apollo command and service module (CSM) and NASA's Space Shuttle Orbiter.

Thanks to a significant amount of research in the field, several types of FCs exist at different stages of development. They can be classified according to their specific characteristics such as species of reagents, the place of formation of fuel ("reformer" external or internal to the stack), the type of electrolyte, and the operating temperature. The most common classification used today is based on the type of electrolyte and the cell operating temperature. The sizing parameters of FC focus on two parameters, that is, the output power that determines the stack sizing and the energy needs that determine the storage size (i.e., reactants). The main FC technologies include the following:

—alkaline fuel cell (AFC),
—phosphoric acid fuel cell (PAFC),
—proton exchange membrane fuel cell (PEMFC)
—direct methanol fuel cell (DMFC),
—molten carbonate fuel cell (MCFC),
—solid oxide fuel cell (SOFC),
—intermediate-temperature solid oxide fuel cell (ITSOFC).

Table 2.24 provides an overview of the main engineering features of the identified FC options. Since DMFC is a subset of PEMFC hence is not repeated in the table. Based on the table, the PEMFC technology stands out because it requires a lower operating temperature and provides a comparatively good power density. This kind of FCs can reach efficiency up to 60% and operate at a relatively low stack temperatures (around 50–100 °C). For regenerative H_2-O_2 PEMFCs, two types of reactants storage can be considered such as the high pressurized or cryogenic storage. The cryogenic storage of reactants reduces the mass to 1/2 and the volume to 1/4 compared with the high-pressure storage [147]. A regenerative PEMFC using cryogenic storage has the potential to provide > 1000 W h kg^{-1}, but the current TRL is only about 3–4 that requires significant technology development. The

Table 2.24 Fuel cell technologies and typical characteristics [148].

Parameter	AFC	PAFC	PEMFC	MCFC	ITSOFC	SOFC
Electrolyte	30–50% KOH	Conc. H_3PO_4	Polymeric membrane	Molten Li_2CO_3-K_2CO_3	To be developed	Yttria stabilized ZrO_2
Temperature (°C)	90–100	150–200	50–100	600–700	600–800	800–1000
Electrical efficiency (%)	55–60	35–40	50–60	45	–	35–45
Power density (mW cm^{-2})	300–500	150–300	300–900	150	–	150–300
Start-up	Min	1–4 h	Min	5–10 h	Hours	5–10 h

regenerative PEMFCs with pressurized reactants are expected to reach a specific energy at around 550 W h kg^{-1} in the near future.

In general, cryogenic FCs are more promising in limiting the mass of the system but also pose more technological challenges. For example, long-term management of cryogenic fluids (such as hydrogen and oxygen) can be difficult. Moreover, additional power and additional equipment are required in order to maintain the reactants in their form as well as to liquefy them after electrolysis, which can significantly reduce the gain from using cryogenics. Miniaturization of the technology is also not trivial. The Japanese Aerospace eXploration Agency (JAXA) is currently developing a 100 W class regenerative polymer electrolyte FC for lunar and planetary missions [149]. An unitized regenerative FC (i.e., FC stack and electrolyzer are combined) was designed and tested in a laboratory environment. During the operation of the 100 W device, a stable FC and electrolysis reaction was observed which help raise the technology to TRL4.

The regeneration of the reagents provides a valuable feature for planetary exploration where solar energy can also be used over the course of the mission. The FC can then be used when illumination is irregular or when extended periods in darkness are expected (e.g., exploration of caves, and dark craters). The FC can be regenerated by the energy from conventional solar arrays when the robot returns to full illumination [150].

2.7.2
Thermal Subsystem

The scope for exploration can be found on all the planets and moons of the solar system, each of which has a different thermal environment due to the local insolation, the presence or absence of the atmosphere, and its typical daily cycle. To operate and survive in these environments and fluctuating thermal conditions, design of the thermal control system (TCS) is critical to keep every components

Table 2.25 Typical temperature ranges for robotic components.

Component	Operating temperature (°C)	Survival temperature (°C)
Digital electronics	0–50	−20 to 70
Analog electronics	0–40	−20 to 70
Batteries	10–20	0–35
Solar panels	−100 to 125	−100 to 125
Motors (e.g., MER [35])	−70 to 45	−120 to 110

of the robotic platform within the survival and operating temperatures over the course of the mission. Table 2.25 summarizes the temperature ranges of some key robotic components. To manage both extremes, the TCS must be able to generate heat and conserve the heat in the *cold case* as well as dissipate efficiently the heat in the *hot case*.

Figure 2.26 illustrates the major functions and design options for the TCS. Each function of the TCS can be design as passive thanks to the inherent properties of materials (e.g., multilayer insulation or coating), or as active using actuation, powering up or control of the TCS (e.g., electric heater or loop heat pipe). Typically, the active designs provide greater temperature control, but are heavier than the passive ones and also require power.

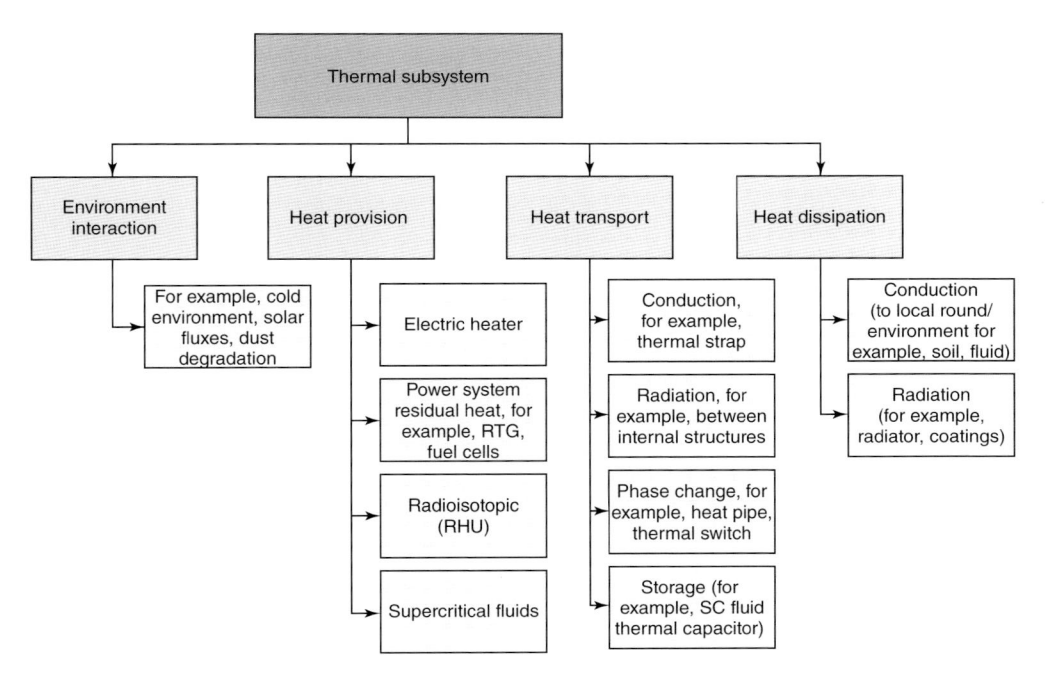

Figure 2.26 Thermal subsystem design options.

2.7.2.1 Sizing Warm/Cold Cases

The TCS design requires identification of some critical factors that influence the design of the system. For typical robotic systems and by extension to a wider range of concepts, the thermal subsystem is driven by the two cases that provide opposite operational requirements. The two cases significantly drive the thermal subsystem design as well as the overall configuration of the robotic system.

- **Warm Case** This case is used during solar illumination and requires the system to deal with excess heat coming from the environment or from its internal processes. It represents the radiator sizing case. Therefore, the system must
  —provide efficient heat rejection from radiator surfaces,
  —protect the radiators from solar flux and ground heating,
  —couple efficiently the dissipating units to the radiator surfaces,
  —insulate the robot body to protect it from solar and ground heating,
  —reduce the robot dissipation to minimize the required radiator area,
  —maximize the unit allowable temperatures to reduce the required radiator area.
- **Cold Case** This case is representative of a scenario where the robotic system is in darkness without solar illumination. In this instance, the robot must conserve the energy and minimize heat loss to the environment. It represents the heat input sizing case. Therefore, the system must
  —retain heat within the robot body,
  —decouple the radiator surfaces from the robot and/or space,
  —insulate robot body to retain heat,
  —provide additional heat to maintain units above their lower temperature limits.

2.7.2.2 Heat Provision

The provision of heat to the robotic system can be made through a number of methods:

- **Electric heaters:** Electric heaters represent the simplest way to convert electric power into heat by passing electricity through a resistance to make it dissipate heat. It is used extensively in space and exploration missions due to low mass and ease of actuation. They are typically used with thermostats or solid-state controllers to regulate precisely the temperature of a particular component. Similarly, they can be used to warm up components to their minimum operating temperatures before the components are turned on (e.g., locomotion subsystem in the morning). Due to the dependency on the power subsystem to provide the energy, this method is limited by the amount of power it can generate and maybe impractical for some cases. For example, if a rover requires 15 W through extended periods of darkness (e.g., lunar night of 200 h), this equals to 3000 W h thermal energy, hence requires a battery mass in the range of tens of kilograms.
- **Power System Residual Heat:** When the power subsystem cannot readily supply power to electric heaters, the residual heat from the power subsystem can

be used. For example, the nuclear or radioisotope power source can generate a significant amount of extra heat that the robotic system needs to deal with in the hot case. During the cold case, this extra heat can be used to keep the robot suitably warm to survive the extreme cold. Similarly, the FC power source dissipates heat of around 1.3–1.5 W per watt of electrical energy produced. A 100 W FC can thus dissipate around 130–150 W of heat.

- **Radioisotopic Heater Units:** RHUs are small heating devices that generate heat through radioactive decay such as RTGs, but do not generate electric power. In the simplest form, a small core of radioactive material (e.g., plutonium or polonium) is embedded into a protective casing that can be fitted where necessary on the spacecraft to provide known quantities of heat locally. The heat generated is continuous hence needs to be managed and dissipated during the hot case. These units run at around 300 °C, making the mechanical mounting and thermal design particularly important. Though being mechanically simple, the use of RHU raises other issues that are mainly related to radioactive safety during the robotic system integration and launch. Furthermore, the heat management during launch and cruise is paramount to the safety and survival of the system and mission.

  To date, only Russia, United States, and recently China have developed RHUs, all of which use plutonium. The Russian model "Angel" delivers 8.5 W of heat, and the American lightweight radioisotope heater unit (LWRHU) delivers 1 W. The small device is typically used in 1 W increment to supply necessary heat to a system. The MER rovers implemented 8 RHUs each. A wide range of space and planetary missions have used RHUs such as the Galileo, Cassini/Huygens, MPF, and MER. The technology has been validated to provide reliable heat generation throughout the mission lifetime. It is worth noting that in addition to the technical aspects concerning the use of radioisotopes, the political aspect is equally practical. To date, the launch of RHU is only possible by the country of origin, that is, only US-made RHUs are used in the US launchers, and similar to Russia.

- **Supercritical Fluids:** Recently, significant progress has been made in the development of thermal energy storage based on supercritical fluids (SCFs), which allows the capture and storage of heat by taking advantage of latent and sensible heat both in the two-phase regime and in the supercritical regime while at the same time reduces the required volume by taking advantage of the high compressibilities. The storage performance and pressure can be optimized by judicious selection of the working fluid with the key properties such as the high latent heat of vaporization, high specific heat, and low vapor pressure [151].

  Development of this relatively new technology units currently stays at a low TRL (i.e., TRL4), although the fundamental principle is well known and research institutes such as UCLA/JPL have been working on 5 kW h prototypes for terrestrial applications and aiming at advancing toward 10–30 kW h [151]. The space-oriented products are neither yet out of the research labs nor ready for the field; one example is the work focused on Brayton cycles requiring high pressure and temperature reaching 1000–1200 K [152].

2.7.2.3 Heat Management (Transport and Dissipation)

Heat is managed through three main principles: (i) convection involving the heat transfer via working fluids, (ii) conduction transferring heat through materials other than working fluids, and (iii) radiation where the heat is transferred via electromagnetic waves (e.g., infrared). The management of heat is applicable to both inward and outward fluxes depending on the specific thermal case.

- **Multilayer Insulation:** MLI provides an efficient means to reduce the radiative heat transfers from the body to the environment and vice versa. They consist of a number of thin layers made out of Kapton or Mylar metalized with aluminum or silver. A thin spacer mesh or scrim is used to prevent different layers from touching each other and to minimize conduction between layers. The MLI is particularly suited for vacuum operation such as exploration of airless bodies, hence is not as efficient in environments where convection is more important (e.g., planet with atmosphere). The design of MLI blankets depends upon the specific environment of the mission. For example, in order to survive the hot and cold extremes of a lunar environment, a rover can use an outer insulation layer with a low solar absorptivity to reduce the heat absorbed during periods of long and sustained illumination, and a low IR emissivity to reduce the heat lost during long periods of darkness. Previous lunar and Martian missions have taught us that the dust is extremely effective in coating external surfaces of the robotic system. Over the mission duration, dust deposition tends to increase the absorptivity and emissivity, which degrades the performance of MLI.
- **Radiators:** Radiators are used to radiate the heat to a cold environment (e.g., a view to deep space). Though they can take different shapes and sizes, they are typically designed as highly reflective surfaces fitted on a surface in the shade. Taking the rover platform as an example, the operation on sloping terrain potentially drives the TCS design and the rover configuration. If the rover targets the south pole of the Moon where the solar angle is very low (meaning that when the rover operates on a flat ground, only the sides of the body facing the Sun are significantly illuminated; but if the rover operates on a slope, the angle of the solar flux relative to the rover increases). The radiator design (e.g., the shape and shielding) must take these factors into consideration so that the impact of the solar flux is managed, either through limiting the rover operations to avoid direct solar illumination on the radiator surfaces or modifying the overall design to shadow the radiator surfaces. A combination of loop heat pipes and thermal switches can be used to further manage and control the heat dissipated to the radiator. If lunar dust coats the radiator surfaces, this will negatively affect the radiators' performance. Location of the radiator surfaces along with mitigating actions to reduce dust must be considered in the TCS design; for example, locating radiators on the top surface may avoid dust contamination during rover operation but at the solar terminus some dust can levitate from the lunar surface and get deposited on the horizontal surfaces of the rover and gradually affect any radiator surfaces located such way over time.
- **Heat Switch:** The use of conductive straps to a radiator is a reliable way to transfer the heat; however, survival of the robot during the cold cases would require

minimization of the conduction paths to the radiator as much as possible. Heat switches are devices that present a variable thermal resistance to perform a dual role either as a good thermal conductor or a good thermal insulator [153]. Many different heat switch configurations have been developed for space applications, mainly for cryogenic systems of satellites. Existing developments include the use of phase change material (paraffin) to create a simple "actuator" allowing two plates to be kept apart when the paraffin is solid and transitioning to a contact when the paraffin turns liquid. The Starsys Pedestal Switch provides a 100:1 conductance ratio between the two states (0.73 W K^{-1} max and 0.0075 W K^{-1}) at 100 g [153].

- **Loop Heat Pipes:** LHP was invented in Russia in the early 1980s [154], which is a two-phase heat transfer device that uses the evaporation and condensation of a working fluid to transfer heat and uses the capillary forces developed in fine porous wicks to circulate the fluid. It has gained rapid acceptance in recent years as a thermal control device in space applications, for example, Martian [155, 156] and lunar projects [157], and a number of mission concepts [101] use LHPs at the core of their TCS design. These designs typically allow transfer of heat from key components to the radiators and can be effectively turned off to limit heat rejection during the cold cases. Unlike other simpler methods

Table 2.26 Typical thermal trade-off design options for rovers.

Function	Trade-off design options
Rover body heating and thermal control	—Electrical heating during the night with associated impact on battery sizing —RHUs and the need to reject waste heat —RTGs for electrical power and the waste heat issue —Cruise environment and interaction with lander
Rover body heat rejection	—Location of radiating surfaces —Need for directional radiators to avoid heat flux from surface —Avoidance of dust build up on surfaces —Cruise environment and interaction with lander
Rover body insulation	—MLI outer surface (dust accumulation, electrostatic-sensitive devices) —Decoupling of radiators during night —Covering surfaces with louvers or the solar array —Loop heat pipes with diode (e.g., ExoMars) —Thermal switches (phase change material between unit I/F and radiator) —Rover body gas gap [52]
Appendage thermal control	—Expected temperature ranges of the appendages and equipment acceptability —Insulation from solar flux and methods of heating during the night —Retraction of equipments during the night

such as thermal straps and switches, the LHPs can transport heat effectively over greater distances. Smaller versions have been investigated to provide high efficiency thermal switching in a form factor closer to conventional thermal switches. Such a development was reported with vapor-modulated loop heat pipe (VMLHP) with a conductance ratio of $580:1$ (3.5 W K^{-1} max, 0.006 W K^{-1} min) at 146 g [158].

- **Other Passive Methods:** The design of a TCS is a holistic exercise taking into consideration the environment, the operation of the robotic system and the thermal capabilities of each onboard subsystem. Some simplistic or passive methods should always be investigated as they can be easier to implement and more cost effective. Examples include surface coatings tailored to a specific need (e.g., absorptivity or emissivity) or insulation (e.g., gas gap similar in principle to double glazed windows [60], and aerogel) that can help address some of the thermal constraints for the robotic system.

2.7.2.4 Trade-Off Options

Using a rover system as an example, Table 2.26 provides a list of key functions and trade-off design options for the TCS.

References

1. Doran, G.T. (1981) There's a smart way to write management's goals and objectives. *Management Review*, **70** (11), 35–36.
2. Mannion, M. and Keepence, B. (1995) Smart requirements. *ACM SIGSOFT Software Engineering Notes*, **20** (2), 42–47.
3. Gwynne, K. (2013) *How to Write Smart Requirements*, International Institute of Business Analysis, Columbus, OH.
4. Spencer, D.A., Blanchard, R.C., Thurmann, S.W., Braun, R.D., Peng, C.-Y., and Kallemeyn, P.H. (1998) Mars Pathfinder Atmospheric Entry Reconstruction. NASA Technical Report, American Astronomical Society, 98.
5. Flandin, G., Polle, B., Frapard, B., Vidal, P., Philippe, C., and Voirin, T. (2009) Vision based navigation for planetary exploration. *32nd Annual AAS Rocky Mountain Guidance and Control Conference*.
6. Parreira, B., Vasconcelos, J.F., Oliveira, R., Caramagno, A., Motrena, P., Dinis, J., and Rebord ao, J. (2010) Performance assessment of vision based hazard avoidance during lunar and martian landing. *Proceedings of 7th International Planetary Probe WorNshop, Barcelona*, Spain.
7. Johnson, A.E., Klumpp, A.R., Collier, J.B., and Wolf, A.A. (2002) Lidar-based hazard avoidance for safe landing on mars. *Journal of Guidance, Control, and Dynamics*, **25** (6), 1091–1099.
8. Prakash, R., Burkhart, P.D., Chen, A., Comeaux, K.A., Guernsey, C.S., Kipp, D.M., Lorenzoni, L.V., Mendeck, G.F., Powell, R.W., Rivellini, T.P. et al. (2008) Mars science laboratory entry, descent, and landing system overview. *Aerospace Conference, 2008 IEEE*, IEEE, pp. 1–18.
9. Laspace (2015) Automatic Station Luna 16.
10. Siddiqi, A.A. (2002) *A Chronology of Deep Space and Planetary Probes 1958±2000*, Monographs in Aerospace History, vol. 24, National Aeronautics and Space Administration, Washington, DC.
11. Akyildiz, L.F., Su, W., Sankarasubramaniam, Y., and Cayirci, E. (2002) A survey on sensor networks. *IEEE Communications Magazine*, **40** (8), 104–112.

12. Ezell, E.C. and Ezell, L.N. (2013) *On Mars: Exploration of the Red Planet, 1958–1978 The NASA History*, Courier Corporation.
13. Space Studies Board et al. (1999) *A Scientific Rationale for Mobility in Planetary Environments*, National Academies Press.
14. NASA, Viking Lander Model, NASA/JPL-Caltech/University of Arizona – http://www.nasa.gov/multimedia/imagegallery/image_feature_2055.html
15. NASA (1999) Nasa Fact Sheet - Mars Pathfinder.
16. NASA/JPL (1999) An Artist's Concept of the Mars Polar Lander on Mars.
17. Bonitz, R.G., Nguyen, T.T., and Kim, W.S. (2000) The Mars surveyor '01 rover and robotic arm. *Aerospace Conference Proceedings, 2000 IEEE, vol. 7*, pp. 235–246.
18. NASA (2008) *Phoenix Landing - Mission to the Martian North Pole*, Press Kit, NASA.
19. Lebreton, J.-P. and Matson, D.L. (2003) The Huygens probe: science, payload and mission overview, in *The Cassini-Huygens Mission*, Springer-Verlag, pp. 59–100.
20. Schilling, K., De Lafontaine, J., and Roth, H. (1996) Autonomy capabilities of European deep space probes. *Autonomous Robots*, **3** (1), 19–30.
21. Clausen, K.C., Hassan, H., Verdant, M., Couzin, P., Huttin, G., Brisson, M., Sollazzo, C., and Lebreton, J.-P. (2003) The Huygens probe system design, in *The Cassini-Huygens Mission*, Springer-Verlag, pp. 155–189.
22. Clemmet, J. (2015) Beagle 2 discovered - entry, descent & landing and mission outcome. *UK Space Conference*.
23. Bridges, J.C., Seabrook, A.M., Rothery, D.A., Kim, J.R., Pillinger, C.T., Sims, M.R., Golombek, M.P., Duxbury, T., Head, J.W., Haldemann, A.F.C. et al. (2003) Selection of the landing site in Isidis Planitia of Mars probe Beagle 2. *Journal of Geophysical Research: Planets (1991–2012)*, **108** (E1), 1.
24. Wright, I.P., Sims, M.R., and Pillinger, C.T. (2003) Scientific objectives of the Beagle 2 lander. *Acta Astronautica*, **52** (2), 219–225.
25. Sims, M.R., Pillinger, C.T., Wright, I.P., Morgan, G., Praine, I.J., Fraser, G.W., Pullan, D., Whitehead, S., Dowson, J., Wells, A.A. et al. (2000) Instrumentation on Beagle 2: the astrobiology lander on ESA's 2003 Mars Express mission. *International Symposium on Optical Science and Technology*, International Society for Optics and Photonics, pp. 36–47.
26. Bibring, J.-P., Rosenbauer, H., Boehnhardt, H., Ulamec, S., Biele, J., Espinasse, S., Feuerbacher, B., Gaudon, P., Hemmerich, P., Kletzkine, P. et al. (2007) The Rosetta lander ("philae") investigations. *Space Science Reviews*, **128** (1-4), 205–220.
27. Petrov, G.I. (1972) Investigation of the moon with the Lunokhod 1 space vehicle. *Space Research Conference, vol. 1*, pp. 439–447.
28. Malenkov, M. (2015) Self-propelled automatic chassis of Lunokhod-1: history of creation in episodes. *Proceedings of 2015 IFToMM Workshop on History of Mechanism and Machine Science*.
29. Kassel, S. (1971) Lunokhod-1 Soviet Lunar Surface Vehicle. Technical report, R-802-ARPA, DTIC Document.
30. Kemurdjian, A.L., Gromov, V.V., Kazhukalo, I.F., Kozlov, G.V., Komissarov, V.I., Korepanov, G.N., Martinov, B.N., Malenkov, V.I., Mitskevich, A.V., and Mishkinyuk, V.K. (1993) Soviet developments of planet rovers in period of 1964-1990. *Missions, Technologies, and Design of Planetary Mobile Vehicles*, **1**, 25–43.
31. Waye, D.E., Cole, J.K., and Rivellini, T.P. (1995) Mars pathfinder airbag impact attenuation system. *13th AIAA Aerodynamic Decelerator Systems Technology Conference*, pp. 109–119.
32. Matijevic, J. (1996) Mars pathfinder microrover-implementing a low cost planetary mission experiment. *2nd IAA International Conference on Low-Cost Planetary Missions*, CiteSeer, pp. 16–19.

33. Stone, H.W. (1996) *Mars Pathfinder Microrover: A Small, Low-Cost, Low-Power Spacecraft*, Jet Propulsion Laboratory.
34. NASA (2003) NASA Press Kit - Mars Exploration Rovers.
35. Fleischner, R. (2003) Development of the Mars exploration rover instrument deployment device.
36. Ratnakumar, B.V., Smart, M.C., Halpert, G., Kindler, A., Frank, H., Di Stefano, S.D., Ewell, R., and Surampudi, S. (2002) Lithium batteries on 2003 Mars exploration rover. *Battery Conference on Applications and Advances, 2002. The 17th Annual*, IEEE, pp. 47–51.
37. NASA (2011) NASA Launch Kit - Mars Surface Laboratory.
38. Principal Author Billing, R. and Co-Author Fleischner, R. (2011) Mars Science Laboratory Robotic Arm. *14th European Space Mechanisms & Tribology Symposium-ESMATS*.
39. Hammel, T.E., Bennett, R., Otting, W., and Fanale, S. (2009) Multi-mission radioisotope thermoelectric generator (MMRTG) and performance prediction model. *Proceedings of International Energy Conversion Engineering Conference (IECEC 2009)*, pp. 2–5.
40. Woerner, D., Moreno, V., Jones, L., Zimmerman, R., and Wood, E. (2012) The Mars Science Laboratory (MSL) MMRTG in Flight: A Power Update.
41. Ritz, F. and Peterson, C.E. (2004) Multi-mission radioisotope thermoelectric generator (MMRTG) program overview. *Aerospace Conference, 2004. Proceedings. 2004 IEEE, vol. 5*, IEEE.
42. Liu, J.J., Zhang, H.B., Su, Y., Li, H., and Li, C.L. (2015) Te Chang'e 3 misison: one year overview. *46th Lunar and Planetary Science Conference*.
43. Ip, W.H., Yan, J., Li, C.L., and Ouyang, Z.Y. (2014) Preface: the Chang'e-3 lander and rover mission to the Moon. *Research in Astronomy and Astrophysics*, **14** (12), 1511.
44. NASA (2013) The ExoMars Programme, NASA PPS. NASA.
45. Giorgio, V. (2014) ExoMars - two programs, one mission, in *Planetary Exploration Symposium*, TU Delft.
46. Matti, E.A. (2005) Concept evaluation of Mars drilling and sampling instrument.
47. Mühlbauer, Q., Ng, T.C., Paul, R., Schulte, W., Richter, L., and Hofmann, P. (2013) Technologies for automated sample handling and sample distribution on planetary landing missions. *Proceedings of ASTRA*, ESA.
48. Griffiths, A.D., Coates, A.J., Jaumann, R., Michaelis, H., Paar, G., Barnes, D., and Josset, J.-L. (2006) Context for the ESA ExoMars rover: the Panoramic Camera (PanCam) instrument. *International Journal of Astrobiology*, **5** (03), 269–275.
49. Corbel, C., Hamram, S., Ney, R., Plettemeier, D., Dolon, F., Jeangeot, A., Ciarletti, V., and Berthelier, J. (2006) WISDOM: an UHF GPR on the ExoMars mission. *AGU Fall Meeting Abstracts*, vol. 1, p. 1218.
50. Plettemeier, D., Ciarletti, V., Hamran, S.-E., Corbel, C., Cais, P., Benedix, W.-S., Wolf, K., Linke, S., and Roeddecke, S. (2009) Full polarimetric GPR antenna system aboard the ExoMars rover. *Radar Conference, 2009 IEEE*, IEEE, pp. 1–6.
51. Josset, J.-L., Westall, F., Hofmann, B.A., Spray, J.G., Cockell, C., Kempe, S., Griffiths, A.D., De Sanctis, M.C., Colangeli, L., Koschny, D. et al. (2012) CLUPI, a high-performance imaging system on the ESA-NASA rover of the 2018 ExoMars mission to discover biofabrics on Mars. *EGU General Assembly Conference Abstracts*, vol. 14, p. 13616.
52. Baglioni, P. et al. (2013) ExoMars project 2018 mission: rover development status. *Proceedings of ASTRA*.
53. Lopez-Reyes, G., Rull, F., Venegas, G., Westall, F., Foucher, F., Bost, N., Sanz, A., Catalá-Espí, A., Vegas, A., Hermosilla, I. et al. (2013) Analysis of the scientific capabilities of the ExoMars Raman laser spectrometer instrument. *European Journal of Mineralogy*, **25** (5), 721–733.
54. Edwards, H.G.M., Hutchinson, I., and Ingley, R. (2012) The ExoMars Raman spectrometer and the identification of biogeological spectroscopic signatures

using a flight-like prototype. *Analytical and Bioanalytical Chemistry*, **404** (6-7), 1723–1731.

55. Baglioni, P. and Joudrier, L. (2013) ExoMars rover mission overview. *IEEE International Conference on Robotics and Automation Workshop on Planetary Rovers*.
56. Patel, N., Slade, R., and Clemmet, J. (2010) The ExoMars rover locomotion subsystem. *Journal of Terramechanics*, **47** (4), 227–242.
57. Michaud, S., Gibbesch, A., Thüer, T., Krebs, A., Lee, C., Despont, B., Schäfer, B., and Slade, R. (2008) *Development of the ExoMars Chassis and Locomotion Subsystem*, Eidgenössische Technische Hochschule Zürich, Autonomous Systems Lab.
58. Michaud, S., Hoepflinger, M., Thueer, T., Lee, C., Krebs, A., Despont, B., Gibbesch, A., and Richter, L. (2008) Lesson learned from ExoMars locomotion system test campaign. *Proceedings of 10th Workshop on Advanced Space Technologies for Robotics and Automation, ESTEC The Netherlands*.
59. Batteries, S. (2015) Saft Li-ion battery to power the ExoMars rover as it searches for life on the red planet, July 8, 2015.
60. Valter, P., Marco, G., Sergio, M., Renato, M., and Salvatore, T. (2011) Design and test of ExoMars thermal breadboard. *41st International Conference on Environmental Systems*, p. 5118.
61. AlaryF, C. and LapenséeF, S. (2010) Thermal design of the ExoMars rover module.
62. Silva, N., Lancaster, R., and Clemmet, J. (2013) ExoMars rover vehicle mobility functional architecture and key design drivers. *Proceedings of the 12th Symposium on Advanced Space Technologies in Robotics and Automation, ESTEC, Noordwijk, The Netherlands*.
63. McManamon, K., Lancaster, R., and Silva, N. (2013) ExoMars rover vehicle perception system architecture & test results. *Proceedings of ASTRA*.
64. Winter, M., Barclay, C., Pereira, V., Lancaster, R., Caceres, M., McManamon, K., Nye, B., Silva, N., Lachat, D., and Campana, M. *ExoMars rover vehicle: detailed description of the GNC system*, ASTRA Conference 2015 – Noordwijk, ESTEC, ESA.
65. Lancaster, R., Silva, N., Davies, A., and Clemmet, J. (2011) ExoMars rover GNC design and development. *ESA GNC Conference 2011*.
66. Mustard, J.F., Adler, M., Allwood, A., Bass, D.S., Beaty, D.W., Bell, J.F. III, Brinckerhoff, W.B., Carr, M., Des Marais, D.J., Drake, B. et al. (2013) Report of the Mars 2020 science definition team, posted July, p. 154.
67. Allwood, A., Hurowitz, J., Wade, L.W.A., Hodyss, R.P., and Flannery, D. (2014) Seeking ancient microbial biosignatures with PIXL on Mars 2020. *AGU Fall Meeting Abstracts*, vol. 1, p. 07.
68. NASA (2015) NASA Fact Sheet - Mars 2020.
69. Farley, K. Mars 2020 Mission: Science Rover, August 2015.
70. XPRIZE Foundation About Xprize.
71. Mitrofanov, I., Dolgopolov, V., Khartov, V., Lukjanchikov, A., Tret'yakov, V., and Zelenyi, L. (2014) "Luna-Glob" and "Luna-Resurs": science goals, payload and status. *EGU General Assembly 2014, held 27 April-2 May, 2014 in Vienna, Austria, ID. 6696*.
72. Novara, M. (2002) The bepicolombo ESA cornerstone mission to mercury. *Acta Astronautica*, **51** (1), 387–395.
73. Wu, X., Bender, P.L., and Rosborough, G.W. (1995) Probing the interior structure of Mercury from an orbiter plus single lander. *Journal of Geophysical Research: Planets (1991–2012)*, **100** (E1), 1515–1525.
74. Bertrand, R., Brueckner, J., van Winnendael, M., and Novara, M. (2001) Nanokhod–a micro-rover to explore the surface of mercury. *6th International Symposium on Artificial Intelligence, Robotics and Automation in Space, Montreal, Canada*.
75. Sagdeev, R.Z., Linkin, V.M., Blamont, J.E., and Preston, R.A. (1986) The VEGA Venus balloon experiment. *Science*, **231** (4744), 1407–1408.
76. Izutsu, N., Yajima, N., Hatta, H., and Kawahara, M. (2000) Venus balloons at

77. Cutts, J.A. and Kerzhanovich, V.V. (2001) Martian aerobot missions: first two generations. *Aerospace Conference, 2001, IEEE Proceedings*, vol. 1.
78. Hall, J.L., Pauken, M., Kerzhanovich, V.V., Walsh, G.J., Fairbrother, D., Shreves, C., and Lachenmeier, T. (2007) Flight test results for aerially deployed Mars balloons. *Proceedings of AIAA Balloon Systems Conference, AIAA Reston, VA*, p. 21–24.
79. Jaroszewicz, A., Sąsiadek, J., and Sibilski, K. (2013) Modeling and simulation of flapping wings entomopter in Martian atmosphere. *Aerospace Robotics*, Springer-Verlag, pp. 143–162.
80. Trost, A. (2001) *Extendable swarm programming architecture*. PhD thesis. University of Virginia.
81. Domaine, H. (2006) *Robotics*, Lerner Publications.
82. Howe, S.D., O'Brien, R.C., Ambrosi, R.M., Gross, B., Katalenich, J., Sailer, L., Webb, J., McKay, M., Bridges, J.C., and Bannister, N.P. (2011) The Mars hopper: an impulse-driven, long-range, long-lived mobile platform utilizing in situ Martian resources. *Proceedings of the Institution of Mechanical Engineers, Part G: Journal of Aerospace Engineering*, **225** (2), 144–153.
83. Powell, J., Maise, G., and Paniagua, J. (2001) The Mars hopper- a mobile, lightweight probe to explore and return samples from many widely separated locations on Mars. *IAF, International Astronautical Congress, 52nd, Toulouse, France*.
84. Shafirovich, E., Salomon, M., and Gökalp, I. (2006) Mars hopper versus Mars rover. *Acta Astronautica*, **59** (8), 710–716.
85. Zubrin, R., Muscatello, T., Birnbaum, B., Caviezel, K.M., Snyder, G., and Berggren, M. (2002) Progress in Mars ISRU technology. *AIAA Aerospace Sciences Meeting & Exhibit, 40th, Reno, NV*.
86. Sanders, G.P. and Peters, T. (2001) Report on development of micro chemical/thermal systems for Mars ISRU-based missions. *AIAA, Aerospace Sciences Meeting and Exhibit, 39th, Reno, NV*.
87. Rice, E.E., Gramer, D.J., St. Clair, C.P., and Chiaverini, M.J. (2003) Mars ISRU CO/O2 rocket engine development and testing. *7th International Workshop on Microgravity Combustion and Chemically Reacting Systems*, p. 101.
88. Elfes, A., Montgomery, J.F., Hall, J.L., Joshi, S., Payne, J., and Bergh, C.F. (2005) *Autonomous Flight Control for a TITAN Exploration AEROBOT*, Jet Propulsion Laboratory, National Aeronautics and Space Administration, Pasadena, CA.
89. Fink, W., Tarbell, M.A., Furfaro, R., Powers, L., Kargel, J.S., Baker, V.R., and Lunine, J. (2011) Robotic test bed for autonomous surface exploration of Titan, Mars, and other planetary bodies. *Aerospace Conference, 2011 IEEE*, pp. 1–11.
90. Hartwig, J.W., Colozza, A., Lorenz, R.D., Oleson, S., Landis, G., Schmitz, P., Paul, M., and Walsh, J. (2016) Exploring the depths of Kraken Mare – Power, thermal analysis, and ballast control for the Saturn Titan submarine. *Cryogenics*, **74**, 31–46.
91. Lorenz, R.D. (2004) Titan: a new world covered in submarine craters? in *Cratering in Marine Environments and on Ice*, Springer-Verlag, pp. 185–195.
92. Oleson, S.R., Lorenz, R.D., and Paul, M.V. (2015) Phase 1 final report: Titan submarine.
93. Friedlander, A.L. (1986) Buoyant station mission concepts for titan exploration. *Acta Astronautica*, **14**, 233–242.
94. Atkinson, D.J. (1999) Autonomy technology challenges of Europa and titan exploration missions. *Artificial Intelligence, Robotics and Automation in Space*, **440**, 175.
95. Ross, C.T. (2007) Conceptual design of submarine to explore Europa's oceans. *Journal of Aerospace Engineering*, **20** (3), 200–203.
96. Reh, K., Elliot, J., Spilker, T., Jorgensen, E., Spencer, J., and Lorenz, R. (2007) Titan and Enceladus $1b Mission

Feasibility Study Report. JPL D-37401B, NASA/JPL.
97. Jones, J.A. and Heun, M.K. (1997) Montgolfiere balloon aerobots for planetary atmospheres. AIAA, Paper No. 97-1445.
98. Herrmann, F., Kuß, S., and Schäfer, B. (2011) *Mobility Challenges and Possible Solutions for Low-Gravity Planetary Body Exploration*, ESA/ESTEC, Noordwijk, The Netherlands.
99. Ulamec, S., Kucherenko, V., Biele, J., Bogatchev, A., Makurin, A., and Matrossov, S. (2011) Hopper concepts for small body landers. *Advances in Space Research*, **47** (3), 428–439.
100. Hager, P.B., Parzinger, S., Haarmann, R., and Walter, U. (2015) Transient thermal envelope for rovers and sample collecting devices on the moon. *Advances in Space Research*, **55** (5), 1477–1494.
101. Barraclough, S., Huston, K., and Allouis, E. (2009) Thermal Design for Moon-Next Polar Rover. Technical report, SAE Technical Paper.
102. Richter, L. (2008) Low temperature mobility and mechanisms, SAE International.
103. Phillips, N. (2001) Mechanisms for the Beagle 2 lander. *9th European Space Mechanisms and Tribology Symposium*, vol. 480, pp. 25–32.
104. Briscoe, H.M. (1990) Why space tribology? *Tribology International*, **23** (2), 67–74.
105. Jones, W.R. Jr. and Jansen, M.J. (2000) Space tribology. NASA TM, 209924.
106. Patterson, R.L., Hammoud, A., and Elbuluk, M. (2008) Electronic components for use in extreme temperature aerospace applications. *12th International Components for Military and Space Electronics Conference, San Diego, CA*.
107. Bruhn, F., von Krusenstierna, N., Habinc, S., Gruener, G., Rusconi, A., Waugh, L., Richter, L., and Lamoureux, E. (2008) Low temperature miniaturized motion control chip-enabled by MEMS and microelectronics. *Proceedings of ASTRA*, vol. 2008, CiteSeer, pp. 11–13.
108. Farrell, W.M., Stubbs, T.J., Vondrak, R.R., Delory, G.T., and Halekas, J.S. (2007) Complex electric fields near the lunar terminator: the near-surface wake and accelerated dust. *Geophysical Research Letters*, **34** (14), L14201, doi: 10.1029/2007gl029312.
109. Stubbs, T.J., Vondrak, R.R., and Farrell, W.M. (2006) A dynamic fountain model for lunar dust. *Advances in Space Research*, **37** (1), 59–66.
110. Stubbs, T.J., Vondrak, R.R., and Farrell, W.M. (2007) Impact of Dust on Lunar Exploration, 4075, http://hefd.jsc.nasa.gov/files/StubbsImpactOnExploration.
111. Williams, D.R. (2015) Nasa Planetary Fact Sheets.
112. Vasavada, A.R., Paige, D.A., and Wood, S.E. (1999) Near-surface temperatures on Mercury and the moon and the stability of polar ice deposits. *Icarus*, **141** (2), 179–193.
113. Geelen, K. (2012) Miniaturisation needs of the Mars network landers. *ESA Workshop on Avionics Data, Control and and Software Systems (ADCSS)*.
114. NASA Advanced Space Transportation: Paving the Highway to Space, NASA, 2010.
115. Way, D.W., Davis, J.L., and Shidner, J.D. (2013) Assessment of the Mars Science Laboratory entry, descent, and landing simulation. *23rd AAS/AIAA Space Flight Mechanics Meeting*, pp. 2013–0420.
116. Beaty, D., Grady, M., May, L., Gardini, B. et al. (2008) Preliminary planning for an international Mars sample return mission. *Report of the International Mars Architecture for the Return of Samples (iMARS) Working Group*.
117. Elie, A., Jorden, T., Patel, N., and Ratcliffe, A. (2011) Sample fetching rover - lightweight rover concepts for Mars sample return. *Proceedings of ESA ASTRA*.
118. Squyres, S.W., Arvidson, R.E., Bollen, D., Bell, J.F., Brueckner, J., Cabrol, N.A., Calvin, W.M., Carr, M.H., Christensen, P.R., Clark, B.C. et al. (2006) Overview of the opportunity Mars exploration rover mission to Meridiani Planum: eagle crater to purgatory ripple. *Journal of Geophysical*

Research: Planets (1991–2012), **111** (E12), doi: 10.1029/2006je002771.
119. Beauchamp, P., Ewell, R., Brandon, E., and Surampudi, R. (2015) Solar power and energy storage for planetary missions, in NASA, editor, Outer Planet Assessment Group - August 2015, NASA - JPL, LP.
120. Lemmon, M.T., Wolff, M.J., Bell, J.F., Smith, M.D., Cantor, B.A., and Smith, P.H. (2015) Dust aerosol, clouds, and the atmospheric optical depth record over 5 Mars years of the Mars exploration rover mission. *Icarus*, **251**, 96–111.
121. Mazumder, M.K., Biris, A.S., Sims, R., Calle, C., and Buhler, C. (2003) Solar panel obscuration in the dusty atmosphere of Mars. *Proceedings ESA-IEEE Joint Meeting on Electrostatics*, pp. 208–218.
122. Mittal, K.L. (2006) *Particles on Surfaces: Detection, Adhesion and Removal*, vol. 9, CRC Press.
123. Biris, A.S., Saini, D., Srirama, P.K., Mazumder, M.K., Sims, R.A., Calle, C.I., and Buhler, C.R. (2004) Electrodynamic removal of contaminant particles and its applications. *Industry Applications Conference, 2004. 39th IAS Annual Meeting. Conference Record of the 2004 IEEE*, vol. 2, IEEE, pp. 1283–1286.
124. Horenstein, M.N., Mazumder, M., and Sumner, R.C. (2013) Predicting particle trajectories on an electrodynamic screen–theory and experiment. *Journal of Electrostatics*, **71** (3), 185–188.
125. Sharma, R., Wyatt, C., Zhang, J., Calle, C., Mardesich, N., Mazumder, M.K. et al. (2009) Experimental evaluation and analysis of electrodynamic screen as dust mitigation technology for future Mars missions. *IEEE Transactions on Industry Applications*, **45** (2), 591–596.
126. Surampudi, S. (2011) Overview of the space power conversion and energy storage technologies. *Technology Forum on Small Body Scientific Exploration*, ibid.
127. Stella, P., Mueller, R., Davis, G., and Distefano, S. (2004) The environmental performance at low intensity, low temperature (LILT) of high efficiency triple junction solar cells. 2nd International Energy Conversion Engineering Conference, p. 5579.
128. Schmidt, G.R., Dudzinski, L.A., and Sutliff, T.J. (2011) *Radioisotope Power: A Key Technology for Deep Space Exploration*, INTECH Open Access Publisher.
129. Dudzinski, L.A., McCallum, P.W., Sutliff, T.J., and Zkrajsek, J.F. (2013) NASA's radioisotope power systems program status. *11th Energy Conversion Engineering Conference*.
130. Summerer, L., Roux, J.P., Pustovalov, A., Gusev, V., and Rybkin, N. (2011) Technology-based design and scaling for RTGS for space exploration in the 100W range. *Acta Astronautica*, **68** (7), 873–882.
131. O'Brien, R.C., Ambrosi, R.M., Bannister, N.P., Howe, S.D., and Atkinson, H.V. (2008) Safe radioisotope thermoelectric generators and heat sources for space applications. *Journal of Nuclear Materials*, **377** (3), 506–521.
132. Watkinson, E.J., Ambrosi, R.M., Williams, H.R., Sarsfield, M.J., Tinsley, T.P., and Stephenson, K. (1798, 2014) Americium oxide surrogates for European radioisotope power systems. LPI Contributions.
133. Thieme, L.G., Qiu, S., and White, M.A. (2000) Technology development for a stirling radioisotope power system. *AIP Conference Proceedings, Number 2*, IOP Institute of Physics Publishing Ltd, pp. 1260–1265.
134. Dudzinski, L.A. (2014) NASA's radioisotope power systems program. *Presentation to NRC - Committee on Astrobiology and Planetary Science*.
135. Bennett, G.L. (2006) Space nuclear power: opening the final frontier. *4th International Energy Conversion Engineering Conference and Exhibit (IECEC)*, pp. 26–29.
136. Ambrosi, R.M., Williams, H.R., Samara-Ratna, P., Bannister, N.P., Vernon, D., Crawford, T., Bicknell, C., Jorden, A., Slade, R., Deacon, T. et al. (2012) Development and testing of Americium-241 radioisotope thermoelectric generator: concept designs and breadboard system. *Proceedings of*

Nuclear and Emerging Technologies for Space, The Woodlands, TX, pp. 21–23.

137. Williams, H.R., Ambrosi, R.M., Bannister, N.P., Samara-Ratna, P., and Sykes, J. (2012) A conceptual spacecraft radioisotope thermoelectric and heating unit (RTHU). *International Journal of Energy Research*, **36** (12), 1192–1200.

138. Sutliff, T.J. and Dudzinski, L.A. (2009) NASA radioisotope power system program–technology and flight systems. paper AIAA, 4575, pp. 2–5.

139. Hughes, T.G., Smith, R.B., and Kiely, D.H. (1983) Stored chemical energy propulsion system for underwater applications. *Journal of Energy*, **7** (2), 128–133.

140. Ramanarayanan, C.P. (2013) Avenues for underwater propulsion. *Defence Science Journal*, **42** (3), 205–208.

141. Michael, P., Sal, O., Steve, O., Miller, T., and Lee, M. (2015) SCEPS in space: non-radioisotope power for sunless solar system exploration missions, in *Outer Planets Assessment Group (OPAG)*, Lunar and Planetary Institute.

142. Michael, P. (2015) SCEPS in space - non-radioisotope power systems for sunless solar system exploration missions.

143. NASA (2015) NASA Technology Roadmaps - TA 3: Space Power and Energy Storage. Technical report TA-3, NASA.

144. Ratnakumar, B.V., Smart, M.C., Kindler, A., Frank, H., Ewell, R., and Surampudi, S. (2003) Lithium batteries for aerospace applications: 2003 Mars exploration rover. *Journal of power sources*, **119**, 906–910.

145. Lyons, V.J., Gonzalez, G.A., Houts, M.G., Iannello, C.J., Scott, J.H., and Surampudi, S. (2010) *Draft Space Power and Energy Storage Roadmap*, National Aeronautics and Space Administration.

146. Allan, P. (2011) Lithium ion batteries: going the distance, in *Electric & Hybrid Vehicles and Fuel Cell Network Seminar*, Imperial College London.

147. Lu, C.-Y. and McClanahan, J. (2009) NASA JSC lunar surface concept study lunar energy storage. *US Chamber of Commerce Programmatic Workshop*, vol. 26.

148. EG&G technical services (2004) *Fuel Cell Handbook*, 7th edn, Inc., Albuquerque, NM, DOE/NETL-2004/1206, EG&G Technical Services, Inc.

149. Sone, Y. (2011) A 100-W class regenerative fuel cell system for lunar and planetary missions. *Journal of Power Sources*, **196** (21), 9076–9080.

150. Lyons, V.J. (2007) NASA energy/power system technology. *Army Research Office Base Camp Sustainability Workshop*.

151. Ganapathi, G.B. (2013) High density thermal energy storage with supercritical fluids (SUPERTES).

152. Rousseau, I.M. and Driscoll, M. (2007) Analysis of a high temperature supercritical Brayton cycle for space exploration.

153. David, G.G. (2002) *Spacecraft Thermal Control Handbook: Fundamental Technologies*, vol. 1, AIAA.

154. Ku, J. (1999) Operating Characteristics of Loop Heat Pipes. Technical report, SAE Technical Paper 1999-01-2007, SAE International.

155. Tosi, M.C., Mannu, S., Jones, G., Quinn, A., and Clemmet, J. (2008) ExoMars Rover Module Thermal Control System. Technical report, SAE Technical Paper 2008-01-2003, SAE International 2008.

156. Molina, M., Franzoso, A., Bursi, A., Fernandez, F.R., and Barbagallo, G. (2008) A Heat Switch for European Mars Rover. Technical report, SAE Technical Paper 2008-01-2153, SAE International.

157. Christopher, J.P., John, R.H., Anderson, W.G., Tarau, C., and Farmer, J. (2011) Variable conductance heat pipe for a lunar variable thermal link. *41st International Conference on Environmental Systems (ICES 2011)*, Portland, OR.

158. Mishkinis, D., Corrochano, J., and Torres, A. Development of miniature heat switch-temperature controller based on variable conductance LHP, September 2014.

3
Vision and Image Processing

Gerhard Paar, Robert G. Deen, Jan-Peter Muller, Nuno Silva, Peter Iles, Affan Shaukat, and Yang Gao

3.1
Introduction

The use of autonomous robotic platforms has experienced a tremendous growth over the past five decades of planetary exploration missions with a diversity of technologies such as planetary landers and rovers. Within the variety of sensor concepts on board these spacecraft, machine vision has played a significant role in gathering planetary data, and extracting and analyzing useful information for increased autonomy and scientific gain. Camera images and related vision sensors are used both directly on board to contribute to autonomy and on ground to aid operations and science decisions. Software on board the rovers and landers is used for image compression, mapping and 3D scene reconstruction, navigation, and – as a topic getting more and more important – for scene interpretation to feed decisions and raise the level of autonomy of robotic assets on planetary bodies.

Robotics vision is a method of perception by machines, specifically those that react on the perceived information (see Figure 3.1 for a simplified scheme). It is typically done with the help of vision sensors that measure the (reflected) radiation of the environment in specific parts of the electromagnetic spectrum (e.g. light of the visual spectrum typically between 400 and 700 nm wavelength). Most representative are digital cameras that are equipped with a lens and many individual detector components (e.g., a regular array arranged in a square grid on a CCD or CMOS chip) to measure also the spatial distribution of light coming in.

Perception, mapping, navigation, and monitoring of activities rely heavily on vision as a major component of robotic assets. This chapter addresses components, sensors, software, and visual cues that are important to fulfill the needs of robotic activities on the surface of planets, asteroids, comets and moons. The main users of robotic vision solutions are planetary rovers, landers and astronauts.

This chapter shows that various passive vision sensor technologies have been successfully implemented in more than a dozen robotic planetary missions, while

Contemporary Planetary Robotics: An Approach Toward Autonomous Systems, First Edition.
Edited by Yang Gao.
© 2016 Wiley-VCH Verlag GmbH & Co. KGaA. Published 2016 by Wiley-VCH Verlag GmbH & Co. KGaA.

Figure 3.1 Image processing/computer vision gives planetary rovers "eyes" to interpret their environment.

active vision such as laser ranging has not been emphasized but is envisaged a growing field for specific applications and environments such as the lunar poles. Stereo vision is the primary concept, opening up the ability to capture scenes in three dimensions whose immediate products range from 3D maps of the environment to accurate localization in a global planetary context when using remote sensing imagery as a reference. Calibration is important to allow proper interpretations of camera-based observations in a quantitative sense. The knowledge about possible error influences in this context is vital for the usability of the captured data. Effective visualization closes the gap between bytes and software in the computational chains and interpretation by humans. Already now, robotics vision in many cases is the only available source of perception during large parts of planetary surface missions, which makes it such an essential component.

It will become clear from the chapter that vision sensors and software in recent years have been improved to raise robustness and accuracy and cope with the harsh and stringent space environment. Onboard processing of planetary robots is limited by computational resources, development life-cycles, and space qualification issues. Processing of images on ground profits from rapid progress in computer vision software technology, which has also been pushed by industry through the availability of digital consumer cameras shortly before the beginning of the 21st century.

This chapter also sheds some light on robotic use cases envisaged in the future that will be served by vision, such as science autonomy, soil characterization for locomotion monitoring and planning, search for a deposited sample cache, or the monitoring of robotic service and construction tasks. New techniques and use cases such as vision supported by saliency models, artificial intelligence, and cybernetics vision are underway.

Table 3.1 gives a list of important terms and acronyms used throughout this chapter and provides a simple explanation for each. The rest of the chapter focuses on rover and lander operation scenarios.

Table 3.1 Important terms and acronyms in the field of planetary robotics vision.

Term	Acronym	Meaning/description
Camera	… Cam, … cam	Vision sensor that is able to capture a one- or two-dimensional array of light measurements
Bidirectional reflectance distribution function	BRDF	Function that defines how light is reflected at an opaque surface
Charge-coupled device	CCD	Light-sensitive electronic element used in cameras
Complementary metal–oxide–semiconductor	CMOS	Light-sensitive electronic element used in cameras
Digital elevation model/digital terrain model	DEM/DTM	3D data structure that is able to store and represent the morphology of a 3D surface (heights are stored in grid values)
Experiment data record	EDR	Raw data as directly received on Earth from a space instrument
Field of view	FoV	Angular extension of a vision sensor pixel
Focal plane		Plane in the camera where images are focused to (e.g., at the plane of the CCD array)
Field programmable gate array	FPGA	Integrated circuit designed for execution of specific software tasks, for example, image processing algorithms; in many cases directly attached to sensors for gaining speed
Ground truth		Information obtained from direct observation or simulation
Guidance navigation and control	GNC	System to control the motion of vehicles (in this context: Robotic, i.e., driven by computers and fed by sensors)
Infrared	IR	Part of the electromagnetic spectrum; longer wavelength than visible light
Lens		Optical component in a camera that focuses light beams onto the light detector (the latter being an array in the case of a camera)
Panoramic camera	PanCam; Pancam	Single or multiple camera system that is able to capture a wide FoV

(*continued overleaf*)

Table 3.1 (Continued).

Term	Acronym	Meaning/description
Panchromatic camera/ Black-and-white camera	B/W	Camera with single color channel that collects a major part of the visible (and in some cases near infrared) spectrum
Panorama		Two or more images stitched together into a mosaic. A full panorama covers 360° in azimuth
Pan-tilt unit	PTU	Movable device with two axes to actively point a camera or a set of cameras into different angles
Planetary data system	PDS	System available to the international space research community, hosted by NASA, to archive and distribute scientific data from planetary missions, astronomical observations, and laboratory measurements
(Rover or camera) pose		Position and pointing angles or orientation of a device or object
Reduced data record	RDR	Processed, often calibrated products derived from immediate sensor data (e.g., lens distortion corrected image, or XYZ point cloud)
Sol		Planetary day (about 24 h 38 m for Mars)
Super resolution (restored)	SR(R)	Process/product that uses images taken at slightly different viewing angles to compile a single imaging product with higher resolution than the source images
Stereo imaging		Image-based perception of a target from at least two spatially distinct positions (example: two human eyes)
Visual odometry	VO	Method for determining ego-motion by viewing the environment at short time intervals, using tracked landmark features

3.2
Scope of Vision Processing

The different mission phases, as pointed out in Table 3.2, have different requirements to the vision system components; nevertheless, maximum synergy needs to be addressed to them in order to save resources (mainly mass, power, and space, also downlink and uplink rate, as well as complexity, which impacts reliability and development effort).

In this chapter we use Mars rovers as the primary examples and use cases to describe robotic vision concepts. Most concepts apply equally well to other planets, with the possible exception of the Moon (where some forms of limited

Table 3.2 Mission phases where robotics vision is required.

Phase/individual step	Requirements/requests to robot vision	Comments/technique/ purpose/typical product
Descent	Maintenance of lander's attitude	Star tracking, landmark tracking
Descent	Landing spot detection	Comparison to satellite-generated images
Landing	Landing spot tracking	Landmark tracking
Landing	Surface-relative speed determination	Landmark tracking
Landing	Hazard avoidance	Roughness measurement; classification; 2D/3D pattern recognition
Rover egress	Initial mapping from lander	Safety assessment; decision (how) to egress
Rover navigation	Mapping, localization, measurement of motion	DEM generation, VO, image comparison, slope and hazard map
Rover science	Geologic context provision and navigation	DEMs, ortho images, panoramas, placement of science sensors
Rover science	Geology, morphology, geochemistry, atmospheric science	Spectroscopy, radiometry, geologic maps, base for other sensor's spatial context
Sample caching and cache collection	Sample and cache search	3D recognition and localization of samples and sample cache
Education, public outreach	Highly informative/beautiful pictures	Important component to justify robotic missions to other planets

tele-operation from Earth become feasible). Also, most non-mobility concepts apply to stationary landers as well as rovers.

The MER and Mars Science Laboratory (MSL) missions represent the first implementations of a so-called "tele-robotic field geology" paradigm on Mars in which geologists, geochemists, and mineralogists on the ground use a remotely operated mobile asset on another planet to systematically study the geology at the rover site in a similar way as they would do when in the field themselves. This includes[1]

- the ability to move around (provided by the rover mobility system),
- the ability to observe, survey, and map the scene to establish scientific context and for planning of movements, of sampling locations and close-up investigations by cameras,
- the ability to approach, "touch" and sample targets of interest identified from stand-off remote observations (provided by robot-mounted *in situ* analysis),

---

1) A similar approach is followed by ESA's ExoMars mission, extended by exobiological objectives and a sophisticated drill function.

and microscopic instruments that determine elemental and mineralogical composition as well as close-up texture.

In a scientific rover mission, imaging therefore has **a twofold function**:

- an **engineering function**: providing a basis for building three-dimensional (3D) models of the rover surroundings to support path planning and execution onboard the vehicle, and support various manipulation tasks,
- a **scientific function**: providing high fidelity imaging in selected wavebands (or other signatures such as 3D surface roughness) and at appropriately high resolution – including a precise fusion of different sensors – to support interpretations of the local geology as well as to support selection of promising targets for close-up study.

From this distinction, the primary requirements (the most important one being restricted downlink data rate) are established in the following two sections. In addition, Section 3.2.3 points out the main physical conditions that vision on the surface of planets must endure.

Onboard and on-ground functions have associated implications. On ground there is a very large processing capability with supercomputers, no real-time constraints (no other functions must be executed at key periods implying functions would have a guaranteed limited maximum time) – this allows employing any available algorithm since complexity and computation time are not an issue. More valuably, there is the possibility (realistically, mandatory) that a human checks the results before any command is sent to the rover. This ensures that the commands are sensible and safe with the full control of the operators. However, there are very few opportunities to receive and send information to the rover in small closed-loop intervals, which highly limits how far the rover can drive based on a ground-command: very slow net motion on the surface will be the effect.

Onboard there is the possibility of sending drive commands at any point in time where there is enough energy to drive: this allows for an almost continuous motion on the surface. However, because of the vision processing limitations of space qualified computers, current-generation rovers regularly stop (after 2.3 m in average) for a certain time (in the range of minutes) in order to compute a new safe and efficient path, making autonomous driving less efficient. There is no human intervention, which means that the onboard functions must be highly reliable, safe, and efficient. The effort put into qualification is high and operators must regularly check its correct behavior.

3.2.1
Onboard Requirements

There are two key engineering-driven functions onboard contemporary rovers that require vision and image processing: perception of the surrounding terrain in order to plan safe and efficient paths (Navigation) and detecting and tracking of landmarks such that the rover egomotion, position and angles ("pose") can be determined at any time (Visual Localization). There are generally two daily opportunities for Earth to contact the rover. It is possible (and likely due to the typical sun-synchronous orbits of relay satellites) that no closed-loop ground contact actually occurs during the period where it is day and the rover has energy and electronics warm enough to drive. In addition, because of the distance between Mars and the Earth, the information may take more than 20 min to get from one planet to the other. Considering that the communication window is just above such duration, it is not practical to implement mobility functions on the ground. Finally, considering the need for almost continuous egomotion estimation, no actual practical mobile mission could have such function implemented on the ground.

Implementing the functions onboard implies a high level of autonomy. In addition to a reliable, robust, accurate, and adequately verified onboard implementation, sufficient data shall be gathered and transmitted every day such that ground can regularly confirm the appropriate behavior of the onboard functions. To name a practical example, the ExoMars Navigation and Localization cameras have a resolution of $1024 \times 1024 \times 8$ bits pixel. The Navigation function, at each stop/waypoint, requires three pairs of images. The Visual Localization function requires a pair every 10 s, each only a few centimeters apart. It has been concluded that full resolution is needed for navigation and that half resolution is sufficient for Visual Localization. Hence, for a 3 m path, this requires 180 Mbit of imagery. Since the rover has on average only 150 Mbit/sol of downlink bandwidth, it is clear that it would not be possible to transmit this amount of data to the ground. It is true that compression will be applied, but other data must also be exchanged with ground, not just mobility imagery – and drives cover much longer distances.

The cameras must obtain good quality images of the Martian environment. Knowledge of the visual environment has been used to derive the camera driving requirements, both for the optics and the sensor. In addition, though the rover speed is low, when its wheels slip off rocks, its vertical speed is high and images taken at that point in time can be compromised. Hence, the cameras typically have a global shutter and a CCD sensor, rather than a rolling shutter (such as a CMOS sensor).

The largest onboard constraint is the computer. For the ExoMars example, there is a dedicated coprocessor for the visual functions, but it remains a challenge to meet the requirements of 180 s for all navigation-related functions and 4.5 s for Visual Localization according to the ExoMars mission layout. These current computing constraints may lead in future missions to a dedicated computer for these functions and to move a maximum of functions to the camera's field-programmable gate array (FPGA). Compromises have to be made on algorithms such that sufficient accuracy and robustness is obtained together with acceptable computational time. This is applicable to all vision-based functions without exception. Using a dedicated computer also enables overcoming most of the real-time constraints. Indeed, the traditional Guidance, Navigation, and Control (GNC) functions can be allocated to the main processor where real-time constraints apply. All time-critical functions are also allocated to that processor making it a "conventional" computer with "conventional" onboard software. As the rover stops to perform Navigation, the function is able to take several minutes and the rover can wait until it completes – time-critical functions keep running on the main processor. Visual Localization also runs on the dedicated processor, but its pose estimate in the ExoMars case is required every 10 s such that the assumed safety margins (linked to the localization accuracy) remain valid. Therefore, though it can run uninterrupted for several seconds, it must complete within its maximum allocated time.

3.2.2
Mapping by Vision Sensors: Stereo as Core

Computer vision, and particularly stereo vision, is at the core of all current, recent, and near-future Mars missions: MSL (Curiosity Rover), the two MERs, Chang'E 3 (Yutu Rover), Phoenix, Mars Pathfinder, ExoMars, Mars 2020 and InSight. It is the only mechanism the rovers have of sensing their environment *a priori*. Sensors on the rocker bogies, wheels, and arms can determine where the ground is after contact is made, but without *a priori* knowledge, there is no way to make sure the rover gets where it needs to go without running into anything.

The Mars Pathfinder Rover Sojourner had a simple laser stripe system that could provide a low-resolution elevation map. But even that required the use of a camera to see the laser stripes [1]. The Mars Pathfinder lander included a mast-mounted stereo camera (IMP). Processing IMP images on Earth yielded excellent terrain maps, validating the use of stereo vision. When the time came to design MERs, the primary trade-off was between stereo vision and laser rangefinding. Stereo vision got the nod for several reasons. First, there was less hardware – no laser or scanning mirrors were needed. Second, power requirements were much lower, with no active laser to power. Third, the Pathfinder results showed that there was likely to be sufficient texture to correlate imagery everywhere on Mars. Finally, images were needed anyway for human consumption [2].

In all relevant missions since, stereo vision has proven to be robust and reliable. It enables operators to navigate in an unknown environment with confidence. It lets the rover determine precisely how far it has driven (visual odometry, VO), track targets, and detect obstacles and choose its own path around them (autonav). It enables scientists to understand the environment around the rover and perform geospatial and geomorphological analysis, photometric studies, strike/dip derivation, bedding relationships, mapping, and more. It enables the public to experience Mars in immersive ways not possible with nonstereo imagery – culminating in virtual or augmented reality systems in the near future [3].

In short, stereo vision is what makes the Mars rover missions possible.[2] Section 3.5.2 gives more details about 3D mapping from stereoscopy.

---

2) This is not to say monocular vision does not have a place; indeed it is useful for a number of scientific purposes (see also Sections 3.8.1 and 3.8.2). But stereo is at the core of rover operations.

3.2.3
Physical Environment

In the planetary environment, vision instruments undergo a set of conditions that need to be either overcome by specific technological concepts or avoided in the operational scenario. The following is a list of such conditions:

- Illumination: Most vision sensors rely on proper illumination of a scene; therefore, most observations and the derived robotic activities can only be made during daytime.[3] The power requirements to adequately light more than a small area (e.g., for a microscopic imager) are enormous.
- The presence of an atmosphere (such as on Mars) has advantages and disadvantages. With atmosphere and wind, shadow areas are indirectly illuminated by the straylight, providing a signal on a sensor even for those areas. On the other hand, visibility (measured by the so-called "optical depth" [5]) is limited by dust, which makes far-range observations hard or impossible. Furthermore, atmospheric dust gradually generates layers on optical devices (lens, baffles), which degrades the optical signal observed by the sensor. Wind events occasionally clean off the dust, however.
- There are obvious physical constraints for a vision system such as depth of field (particularly for close ranges up to a few meters distance), which makes active focusing necessary, and hence adds complexity. Other such constraints include a limited bandwidth (starting with ultra violet - UV to the near infrared) covered by low-mass sensors that do not require special cooling or active illumination, the avoidance of vehicle motion during image acquisition (to prevent motion blur), and the sensitivity to straylight coming through the optical path in case the Sun is in or close to the field-of-view. The latter means the necessity for complex and mass- costly baffle constructions, or operational restrictions to avoid sensing into specific directions.
- A major restriction for any vision asset in a planetary environment that communicates with ground operations is downlink data rate: images are very rich in data content and, therefore, consume the vast majority of available bandwidth for most missions. This requires limiting the amount of data collected, although this can be mitigated somewhat via efficient compression strategies, or intelligent/autonomous ways to select the most effective subset of data to be sent to Earth.
- Temperature is an important driver of image quality: image (electronic) noise rises substantially with temperature; therefore, high temperatures (e.g., over 20 °C) need to be avoided. Fortunately Mars is cold, but electronics generate heat (which can raise temperatures above limits) and require additional heat (when operating at night or early morning). The presence of an atmosphere helps even out temperature variations. On a body such as the Moon, extremes

---

3) There are other implications of day/night scenario such as energy supply in case of dependence on solar panels or thermal conditions.

of heat and cold occur simultaneously on lit and shadowed sides of a spacecraft, making thermal design very tricky.
- A severe constraint is computing power for resource-intensive image-processing tasks. Computers onboard of spacecraft need thorough development, space qualification, and testing/verification time, which, together with launch preparation and cruise time, results in a lag of processing capabilities in the range of 10–20 years, compared with regular consumer hardware. Image processing and analysis tasks (including the ones required for mapping and navigation), therefore, must be heavily optimized in runtime, and are nevertheless often the bottleneck for quick turnaround loops and onboard autonomy.

3.3
Vision Sensors and Sensing

Various types of image sensors have already been landed and operated on the surface of Mars, the Moon, and other planetary bodies/asteroids and comets. Table 3.3 provides an overview. [4]

Imaging sensors may have a variety of features, depending on the requirements, technical implications, implementation costs, and restrictions in the planetary context. The following is a list of such features without claim of completeness, in order to show the various degrees of freedom in design and realization.

- Sensors may be passive or active.
- The detector can be a CCD or CMOS array or a different concept in case of specific spectral and radiometric requirements. Some older cameras used rotating/scanning mirrors.
- The shutter can be global (i.e., exposure is made synchronously across the whole sensor array) or rolling (i.e., exposure and/or readout is made dynamically across the sensor, which can cause artifacts on moving objects or camera).
- Vision sensors have various fields of view (wide or narrow angle) and operate on specific distance ranges (e.g., starting with microscopes, close-range navigation sensors, or sun and star sensors at the far extreme).
- Sensors can be full 2D sensors (frame scan) or line scanning devices.
- The sensor can be monochrome (or black-and-white, b/w), color (red–green–blue, RGB), or multispectral.
- Color or multispectral layout can be provided on single-pixel level (i.e., each pixel has its individual color mask, such as a Bayer pattern) or by a filter wheel that changes the filters on a monochromatic (or – as in the case of MSL MastCam – RGB pattern) sensor array. A mixture of both is implemented for ExoMars PanCam HRC with an RGB filter pattern across the camera's FoV and a necessary pan change to assemble a full RGB image.

4) Specific features such as mixed resolutions (e.g., panchromatic channel with different spatial resolution compared with specific wavelength), type of the sensor, pixel size on the sensor, or temporal behavior of the image capturing process are not mentioned [6].

Table 3.3 Vision sensors as realized on planetary robotic missions by 2015.

Asset	H/R/L[1]	Engineering	Science
Apollo 11–17	H/R	n.a.	Hasselblad (b/w and color film) handheld and tripod; video handheld and on rover (Apollo 15–17)
Mars3	L	n.a.	Stereo (120 mm base) Cycloramic (vertical scanning) 360° × 29° 6000 × 500 pixels, panchromatic
Luna 16 Luna 20	L	Stereo pair, focused on drilling site, 300 × 6000	Panoramic television b/w
Lunokhod 1 (Luna 20)	R	2 TV cameras	Four panoramic line scan cameras using scanning mirrors b/w
Lunokhod 2 (Luna 21)	R	3 TV cameras	Four panoramic line scan cameras using scanning mirrors b/w
Viking 1 and 2	L	n.a.	Stereo (800 mm base) cycloramic (vertical scanning) based on photodiodes. 6000×500 pixels; RGB and IR
Mars Pathfinder	L	Rover navigation. (IMP was used to help drive the rover.)	Stereo fed into one CCD monochrome camera, filter wheel 250×256 pixels CCD each
Sojourner (Pathfinder)	R	Stereo b/w 768 × 484 pixels in conjunction with laser stripe projectors	RGB 768×484 pixels arranged in 4 × 4 colored sub arrays
Beagle 2	L	n.a.	Stereo CCD (209 mm base) 1024×1024 pixels with filter wheels
MER-A&B	R	Navcam stereo (200 mm base) pair 1024 × 1024 pixels b/w 2 HazCam stereo pairs 1024 × 1024 pixels b/w DIMES Descent Imager 1024 × 1024 pixels b/w	Pancam: Stereo (300 mm base) 1024 × 1024 CCD, 8 filters each on 2 wheels MI: microscopic imager 1024 × 1024 b/w
Phoenix	L	Arm operations	Stereo (150 mm base) 1024 × 1024 CCD, 12 filters each on 2 wheels. Robot arm-mounted camera.
Curiosity (MSL)	R	MARDI Mars Descent Imager 1600×1200 color; 4 stereo HazCam pairs, 2 stereo Navcam pairs, each 1024 × 1024 pixels b/w	Mastcam: stereo pair with different FoV (ratio about 1 : 3), RGB 1200 × 1200 pixels, 12 filters each on 2 wheels, variable focus MAHLI: Mars Hand Lens Imager 1600×1200 pixels color, variable focus ChemCam remote micro-imager (RMI) 1024 × 1024 CCD b/w
Yutu (Chang'e 3)	R	NavCam stereo pair, (270 mm base) 1024×1024 pixels; HazCam stereo pair	Pancam: RGB stereo pair
ExoMars	R	NavCam stereo pair (150 mm base) 1024×1024 pixels, LocCam stereo pair 1024 × 1024 pixels	PanCam Wide Angle Cameras stereo pair (500 mm base) 1024 × 1024 pixels, 34° FoV; high-resolution camera 4.8° FoV RGB in 3 filter stripes

1) H, Human; L, Lander; R, Rover.

- Imaging sensors can be fixed or mounted on a movable asset such as a pan-tilt unit (PTU) or a robot arm. There is a trade-off between moving assets versus restrictions in field-of-view: moving parts are always a risky component and prone to failure or deterioration over time, and the location of movable cameras is less precisely known. However, reducing the degree of freedom for sensor pointing means lack of information in areas not covered by the cameras.

In the following sections the main classes are explained in more detail.

3.3.1
Passive Optical Vision Sensors

Since the first robotic missions to planetary bodies took place in the 1960s, cameras or imaging scanners have played a critical role both in the scientific interpretation of the environment that the robotic lander or rover is located in and as a source of unique information for decision making in navigation planning as well as for making "real-time" optical navigation decisions when encountering obstacles in the path of the rover or lander arm. Owing to bandwidth limitations, color imagery is often considered as a secondary objective. However, from the start of Mars exploration with the Viking Landers in 1976, color filter imagery was always employed to generate Public Information Office materials [7], as well as help in the interpretation of rock types, such as the discovery of meteorites [8], the location of frost in Viking Lander 2 [9], and more recently from MERs, the interpretation of the so-called "blueberries" [10], and the discovery of surface ice from underneath the Phoenix lander [11]. For the two Viking Landers [12], line-scan systems were employed. Mars Pathfinder [13] and Phoenix [14] used frame cameras. Rovers generally employ miniature black and white cameras with around 1 Mpixel for navigation (NavCam) and hazard avoidance (HazCam) for rovers such as the MER set [15] and its copy on MSL [16] or "science cameras" such as Pancam on board the MER [17], Mastcam on board the MSL [18] and PanCam on board the ExoMars [19]. However, the trend is toward higher resolution cameras; the Mars 2020 rover for example is likely to have significantly higher resolution navigation cameras.

The cameras tend to have higher dynamic range (12–16-bit) and signal-to-noise ratio (SNR) than consumer units but much lower numbers of pixels than their equivalent even on a mobile device, such as a smartphone. This is primarily due to the very limited bandwidth and onboard solid-state memory. This bandwidth limitation has restricted the application of more up-to-date spectral cameras, such as hyperspectral imagers [20], which have well-known spectral absorption features to map surface mineralogy in the short-wave infrared (SWIR) and thermal infrared (TIR). Instead, either filter-wheels (e.g., Pancam on MER or PanCam on ExoMars) or Bayer filters (e.g., Mars Hand Lens Imager–MAHLI [21] and Mastcam onboard MSL) are employed.

In the future, besides the obvious increase of image resolution, SWIR cameras, using InGaAs cameras may be employed with low temperature-resistant devices

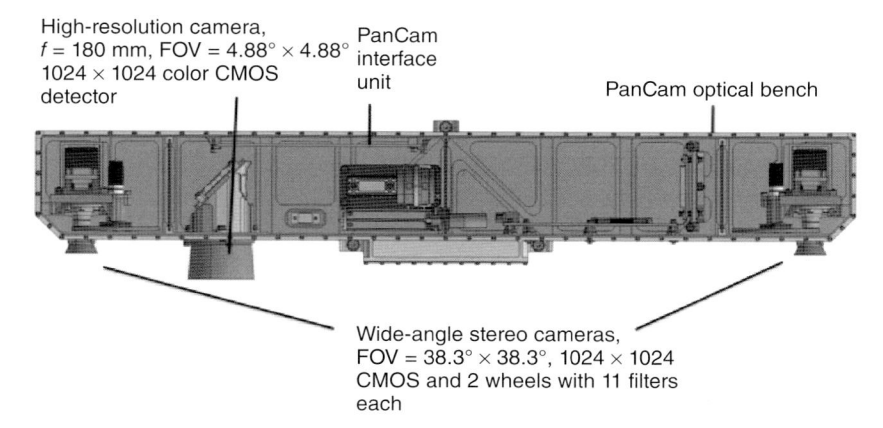

Figure 3.2 CAD drawing of PanCam and its optical bench on the pan-tilt unit. (Courtesy of Mullard Space Science Laboratory, UCL.)

Labels in figure: High-resolution camera, f = 180 mm, FOV = 4.88° × 4.88°, 1024 × 1024 color CMOS detector; PanCam interface unit; PanCam optical bench; Wide-angle stereo cameras, FOV = 38.3° × 38.3°, 1024 × 1024 CMOS and 2 wheels with 11 filters each.

similar to liquid crystal tunable filters (LCTF) [22] such as acousto-optical tunable filters (AOTF) as proposed for a future Chinese lunar rover [23].

Figure 3.2 shows the ExoMars PanCam optical bench including a high-resolution camera bore-sighted [5] with the wide-angle camera with a factor of 7.6 difference in resolution between them [24].

Although a variable focal length (=zoom) camera was proposed for MSL Curiosity, its success-oriented schedule resulted in it not being able to be completed in time. In the future, the Mastcam-Z camera on the Mars 2020 rover [25] will include a zoom capability of 3.6 : 1.

Existing cameras can be employed to achieve zoom factors of up to 1.75 using "super-resolution" by repeatedly imaging the same scene; motions of 1–2 pixels allow subpixel resolution to be obtained. This has been employed on MERs [26]. However, this makes the data rate requirements even worse. Super-resolution works better from orbit, where larger pixel displacements can result in an improvement factor of up to five [27].

3.3.2
Active Vision Sensing Strategies

Only a few contemporary missions use vision that is supported by its own illumination, pattern projection, scene-dependent movement of the sensor, or changing the optical properties of the sensor depending on the content of the scene. Many of these techniques are targeted at 3D sensing.

A class of active vision sensor that has been particularly exploited for precise long- and short-range target detection, identification, and depth estimation is

---

5) Here: Looking into the same direction.

laser-illuminated detection and ranging (LiDAR) technology. LiDAR can be broadly classified into three major types: range finders (depth estimation, digital elevation models, and environmental perception), differential absorption LiDAR (DiAL; for the measurement of atmospheric temperature, trace gases pressure and density) and Doppler LiDAR (velocity measurement). The latest in LiDAR technology for current and future space missions is used for assisting spacecraft with rendezvous and docking, depth estimation and mapping, scientific analysis, and geological surveying. For example, the OSIRIS-REx spacecraft planned to be launched in 2016 will attempt to map the asteroid 1999 RQ36 while flying close to it using the Canadian LiDAR built by MDA and Optech (OSIRIS-REx laser altimeter, OLA). An advanced version of the Laser Camera System by Neptec Design Group; Triangulation + LiDAR (TriDAR) automated rendezvous and docking system was tested on the STS-128, STS-131, and STS-135 missions to track the ISS during docking, undocking, and fly-around operations. Due to the short- and long-range sensing capabilities of TriDAR, the system is undergoing further development as a navigation system for lunar rovers.

In addition to range sensing, surface mapping, and navigation, recent space missions have successfully used laser sensors for scientific analysis, such as the Chemistry Camera (ChemCam) onboard NASA's MSL (November 2011) rover Curiosity [28]. ChemCam is a suite of two distinct instruments: laser-induced breakdown spectroscopy (LIBS) and a Remote Micro-Imager (RMI) telescope. LIBS uses a solid-state laser that produces highly energetic laser pulses as the atomization and excitation source along with a highly sensitive spectrometer for rapid analysis and identification of rocks and soil compositions [29] on the Martian surface. Similarly, the Raman laser spectrometer (RLS) of the ExoMars rover instrument suite is a powerful tool for identifying and characterizing minerals and biomarkers, such as organic compounds, water-related processes, and oxidation that may be related to past signatures of life on Mars [30]. Scientific measurements of the Martian environment were also performed by the Canadian weather system onboard the Mars Phoenix lander (May 25, 2008) using the LiDAR system developed by Optech and MacDonald Dettwiler and Associates (MDA). The LiDAR instrument was capable of detecting snow fall from clouds and obtaining measurements of atmospheric dust and clouds from the surface in the Arctic region of Mars [31].

Other sensors with active illumination include the RAC on Phoenix [32] and MAHLI on MSL [21], which both use LED arrays to illuminate the area, primarily for nighttime imaging.

3.3.3
Dedicated Navigation Vision Sensors: Example Exomars

Vision is one of the most important information sources for navigation, with lots of implications to the operational scenario, including safety. For example, determining orientation is critical for antenna pointing, and path planning requires a reliable 3D map. This section complements Chapter 4 in terms of a brief description of star trackers, sun sensors, and other vision sensors dedicated to rover navigation and localization.

In ExoMars, for example, there are three key functions enabled by the rover engineering camera images: two core functions and an opportunistic one. The two core functions are Navigation (Perception/stereo vision) using the NavCams and Visual Localization using the LocCams. The opportunistic function is locating the Sun using the NavCams and due to specific ExoMars mission design it is not a priority in the design of the cameras. For performance reasons, embedded in the cameras are functions that correct image distortion and that align left and right images, saving precious time in the onboard computer.

3.3.3.1 Navigation (Perception/Stereo Vision)

The navigation of the rover is composed of three key functions: Perception, Navigation, and Path Planning. Perception uses stereo vision in order to obtain a high-quality and reliable disparity map of the region imaged by the cameras. On ExoMars, there are three pairs of NavCam images per stop and Navigation initially uses the three corresponding disparity maps in order to produce a 3D model of the terrain surrounding the rover. It then assesses that terrain with regard to the locomotion capabilities and produces a navigation map containing several assessments: forbidden areas, difficulty levels of driveable areas, and so on. Path Planning uses the navigation map together with the maneuverability information of the rover to find a path that is safe (avoids forbidden areas), efficient (keeps the rover far from difficult areas when possible), and driveable (smooth path avoiding step changes of curvature, for example). Since the terrain includes rocks and slopes, if the NavCams were close to the ground, it would be impossible to see behind those obstacles. This would represent large areas of unknown terrain (considered unsafe), which would greatly reduce the number of possible viable paths. Therefore, the NavCams have been mounted on top of a 2 m high mast, and this height is mainly constrained by mass and volume limitations. In order to gather the three image pairs (left, center and right), the NavCams are mounted on top of a pan-tilt unit.

3.3.3.2 Visual Localization and Slippage Estimation

The ExoMars Rover has closed-loop control around its trajectory ensuring at all times that the rover remains within the safety margins of the commanded path. The closed-loop control takes advantage of the high maneuverability of the ExoMars Rover characterized by the capability of steering all six wheels simultaneously, optionally while driving. In order to correct deviations from the commanded trajectory, the rover must know where it is: this is the function of Localization. Since trajectory control is performed at 1 Hz, Localization provides information at the same frequency. However, because of onboard computer limitations, Visual Localization only provides a complete pose estimate every 10 s. In-between visual frames, localization uses gyro and wheel odometry information to propagate the previous VisLoc estimate. Visual Localization actually runs much faster than 10 s, but one must consider processing margins and other functions that must also run on the coprocessor. The function uses the stereo images from the LocCams, detecting and tracking features in the scene by estimating the linear and angular displacement from frame to frame. Long-range features are important for attitude estimation while short-range features are key for position estimation. While the NavCams are used when the rover is at rest, the LocCams are used when the rover is in motion and, therefore, blur from motion is possible. Hence, it is important to minimize the motion of the LocCams, and this is one of the reasons not to have them on top of the mast. The other reason is to enable a good track of close-range features in front of the rover for accurate position estimation. A key advantage of the ExoMars architecture is the fact that it allows obtaining a VisLoc pose estimation every 10 s while the rover moves. This information is used for estimating the rover slippage during the move allowing an essential safety monitor, as well as making operations more efficient by ensuring that the rover reaches the desired target.

3.3.3.3 Absolute Localization

The ExoMars Rover does not have an X-band antenna directly to Earth. Hence, unlike the MER and MSL US rovers, it does not need to point an antenna – it only uses the UHF link to the orbiter. Nevertheless, the need for an inertial (absolute) attitude remains because pitch and roll are necessary for Navigation on the 3D map of the terrain and for safety (tilt angle monitor); heading is necessary since it is required to command the rover in a Mars-associated frame (for energy and thermal reasons angle with the Martian north is important). The NavCams on top of the mast are used to take pictures of the sky and detect the Sun direction. It is important to point out that no filters and features that could have a negative impact on the Navigation functions have been implemented for this purpose. The Sun image from the NavCams presents a series of artifacts, which have been minimized but never at the price of degrading navigation. The onboard algorithms cope with those artifacts and estimate an accurate Sun direction.

3.4
Vision Sensors Calibration

Calibration is required to understand the sensor properties in a quantified way so that data can be interpreted in a precise and objective manner. For the planetary robotics case, it is not enough if images and their products "look good" (e.g., such as tourist photo panoramas), but each pixel and measurement (e.g., gray level) needs to be known exactly in terms of its projection in 3D space (geometric) and its spectral content (radiometric).

3.4.1
Geometric Calibration

Geometric calibration is used to determine the precise vector for each pixel in the camera(s) with respect to 3D space. This is particularly important for stereo cameras as this forms the basis for quantitative stereo processing. Good calibration allows for "epipolar alignment" between cameras: meaning that features in one camera appear in the same row (i.e., vertically) in the other camera. This simplifies the disparity-matching process used to locate features in both images, as a one-dimensional (horizontal) search can be conducted. This allows for a point cloud (in Cartesian coordinates) to be produced by the cameras.

Geometric calibration includes the estimation of both intrinsic (or internal) and extrinsic (or external) parameters of a camera system. Intrinsic parameters model the internal properties inherent to a camera, independent of how it is pointed in space (such as focal length and pixel scale). Extrinsic parameters model the relationship between cameras, or between cameras and another system component or coordinate frame (in other words, camera pointing). A camera model describes the relationship between pixels in the image and the location in Cartesian space of the object being imaged, combining the intrinsic and extrinsic parameters. A camera model is bidirectional: given an XYZ point, it can specify where in the image that point appears, or given an image pixel, it can specify the ray (origin and unit direction vector) in 3D space along which the light hitting that pixel must have originated. Camera models often consist of a mathematical model with parameters, although it is also possible to implement them using look-up tables. Typical models include the Heikkil/Silven model [33], the popular Matlab camera calibration toolbox [34], and the CAHV/CAHVOR/CAHVORE model family used by many robotics practitioners, including all of the NASA *in situ* missions [35] (except Viking). The Heikkil/Silven model includes intrinsic parameters for focal length, principal point (optical center), skew coefficient, and radial/tangential distortions. Typically, intrinsic calibration of a camera or a stereo pair requires the collection of many images of a calibration board (e.g., a checkerboard or dot grid) at various distances, orientations, and placements in each camera field of view.

Extrinsic calibration is used to determine the relative transformation between the frames of reference of the two cameras in a stereo pair, or more generally, between a camera and another object or coordinate frame. Basically, extrinsic parameters describe the position and pointing of a camera. For both cases, the parameters are simply the rotation and translation between the objects, typically represented by a 4×4 transformation matrix, or translation plus roll-pitch-yaw angles, or translation plus a quaternion. As discussed earlier, extrinsic parameters between cameras must be known to perform stereo vision. It is also important to know a camera's position relative to the rest of the rover system, for example, in order to apply visual odometry measurements in the proper orientation relative to the rover, or to accurately perform obstacle detection. Cameras sometimes have calibration cubes installed on them, which provide a reference point to the camera reference frame (as shown in Figure 3.3). Often complicating extrinsic calibration

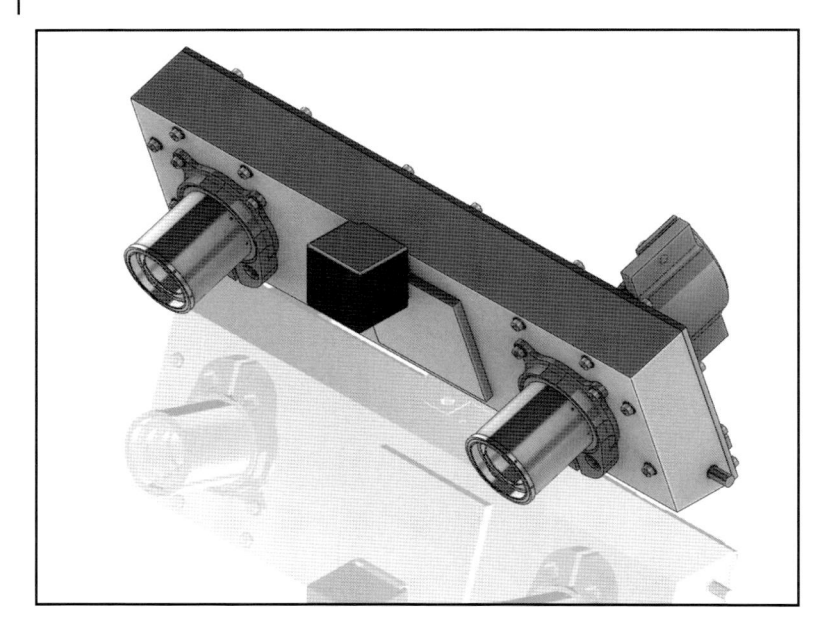

Figure 3.3 Models of ExoMars Navigation and localization cameras, showing calibration cube. (Courtesy Neptec Design Group.)

between a camera and another rover component is an articulating component between them, such as a PTU or a robotic arm. The motion of these components needs to be modeled using kinematics and the results factored in to the camera model parameters, a process sometimes called "pointing" the camera model. Since knowledge of the articulation state is necessarily imprecise, movable cameras will add larger errors in their pointing to the extrinsic calibration parameters, which must be taken into account for the intended usage. For example, a stereo pair rigidly mounted to a PTU will suffer little range error, because the baseline between the cameras is fixed (although thermal effects on the camera bar can sometimes be an issue). But translating that range to XYZ coordinates in a rover frame adds error due to the articulating PTU. By contrast, when a single arm-mounted camera (such as InSight IDC or MSL MAHLI) takes stereo images by moving the arm, the baseline between the cameras is not fixed, and thus the range error is significantly greater (in addition to the XYZ translation error).

A challenge with calibrating cameras for rovers is that calibration performed on the ground before a mission may not be sufficient to maintain the state of the camera system calibration when used in-flight, since geometric calibration status may change, due to, for example, mechanical impact, varying temperature, or degradation of mechanical parts. Experience with MER and MSL shows that temperature effects can be significant, while other effects are possible but rare. One way to mitigate the temperature issue is by calibrating the cameras at different ambient temperatures, and generating different parameters for different temperatures. Calibration should be monitored during the mission, and if necessary, adjusted using a reduced calibration process. One method is to view calibration targets on the deck of the rover or lander and comparing their position in the images with the expected ones as calculated using the ground calibration. Second, natural targets at large (i.e., "infinite") distances (such as sun, stars, or horizon) can be imaged using stereo imagery and thus can be used to measure stereo disparities as expected for infinity. This particularly can be used to correct errors in stereo toe-in angle, which may cause significant errors in distance measurement. Finally, one can use multiple image observations of a scene to apply adjustment procedures (e.g., bundle adjustment) to regain calibration parameters from a mathematically overdetermined system.

This actually happened on MSL after the sol 200 anomaly [36] required switching to the B-side computer and cameras. Checkout showed increased range error compared with the A-side cameras. After some investigation, it was determined that temperature variations were causing the calibration to vary. The effect was mitigated by performing self-correlation of the same scene taken at different temperatures in order to determine how the pixels moved in image space with temperature. This knowledge was then applied to the preflight camera calibration data to create temperature-dependent camera models, which solved the problem.

Planetary rovers use a variety of approaches for geometric camera calibration. MER and MSL have used flat calibration targets and use the CAHVORE formulation [35]. The ExoMars navigation and localization cameras use a multistep process for calibration [37], which is summarized here. The first step is a detailed calibration that is done at ambient temperature, which uses an array of theodolites, a flat target board, and the alignment cube to calculate the stereo reference frame with respect to the alignment cube. The rotation and translation accuracy of the estimated transformation is known within an accuracy of 60 arcsec and 0.1 mm, respectively. Next, to allow the cameras to stay in calibration over the wide operational temperature range (-50 to $+40\,°C$), further calibration is done by mounting the cameras to a small, thermally stable target board. The cameras and board are then temperature-cycled, and the calibration values are adjusted for each temperature step. During the mission, the calibration will be applied onboard the camera FPGA, so that the main rover processor gets distortion-free images.

Techniques exist to "self-calibrate" the camera by taking many images of a moving object, or moving the camera around a static object, without using surveyed or measured calibration targets (e.g., Ref. [38]), and these can be used to derive a camera model. However, surveying techniques tend to be more robust and generally require fewer images (spacecraft time during assembly, test, and launch operations (ATLO) is extremely valuable) and have thus been used in most rover missions to date.

There are specific calibration aspects for 3D sensors that significantly differ from those of cameras. For example, 3D active sensors such as LiDARs have different calibration approaches (e.g., a target array at different ranges), but the goal is the same: map the sensor measurements into Cartesian space with a defined origin and avoid (or at least characterize) errors in each single measurement.

3.4.2
Radiometric Calibration

While geometric calibration is essential for fulfilling the operations-related vision tasks in a robotic chain, radiometric calibration is mainly necessary for scientific purposes. However, a minimum level of radiometric calibration (including at least exposure time compensation) is needed for engineering uses as well, to avoid large differences in contrast or brightness between neighboring frames of a mosaic or mesh, which can interfere with proper interpretation of the surroundings.

In general, the main aspects of radiometric calibration are the following:

- flat field compensation
- dark and hot pixel removal,[6]
- exposure time compensation,
- temperature responsivity of the sensor,
- dark current removal,
- blooming/smearing compensation,
- multispectral calibration [39],
- obtaining true radiometry,
- using in-flight calibration targets for daily in-flight radiometric calibration (MER Pancam, MSL Mastcam, ExoMars PanCam),
- tracking and compensating radiometric degradation (by dust and radiation).

In order to transform rover images into science data, we require absolute radiometric calibration of rover images [40] so that we can exploit these images to yield information on (a) surface anisotropic scattering (through a BRDF – bidirectional reflectance distribution function); (b) absolute reflectances to compare against stored spectral reflectance libraries of different mineral types in order to map the mineralogy of the surface; (c) quantify how much atmospheric dust is suspended in the air through atmospheric opacity measurements of the Sun, or (d) compare measurements made from different rovers and landers in different areas on the surface of Mars. Absolute radiometry is very difficult to achieve outside of the laboratory, so radiometric calibration systems are usually developed to measure the change in radiance as a function of time using onboard radiometric calibration (known hereafter as *radcal* targets). For rovers, this is compounded by the fact that atmospheric dust will tend to settle on these radiometric calibration targets, rendering it even harder to measure the sensor degradation. Nevertheless, radiometric calibration has been achieved by employing a "post and bull's eye" calibration target modified from the one originally developed for the Pathfinder IMP camera [17]. An example can be observed from MER in Figure 3.4 of the impact of dust over an Earth year for Spirit.

As Mars has no stratospheric layer of UV-absorbing ozone or any equivalent, any materials placed on the surface will degrade from UV damage. To minimize this, all materials used for the radcal target were UV-irradiated for the

---

6) Single pixels with large radiometric errors or transmitting no meaningful signal at all.

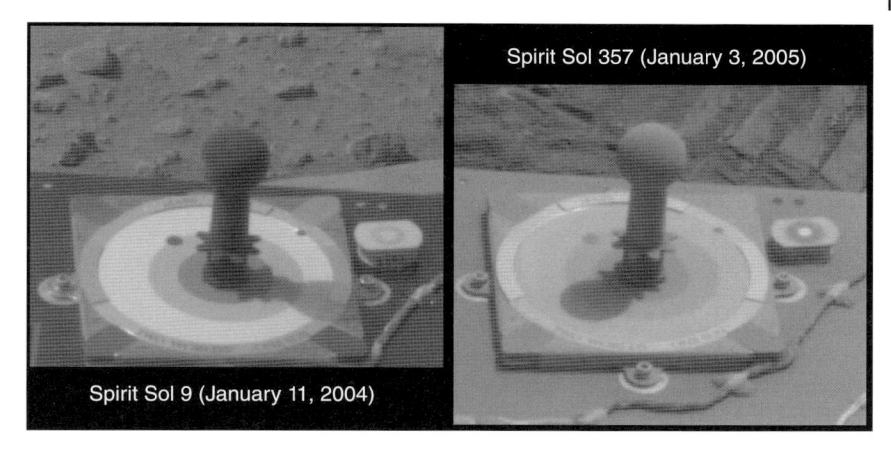

Figure 3.4 Example of the impact of dust on the MER-A (Spirit) rover PanCam over approximately 1 year. (Original taken from http://mars.nasa.gov/mer/gallery/press/spirit/20050125a/Spirit_dust_comparison-A379R1.jpg. Courtesy NASA/JPL-Caltech.)

equivalent of 30 sols prior to calibration. The CCDs used had their quantum efficiency (QE) defined along with their gain and bias at the three temperatures of interest ($-55\,°C$, $-10\,°C$, and $+5\,°C$) as well as their blooming characteristics, CCD frame transfer smear and above all, dark current to take into account a temperature-dependent current due to thermal noise. Dark current (shutter closed) is measured both in the active and storage sections of the frame-transfer CCD and is continuously monitored during rover operations. Each spectral filter was also characterized using a monochromator for "out of spectral band" leakage and this was found to be negligible for MER geology filters but potentially significant for the solar filters. Flat-field images were acquired from an integrating sphere for each filter and for each camera with the sphere itself being calibrated with a NIST-traceable diode radiometer. Spectral responsivity curves were then produced for each filter preflight and using a NIST-calibrated tungsten lamp an independent check was carried out on the radiometric calibration coefficients. These were found to lie within 5–10% for all filters. Noise in remote sensing systems can be expressed in terms of the equivalent amount of power in a hypothetical "noise signal," also known as the noise equivalent spectral radiance (NESR) and for the Pancam these were measured to yield an SNR of 200 for most images acquired using the geology filters. A BRDF model was generated for each of the seven reference areas on the target (white, gray, black, blue, red, yellow, and green) using goniometers in the United States and France [41]. A similar but less extensive radcal was performed using a MER flight-spare calibration target [18].

3.4.3
The Influence of Errors

Accurate knowledge about the environment is essential for the operations of robotic devices in unknown terrain. In terrestrial applications, redundancies (e.g., many overlapping images from the same scenery taken from many different viewpoints) often allow a so-called "structure-from-motion" reconstruction, which simultaneously derives camera positions and reduces reconstruction noise by overconstraining the data. However, in the planetary context, redundancy is avoided to the largest possible extent in order to save data downlink resources. This leads to more stringent requirements on calibration, requiring additional *a priori* knowledge about the system. One important example is knowledge of camera pointing (e.g., from a PTU, or in particular the pointing of two stereo cameras with respect to each other). Even small errors in pointing lead to large errors in larger distances, which may make robot-driven contact instrument placement impossible or even hazardous to the entire device in case of unexpected collisions on the way. One important distinction here is between where the camera was intended to point, and where it actually pointed. In general, for stereo analysis, it is not really essential to point exactly where it was intended (within some rough limits); what is critical is precise knowledge about the final pointing result.

Bad pixels in images represent another source of error that may lead to false correlations, which is especially true for long-baseline or arm camera scenarios where only one camera is used and thus the same camera noise pattern may spoof the correlator into finding false matches.

Various types of errors are particularly dependent on imaging geometry. Stereo range accuracy drops with the square of the viewing distance. Figure 3.5 shows different examples of range error behavior of typical sensors, with a given typical stereo-matching error in the range of 1/4 pixel. Current rover stereo image products are used up to about 20 m routinely, sometimes up to 50 or 100 m with rapidly increasing noise level. Pointing errors in this scenario influence the measurement results as a whole (i.e., resulting in a systematic error), whereas image-processing deficiencies and image noise result in random noise in the products. This is demonstrated in Figure 3.6 using MER from a viewing distance of about

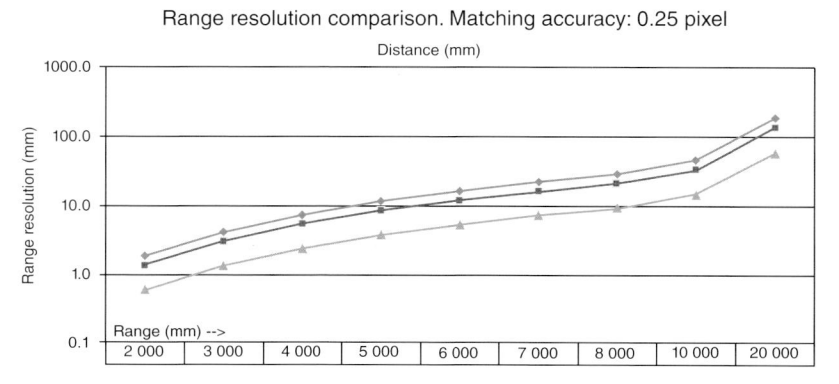

Figure 3.5 Expected stereo range accuracy for different stereo configurations on MER Pan-Cam (highest curve), MSL MastCam (middle), and NASA Mars 2020 MastCam-Z (lowest curve).

(a)   (b)

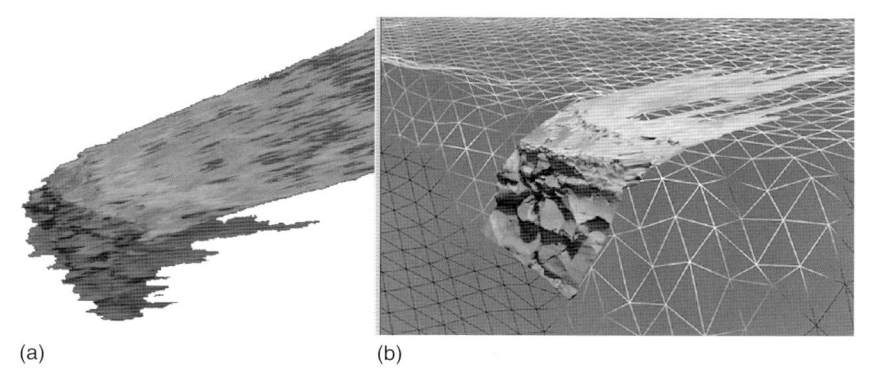

Figure 3.6 3D rendering of a similar portion of Victoria crater (Cabo Corrientes [42]) reconstructed from the rover stereo images (a) and from two different rover poses stereo (embedded in the geometric context of the HiRISE-derived DTM) (b).

50 m toward adjacent cliffs of Victoria crater at Cabo Corrientes. Using the fixed-base length stereo configuration of the MER-B PanCam, the 3D reconstruction result is not usable for geologic interpretation. If a stereo pair taken at different positions of the rover is used (in this case about 3 m apart from each other), otherwise known as long-baseline stereo, the 3D structure is sufficiently represented to obtain geologic measurements.

Errors or deficiencies in 3D reconstruction are often caused by regions that do not show enough texture for stereo correlation. Figure 3.7 shows the input image with large areas without useful texture, and the respective slope (from stereo) calculation result, which shows many unknown regions due to failure of stereo

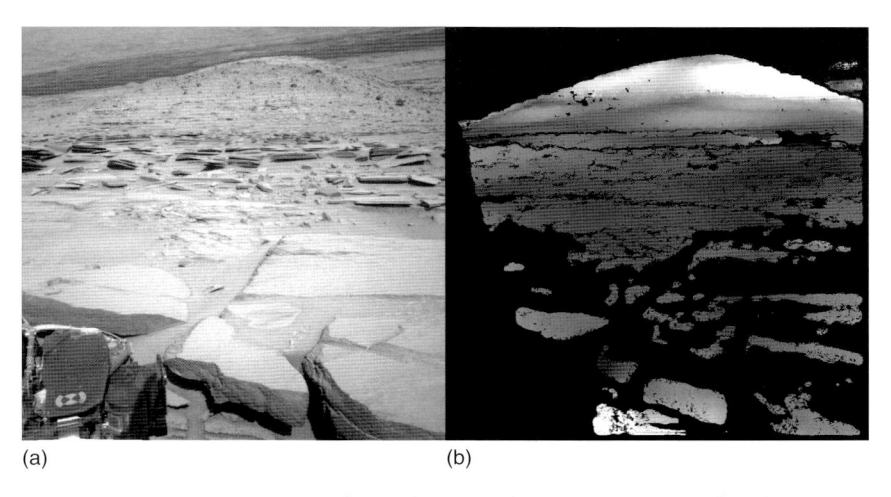

(a)   (b)

Figure 3.7 (a) MSL Navcam image from Sol 581. (b) Slope image (white = 20° and more); note the lack of data in the foreground where texture is insufficient for correlation, indicating unknown result. (Courtesy NASA/JPL-Caltech.)

matching. Although correlation parameters can be tuned to get better coverage (at the expense of accuracy), such holes in some terrains are an inevitable result of using passive optical sensors.

3.5
Ground-Based Vision Processing

In contrast to onboard processing, the processing power available on the ground as well as the availability of humans allow for more sophisticated, robust, and "interactive" tools. Once downlinked images and their respective metadata are received, they are distributed to several processing channels for the production of different kinds of products for science and operations as well as data analysis and visualization. In most cases, this starts with a batch chain that performs low-level processing in a fully automatic way, followed by standard processing that is still automatic for most datasets, and finally an interactive data analysis and interpretation chain with scientists and automated processes jointly working together to set-up decisions how to further proceed in the mission. The final round after proper data archiving is scientific exploitation, which may go on for years after data have been captured.

The whole operations cycle of contemporary Mars rover missions is depicted in Figure 3.8, with the generation of data products (feeding both analysis and simulations) being a central task heavily relying on vision. In the following sections some important steps of the ground vision processing chain are explained.

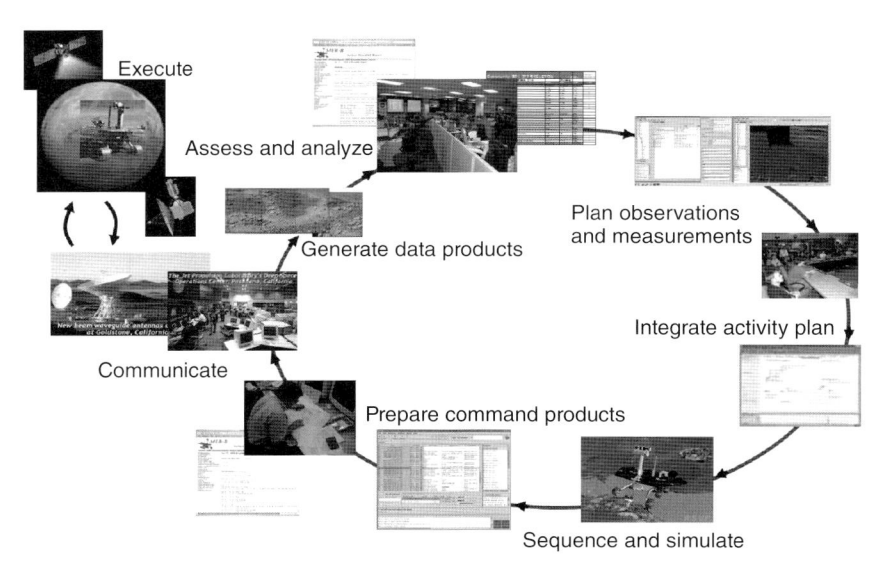

Figure 3.8 Operations sequence cycle for planetary rover mission control (MER paradigm). (Courtesy NASA/JPL-Caltech.)

3.5.1
Compression and Decompression

Due to severe constraints on communication bandwidth, data compression and decompression processes are imperative to space missions. The use of data compression in space missions dates back to the Apollo program [43]. Image compression is extensively used onboard the planetary rovers in order to make the most effective use of the downlink resources. For MER and the MSL engineering cameras, most of the images are compressed using the ICER image compression software [44] and a modified version of the low complexity lossless compression (LOCO) software [45]. The MSL science cameras, as well as InSight, use mostly JPEG compression. Lossy image compression and decompression methods are developed such that they allow for a graceful trade-off among compression and degradation of image quality (induced artifacts, added noise, and loss of certainty in the content of the originally captured image), associated software and hardware requirements, reliability and data integrity, compatibility with the transmission medium and ground stations, robustness in the face of lost data packets, and most importantly the associated implementation cost.

3.5.2
3D Mapping

Cartography of planetary surfaces dates back to the early 1960s when the United States and USSR sent orbiting probes with imagers onboard to Moon and Mars. Due to sensor maturity and orbit distance, the ground resolution was quite sparse, in the range of several tens of meters in the best case. Since then, mapping from orbit (particularly in 3D) has evolved considerably: the best achievable resolution on Mars is the HiRISE instrument on Mars Reconnaissance Orbiter (MRO) [46], which allows to acquire images with a resolution down to 30 cm. Executing two adjacent orbit passes across the same site from different viewing angles allows for stereo imaging and photogrammetric processing, to come up with digital elevation models (DEMs) 0.6 m in resolution and a proven height resolution in the range of 0.2 m [47]. However, to map the environment of a landed/roving probe in sufficient level of detail – for example, for rover navigation or science – target selection requires much higher resolution and often also different viewpoints compared with orbiter imagery. Although earlier landers such as Luna 9 [48] and Surveyor-1 [49] for Moon in 1966 had already successfully transmitted camera images from the surface of Moon and Mars, the first opportunity to perform 3D mapping of the environment in stereo using instruments on board of a planetary surface asset was offered by the Viking missions in 1976 (see Section 3.7.2). Since then, around 10 missions have used the stereo imaging technique for close-range mapping in 3D.

Apart from the images themselves, 3D mapping requires detailed knowledge about imaging geometry (i.e., a description of interior camera parameters and the position and pointing of cameras embedded in a unique geometric context), in order to obtain proper scaling and to be able to place the mapping results with respect to the data acquisition platform (e.g., lander, rover) and the outside world (e.g., in a geographic coordinate system of the planet, or a local coordinate system of the landing site).

Figure 3.9 displays a general scheme for mapping from a rover, based on stereoscopic images, which is the most frequently used technique nowadays. The stereo imaging system is mounted on a mast or the rover/lander chassis itself. To obtain a panoramic view of large parts of the surroundings, devices are used that allow pointing of the cameras into different directions – so-called "pan-tilt units" (PTU). With knowledge about the geometry of the PTU angles and axes with respect to the cameras' geometry, each image pair can be accurately assigned to its correct geometry (see Figure 3.10 for a monoscopic panorama) and, in case stereo images have been captured, the mapping results obtained from each individual pointing state can be seamlessly combined into a unique mapping result covering the full surroundings of the rover/lander. Alternative flexible imaging geometries are realized by a robotic arm (e.g., during the Phoenix mission [50], still enabling useful 3D reconstructions [51]), by using mirrors (as planned for the ExoMars mission and its rover inspection mirror – RIM [52]), or mounting the cameras on other movable parts of a platform (e.g., the ExoMars Close-Up Imager CLUPI, being mounted on the ExoMars drill box [53]).

The stereo mapping case starts with two images from a stereo camera system (left and right image). Each pixel in the left camera is sought for a corresponding location in the right image (i.e., the same object point occurring in both images). This generates a so-called (dense) disparity map, which contains the parallaxes for each pixel (see Figure 3.15 for a more detailed flowchart of this process). These are used to reconstruct a 3D location. Since the camera pointing is known from the PTU angles and the platform (rover) pose, these 3D locations of object points can be obtained in an overlaid coordinate system (in a Mars-centric coordinate system, or in a local one such as lander-centric). An important by-product of this process is the texture taken from one or both images at the source location in the image, to colorize the respective object point. The combination of object points is projected into a gridded data structure, either in Cartesian,

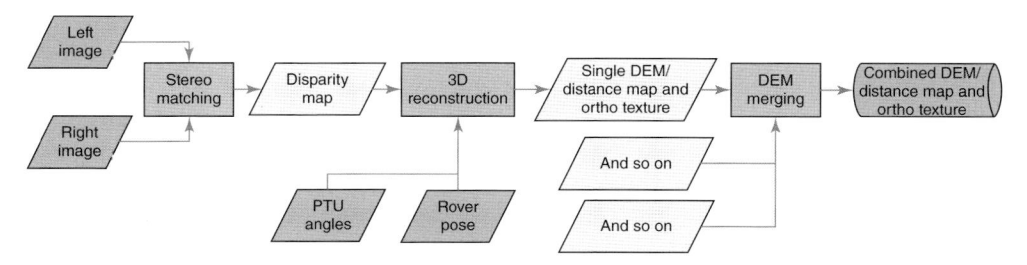

Figure 3.9 General scheme for mapping based on stereoscopy from a rover.

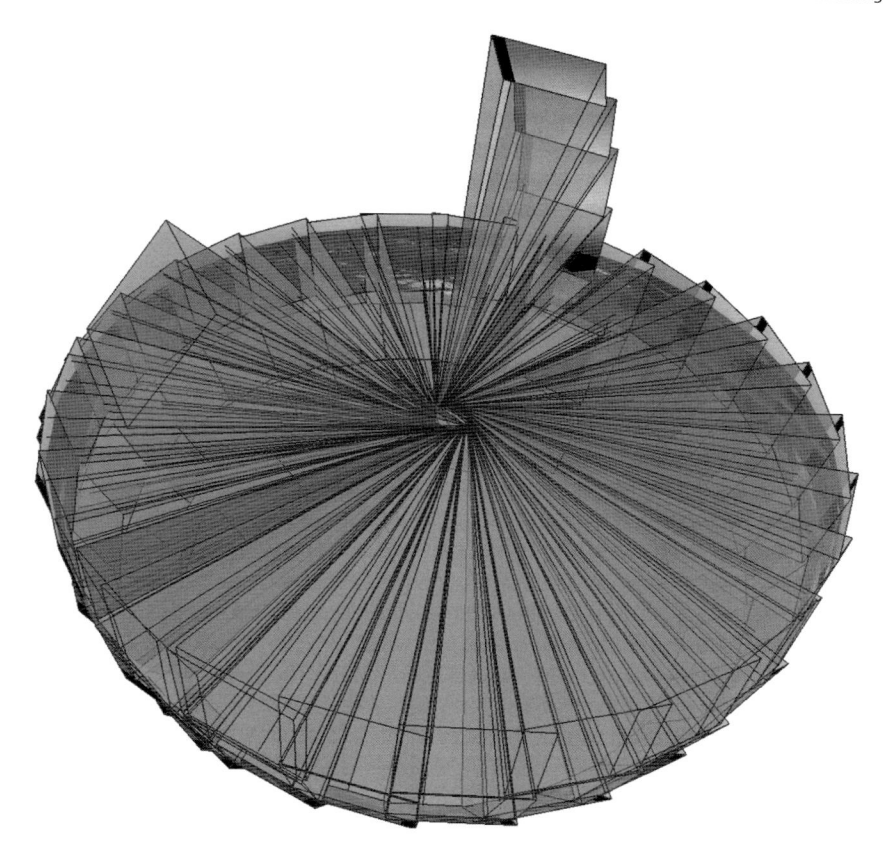

Figure 3.10 Imaging geometry for panorama (MER-B, 146 images taken in Sols 652,653,655,657–661,663,665,667–669,672, 673,686,697, and 703). Note some images taken from the sky region used for aerosol optical depth determination and/or check-/determination of camera vignetting.

spherical, or cylindric coordinate space. Finally, in this representation (or as simple textured 3D meshes), different views can be combined to a merged 3D reconstruction.

A typical mapping case from an ESA study (EUROBOT Ground Prototype Vision System [54]) is depicted in Figure 3.11. First a mapping stereo image sequence is taken from a single static viewpoint leading to a circular region with radius 10 m (maximum selected mapping distance) around the rover station. Then, the rover is moved to the right in stop-and-go mode every 1–2 m with subsequent triple stereo mapping sequences pointing to the right in motion direction. Knowledge about rover pose from visual odometry allows for merging the individual stereo constructions to a unique DEM and ortho image.

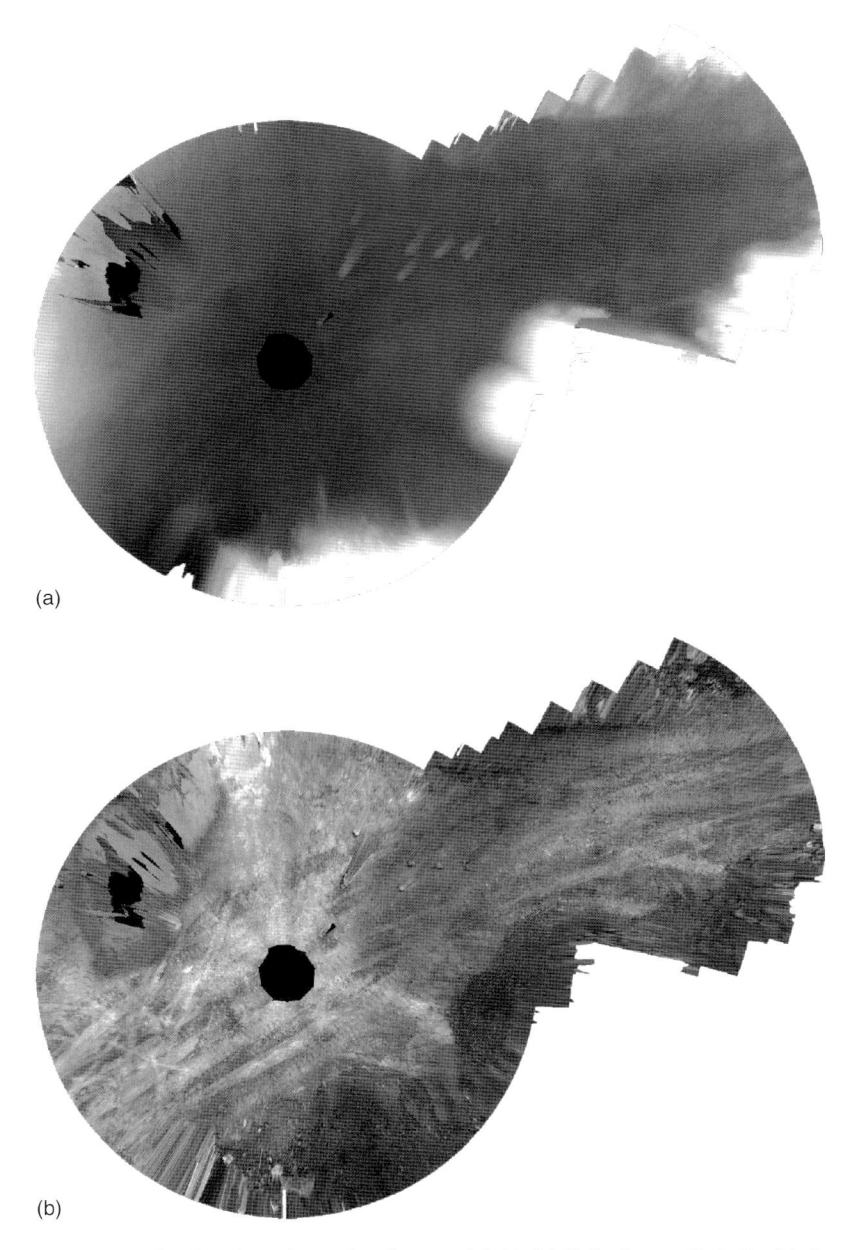

Figure 3.11 Combination of mapping from a single viewpoint (circular region) and rover motion to the right. (a) DEM with low regions coded in dark and high regions in bright; (b) Ortho image. Note the black center in the circular region which stems from occlusion caused by the vehicle itself when image capture of panorama took place.

3.5.3
Offline Localization

Rover localization is the process of determining where the rover is. This is done both onboard (to ensure the vehicle goes where commanded) and on the ground (to generate a more accurate solution and to be able to use orbital navigation maps). In addition to measuring orientation using an Inertial Measurement Unit (IMU), both MER and MSL have two onboard localization systems, both concerned with determining the motion for each drive. The simplest is based on integrating wheel revolutions. This is simple to do and available on every drive, but is inaccurate—generally 10% error (of distance traveled) as a rule of thumb, much more if the wheels slip (e.g., in sand). The second, called visual odometry (VO), takes images after every drive arc and uses vision techniques to derive the rover motion. This is much more accurate (1% of range rule-of-thumb error, depending on the initial pose of the rover), but it is very resource intensive on current missions, slowing down driving tremendously. As a result, it is used only when high slip or precision driving is required. On future rovers with more computing power, VO could be used on every drive, as it will be done on the ExoMars rover. For a more thorough explanation of onboard VO, see Section 3.6.4.

Ground-based localization using imagery takes advantage of essentially unlimited processing power as well as other information sources such as orbital imagery to improve the localization of the rover. These techniques include incremental bundle adjustment (IBA) of rover imagery and rover-to-orbiter imagery matching.

An IBA technique was used early in the MER missions to link all rover imagery together from a traverse, by matching common features in the imagery [55]. The technique batch-processes the linked set of images to generate optimal estimates of rover positions given for each image. This is described further in Section 4.4.

MSL and current MER operations use rover-to-orbiter imagery matching. Using orbital imagery, the absolute position of the rover can be estimated after a day's drive is complete. Images from the rover are projected into an orthorectified mosaic, which uses the XYZ location of each pixel to create a true overhead view (early on in MER, vertical projection mosaics were used instead, but they suffer from parallax and layover effects). These mosaics are then compared with orbital views of the terrain to pinpoint exactly where the rover is in terms of latitude and longitude on the Martian surface. This comparison typically achieves an accuracy of 1–4 (orbital) pixels, depending on the scene activity. With the standard HiRISE 0.25 m mosaics, this means the rover's absolute position on Mars (with respect to HiRISE) is determined to within 25 cm in most cases, with 1 m worst case [56]. Current operational techniques require human operators to do the ortho-orbital comparison.

Improved rover-to-orbiter imagery matching can be achieved by using super-resolution restored (SRR) images down to 5 cm resolution generated from 5 or more 25 cm resolution HiRISE images using the methods described in Ref. [27]. These SRR images can also be employed to update the rover traverse from the accumulated errors from optical navigation using the methods described in Ref. [57]. Automation of rover-to-orbiter localization using image-based correlation techniques has been investigated. An automated processing system has been developed by UCL-MSSL to ensure that rover transects fit features, which are discernable in orbital high-resolution imagery using landmarks visible in both the rover imagery and from space [57]. The tracks can be validated using observed rover wheel tracks, where visible. The UCL-MSSL processing system starts with an orthorectified image generated either from 25 cm HiRISE imagery or from an orthorectified 5 cm SRR imagery [27]. In both cases, the HiRISE orthoimage is itself co-registered to CTX and thence to orthorectified HRSC images, which are in turn co-registered to the global coordinate system generated by the MOLA instrument [58]. This means that features located by the rover can be assigned in absolute aerographic coordinates.

Scale-invariant feature transform (SIFT) is employed for the generation of feature points and matching of rover-to-orbital images. As far as feasible, panoramic orthorectified image mosaics are employed from Navcam and Pancam data. Experimental results for all three NASA rovers (MER-A,B, MSL) show that errors are minimal except for some long transects for MER-B where small shifts can be observed. An example of the automatically derived rover transects and a comparison with HiRISE-SRR rover tracks is shown in Figure 3.12. SRR images are now being employed by the MSL science team to help with science target selection and with rover navigation planning.

For the Lunar case, such techniques have been applied for the Chang'E 3 lander and its rover Yutu [59].

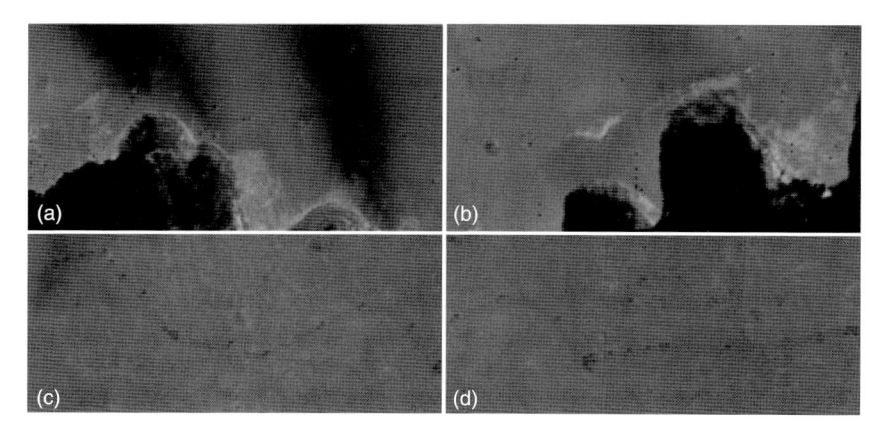

Figure 3.12 Localized rover positions (marked as dots) on HiRISE ORI for MER-B (a, b) and MSL (c, d) showing compliance between the automated localized rover traverse and the actual tracks recorded by a HiRISE image of 25 cm from orbit.

Other rover-to-orbiter image-matching techniques have been conceptualized, but not yet operationally applied by a flight rover. These include rover-to-orbiter horizon matching and rover-to-orbiter digital elevation model (DEM) matching. For the former, a DEM of the terrain around the rover is used to produce a simulated skyline view from multiple candidate rover positions. The position of the rover is estimated by finding the simulated skyline that best matches the rover's actual skyline view. In the latter case, a DEM is matched directly to a 3D model of the terrain generated around the rover, which can be generated via stereo imagery or, possibly, by a 3D sensor such as a LiDAR.

3.5.4
Visualization and Simulation

Being host organization of the most successful and still ongoing Mars rover missions, JPL has realized various ways to visualize the image data and vision processing results. The three most important to daily rover operations are the following.

- Marsviewer [60] is an image display and Reduced Data Record (RDR) visualization program developed by Multimission Image Processing Lab (MIPL) at JPL. It was originally developed as a quality control (QC) tool for MIPL. As such, it provides the ability to directly visualize all RDR types. Most of the RDRs are not traditional images: pixels may be XYZ coordinates, slope values, arm reachability codes, and so on. These are interpreted by Marsviewer and visualized for display, most often either using a color ramp to indicate the value (blue = "good" to red = "bad") or a contour stretch where values are highlighted every so many meters (default 0.1 m). Marsviewer provides the most direct "hands-on" view of the image data itself. The overlay visualizations shown in Figure 3.18 capture output from Marsviewer. Marsviewer can also view point clouds or meshes in 3D.

  Marsviewer comes in three versions: a desktop Java client, a Web version (JavaScript), and an iOS version for iPads and iPhones. All communicate with servers to access the data, or the Java client can read the file system where images are stored directly. One of the key features of Marsviewer is that it hides the complexities of where to find data behind an interface that is common across missions.

  Marsviewer is available for the general public to use to access image data in planetary data system (the PDS) for the MER, Phoenix, and MSL missions [60].

- The second tool is the science planning tool. It goes by different names on different missions: Maestro for MER [61], PSI for Phoenix [62], and MSLICE for MSL [63]. These tools are used by scientists and instrument operators to plan their observations and write commands. In order to do this, they display imagery in various forms. Users can define targets on the imagery and plan observations using the displayed imagery. The tools do not show all RDR types, but do show the ones most important for instrument commanding, including mosaics, range, and arm reachability.

- The third tool is the Robot Sequencing and Visualization Program (RSVP) [64]. This tool is used by the rover planners to plan drives and arm motions, and to build the entire command sequence package. A key feature of RSVP is its ability to simulate the command sequence, using 3D rendering to show what the rover will do and how it will interact with the surface. This is key for safety. RSVP primarily uses terrain meshes derived from stereo image analysis, along with "height maps" (DEMs). However, it also has modes to display the images directly.

Science teams are using tools that are dedicated to their instruments, such as *Viewpoint*, which is designed to provide the MER science team with easy-to-use 3D planning tools for rover operations [65]. ExoMars PanCam will use a tool called "PRo3D" (Planetary Robotics 3D Viewer) [66] for immersive 3D presentation of the PanCam 3D vision products, and their geological interpretation (Figure 3.13). PRo3D is supported by PRoViP (Section 3.7.7) being the backbone processing engine for ExoMars PanCam. For the operations part of ExoMars to be run in the Rover Operations Control Centre (ROCC), rover system simulation environments such as 3DROV [67] or VERITAS [68] will converge to the ROCS RVP, which stands for ROCC Operations Control System, Rover Visualization and Planning, being an analogue to the JPL RSVP [64].

An important tool for direct visualization of stereo images are viewers that present the left and right images of stereo views (or combinations thereof, such as panoramas) to the left and right eye of the spectator. Inspired by the functionality and availability of the JADIS open source Java library developed at JPL [71, 72] to allow the display of stereo images on different stereo display technologies, a generic stereo workstation called StereoWS has been developed at UCL-MSSL and is available as open source through sourceforge [73] including full user and programmer documentation. This allows the display of stereo images from rover cameras, after epipolar alignment, along with tools to permit 3D measurement using a standard "floating crosshair cursor" displayed over the image. 3D measurements can be performed on standard flat-screen displays or specialist stereo displays or consumer unit stereo displays. Different stereo-matching disparities can be displayed including standard PDS products using a Java I/O class from JPL. StereoWS was employed in several investigations of stereo-matching accuracy of different stereo images from different rovers in some cases under laboratory controlled conditions with correlative laser point clouds [74]. By holding the left point position fixed from previously selected stereo-matching results, the user is able to modify just the apparent 3D position of the cursor and place this 3D cursor onto the surface of the same feature. The positions are then selected either as being both obvious stereo landmarks or difficult points suffering from partial occlusion. Screen shots of StereoWS are shown in Figure 3.14. The reader is suggested to download and install the JADIS library along with the tool from sourceforge.

3.5 Ground-Based Vision Processing | 137

Figure 3.13 Screenshot of PRo3D showing a geologic interpretation session in the Shaler area (Gale Crater, MSL mission). This detailed interpretation of the stratigraphy shows the main stratigraphic boundaries as gray lines, bedset boundaries as thick white lines, and laminations within those bedsets as the thin white lines (note that the original image is in color). The dip and strike values are available directly in PRo3D color coded by dip value, and generally dip 15–20° to the southeast; however, this requires validation. The findings are consistent with those in Refs [69, 70] in that the outcrop represents a changing fluvial environment, with recessive, fine-grained units interlayered with coarse, pebbly units. (Data courtesy NASA/JPL, image courtesy: Imperial College London, Robert Barnes/Sanjeev Gupta; www.provide-space.eu.)

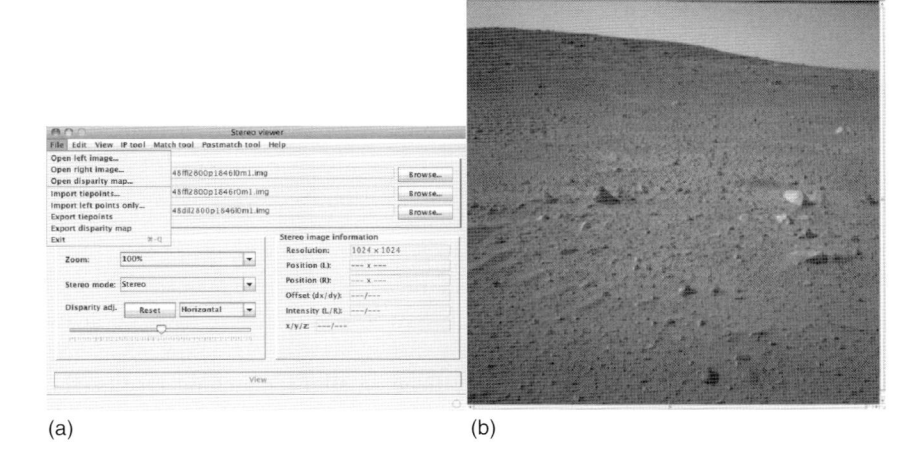

(a) (b)

Figure 3.14 Screenshots taken from the StereoWS tool showing images from MER-A in stereo anaglyph on a standard flatscreen display (originally appearing in red/blue to be viewed with anaglyph spectacles). The left panel, the control panel, is shown along with the pull-down menu, which shows inputs and outputs including the importing of left points only for subsequent 3D stereo measurements.

3.6 Onboard Vision Processing

3.6.1 Preprocessing

There are no such things as perfect cameras. Indeed, optics will distort the image and distortion increases with the field of view – ExoMars NavCam uses a 65° horizontal and vertical field of view, a fairly wide angle. Sensors and associated electronics are also imperfect and add effects such as noise and quantization error. Because of the onboard processing constraints, the ExoMars stereo vision assumes that it is known which row on the right (left) camera corresponds to each row on the left (right) camera. This requires a very stable stereo-bench and/or a very accurate calibration method. Both are used in ExoMars and the necessary image shift to align right and left images is performed in the cameras. The pixel size is very small (in the range of 5 μm) and full resolution is used – optical distortion must be corrected to an accuracy of one-fourth of a pixel, which is again a difficult challenge (note that on Mars the temperature ranges from −120 to +50 °C). Once again, a very stable stereo-bench and/or a very accurate temperature-dependent calibration method are employed. Though calibration data are stored in the onboard computer, the actual image correction process is performed inside the cameras. Indeed, the cameras include an FPGA, and this process can be highly parallelized, making this a very fast operation. However, this remains a challenge for space-qualified FPGAs because of the amount of data to be processed. The electronic imperfections are dealt with by the Perception and Visual Localization algorithms. No dedicated algorithm is applied; this is addressed by thresholds and tuning of those algorithms such that the final performance is acceptable. If it becomes too inaccurate, dedicated calibration might be necessary but this is unlikely to be needed for GNC purposes.

3.6.2 Compression Modes

It was established in Section 3.5.1 that the compression strategies are required for downlink to ground stations for Earth-based analysis of the data. The data compression most importantly applies to the high bandwidth type of instruments, for example, cameras and scientific payloads, such as spectrometers. Contemporary missions, such as MSL Curiosity and the ESA ExoMars, have the capabilities to perform some scientific processing onboard, which saves bandwidth costs compared with downlinking the raw data. For example, the MAHLI mounted on the turret at the end of Curiosity's robotic arm uses a 2-Mpixel color camera with adjustable-focus macro lens for the observation of morphological, mineralogical, structural, and textural properties in geologic materials [21]. The adjustable-focus capability enables MAHLI to acquire images of the Martian surface at varying resolutions and distances. Much of the image processing is done onboard, such as z-stacking (focus stack merging) for best-focus views, and creating depth (range) maps of the target, as a measure of its microtopography. Given the large data storage available for MAHLI, it is possible to store data in uncompressed format, and send their respective thumbnail images to Earth. A specific selection of images from a stack can be downlinked to Earth upon request at a later stage, thus saving on downlink bandwidth.

MAHLI instruments can also perform compression of 12-bit images to 8-bit images using a square-root companding look-up table. Most MAHLI images use a lossy compression format (JPEG), which results in some level of degradation. However, it is also possible to use a lossless compression mode, to achieve highest quality. Often the original image is downlinked as JPEG, and then the most interesting ones are selected for retransmission using lossless compression. The MSL Mastcam also has most of these same capabilities [18].

3.6.3
Stereo Perception Software Chain

The role of the stereo perception system is to analyze a pair of stereo images and produce a disparity map. A disparity map describes the apparent shift in corresponding pixels between a pair of stereo images, as illustrated in Figure 3.15a. As an example, the following points summarize some key features of the ExoMars perception system (Figure 3.15b):

- Optical distortion effects are rectified within the rover cameras and are, therefore, not considered within the perception system.
- A multiresolution approach is used to maximize the amount of terrain covered by the disparity maps while mitigating the adverse processing time implications of using high-resolution images.
- When calculating the low-resolution section of the disparity map, the 1024 by 1024 pixel camera images are reduced to 512 by 512 pixels.
- A Laplacian of Gaussian preprocessing filter is used to produce gradient images before the stereo-correlation algorithms attempt to calculate disparity values.
- A sum of absolute differences correlation algorithm is used to scan the stereo images horizontally to detect disparity values to the nearest pixel.
- A linear subpixel interpolation method is used to calculate decimal disparity values.
- Disparity map filtering is used to remove any erroneous values from the raw disparity map to produce the final, filtered, multiresolution disparity map.

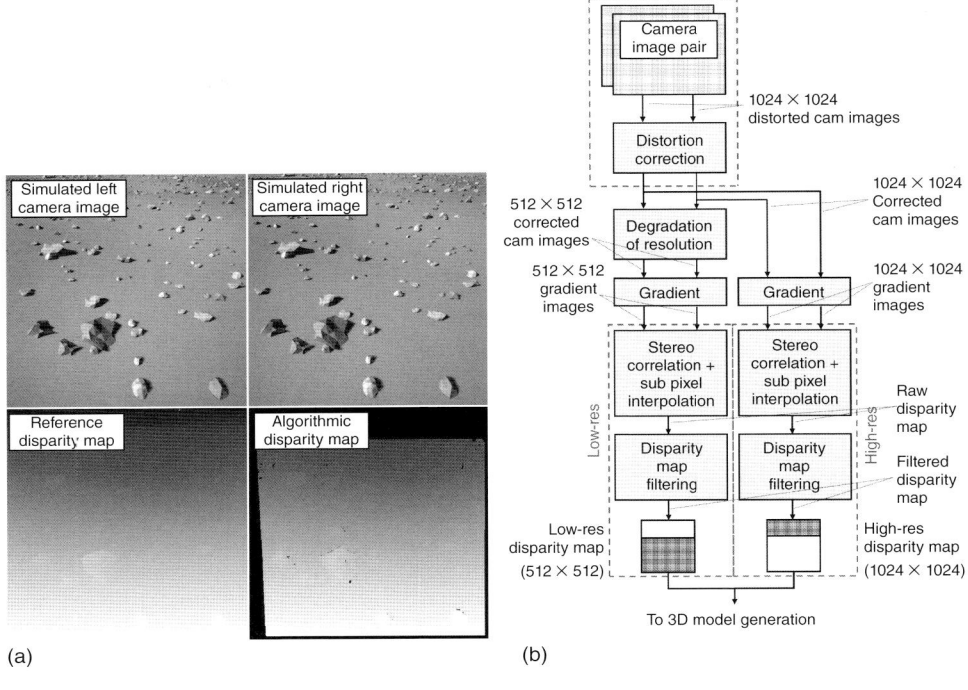

Figure 3.15 (a) Reference and algorithmic disparity maps from a ground-truth simulation (PANGU [75]). (b) Scheme of Perception architecture (ExoMars perception system examples).

3.6.4
Visual Odometry

The role of the visual odometry (VO) is to use visual information to estimate the linear and angular displacement of the rover as it moves (Figure 3.16).[7] There are a large number of different approaches to visual odometry. However, they can be structured using the following key steps:

- Calibration: Calibration data are obtained on ground before launch. It can be split into intrinsic (e.g., distortion correction) and extrinsic (where the camera is located and pointing). In ExoMars, this is performed outside the Visual Localization function. See Section 3.4.1 for calibration.
- Motion prediction: This is based on previous frames and it allows estimating where corresponding features are expected to be on the new frame – this accelerates the process.

7) In addition to ground navigation, the scheme is applicable to support the navigation in other mission phases. For example, MER used the descent imager to determine its horizontal speed in order to know how to fire the rockets when landing. Supporting the flying, hopping, and so on of robots via a VO-like mechanism is also feasible.

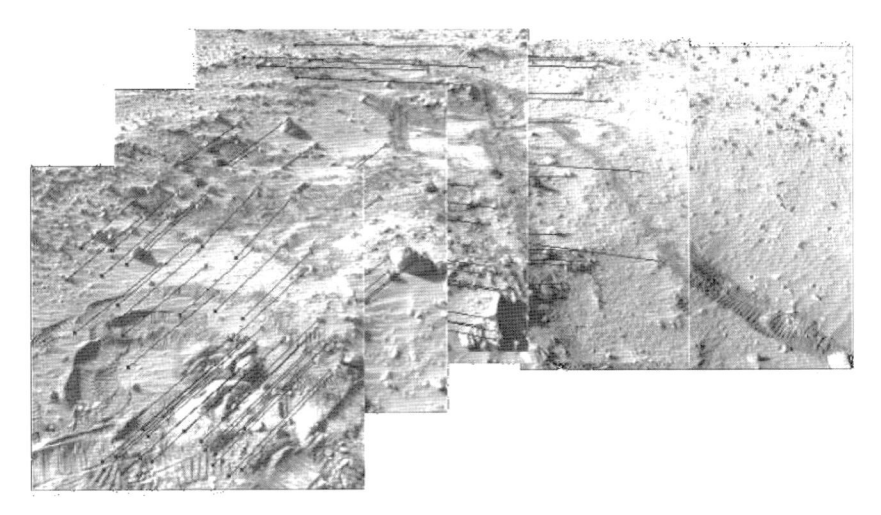

Figure 3.16 MER example of vision-based odometry with feature detection and matching shown [76] (contrast modified). (Courtesy NASA/JPL-Caltech.)

- Feature detection: There are numerous feature detection techniques. Essentially, this corresponds to detecting strong features in the image (e.g., corners such as vertex of rocks) that can more "easily" be detected and matched in subsequent frames.
- Feature matching: Based on the characteristics of each feature previously detected (surrounding visual environment), a match of the newly detected features to the previous ones is made.
- Motion estimation: Again there are numerous possibilities for this step. For example, a Kalman filter can be used in order to, based on the motion of the features, estimate the motion of the rover. It is a key assumption (valid on Mars) that the scene does not evolve and, therefore, if features have moved, this is solely due to the motion of the rover.

3.6.5 Autonomous Navigation

Autonomous navigation (and other autonomous abilities such as the search for scientific targets and collection of information and samples) relies heavily on vision components being the main perception asset, supporting duties such as mapping, terrain assessment, obstacle avoidance, target detection, and path planning.

The MER, MSL, and ExoMars rovers have similarly structured autonomous navigation pipelines, and all use stereo vision data as the main sensory input. The primary cameras are the mast-mounted Navcams for all the rovers, but the MERs and MSL also have wide-angle, body-mounted Hazcams for obstacle detection. Each of the rovers generates local maps from the stereo inputs, and then merge these maps into a larger map constructed from previous stereo data collections.

These maps are then used to create traversability maps, which indicate safe areas for the rovers to travel by analyzing the map for hazards such as rocks or excessive slopes. Path planning to a goal location is then performed on the traversability map, and the rover is commanded to follow the path. The rover travels a short distance and these steps are performed again. These autonomous navigation steps are described further in Section 4.5.

Orbital imagery also plays a key role in the navigation abilities of the MERs, MSL, and ExoMars rovers (see Section 3.5.3). This imagery is (and will be) used to select goal locations for the rover to reach, which the rovers then can autonomously travel to. However, at present this long-term/global localization is performed on the ground.

The MERs also have the ability to specify a goal location by its appearance, rather than by its coordinates [77]. For example, a rock of interest can be selected as a goal, and the rover will use image-based techniques to track the rock and guide the rover there.

3.7
Past and Existing Mission Approaches

This section introduces a set of solutions coming from past and present missions highly related to planetary robotics vision, relevant field trial assets, and trials themselves, as well as experimental setups. It does not claim to be complete but rather to capture a representative sampling of planetary bodies, technical approaches, maturity, and use cases.

3.7.1
Lunar Vision: Landers and Rovers

The Luna 9 spacecraft of the Soviet Union's Luna program was the first spacecraft to achieve soft landing on the Moon (February 3, 1966) and transmit photographic data of the lunar surface back to Earth. Luna 9 was equipped with a lightweight panoramic television camera system and an active rotating and tilting mirror mounted on an 8 cm turret above the camera for a 360° coverage of the lunar surface. Follow-up missions Luna 13 (December 24, 1966) used a television system to transmit cycloramas of the surrounding lunar landscape, while Luna 16 (September 20, 1970) consisted of two panoramic optical-mechanical cameras with better light sensitivity compared with its predecessor (Luna 9), a wider aperture lens, and artificial illumination instruments. The Luna 17 lander carried onboard the Lunokhod 1 robotic rover for the exploration of the lunar surface, which landed on November 15, 1970. Lunokhod 1 was a successful remote-controlled rover deployment on the Moon. Equipped with two vidicon-tube television cameras, it returned 250-line images at 10 frames/s and four panoramic telephotometers; two oriented for 180° horizontal panoramic images at 500 × 3000 pixels and two oriented for 360° vertical panoramic images at 500 × 6000 pixels [78]. Luna

20 was a lunar sample return mission, which landed on the Moon on February 21, 1972. It was equipped with a stereo pair of optical-mechanical panoramic cameras imaging the lunar surface, spacecraft, and the sky at four lines (300 pixels)/s. Lunokhod 2 rover onboard the Luna 21 (January 8, 1973) was a continuation of its predecessor, Lunokhod 1, which consisted of three television cameras (one used for navigation) and four panoramic cameras for surface observation. This mission returned 86 panoramas and over 80 000 pictures from navigation videos. The Luna program eventually ended with its final lander; Luna 24 on August 18, 1976.

Most of the images acquired from the Lunokhod rovers have missing relevant information, such as operational and extrinsic parameters, timing, and positional information. Furthermore, these images have noise in them [79]. However, recent advances in photogrammetric image processing along with complementary data from the Lunar Reconnaissance Orbiter's (LRO) narrow-angle camera (LRO NAC: up to 0.3 m/pixel) may be used to exploit the Lunokhod dataset. Data fusion can be applied to DEMs generated using the LRO NAC images, stereo photogrammetric processing of the Lunokhod panoramas and orthorectified panoramic images in order to provide morphology and morphometry information of the Lunokhods' surroundings [79, 80].

Almost 40 years later, on December 14, 2013, the next soft landing mission on the Moon was the Chinese Chang'E 3 mission, which deployed the Lunar rover Yutu. It is equipped with a stereo panoramic camera [81] and a stereo navigation camera on a mast, as well as two hazard cameras also used for stereo imaging. For monitoring rover activities (among other operations), the lander also carried a panoramic camera system [82].

3.7.2
Viking Vision System

The Viking Lander camera system [83] was very different from current solid-state camera systems. Rather than having an array of photosensitive pixels, it had just one single pixel (consisting of 12 photodiodes at different wavelengths). Images were formed by physically rotating mirrors both in elevation and in azimuth. The elevation mirror had 512 steps in each cycle, meaning images are 512 pixels high. The camera could see from 60° below the horizon to 40° above it. Two stepping modes allowed for different fields of view: low resolution (0.12°) and high resolution (0.04°). Regardless of stepping mode, there were always 512 steps. The camera could see 342.5° of azimuth, and the azimuth range varied per image [84].

There were two of these cameras per lander. This allowed for stereo, but only in some directions. When looking perpendicular to the baseline between the images, full stereo could be achieved. When looking parallel (i.e., through the other camera), no stereo was possible since there was effectively no baseline. In between, the baseline varied with the angle.

Processing various stereo sets and panoramas was successful already in 1977 [85].

3.7.3
Pathfinder Vision Processing

The Imager for Mars Pathfinder (IMP) [13, 86] was a stereo imaging system with color capability provided by a set of selectable filters for each of the two camera eyes. It consisted of three physical subassemblies: camera head (with stereo optics, filter wheel, CCD and preamp, mechanisms, and stepper motors); extendable mast with electronic cabling; and two plug-in electronics cards (CCD data card and power supply/motor drive card), which plug into slots in the warm electronics box within the lander.

Azimuth and elevation drives for the camera head were provided by stepper motors with gear heads, providing a field of regard of $\pm 180°$ in azimuth and $+83°$ to $-72°$ in elevation, relative to lander coordinates. The camera system was mounted at the top of a deployable mast. When deployed, the mast provided an elevation of 1.0 m above the lander mounting surface. The focal plane consisted of a CCD mounted at the foci of two optical paths where it was bonded to a small printed wiring board, which in turn was attached by a short flex cable to the preamplifier board. The CCD was a front-illuminated frame transfer array with 23 μm^2 pixels. Its image section was divided into two square frames, one for each half of the stereo FOVs. Each had 256×256 active elements.

The stereoscopic imager included two imaging triplets, twofold mirrors separated by 150 mm for stereo viewing, a 12-space filter wheel in each path, and a fold prism to place the images side-by-side on the CCD focal plane. Fused silica windows at each path entrance prevented dust intrusion. The optical triplets were an f/10 design, stopped down to f/18 with 23-mm effective focal lengths and a $14.4°$ field of view. The pixel instantaneous field of view was 1 mrad. The filter wheel contained 4 pairs of atmospheric filters, 2 pairs of stereo filters, 11 individual geologic filters, and 1 diopter or close-up lens, designed to acquire images of magnetic, wind-blown dust, which adhered to a small magnet located on the IMP tip plate.

Pathfinder proved the concept of passive stereo imaging on Mars that has been adopted by all later missions. IMP captured thousands of images that could be used for cartographic panoramas, local DEMs, observations of the Pathfinder Rover Sojourner as well as change analysis of the Martian surface [87]. The first super-resolution datasets and processing from a Mars surface probe were captured and processed in this mission [88].

On Pathfinder's rover Sojourner, two front monochromatic camera/laser systems provided stereo viewing and ranging, while the rear-mounted camera was discriminant in the red, green, and IR spectral bandwidths to produce color images [89]. The laser stripe system was used for ranging. Whereas the IMP camera viewed the Martian scenes from a fixed platform at spatial resolutions that diminished with distance, the mobile rover was able to drive up to objects, including those hidden from the IMP, to image them at higher resolution. Vision processing for the Pathfinder rover included mapping and localization [2].

3.7.4
MER and MSL Ground Vision Processing Chain

While onboard processing and autonomy is playing an increasingly important role in rover and lander operations, most of the planning for current and contemplated missions is still done by operators on the ground. Doing this efficiently and safely requires a large variety of image-based datasets, which together provide situational awareness, both qualitative and quantitative, to the human operators. This situational awareness is largely based on images, and image processing used to derive higher-order products. For example, mosaics provide an overall view of the area (supporting navigation and targeting), terrain meshes provide a quantitative 3D model for rover simulations to interact with, slope and solar energy maps help determine where it is safe to drive, and reachability products show where arm-mounted instruments can be used in the arm workspace.

Understanding the environment with which the rover or lander is interacting is critical to safe operations: reliability is one of the most important aspects for any application in a space environment. This specifically applies to products generated to support operations, whether onboard or on the ground. When driving a multi-billion-euros asset with no possibility of rescue in the case of failure, such a system needs to be "bulletproof" in the sense of reliability of processing results. As an example it is critical to know where unknown areas are located, rather than, for example, interpolating over them to make a map complete. Decision upon failure has to be taken conservatively: when faced with some remaining level of uncertainty a result should be flagged as "unknown" rather than guessing.

The processing needed to create these products must be (at least mostly) automated, in order to support the rapid turnaround needed for tactical surface operations (especially on Mars). It also needs to be highly robust, and fail-safe in that it reports only what it knows, with holes or gaps indicating unknown areas.

In addition to tactical products (used in the critical path of daily operations), strategic products also need to be considered. These are made outside the tactical timeline, and are used for long-term planning (days to weeks), science analysis, and public release. While the focus in this book is on tactical products as they relate to robotic operations, it should be noted that strategic products are just as important for the overall success of a mission, even if they do not contribute to daily operations.

One example of a system capable of this processing is the one created and operated by the Multimission image processing lab (MIPL) at JPL, which supports image processing for all NASA *in situ* Mars missions. A common code base, initially written for Mars Pathfinder, has supported MER, MSL, Phoenix, Mars Polar Lander, and the upcoming InSight and Mars 2020 missions. This section provides some details on the MIPL system, as an example of what a comprehensive ground-based image processing system for planetary rovers and landers could look like.

The MIPL system starts with telemetry data, reconstructing that into usable image (and non-image instrument) data. Images are decompressed and reformatted into usable formats, and telemetry about the image is decoded to metadata "labels" that go along with the image. This metadata describes the conditions under which the image was taken (pointing, temperatures, exposure time, arm state, etc.) and is critical to being able to understand the image. This metadata is what makes the difference between scientifically or operationally useful data, and just a pretty picture. These images and their metadata, called experiment data records (EDRs), are the starting point for all image processing, and are the most important image products in the mission archive (such as PDS).

From the EDRs, the MIPL system makes RDRs containing higher-order analysis products. Over two dozen distinct products are made from each stereo pair of images, or a combination of several stereo pairs, which are described (for MER) in [90]; see also software interface specifications for MER [91], Phoenix [92], and MSL [93]. The key products include:

- radiometrically corrected images
- geometrically rectified images (epipolar aligned)
- disparity maps
- XYZ images
- surface normals
- slope maps
- reachability/preload maps (for arm instruments)
- roughness maps
- range error maps
- terrain meshes (including meshes from orbital images)
- mosaics.

Some of these RDRs are shown in Figures 3.17–3.19.[8]

Stereo correlation is done in a two-stage process. The first stage works with epipolar aligned, reduced-resolution, linearized images using a one-dimensional correlator (the same correlator as is used onboard the rovers). The second stage uses a modification of the Gruen algorithm [94], which implements a two-dimensional correlation [95]. Even though the images are epipolar aligned, calibration is not perfect, so a 2D correlator helps achieve better results. In addition, the 2D correlator can work with nonlinearized images, as well as unusual stereo pairs, such as the MSL Mastcams, which differ by a factor of 3 in zoom.

The XYZ generation step uses *a priori* camera models based on ground calibration. While self-calibration of models is possible, more quantitative accuracy can be obtained by preflight calibration. The XYZ generation includes a large number of filters and consistency checks that help to remove bad points (conservatism and safety dictates removing any points that are questionable, rather than interpolating over them) [95].

The arm reachability products use arm kinematics and collision models to determine which pixels can be reached by each arm instrument, and in what arm

---

8) Such informative RDR products for human interpretation and operations planning are actually produced in color.

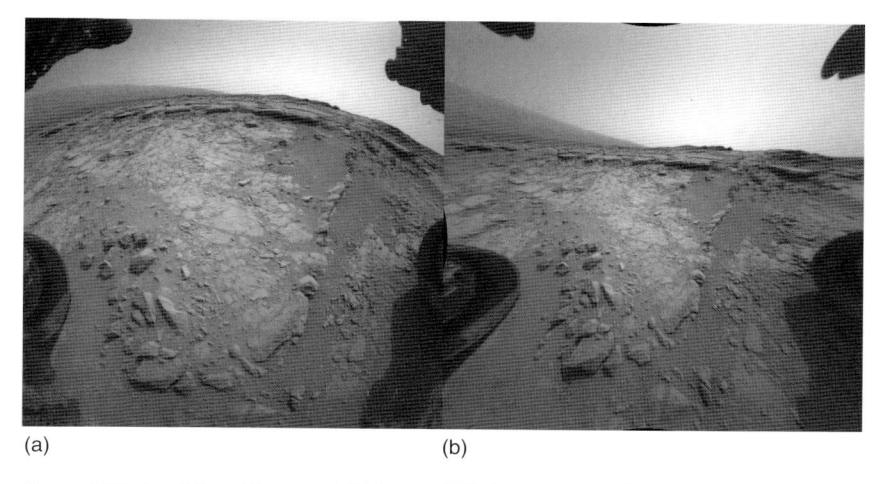

Figure 3.17 Raw (a) and linearized (b) image (MSL front HazCam from sol 151). Note fisheye distortion on the raw image. (Courtesy NASA/JPL-Caltech.)

Figure 3.18 Overlays on image from Figure 3.17. (a) Slope; dark = low slope; bright = high slope. (b) *XYZ* image where white = lines of constant *X*; gray = *Y*. (Courtesy NASA/JPL-Caltech.)

mode (shoulder in or out, elbow up or down, etc.). The same algorithms are used in the flight software for collision checks as an extra safety mechanism.

Mosaics are complicated by the fact that for these missions, the camera moves as it is taking pictures; it does not rotate around the camera entrance pupil but rather around the center point between the stereo pair of cameras. That introduces parallax into the imagery, which causes geometric seams in the mosaic [96]. Pointing knowledge and camera modeling errors also contribute to seams. Techniques have been developed to minimize these seams [97], but in general parallax seams are uncorrectable without introducing unacceptable image warping.

Figure 3.19 360° drive mosaic from MSL Navcam data, sol 169, cylindrical projection. (Courtesy NASA/JPL-Caltech.)

There are several important principles that MIPL adheres to regarding mosaics. Maintaining scientific integrity of the results is paramount. It is critically important to maintain traceability of results; each pixel can be mathematically traced to the source image it came from. This leads to no use of unconstrained warping. The image is treated as a single entity, with seam correction achieved by adjusting the pointing of the entire image. Warping distorts the scene in ways that make it useless for scientific measurement. For similar reasons, seams are not blended. Blended seams can create artifacts or "fuzzy" areas that could lead to misinterpretation of the data. *A priori* metadata, especially regarding camera pointing, is used. Although techniques exist to determine camera pointing solely from camera imagery, that does not always work, and MIPL has seen examples where frames were swapped with no visual cue this had occurred. The risk to a mission from a blunder of that magnitude is not acceptable. Finally, while great attempts are made to minimize seams, a hard-edged, straight seam is preferable to something that could be misinterpreted as a feature in the image.

These principles lead to uncorrectable parallax errors. For example, Figure 3.20 shows two adjacent frames with a portion of the MER solar panel and the ground. On the left, the ground has been corrected to be seamless, but the solar panel has a seam. On the right, the panel is seamless, but the ground is not. It is simply not possible to eliminate both seams at once without introducing gaps in the image or removing data. In general, MIPL opts for correcting the ground, except for self-portraits. It should be noted that the MSL self-portraits taken by MAHLI have no parallax, because the arm-mounted MAHLI camera has enough degrees of freedom to rotate around the entrance pupil of the camera. Mosaicking those images is thus almost a trivial effort.

One of the most significant enhancements that MSL Mastcam has brought to Mars exploration in general and MSL science target selection is the creation of gigapixel mosaics [98], the first one of which consists of 850 individual megapixel Mastcam color frames mosaicked into a cylindrical projection to allow exploration at a variety of different distances. A small section of the first Mastcam mosaic at

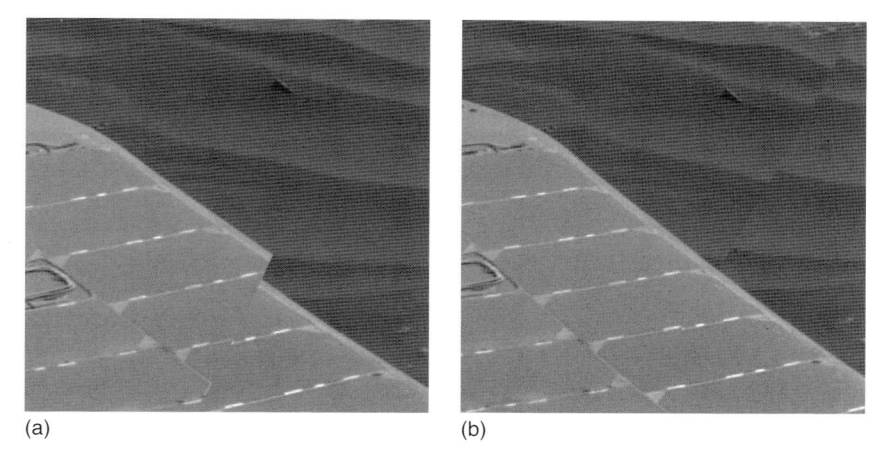

Figure 3.20 Parallax effects on mosaics. (a) Correcting on the ground, (b) correcting on the foreground solar panel. (Courtesy NASA/JPL-Caltech.)

Figure 3.21 Mosaic of Mastcam images of Mt Sharp taken on MSL Sol 45 (September 12, 2012; originally in color) using white balance to show an Earth-like sky. (Courtesy NASA/JPL-Caltech.)

location "Rocknest," showing the top of Mount Sharp is shown in Figure 3.21.[9] Large-FOV mosaics of Mastcam have become the operational norm for MSL and should be considered as the new default mode for future rovers such as the ESA ExoMars rover 2018.

The processing described earlier is implemented in dozens of separate application programs using the VICAR image processing system [99]. Each program does one part of the process—first stage disparity, second stage disparity, XYZ generation, mosaic, pointing correction, and so on. This allows great flexibility as compared with a monolithic or GUI-based program with few options—flexibility that is often needed in the dynamic world of rover operations. A pipeline glues all these programs together into an automated processing system.

Finally, the application programs are written using a multimission library, which abstracts out all the mission-specific details. All the missions MIPL supports use the same code base, with 130 000 lines (and growing) of multimission code and only about 5000 or less lines per mission. This means the software is easily adapted for each new mission, at great cost savings [100]. It also means that improvements made for a later mission (e.g., MSL) are easily usable in older missions (e.g., MER).

9) http://photojournal.jpl.nasa.gov/catalog/PIA16768.

3.7.5
ExoMars Onboard Vision-Based Control Chain

ExoMars uses fast state-of-the-art cameras with embedded image correction. This advance in technology enables performing Visual Localization as the rover drives without having to stop to capture images, which indirectly allows checking the rover slippage levels every 10 s. Though its Navigation strategy is similar to MER/MSL (regular planning stops), ExoMars is able in equivalent timescales to produce a more accurate 3D model of the terrain, to find more possible paths for driving, and to further optimize these paths. The advances in computing power and the tailored onboard computer with dedicated coprocessor for image-processing functions are key for enabling these advances. The disparity maps produced using the NavCam images allow coverage of terrain with high accuracy at least 6 m in front of the rover. A multiresolution strategy is used to cope with computation timing constraints. The disparity maps are also checked onboard for inconsistencies with what we can expect from the environment removing eventual artifacts. The 3D model of the terrain merges the three disparity maps (left, center, right) produced at each stop and is then used to assess where the rover can drive, and where it should drive. The navigation maps produced use information from previous stops such that the rover always has an efficient long-term planning map and avoids known dead-ends for example. The path planning has a long-term and a short-term component, the latter being optimized to take into account terrain difficulty, exploiting high rover maneuverability, and minimizing tight trajectories.

The Visual Localization function on ExoMars is one of the most advanced techniques of its kind, being a good compromise between computational speed, pose estimation accuracy, and robustness to challenging environments (e.g., shadowed terrain at low optical depth, blended sand without rocks). Having a regular pose estimation, together with a more maneuverable locomotion, enables continuous trajectory closed-loop control, which ensures that the commanded path is always followed with good accuracy. This again has allowed decreasing safety margins, which enables finding more driveable paths, being a key mission enhancer for challenging terrains where ground does not need to be in the control loop.

The ExoMars engineering cameras have a stereo baseline of 150 mm, 1024×1024 pixels at 8 bits full resolution (note that the sensor is 12-bit but, for computation timing reasons only 8 bits are used), and a 65° horizontal and vertical field of view: the same equipment (design) is used for both Navigation (NavCam) and Visual Localization (LocCam) – there are no such thing as Hazcams on ExoMars. The stereo baseline corresponds to an empirical optimum found in accuracy and computation time for the ExoMars mission, in particular for the perception function (first function in the Navigation computing a disparity map of the terrain surrounding the rover). While the NavCams are on top of a 2 m mast mounted on a pan-and-tilt mechanism, which also accommodates the science cameras, the LocCams are on top of the rover body (1 m above ground) minimizing motion and being sufficiently close to the terrain such that good positional accuracy can be obtained. Accommodation on the ExoMars Rover is challenging, and the LocCams' field of view is partly obstructed by the drill box (payload) and the locomotion legs and wheels.

In order to have an absolute heading reference on Mars, the NavCam images are also used for determining the Sun direction. This is a secondary function of the equipment and it is possible, thanks to the good quality of the optics and the CMOS sensor, which returns a useable circle of light where the Sun is (no rows or columns completely saturated). Though artifacts exist in the image (mainly due to optics), a good estimation of the Sun direction is still obtained. These functions are organized as per Figure 3.22.

3.7.6
ExoMars Onboard Vision Testing and Verification

The ExoMars GNC is verified in a multitude of test benches. While accuracy is assessed in high-fidelity simulators (Figure 3.23), timing performance is assessed during development on a Leon2 and verified on the onboard computer. In addition to these formal verification benches, all visual functions are also tested with real imagery for robustness and de-risking: from the Airbus DS Mars yard, from outdoors field testing (e.g., Atacama Desert or outdoors Mars yard). A specific test using the cameras' qualification models, the flight software, and a high-accuracy ground-truth system will be also used to ensure nothing important has been forgotten or misrepresented in the modeling. Finally, all visual functions run on the development rovers as they will run on Mars on the actual flight rover. They are

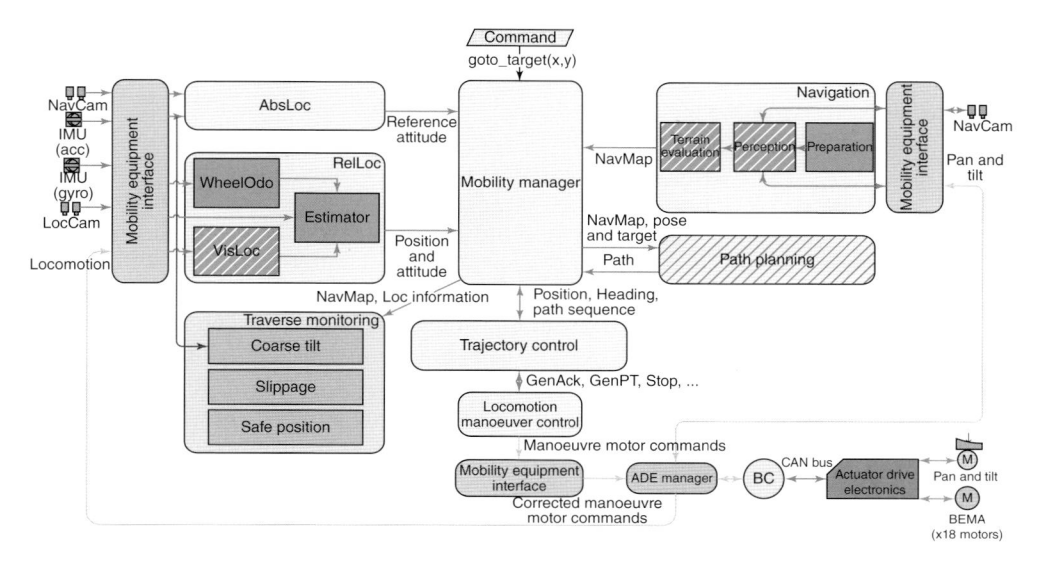

Figure 3.22 ExoMars Rover Functional Architecture depicting the use of visual information via NavCam and LocCam.

Figure 3.23 (a) Comparison of the Mars images with the ExoMars simulator generated images. (b) Development and testing rover in the Mars yard.

integrated in a completely representative manner in the GNC using representative sensors and actuators.

3.7.7
ExoMars PanCam Ground Processing

3D scene reconstruction for ExoMars PanCam, delivered as "PRoViP" (Planetary Robotics Vision Ground Processing), produces DEMs and true orthophotos (including Panoramas) in various geometries (Cartesian, spherical, general plane) from PTU-acquired stereo imagery [101]. 3D data export is provided to known standards (GeoTiff, VRML, and PDS [102]) as well as further visualization for PanCam by the PRo3D visualization tool [66]. Figure 3.24 displays a typical workflow. Starting with images from the instrument available via PDS (or similar sources – during the mission directly from downlink sources in the Rover Operation Control Center – ROCC), stereo matching is performed, followed by 3D reconstruction into DEMs in various geometry, generation of an intermediate dataset (GPC: Generic Point Cloud), combination of the DEMs into unique consistent mosaicked products, and finally the export into products to be exploited by scientists and operations personnel. The workflow has been extensively tested with various Mars mission datasets from sensors similar to ExoMars, including MSL Mastcam, see Figure 3.25.

3.7 Past and Existing Mission Approaches | 153

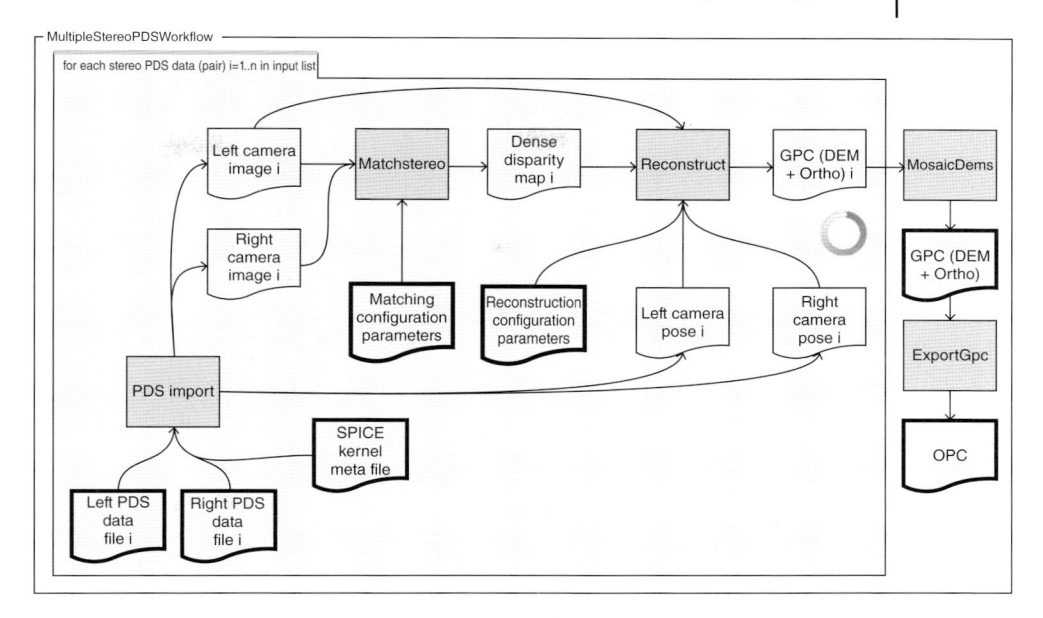

Figure 3.24 ExoMars PanCam 3D vision processing workflow scheme.

Figure 3.25 Result of MSL Mastcam processing of Garden City outcrop area taken at MSL Sol 926 and 929 (DEM, rendered by PRo3D [66]) making use of ExoMars PanCam 3D vision processing workflow PRoViP.

Working on top of the traditional data pipeline, ExoMars PanCam ground data processing chains are also envisaged to adopt an innovative software framework called mission-specific data processing (MSDP), which automates and simplifies the overall operations of the ExoMars rover [19]. The framework supports multichannel input and processes data according to user-defined scenarios on the

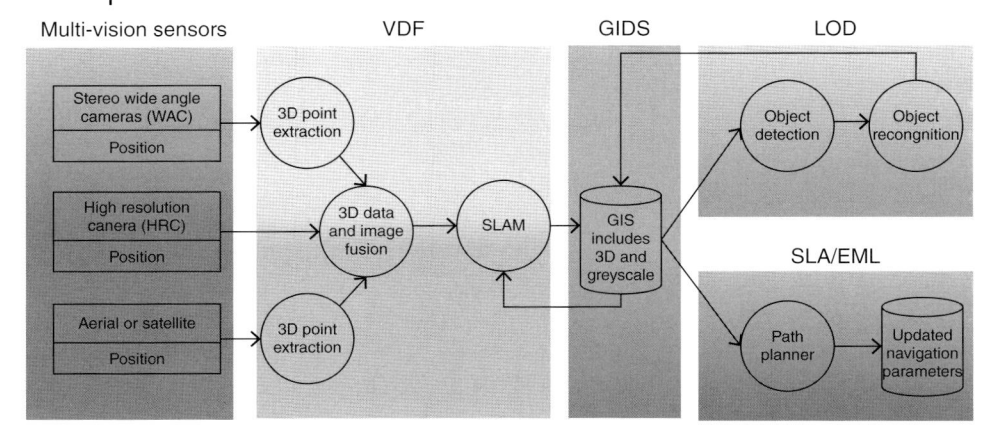

Figure 3.26 The MSDP framework.

ground by default but with the potential and flexibility for extension to support real-time data processing on board the rover. As shown in Figure 3.26, MSDP represents a multitasking process and system architecture for pipelined/concurrent multicore CPUs and multinodes. It supports several databases including a multidisciplinary *Commands Database* that handles sets of commands, a *Scenarios Telemetry Database* that provides information on the data/parameters, and a *Process Functions Database* that supplies the data processing algorithms including data representation and transfer functions. At each level of the output, a data logger tracks the data that passes through the current level. A synchronization process will safeguard the integrity of the databases.

The major processes within the MSDP include the following:

- **Visual Data Fusion (VDF)**: VDF is a software engine that ingests multiple visual inputs to produce accurate and complete 3D data products, such as nested multiresolution digital terrain models (DTMs), texture maps, and a constantly updated location map. For most ExoMars-type planetary rover missions, the visual information is a vital clue to compensate for errors involved in mechanical odometers for navigation and localization tasks (as discussed previously in Sections 3.5 and 3.6). The state-of-the-art approach based on simultaneous localization and mapping (SLAM) [103] will benefit from such fused data within the VDF. The process is also scalable to include sensors from multiple platforms such as Mars rovers and aerobots.
- **Geographical Image Database Server (GIDS)**: GIDS stores the sensed data and its derived data products. The data fusion process is expected to refine a current 3D scene reconstruction by combining previous sensing data at different locations (e.g., rear and front views from the rover) such as using 2D image matching with 3D data products. Therefore, fast 3D model-based searching within a large GIDS database is desirable.
- **Learning-Enabled Object Detection (LOD)**: The ability to perceive surrounding objects by extracting meaningful information from PanCam data is key to

both scientific and engineering operation of the rover. Objects can be classified as *hazards* that potentially obstruct the rover planned path, *landmarks* that are reference points for localization and building maps, or *regions of interest (ROI)* such as scientifically important rocks on Mars. Current vision-based methods of detecting and classifying objects are based on the object's geometric features and appearance. The dominant approaches for fast object detection use either ROI selection (such as bottom-up saliency models [104] discussed in Section 3.8.1), or holistic generative and discriminative models [4]. Generative holistic models are suitable for recognizing objects with relatively uniform geometric properties whereas discriminative models work best with previous knowledge of the appearance of the object that is stored in a template during training.

- **Self-Learning Agent (SLA) and Environment Model Library (EML)**: SLA employs a hierarchical approach where object class descriptions are learned based on the complexity of the classification method. A template of each object class is stored and updated in the EML upon successful detection of a target object to increase the processing speed (particularly relevant if these processes are required to work on board the rover). The SLA is capable of inferring objects from partial knowledge of the detected objects by utilizing a multitude of data mining and machine learning techniques. The EML is a result of the multiple pipeline processes and consists of detected objects and environment properties that are mapped with the location information from the GIDS. Objects are stored in various classes based on human-like interpretation of the environment such as geometry, appearance, and material attributes. This in turn allows human-like decision making such as producing a path cost map that is optimized based on the EML and can be used for path planning of the rover. If required, data from the EML can also be displayed from a 3D virtual reality simulator at the ROCC, which help increase learning accuracy by incorporating human feedback in a timely manner.

3.7.8
Additional Robotic Vision Systems

In addition to probes that have landed and moved on the surface of a planetary body, a number of other space missions have used vision systems for the surveying of minor planets, comets, and asteroids from close range.

Deep Space 1 (DS1) of the NASA New Millennium Program (launched October 24, 1998) performed a flyby of the asteroid 9969 Braille for scientific study, and with the Comet 19P/Borrelly for further engineering testing. The main vision payload onboard the DS1 consisted of the miniature integrated camera and spectrometer (MICAS); however, it achieved only partial successes in the assigned mission objectives due to problems within the system.

Dawn is a spacecraft which was launched by NASA on September 27, 2007 in order to survey the two proto-planets Vesta and Ceres. In order to capture detailed optical images of Vesta and Ceres for navigation and scientific study,

Dawn is equipped with a framing camera instrument. The probe carries two similar cameras with separate optics, electronics, and internal data storage and are capable of using seven color filters for the study of minerals on the surface of Vesta. The cameras can also record NIR images in addition to the visible spectrum.

The European Space Agency's space probe Rosetta (launched March 2, 2004) is on a mission to perform a detailed study of the comet 67P/Churyumov-Gerasimenko (67P) [105, 106]. The spacecraft reached its designated science target on August 6, 2014, while performing flybys of the asteroids 21 Lutetia, 2867 Steins, and the planet Mars during its journey. In addition to other onboard instruments, Rosetta carried the Philae lander, which made a soft landing on the comet 67P on November 12, 2014. The main camera instrument onboard Rosetta is the optical, spectroscopic, and infrared remote imaging system (OSIRIS). OSIRIS consists of two cameras; a narrow-angle camera (NAC) and a wide-angle camera (WAC). NAC is designed for capturing high-resolution images of the surface of the comet 67P/Churyumov-Gerasimenko, using 12 discrete filters covering a wavelength range 250–1000 nm and angular resolution of 18.6 μrad/pixel. WAC is designated for mapping gases and dust in the surroundings of the target 67P using 14 discrete filters at an angular resolution of 101 μrad/pixel [107].

The Near Earth Asteroid Rendezvous (NEAR) Shoemaker, launched on February 17, 1996, was an earlier mission consisting of a robotic space probe to the study the near-Earth asteroid Eros. The spacecraft successfully achieved its mission goals, finally terminating upon touch down on the asteroid on February 12, 2001. In addition to other instruments, the spacecraft was equipped with an NIR imaging spectrograph, and a multispectral camera using a CCD imaging sensor for imaging the surface of Eros. NEAR captured 160 000 images, which helped spot more than 100 000 craters on the surface of the asteroid.

Other follow-up missions have used high-resolution cameras for imaging asteroids, such as Chang'E 2 – the Chinese lunar probe (launched October 1, 2010) – which captured images of the asteroid 4179 Toutatis at a resolution of 10 m/
pixel.

Galileo was another robotic NASA space probe (launched October 18, 1989), which surveyed Jupiter and its moons along with other celestial bodies, such as the asteroid 243 Ida. Galileo's imaging instruments included an NIR mapping spectrometer for multispectral imaging, a UV spectrometer for studying gases and a photopolarimeter radiometer. The camera system was capable of capturing images of Jupiter's satellites at very high resolutions and a broad color detection band.

None of these instruments helped immediate robotic interaction with the observed surfaces; however, the concepts are valuable to learn more about the physical, chemical, mechanical, and morphological setup on the respective bodies. In addition, well-proven concepts are often reused for further exploitation in similar and even different environments.

3.8
Advanced Concepts

Space robotics is a fast growing field, and particularly the vision components undergo massive improvement in various streams of research. This section lists some of the future trends and sketches their technical, scientific, and mission-related implications.

3.8.1
Planetary Saliency Models

In an effort to reduce computational load onboard planetary rovers, recent studies have proposed the use of visual saliency methods to detect landmark features for autonomous navigation on extraterrestrial planetary surfaces [4, 104]. Visual saliency models are mostly inspired by the information selection property of biological visual systems and their underlying paradigms can either be based on computational models or cognitive research. Applications of visual saliency models cover a range of different areas: from low-level object detection and tracking [108, 109] to more complex robot localization and navigation [110]. Models of visual attention can vary on the basis of their processing characteristics [111]. The following discussion specifically focuses on the bottom-up attention models. Further classifications exist in the literature, such as *object-* and *space-based* models [111]. Within the context of the planetary navigation, models that are bottom-up, space-based, and are able to generate topographic saliency maps of the input visual scene seem to fit the problem.

A quantitative assessment of a saliency model for interesting objects detection has been performed on real-world and simulated datasets [104] (e.g., Figure 3.27) that are representative of a homogeneous planetary surface in order to identify the types of stimuli that can help delineate rocks or boulders in Mars-like environments with minimum computational load.

The work by Itti *et al.* [112] was the forefront in bottom-up visual saliency modeling that relates to human visual search strategies. It uses center-surround differences across multiscale image features within three topographical feature maps (*color, intensity* and *orientation*) for identifying conspicuous regions. The three feature maps are combined into a single saliency map for local conspicuity over the entire visual scene. Walther and Koch [113] extend this concept toward modeling visual attention in terms of proto-objects (the generated saliency maps are used for the deduction of a proto-object, via a winner-take-all neural network framework, at the attended location), (see Refs [114, 115] for further literature on proto-objects). Harel *et al.* [116] uses the computational power, topographical structure, and parallel nature of probabilistic graphical models in order to describe visually salient regions in an image. The equilibrium distribution of Markov chains along with a dissimilarity measure is used to compute saliency values in the visual scene. Hou and Zhang [117] proposed the use of log spectrum of an input image along with the average Fourier envelope in order to extract the spectral residual in the

Figure 3.27 Detecting interesting objects on planetary surfaces (such as rocks) using visual saliency method. (a) Image capture, (b) detect salient objects, (c) segmentation, and (d) track detected objects.

frequency domain to generate a saliency map. Seo and Milanfar [118] proposed a unified framework for (static and space-time) saliency detection that defines visually conspicuous regions in a local way. The model utilizes nonparametric local regression kernels to estimate the likelihood of pixel to its surroundings. A *"self-resemblance"* map (that is used to estimate saliency likelihood) is generated, which measures the statistical likelihood (similarity) of a feature matrix at a given pixel to its surrounding feature matrices using a matrix cosine similarity measure. Guo and Zhang [119] introduced a multiresolution spatiotemporal-based visual saliency detection model called phase spectrum of quaternion Fourier transform (PQFT) that represents an image in terms of color, intensity, and motion features (texture pop-out). This model captures the temporal characteristic (an additional motion dimension) of conspicuous regions within a visual scene in addition to its spatial conspicuity and has been proven to have very low computational complexity with little compromise on performance. Hou *et al.* [120] model is based on the concept of figure-ground separation for separating objects from the background using a binary, holistic image descriptor, called *"image signature"* (performed within the framework of sparse signal mixing).

These models could potentially be used for rock detection on homogeneous planetary surfaces, such as Mars for autonomous navigation, hazard detection, and scientific study. The models are tested using simulated and real-world datasets mentioned in Refs [103, 104] (more formally known as PANGU simulated, RAL

Space SEEKER, SSC lab-based data), that closely replicate a homogeneous planetary surface. The objective of these experiments is to examine the performance of the saliency models in terms of their ROI detection capabilities, which in the current case are surficial rocks and boulders. An example of the visual saliency-based rock detection is illustrated in Figure 3.27. The object detection performance is tested using quantitative evaluation metrics and protocols set out in Refs [121, 122]. For any given frame t, the number of "*false positives*" (fp_t), "*misses*" (ms_t), and "*true positives*" (tp_t) is calculated by measuring the spatial overlap between the ground-truth objects and the detector/tracker outputs. If for a given frame t, G_i^t is the ith ground-truth object and D_i^t is the ith detected object then the spatial overlap ratio (OR_i^t) is calculated as

$$OR_i^t = \frac{|G_i^t \cap D_i^t|}{|G_i^t \cup D_i^t|} \quad (3.1)$$

$$N - MODA = 1 - \frac{\sum_{t=1}^{N_{frames}} (c_{ms}(ms_t) + c_f(fp_t))}{\sum_{t=1}^{N_{frames}} N^t} \quad (3.2)$$

where

$$\forall t, N^t = \begin{cases} N_G^t & \text{if } N_G^t \geq N_D^t \\ N_G^t & \text{if } N_G^t < N_D^t \end{cases}$$

The detected object is considered a *true positive* for $OR_i^t \geq 0.2$ and *false positive* for $OR_i^t \leq 0.2$, whereas any unmatched objects in the ground-truth set are considered misses. The normalized multiple object detection accuracy (N-MODA) is computed for the entire image sequence of each dataset. For $\sum_{t=1}^{N_{frames}} N^t = 0$, $N - MODA = 0$. The parameters, c_{ms} and c_f are weighting parameters that can be varied according to the specification of the problem (here, $c_{ms} = c_f = 1$), N_G^t is the number of ground-truth objects and N_D^t is the number of detected objects.

In the case of the simulated dataset, it can be seen that most of the saliency models scored relatively well in terms of ROI predictions. The simulated images from PANGU were relatively simple in terms of complexity of textures. The background is flat with a constant grayscale value across the entirety of the image. Rocks are defined in terms of varying grayscale values with sharp boundaries. The majority of the saliency methods tested here showed good performance against the annotated ground truth (see Table 3.4); however, two models based on the "feature integration theory" (FIT), such as STB and GBVS, achieved the least performance scores in terms of N-MODA. The STB model generated local density of saliency fixations around the higher intensity pixels within the simulated rocks. Since simulated rock textures have uniformly distributed clusters of grayscale intensity pixels with hard boundaries, they act as local conspicuous regions deviating from the surrounding pixels. It naturally attracted the proto-objects-based attention toward these small areas on the rock's surface, resulting in smaller fixations, which did not satisfy the evaluation criterion, causing a lot of miss detections. This anomaly is clearly noticeable from the very high miss rate and false negatives for STB. In

Table 3.4 Normalized performance results for various saliency algorithms on chosen datasets.

Saliency model	Dataset	N-MODA	Miss rate	TPR
Itti	LAB	0.80	0.15	0.85
	PANGU	0.84	0.11	0.89
	SEEKER	0.32	0.56	0.44
GBVS	LAB	0.51	0.43	0.57
	PANGU	0.55	0.34	0.66
	SEEKER	0.14	0.75	0.22
PQFT	LAB	0.76	0.17	0.83
	PANGU	0.78	0.04	0.96
	SEEKER	0.49	0.13	0.87
SDSR	LAB	0.66	0.27	0.73
	PANGU	0.83	0.10	0.90
	SEEKER	0.47	0.08	0.92
SigSal	LAB	0.62	0.33	0.67
	PANGU	0.78	0.15	0.85
	SEEKER	0.29	0.51	0.48
SRA	LAB	0.74	0.18	0.82
	PANGU	0.83	0.01	0.99
	SEEKER	0.45	0.15	0.85
STB	LAB	0.49	0.42	0.58
	PANGU	0.47	0.30	0.70
	SEEKER	0.39	0.57	0.43

the case of GBVS, classification of pixels as salient rocks was performed in terms of probability distributions; the simulated textures (similar to STB) of the rocks were adding a slight center-bias toward specific regions within the rock. As a result of this behavior, the equilibrium distribution of Markov chains selectively increased the conspicuity of the larger rocks located at the center of the image, where the surrounding smaller rocks were given negligible weightings. This resulted in a relatively higher number of false negatives. It is worth noting, however, that the performance of GBVS was still better than STB.

Real-world images are more challenging compared with the simulated PANGU images. On average, a performance drop would naturally occur; however, this still depends upon the visual scene complexity within the real-world images. The images from the indoor lab-based dataset were relatively less challenging among all real-world datasets tested. Pixels in the image that define the rocks are highly conspicuous against the flat approximately uniform textured background floor. One important feature of the lab-based images is that the rock textures have a more uniform distribution of grayscale intensities compared with the simulated PANGU rock surface textures. Itti-98, based on the FIT model, showed very good performance in generating saliency fixations on rocks, with N-MODA score nearly as good as with the simulated images. Two frequency domain-based models; SRA and PQFT, performed equally good. In the case of SRA, the blob of pixels defining rocks have a statistical distribution that is significantly different from its surrounding, therefore, making it much easier to suppress the highly redundant background pixels. Furthermore, temporal information is very much consistent among subsequent frames in the lab-based dataset. Taking the phase spectrum of the images resulted in highly accurate stimuli. This was one of the main reasons why PQFT performance was really good. SigSal, SDSR, and GBVS had a relatively high miss rate compared with the aforementioned models; however, N-MODA is above 50%, which is still good in terms of object detection.

The SEEKER images [123] very closely replicate a homogeneous planetary surface, for example Mars. The images have sparse scattering of very large boulders of rocks against a nearly homogeneous background surface along with a small number of smaller rocks. Visual stimuli based on the statistical redundancy of specific class of pixels as well as orientation information provided good saliency fixations. Furthermore, sequential images in the dataset exhibit a very smooth motion pattern, therefore, providing highly reliable temporal information. Examining the results in Table 3.4, the best performing models are in fact the ones that tend to rely on these types of visual stimuli. GBVS, as with other real-world datasets, completely failed to satisfy the acceptable criteria for detection. SigSal was successful to a certain degree; however, its N-MODA measure was still below other models. PQFT outperformed other models in this dataset due to good temporal information, which is readily used by this model.

Saliency map computation time (averaged over the total number of frames) is computed to examine the processing requirements (see Table 3.5) of each model. The experimental workstation used for generating these results was a quad-core Intel(R) Core(TM) i5-2400 CPU (3.10 GHz) running Linux (Ubuntu 10.04, 64-bit architecture). Experimental results show that *SDSR* took the longest

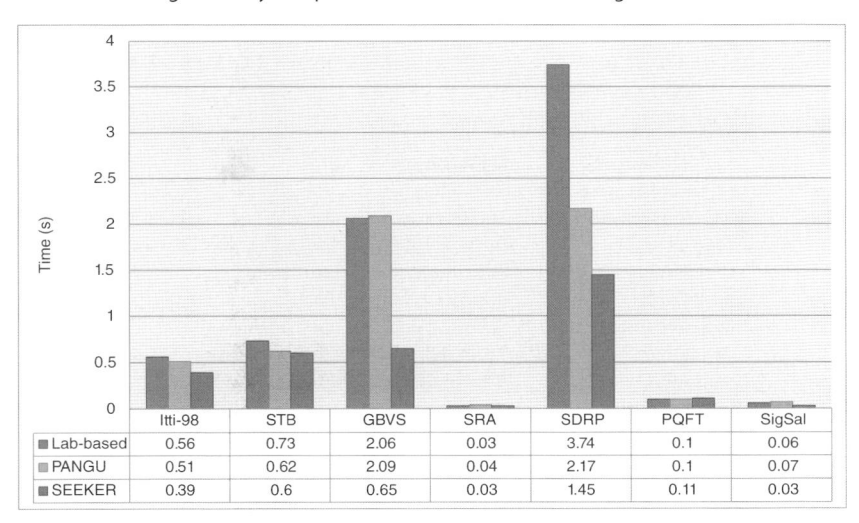

Table 3.5 Average saliency computation time for all models using different test datasets.

	Itti-98	STB	GBVS	SRA	SDRP	PQFT	SigSal
Lab-based	0.56	0.73	2.06	0.03	3.74	0.1	0.06
PANGU	0.51	0.62	2.09	0.04	2.17	0.1	0.07
SEEKER	0.39	0.6	0.65	0.03	1.45	0.11	0.03

time to process an image, whereas *SigSal*, *PQFT* and *SRA* required the lowest processing time.

3.8.2
Vision-Based Rover Sinkage Detection for Soil Characterization

Mars exploration missions over the past few decades have helped us understand the challenges that planetary rovers have to endure during the course of their long-lasting missions. For example, the Mars Exploration Rovers (MERs), Spirit and Opportunity, have undergone several types of difficulties in traversing the Martian terrain. To ensure safety against soft and deformable terrain that can result in mission-threatening situations, such as Spirit's mission, which came to an end when the rover became embedded in soft Martian soil in May 2009 [124], future planetary rover missions would require real-time remote soil sensing methods. One technique proposes to incorporate a small robotic scout rover *in situ* with the main exploration rover that would provide information about soil characteristics ahead of the main rover [125, 126]. This section focuses on the experimental evaluation of the this proposed technique using real-world datasets (see Table 3.6) and the evaluation protocols presented in Refs [127, 128].

The FASTER Scout rover is an agile high-mobility rover with a hybrid legged wheel system [129], which allows for fast traversal and extra agility over highly deformable terrains. The hybrid wheel design of the scout rover allows the use of passive vision-based real-time trafficability analysis of the soil through dynamic leg sinkage detection [128]. The vision system proposed in Refs [127, 128] consists of a single camera fixed below the center of the legged wheel hub, and an image processing algorithm that detects wheel sinkage in different types of soil, sand, or loose gravel.

The algorithm exploits the homogeneous surface textures of the Martian soil by using a color based segmentation technique to delineate the wheel or leg contour from the background deformable terrain. This is achieved by using blue-colored construction material for the wheel, which seems to be a good segregation

Table 3.6 Dataset characteristics and types of terrains [127].

Dataset	Terrain type	Characteristic
1	Very compact sand	Back illumination, gravel
2	Very compact sand	Side illumination, gravel
3	Compact sand	Back illumination, harsh shadows
4	Varying terrain	Side illumination, sand dunes
5	Loose sand	Side illumination, small dunes
6	Medium compact sand	Side illumination, harsh shadows
7	Varying terrain	Front illumination, harsh shadows
8	Loose sand	Top illumination, sun glare
9	Very compact sand	Back illumination, mostly shadowed
10	Compact sand	Front illumination, moving shadows
11	Loose sand	Flat, front illumination
12	Medium compact sand	Sloping down, side illumination

feature against the Martian soil texture [127]. Images captured by the camera are processed for the identification of blue pixels cluster within the image using a color-based segmentation that is based on an *a priori* defined threshold selection criterion.

The resulting binary output image comprises clusters of pixels forming blobs (Regions of Interest – ROIs), which are postprocessed to detect only the largest blob while suppressing the relatively smaller ones. Morphological operators are applied to the resulting image to mitigate noise. The contour of the largest blob is then extracted using a border following method introduced by Suzuki [130]. Image processing is further complemented with an encoder that computes the pose of the wheel or leg such that anything that lies outside the ROI is masked. The sinkage is a measure of the level of occlusion occurred due to a deformable terrain, which is calculated by comparing the size of the detected occluded leg against the actual size of the nonoccluded leg [127], as shown in Figure 3.28.

Multiple tests were conducted during some outdoor field trials presenting a range of challenges to the proposed vision system with varying terrain compactness. Images recorded during these field trials were ground-truth annotated using a manual in house tool for the evaluation of the proposed system. Table 3.7 summarizes the performance evaluation results of the sinkage measurement for the 12 datasets described in Table 3.6. In these results, error is defined as a percentage of the actual sinkage by computing the difference between the detected sinkage and the ground-truth sinkage. Negative error occurs where the algorithm reports

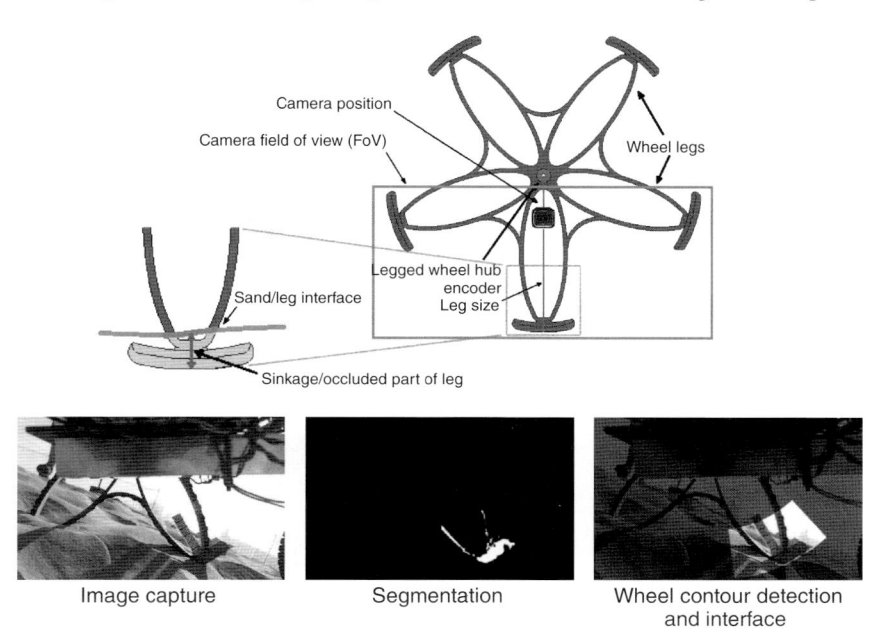

Figure 3.28 Legged wheel sinkage detection in a deformable terrain by measuring the level of occlusion of the locomotion contour [127].

Table 3.7 Average and maximum error for all datasets [125].

Dataset	Maximum positive error (%)	Maximum negative error (%)	RMS error (%)
1	0.58	1.06	0.01
2	0.01	0.29	0.05
3	0.58	0.64	0.03
4	0.61	3.83	0.38
5	0.62	6.17	0.86
6	0.61	0.30	0.08
7	0.93	8.22	0.53
8	0.64	9.82	1.06
9	0.29	1.32	0.24
10	0.01	1.03	0.31
11	0.01	1.79	0.37
12	0.31	0.62	0.18

a deeper sinkage than the actual sinkage and vice versa. Performance evaluations show that the algorithm can detect sinkage with errors below 1% and 10% for positive and negative errors, respectively.

Soil analysis or characterization can be further carried out using the noninvasive imaging techniques, taking the advantage of existing vision system on board the rovers without having to add any dedicated *in situ* soil sensing instruments (such as penetrometers or specially designed wheels). One such technique has been tested on images taken from China's lunar mission Chang'E-3 (CE-3), the former Luna 17 and Apollo 15 landing sites, which uses photoclinometry or shape from shading [4, 131]. The goal in this case is also to retrieve wheel sinkage; hence, the image processing technique applies to rover tracks in order to extract track depth at multiple locations. The output of photoclinometry is a DTM which if viewed in the heading direction of the rover (as shown in Figure 3.29) can be used

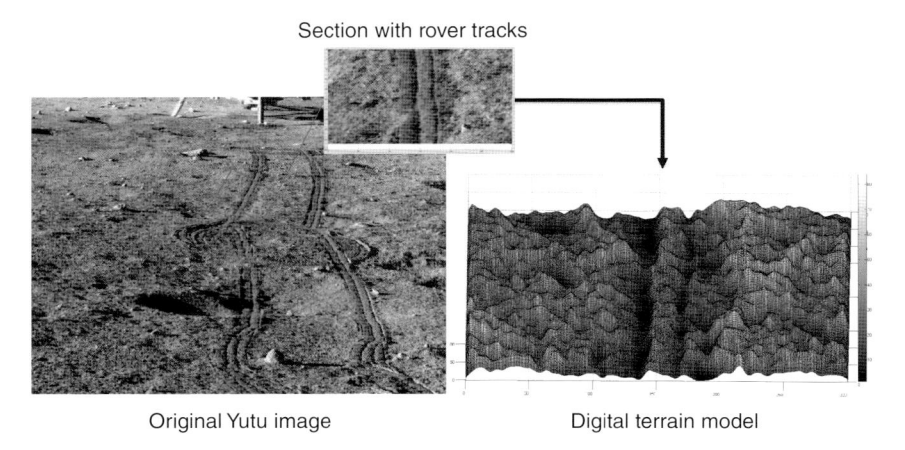

Figure 3.29 DTM generated for image section of the CE-3 Yutu rover tracks.

to calculate wheel sinkage by measuring the distance between the lowest point of the DTM and the surface elevation. The wheel track depth or sinkage value can then be used to estimate the pressure (or stress) versus sinkage relationship (also known as the $\rho - z$ model) where the pressure is induced by the rover weight on the wheels. The $\rho - z$ model can typically represent the soil strength or stiffness, hence is a key measure in soil mechanics that determines whether or not the soil will be stable or how much it will deform. For a planetary rover case, the modern *small wheel model* in terramechanics as shown in Equation 3.3 [132] is applicable, which is a modification of the traditional *Bernstein–Goriatchkin (BG) model* used to estimate pressure-sinkage behaviors of static penetration in soil, for example, based on penetrometer data [133]. This technique has been tested and demonstrated using various track images of the Yutu rover in CE-3, the Lunokhod 1 rover in Luna 17, and the Lunar Roving Vehicle (LRV) in Apollo 15. Using the suggested $n = 0.8$ and $m = 0.39$ values for dry sand in Equation 3.3 applicable to the Moon, the soil stiffness modulus k can be qualified and compared among the three lunar landing sites as illustrated in Figure 3.30

$$\rho = kz^n D^m \tag{3.3}$$

where ρ is the pressure encountered by a rover wheel, z is the wheel track depth or sinkage that retrieved from photoclinometry, D is the wheel diameter with m being the diameter exponent. The parameters k and n are the stiffness modulus and sinkage exponent, respectively.

With regard to Mars missions, vision-based analysis of Martian soil has been previously studied by applying stereopsis to the images from the MER; Spirit and Opportunity, within Gusev crater and at Meridiani Planum. A set of cameras captured wheel tracks left behind by the rovers. Stereo observation of these images manifested wheel sinkage of both rovers, which were studied as indicators of the strength of the soil-like deposits traversed by the rovers [134]. Using wheel-soil theory calibrated to the MER wheel shape characteristics and taking wheel slip into account; the sinkage data were analyzed by performing comparisons with terrestrial soils. Observations of the wheel tracks over the distance traveled provided the basis for studying variations of soil physical properties as function of the spatial scale, surface feature type and local topographical characteristics. The analysis in combination with knowledge gained from studying lunar regolith and

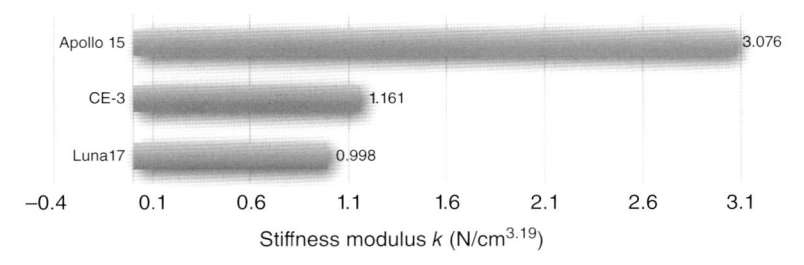

Figure 3.30 Soil stiffness of lunar landing sites based on small wheel model in Equation 3.3 and wheel sinkage data.

dry terrestrial soils allows inference of other properties, such as fine-component thermal inertia and dielectric properties.

3.8.3
Science Autonomy

The robotic exploration of planetary surfaces will be the precursor (or replacement) of human missions within the next few decades. Such exploration will require platforms that are much more self-reliant and capable of exploring long distances with limited ground support in order to advance planetary science objectives in a timely manner. As Steve Squyres (MER Principal Investigator – PI) noted "It takes MER a day to do what a field geologist could do in 45 seconds!" [135]. Autonomous identification and selection of scientifically interesting targets is, therefore, essential to all such missions. The so-called "Science scouting" is required to enhance the science gain of a robotic exploration mission, and contains relevant robotic vision components such as the automatic detection, measurement and assessment of potentially interesting scientific targets, their embedding in the overall onboard map, and important interfaces to feed the onboard decision-making chain.

Technology for performing autonomous commanding of a planetary rover has been successfully demonstrated by the OASIS autonomous science system which provides capabilities for the automated generation of a rover activity plan based on science priorities, the handling of opportunistic science, including new science targets identified by onboard data analysis software as well as other dynamic decision making such as modifying the rover activity plan in response to problems or other state and resource changes [136]. Among others, the MER and MSL missions used and use such techniques for the autonomous detection and recording of clouds and dust devils [137]. The European Framework Programme 7 – Space project PRoViScout realized a demonstrator that implemented science autonomy as a combination between mapping, navigation, science target selection, and a planning agent [138].

3.8.4
Sensor Fusion

Data fusion of different vision sensors into unique data products allows for new ways of data interpretation and rises the scientific value of a mission. Figure 3.6 already showed an example of data fusion between orbiter (HiRISE) and rover (MER-B) image products (3D maps: HiRISE DTM and MER PanCam stereo-derived textured distance map) to demonstrate the ability to fuse sensor products with large differences in resolution and sensor origin. Other contemporary examples are the combination of MER Navcam and Pancam 3D products overlaid with Microscopic Imager (MI) data (Figure 3.31), the embedding of MSL Mastcam 3D products in Navcam reconstructions (Figure 3.32), or the projection of MAHLI texture onto a Mastcam reconstruction (Figure 3.33).

3.8 Advanced Concepts | 167

Figure 3.31 (a) Result of 3D fusion between MER Navcam and Pancam processing, overlaid with Microscopic Imager (MI) Data. (b) 3D rendering of fusion result. (c) Detail showing parts of mesh. (d) Textured result – the MI image shows about 10 times higher surface resolution compared with the Pancam texture. (Data courtesy NASA/JPL-Caltech.)

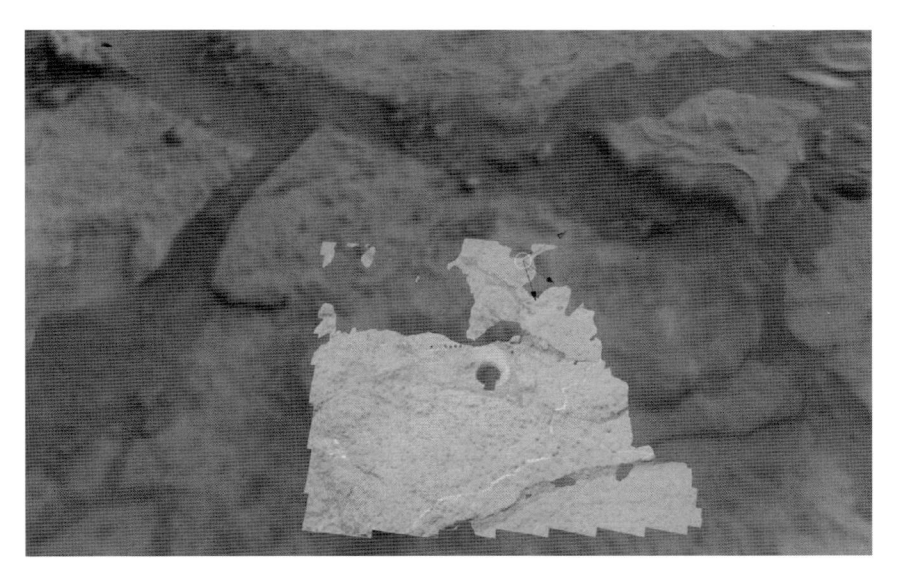

Figure 3.32 Result of fusion between MSL Mastcam and Navcam (taken at MSL Sol 290) processing (textured spherical distance maps, rendered by PRo3D [66]) making use of Exo-Mars PanCam 3D vision processing workflow PRoViP. (Data courtesy NASA/JPL-Caltech.)

Figure 3.33 (Top) Search for corresponding points in MSL Mastcam (a) and MAHLI (b) images. (Bottom) Overlay of co-registered MAHLI image onto Mastcam 3D reconstruction ((c) without overlay, (d) with overlay). (Data courtesy NASA/JPL-Caltech.)

The ExoMars mission will offer further opportunities for fusion between vision sensors and instruments that are mapping features of the scene by means of techniques other than vision: One of the ExoMars sensors is Water Ice and Subsurface Deposit Observation on Mars (WISDOM), a ground penetrating radar that will allow the definition and representation of suitable drilling locations. Utilizing rover motion along a predefined path, it provides 3D profiles of the subsurface containing brightness values that allow a physical interpretation of the subsurface such as individual boulders or humid areas. The PanCam visual data presentation tool PRo3D is able to include 3D visualizations of WISDOM data for a joint presentation in a unique coordinate frame [139], see Figure 3.34. In such way, the interpretation of WISDOM data is directly placed in the structural and textural scene context of the "real" surface obtained from PanCam, which is a valuable immersive basis for drilling decision.[10] One major challenge to such inter-instrument fusion tasks is data co-registration (i.e., to bring the datasets from different sensors into the same unique coordinate frame.). This is accomplished with proper geometric

10) Geologists, physicists, rover operators, and the drill instrument team can use an interactive 3D viewing session making use of the PanCam/WISDOM fusion result to jointly decide upon the drilling strategy such as location, depth and direction.

Figure 3.34 Example for the fusion between 3D surface reconstruction using AU ExoMars PanCam Emulator AUPE [140] and a depth profile from the ExoMars WISDOM mock-up, obtained during the ESA SAFER field trials campaign 2013 in Chile [141] (rendered by PRo3D [66]).

calibration of each sensor individually and with respect to the platform geometric context (Section 3.4), and accurate localization of the platform for each sensing activity (Section 3.3.3.2).

3.8.5
Artificial Intelligence and Cybernetics Vision

Besides the saliency methods pointed out in Section 3.8.1 another promising approach is the use of artificial intelligence and cybernetics vision: in order to achieve complete autonomy for space missions supported by robotic vision without human intervention for complex higher-level decision making and dexterity, future vision systems would require rational-agent based architectures incorporating computational intelligence and learning strategies. Although artificial intelligence is currently implemented in combination with vision systems for autonomous planning, scheduling, or deliberation for navigation, there is a future tendency toward applying more complicated techniques on board, such as pattern recognition, and learning techniques. Most of these processes are currently performed offline on downlinked data. The use of supervised and unsupervised learning techniques used for pattern recognition and object classification [142] can help perform a lot of scientific and geological analysis on board a rover without the need for any Earthbound study. Furthermore, with the increase in computational resources for future planetary rovers, the use of more sophisticated vision systems that allows human-like interpretation of the visual scene, such as cognitive vision and perception-action systems [143–145], would be possible in space missions. To this end, the use of such systems has been limited to terrestrial applications, such as industrial automation, visual communication, and medical imaging. In contrast to classical robotics vision models, where scene representation (perception) is carried out explicitly in geometric terms before actions, cognitive vision utilizing perception-action models build up models of structures relating to perceptual domain to the agent's actions, thus actions of the agent precede perception (or scene representation) [146]. Such vision paradigms have great potential for robustness in unknown environments as well as adaptation to unforeseen situations.

References

1. Matthies, L., Gat, E., Harrison, R., Wilcox, B., Volpe, R., and Litwin, T. (1995) Mars microrover navigation: performance evaluation and enhancement. *Autonomous Robots*, **2** (4), 291–311.
2. Matthies, L., Maimone, M., Johnson, A., Cheng, Y., Willson, R., Villalpando, C., Goldberg, S., Huertas, A., Stein, A., and Angelova, A. (2007) Computer vision on Mars. *International Journal of Computer Vision*, **75** (1), 67–92.
3. *NASA Microsoft Collaboration Will Allow Scientists to 'Work on Mars'* (2015), http://www.jpl.nasa.gov/news/news.php?feature=4451 (accessed 04 April 2016).
4. Gao, Y., Spiteri, C., Pham, M.-T., and Al-Milli, S. (2014) A survey on recent object detection techniques useful for monocular vision-based planetary terrain classification. *Robotics and Autonomous Systems*, **62** (2), 151–167.
5. Lemmon, M.T. (2011) Atmospheric opacity, surface insolation, and dust lifting over 3 Mars years with the Mars Exploration Rovers. *EPSC-DPS Joint Meeting 2011*, vol. 1, p. 1503.
6. Gunn, M.D. and Cousins, C.R. (2013) Mars surface context cameras past, present and future. *Earth and Space Science*, ISSN: 2333-5084, http://dx.doi.org/10.1002/2016EA000166, doi: 10.1002/2016EA000166.
7. Murray, B.C. and Davies, M.E. (1970) Space photography and the exploration of Mars. *Applied Optics*, **9** (6), 1270–1281.
8. Bell, J.F. III,, Calvin, W.M., Farrand, W.H., Greeley, R., Johnson, J.R., Jolliff, B., Morris, R.V., Sullivan, R.J., Thompson, S., Wang, A., Weitz, C., and Squyres, S.W. (2008) *The Martian Surface (Cambridge Planetary Science)*, Chapter 9, Cambridge University Press, pp. 281–314, ISBN: 9780511536076.
9. Svitek, T. and Murray, B. (1990) Winter frost at Viking Lander 2 site. *Journal of Geophysical Research*, **95** (2), 1495–1510.
10. Weitz, C.M., Anderson, R.C., Bell, J.F., Farrand, W.H., Herkenhoff, K.E., Johnson, J.R., Jolliff, B.L., Morris, R.V., Squyres, S.W., and Sullivan, R.J. (2006) Soil grain analyses at Meridiani Planum, Mars. *Journal of Geophysical Research*, **111** (12), 2156–2202.
11. Smith, P.H., Tamppari, L.K., Arvidson, R.E., Bass, D., Blaney, D., Boynton, W.V., Carswell, A., Catling, D.C., Clark, B.C., Duck, T., DeJong, E., Fisher, D., Goetz, W., Gunnlaugsson, H.P., Hecht, M.H., Hipkin, V., Hoffman, J., Hviid, S.F., Keller, H.U., Kounaves, S.P., Lange, C.F., Lemmon, M.T., Madsen, M.B., Markiewicz, W.J., Marshall, J., Mckay, C.P., Mellon, M.T., Ming, D.W., Morris, R.V., Pike, W.T., Renno, N., Staufer, U., Stoker, C., Taylor, P., Whiteway, J.A., and Zent, A.P. (2009) H2O at the Phoenix landing site. *Science*, **325**, 58–61.
12. Arvidson, R., Gooding, J., and Moore, H. (1989) The Martian surface as imaged, sampled, and analyzed by the Viking Landers. *Reviews of Geophysics*, **27** (1), 39–60.
13. Smith, P.H., Tomasko, M.G., Britt, D., Crowe, D.G., Reid, R., Keller, H.U., Thomas, N., Gliem, F., Rueffer, P.,

Sullivan, R., Greeley, R., Knudsen, J.M., Madsen, M.B., Gunnlaugsson, H.P., Hviid, S.F., Goetz, W., Soderblom, L.A., Gaddis, L., and Kirk, R. (1997) The imager for Mars pathfinder experiment. *Journal of Geophysical Research*, **102** (E), 4003–4026.

14. Lemmon, M., Smith, P., Nohara, C.S., Tanner, R., Woida, P., Shaw, A., Hughes, J., Reynolds, R., Woida, R., Penegor, J., Oquest, C., Hviid, S., Madsen, M., Olsen, M., Leer, K., Drube, L., Morris, R., and Britt, D. (2008) The Phoenix surface stereo imager (SSI) investigation. *LPSC08, 39th Lunar and Planetary Science Conference, (Lunar and Planetary Science XXXIX), League City, TX, Number LPI Contribution No. 1391*, p. 2156.

15. Eisenman, A.R., Liebe, C.C., Maimone, M., Schwochert, M.A., and Willson, R.G. (2001) Mars exploration rover engineering cameras. *SPIE 4540 - International Symposium on Remote Sensing, Sensors, Systems, and Next-Generation Satellites V, December, SPIE*, pp. 288–297.

16. Maki, J., Thiessen, D., Pourangi, A., Kobzeff, P., Litwin, T., Scherr, L., Elliott, S., Dingizian, A., and Maimone, M. (2012) The Mars science laboratory engineering cameras. *Space Science Reviews*, **170** (1-4), 77–93.

17. Bell, J., Squyres, S., Herkenhoff, K., Maki, J., Arneson, H., Brown, D., Collins, S., Dingizian, A., Elliot, S., Hagerott, E., Hayes, A., Johnson, M., Johnson, J., Joseph, J., Kinch, K., Lemmon, M., Morris, R., Scherr, L., Schwochert, M., Shepard, M., Smith, G., Sohl-Dickstein, J., Sullivan, R., Sullivan, W., and Wadsworth, M. (2003) Mars Exploration Rover Athena panoramic camera (Pancam) investigation. *Journal of Geophysical Research: Planets*, **108** (12), 8063.

18. Malin, M.C., Caplinger, M.A., Edgett, K.S., Ghaemi, F.T., Ravine, M.A., Schaffner, J.A., Baker, J.D., Bardis, J.M., Dibiase, D.R., Maki, J.N., Willson, R.G., Bell, J.F., Dietrich, W.E., Edwards, L.J., Hallet, B., Herkenhoff, K.E., Heydari, E., Kah, L.C., Lemmon, M.T., Minitti, M.E., Olson, T.S., Parker, T.J., Rowland, S.K., Schieber, J., Sullivan, R.J., Sumner, D.Y., Thomas, P.C., and Yingst, R.A. (2010) The Mars Science Laboratory (MSL) mast-mounted cameras (Mastcams) flight instruments. *41st Lunar and Planetary Science Conference, The Woodlands, TX, Number LPI Contribution No. 1533*, p. 1123.

19. Yuen, P., Gao, Y., Griffiths, A., Coates, A., Muller, J.-P., Smith, A., Walton, D., Leff, C., Hancock, B., and Shin, D. (2013) Exo Mars rover PanCam: autonomy and computational intelligence. *IEEE Computational Intelligence Magazine*, **8** (4), 52–61.

20. Kruse, F.A. (2012) Mapping surface mineralogy using imaging spectrometry. *Geomorphology*, **137** (1), 41–56.

21. Edgett, K., Yingst, R., Ravine, M.A., Caplinger, M., Maki, J.N., Ghaemi, F.T., Schaffner, J.A., Bell, J., Edwards, L.J., Herkenhoff, K.E., Heydari, E., Kah, L.C., Lemmon, M.T., Minitti, M.E., Olson, T.S., Parker, T.J., Rowland, S.K., Schieber, J., Sullivan, R.J., Sumner, D.Y., Thomas, P.C., Jensen, E.H., Simmonds, J.J., Sengstacken, A.J., Willson, R.G., and Goetz, W. (2012) Curiosity's Mars hand lens imager (MAHLI) investigation. *Space Science Reviews*, **170**, 259–317.

22. Wang, W., Li, C., Tollner, E., and Rains, G. (2012) A liquid crystal tunable filter based shortwave infrared spectral imaging system: design and integration. *Computers and Electronics in Agriculture*, **80**, 126–134.

23. He, Z., Shu, R., and Wang, J. (2011) Imaging spectrometer based on AOTF and its prospects in deep-space exploration application. *SPIE 8196 International Symposium on Photoelectronic Detection and Imaging*, vol. 8196, pp. 819625-819625-7.

24. Schmitz, N., Jaumann, R., Coates, A.J., Griffiths, A.D., Leff, C.E., Hancock, B.K., Josset, J.L., Barnes, D.P., Tyler, L., Gunn, M., Paar, G., Bauer, A., Cousins, C.R., Trauthan, F., Michaelis, H., Mosebach, H., Gutruf, S., Koncz, A., Pforte, B., Kachlicki, J., Terzer, R., and the Exo Mars PanCam team (2014) PanCam on the Exo Mars 2018

rover: a stereo, multispectral and high-resolution camera system to investigate the surface of Mars. *International Workshop on Instrumentation for Planetary Missions (IPM-2014), Greenbelt, MD* [Online]. Available: http://ssed.gsfc.NASA.gov/IPM/PDF/1053.pdf (accessed 14 August 2015).

25. Bell, J.F. III,, Maki, J.N., Mehall, G.L., Ravine, M.A., Caplinger, M.A., and the Mastcam-Z Team (2014) Mastcam-z: a geologic, stereoscopic, and multispectral investigation an the NASA Mars-2020 rover. *International Workshop on Instrumentation for Planetary Missions, IPM-2014), Greenbelt, MD*.

26. Bell, J.F., Joseph, J., Sohl-Dickstein, J.N., Arneson, H.M., Johnson, M.J., Lemmon, M.T., and Savransky, D. (2006) In-flight calibration and performance of the Mars exploration rover panoramic camera (Pancam) instruments. *Journal of Geophysical Research*, **111(E)**, E02S03.

27. Tao, Y. and Muller, J.-P. (2016) A novel method for surface exploration: super-resolution restoration of Mars repeat-pass orbital imagery. *Planetary and Space Science*, **121**, 103–114, doi: 10.1016/j.pss.2015.11.010.

28. Grotzinger, J.P., Crisp, J., Vasavada, A.R., Anderson, R.C., Baker, C.J., Barry, R., Blake, D.F., Conrad, P., Edgett, K.S., Ferdowski, B., Gellert, R., Gilbert, J.B., Golombek, M., Gómez-Elvira, J., Hassler, D.M., Jandura, L., Litvak, M., Mahaffy, P., Maki, J., Meyer, M., Malin, M.C., Mitrofanov, I., Simmonds, J.J., Vaniman, D., Welch, R.V., and Wiens, R.C. (2012) Mars Science Laboratory mission and science investigation. *Space Science Reviews*, **170** (1-4), 5–56.

29. Anabitarte, F., Cobo, A., and Lopez-Higuera, J.M. (2012) Laser-induced breakdown spectroscopy: fundamentals, applications, and challenges. *ISRN Spectroscopy*, **2012** (285240), 12.

30. Lopez-Reyes, G., Rull, F., Venegas, G., Westall, F., Foucher, F., Bost, N., Sanz, A., Catalá-Espí, A., Vegas, A., Hermosilla, I., Sansano, A., and Medina, J. (01 2014) Analysis of the scientific capabilities of the ExoMars Raman Laser Spectrometer instrument. *European Journal of Mineralogy*, **25** (5), 721–733.

31. Komguem, L., Whiteway, J.A., Dickinson, C., Daly, M., and Lemmon, M.T. (2013) Phoenix {LIDAR} measurements of Mars atmospheric dust. *Icarus*, **223** (2), 649–653.

32. Arvidson, R.E., Bonitz, R.G., Robinson, M.L., Carsten, J.L., Volpe, R.A., Trebi-Ollennu, A., Mellon, M.T., Chu, P.C., Davis, K.R., Wilson, J.J. et al., (2009) Results from the Mars Phoenix lander robotic arm experiment. *Journal of Geophysical Research: Planets (1991 2012)*, **114**, E00E02.

33. Heikkilä, J. (2000) Geometric camera calibration using circular control points. *IEEE Transactions on Pattern Analysis and Machine Intelligence*, **22** (10), 1066–1077.

34. Bouguet, J.-Y. (2015) *Camera Calibration Toolbox for Matlab*, California Institute of Technology.

35. Gennery, D.B. (2001) Least-squares camera calibration including lens distortion and automatic editing of calibration points, in *Calibration and Orientation of Cameras in Computer Vision* (eds A. Gruen and T.S. Huang), Springer, Berlin Heidelberg, pp. 123–136.

36. Cucullu, G.C., Zayas, D., Novak, K., and Wu, P. (2014) A curious year on Mars long-term thermal trends for Mars Science Laboratory rover's first Martian year. *Proceedings of the 44th International Conference on Environmental Systems*.

37. Cristello, N., Pereira, V., Deslauriers, A., and Silva, N. (2015) ExoMars Cameras - an input to the rover autonomous mobility system. *Planetary and Terrestrial Mining Sciences Symposium*.

38. Lee, S. and Ro, S. (1996) A self-calibration model for hand-eye systems with motion estimation. *Mathematical and Computer Modelling*, **24** (5), 49–77.

39. Barnes, D. (2013) Mars rover colour vision: generating the true colours of Mars. *Proceedings of ASTRA 2013 Conference, ESA, ESTEC, The Netherlands*.

40. Bell, J.F., Savransky, D., and Wolff, M.J. (2006) Chromaticity of the Martian sky as observed by the Mars exploration rover Pancam instruments. *Journal of Geophysical Research*, **111**, E12S05.
41. Johnson, J.R., Sohl-Dickstein, J., Grundy, W.M., Arvidson, R.E., Bell, J., Christensen, P., Graff, T., Guinness, E.A., Kinch, K., Morris, R. et al., (2006) Radiative transfer modeling of dust-coated Pancam calibration target materials: laboratory visible/near-infrared spectrogoniometry. *Journal of Geophysical Research: Planets (1991–2012)*, **111** (E12), doi: 10.1029/2005je002658.
42. Hayes, A.G., Grotzinger, J.P., Edgar, L.A., Squyres, S.W., Watters, W.A., and Sohl-Dickstein, J. (2011) Reconstruction of eolian bed forms and paleocurrents from cross-bedded strata at Victoria crater, Meridiani Planum, Mars. *Journal of Geophysical Research: Planets (1991-2012)*, **116** (E7), E00F21.
43. Meigs, B.E. and Stine, L.L. (1970) Real-time compression and transmission of Apollo telemetry data. *Journal of Spacecraft and Rockets*, **7** (5), 607–609.
44. Kiely, A. and Klimesh, M. (2003) The ICER progressive wavelet image compressor. The Interplanetary Network Progress Report 42-155. Jet Propulsion Laboratory, Pasadena, CA, pp. 1–46.
45. Weinberger, M.J., Seroussi, G., and Sapiro, G. (2000) The LOCO-I lossless image compression algorithm: principles and standardization into JPEG-LS. *IEEE Transactions on Image Processing*, **9** (8), 1309–1324.
46. McEwen, A.S., Eliason, E.M., Bergstrom, J.W., Bridges, N.T., Hansen, C.J., Delamere, W.A., Grant, J.A., Gulick, V.C., Herkenhoff, K.E., Keszthelyi, L. et al., (2007) Mars reconnaissance orbiter's high resolution imaging science experiment (HiRISE). *Journal of Geophysical Research: Planets (1991–2012)*, **112** (E5), doi: 10.1029/2005je002605.
47. Li, R., Hwangbo, J., Chen, Y., and Di, K. (2011) Rigorous photogrammetric processing of HiRISE stereo imagery for Mars topographic mapping. *IEEE Transactions on Geoscience and Remote Sensing*, **49** (7), 2558–2572.
48. Jaffe, L.D. and Scott, R.F. (1966) Lunar surface strength: implications of Luna 9 landing. *Science*, **153** (3734), 407–408.
49. Filice, A.L. (1967) Observations on the Lunar surface disturbed by the footpads of Surveyor 1. *Journal of Geophysical Research*, **72** (22), 5721–5728.
50. Keller, H.U., Goetz, W., Hartwig, H., Hviid, S.F., Kramm, R., Markiewicz, W.J., Reynolds, R., Shinohara, C., Smith, P., Tanner, R. et al, (2008) Phoenix robotic arm camera. *Journal of Geophysical Research: Planets (1991-2012)*, **113** (E3), doi: 10.1029/2007je003044.
51. Havlena, M., Torii, A., Jancosek, M., and Pajdla, T. (2009) Automatic reconstruction of Mars artifacts. European Planetary Science Congress, p. 280.
52. Barnes, D.P. (2007) The ExoMars rover inspection mirror (RIM): new opportunities for Mars surface science. *Geophysical Research Abstracts*, **9**, 10815.
53. Josset, J.-L., Westall, F., Hofmann, B.A., Spray, J.G., Cockell, C., Kempe, S., Griffiths, A.D., De Sanctis, M.C., Colangeli, L., Koschny, D. et al., (2012) CLUPI, a high-performance imaging system on the ESA-NASA rover of the 2018 ExoMars mission to discover biofabrics on Mars. *EGU General Assembly Conference Abstracts*, vol. 14.
54. Medina, A., Pradalier, C., Paar, G., Merlo, A., Ferraris, S., Mollinedo, L., Colmenarejo, P., and Didot, F. (2011) A servicing rover for planetary outpost assembly. *Proceedings of the 11th Symposium on Advanced Space Technologies in Robotics and Automation (ASTRA)*.
55. Li, R., Archinal, B.A., Arvidson, R.E., Bell, J., Christensen, P., Crumpler, L., Des Marais, D.J., Di, K., Duxbury, T., Golombek, M. et al., (2006) Spirit rover localization and topographic mapping at the landing site of Gusev crater, Mars. *Journal of Geophysical Research: Planets (1991–2012)*, **111** (E2), doi: 10.1029/2005je002483.
56. Lourakis, M. and Hourdakis, E. (2015) Planetary rover absolute localization by combining visual odometry with orbital image measurements. *Proceedings of the 13th Symposium on Advanced*

Space Technologies in Automation and Robotics (ASTRA'15), European Space Agency. ESA/ESTEC, Noordwijk, The Netherlands.

57. Tao, Y. and Muller, J.-P. (2015) Automated localisation of Mars rovers using co-registered HiRISE-CTX-HRSC orthorectified images and DTMs. *Icarus*, in review.

58. Abshire, J.B. and Bufton, J.L. (1992) The Mars observer laser altimeter investigation. *Journal of Geophysical Research*, **97** (E5), 7781–7797.

59. Liu, Z.Q., Di, K.C., Peng, M., Wan, W.H., Liu, B., Li, L.C., Yu, T.Y., Wang, B.F., Zhou, J.L., and Chen, H.M. (2015) High precision landing site mapping and rover localization for Chang'e-3 mission. *Science China Physics, Mechanics & Astronomy*, **58** (1), 1–11.

60. NASA/JPL (2015) PDS Marsviewer, http://pds-imaging.jpl.NASA.gov/tools/marsviewer/ (accessed 6 December 2015).

61. NASA (2015) Maestro, https://software.NASA.gov/featuredsoftware/maestro (accessed 28 October 2015).

62. Fox, J.M. and McCurdy, M. (2007) Activity planning for the Phoenix Mars lander mission. Aerospace Conference, 2007 IEEE, IEEE, pp. 1–13.

63. NASA/JPL (2015) MSLICE, http://www.NASA.gov/centers/ames/research/MSL_operations_prt.htm (accessed 6 December 2015).

64. Cooper, B.K., Maxwell, S.A., Hartman, F.R., Wright, J.R., Yen, J., Toole, N.T., Gorjian, Z., and Morrison, J.C. (2013) Robot sequencing and visualization program (RSVP), http://ntrs.NASA.gov/search.jsp?R=20140001459 (accessed 6 December 2015).

65. Proton, J. (2015) Viewpoint, http://www.planetarysciencecommand.com/ (accessed 6 December 2015).

66. Barnes, R., Paar, G., Traxler, C., Muller, J.P., Tao, Y., Sander, K., Gupta, S., Ortner, T., and Fritz, L. (2015) PRo3D – interactive geologic assessment of planetary 3D vision data products. *Proceedings of International Congress on Stratigraphy (STRATI 2015)*.

67. Poulakis, P., Joudrier, L., Wailliez, S., and Kapellos, K. (2008) 3DROV: a planetary rover system design, simulation and verification tool. *Simulation and Verification Tool. iSAIRAS*.

68. Roberts, D.J., Garcia, A.S., Dodiya, J., Wolff, R., Fairchild, A.J., and Fernando, T. (2015) Collaborative telepresence workspaces for space operation and science. Virtual Reality (VR), 2015 IEEE, IEEE, pp. 275–276.

69. Grotzinger, J.P., Sumner, D.Y., Kah, L.C., Stack, K., Gupta, S., Edgar, L., Rubin, D., Lewis, K., Schieber, J., Mangold, N. et al, (2014) A habitable fluvio-lacustrine environment at Yellowknife bay, Gale crater, Mars. *Science*, **343** (6169), 1242777.

70. Anderson, R., Bridges, J.C., Williams, A., Edgar, L., Ollila, A., Williams, J., Nachon, M., Mangold, N., Fisk, M., Schieber, J. et al., (2015) Chemcam results from the Shaler outcrop in Gale crater, Mars. *Icarus*, **249**, 2–21.

71. NASA/JPL (2015) Jadis, http://www.openchannelfoundation.org/projects/JADIS/ (accessed 28 October 2015).

72. Pariser, O. and Deen, R.G. (2009) A common interface for stereo viewing in various environments. Proceedings of SPIE 7237, Stereoscopic Displays and Applications XX, 72371R, International Society for Optics and Photonics.

73. UCL-MSSL (2015) Stereo workstation, http://sourceforge.net/projects/stereows/ (accessed 6 December 2015).

74. Tao, Y., Shin, D., and Muller, J.-P. (2015) An accuracy evaluation of stereo matching results using manual measurements, in preparation.

75. Parkes, S., Martin, I., Dunstan, M., and Matthews, D. (2004) Planet surface simulation with PANGU. *8th International Conference on Space Operations*, pp. 1–10.

76. Cheng, Y., Maimone, M., and Matthies, L. (2005) Visual odometry on the Mars exploration rovers. *Systems, Man and Cybernetics, 2005 IEEE International Conference on, vol. 1*, IEEE, pp. 903–910.

77. Bajracharya, M., Maimone, M.W., and Helmick, D. (2008) Autonomy for

Mars rovers: past, present, and future. *Computer*, **41** (12), 44–50.
78. Vinogradov, A.P. (1971) *Lunokhod 1 Mobile Lunar Laboratory*, translated by Joint Publications Research Service, JPRS. 54525, distributed by National Technical Information Service, U.S. Department of Commerce.
79. Kozlova, N.A., Zubarev, A.E., Karachevtseva, I.P., Nadezhdina, I.E., Kokhanov, A.A., Patraty, V.D., Mitrokhina, L.A., and Oberst, J. (2014) Some aspects of modern photogrammetric image processing of Soviet Lunokhod panoramas and their implemenation for new studies of Lunar surface. *ISPRS Technical Commission IV Symposium on Geospatial Databases and Location Based Service*, ISPRS.
80. Karachevtseva, I., Oberst, J., Scholten, F., Konopikhin, A., Shingareva, K., Cherepanova, E., Gusakova, E., Haase, I., Peters, O., Plescia, J., and Robinson, M. (2013) Cartography of the lunokhod-1 landing site and traverse from LRO image and stereo-topographic data. *Planetary and Space Science*, **85**, 175–187.
81. Yang, J.-F., Li, C.-L., Xue, B., Ruan, P., Gao, W., Qiao, W.-D., Lu, D., Ma, X.-L., Li, F., He, Y.-H. et al., (2015) Panoramic camera on the Yutu lunar rover of the Chang'e-3 mission. *Research in Astronomy and Astrophysics*, **15** (11), 1867.
82. Xiao, L., Zhu, P., Fang, G., Xiao, Z., Zou, Y., Zhao, J., Zhao, N., Yuan, Y., Qiao, L., Zhang, X. et al., (2015) A young multilayered terrane of the northern Mare Imbrium revealed by Chang'e-3 mission. *Science*, **347** (6227), 1226–1229.
83. Mutch, T.A., Binder, A.B., Huck, F.O., Levinthal, E.C., Morris, E.C., Sagan, C., and Young, A.T. (1972) Imaging experiment: the Viking lander. *Icarus*, **16** (1), 92–110.
84. Huck, F.O., Taylor, G.R., McCall, H.F., and Patterson, W.R. (1975) The Viking Mars lander camera. *Space Science Instrumentation*, **1**, 189–241.
85. Levinthal, E.C., Green, W., Jones, K.L., and Tucker, R. (1977) Processing the Viking lander camera data. *Journal of Geophysical Research*, **82** (28), 4412–4420.
86. NASA/JPL (2015) IMP: Imager for Mars Pathfinder, http://mars.nasa.gov/MPF/mpf/sci_desc.html (accessed 27 February 2016).
87. Oberst, J., Hauber, E., Trauthan, F., Kuschel, M., Giese, B., Roatsch, T., and Jaumann, R. (1998) Mars pathfinder: photogrammetric processing of lander images. *International Archives of Photogrammetry and Remote Sensing*, **32**, 436–443.
88. Kanefsky, B., Parker, T.J., and Cheeseman, P.C. (1998) Super-resolution results from pathfinder IMP. *Lunar and Planetary Science Conference*, **29**, 1536.
89. NASA/JPL (2015) Pathfinder Rover Instrument Description, http://pdsimage.wr.usgs.gov/data/mpfr-m-apxs-2-edr-v1.0/mprv_0001/document/rcinst.htm (accessed 6 December 2015).
90. Alexander, D.A., Deen, R.G., Andres, P.M., Zamani, P., Mortensen, H.B., Chen, A.C., Cayanan, M.K., Hall, J.R., Klochko, V.S., Pariser, O. et al., (2006) Processing of Mars Exploration Rover imagery for science and operations planning. *Journal of Geophysical Research: Planets (1991–2012)*, **111** (E2), doi: 10.1029/2005je002462.
91. Deen, R., Chen, A., and Alexander, D. (2014) Mars Exploration Rover (MER) Software Interface Specification; Camera Experiment Data Record (EDR) and Reduced Data Record (RDR) Operations and Science Data Products, version 4.4, NASA Planetary Data System, http://pds-imaging.jpl.nasa.gov/data/mer/opportunity/mer1do_0xxx/document/CAMSIS_V4-4_7-31-14.PDF (accessed 6 December 2015).
92. Zamani, P., Alexander, D., and Deen, R. (2009) Phoenix Project Software Interface Specification (SIS); Camera Experiment Data Record (EDR) and Reduced Data Record (RDR) Data Products, version 1.1.2., NASA Planetary Data System, http://pds-imaging.jpl.nasa.gov/data/Phoenix/phxmos_0xxx/document/cam_edr_rdr_sis.pdf (accessed 6 December 2015).

93. NASA/JPL (2015) Mars Science Laboratory Project Software Interface Specification(SIS); Camera & LIBS Experiment Data Record (EDR) and Reduced Data Record (RDR) Data Products, version 1.1.2., NASA Planetary Data System, http://pds-imaging.jpl.nasa.gov/data/msl/MSLNAV_0XXX/DOCUMENT/MSL_CAMERA_SIS.PDF (accessed 6 December 2015).
94. Gruen, A. (1985) Adaptive least squares correlation: a powerful image matching technique. *South African Journal of Photogrammetry, Remote Sensing and Cartography*, **14** (3), 175–187.
95. Deen, R.G. and Lorre, J.J. (2005) Seeing in three dimensions: correlation and triangulation of Mars Exploration Rover imagery. *Systems, Man and Cybernetics, 2005 IEEE International Conference on, vol. 1, IEEE*, pp. 911–916.
96. Deen, R.G. (2012) In-situ mosaic production at JPL/MIPL. *Proceedings of Planetary Data: A Workshop for Users and Software Developers, Flagstaff, AZ*.
97. Deen, R.G., Algermissen, S.S., Ruoff, N.A., Chen, A.C., Pariser, O., Capraro, K.S., and Gengl, H.E. (2015) Pointing correction for Mars surface mosaics. *Proceedings of the 2nd Planetary Data Workshop, vol. 1846, Flagstaff, AZ*.
98. NASA/JPL (2013) Billion-Pixel View from Curiosity at Rocknest, http://mars.nasa.gov/multimedia/interactives/billionpixel/ (accessed 6 December 2015).
99. NASA/JPL (2015) VICAR Open Source, http://www-mipl.jpl.nasa.gov/vicar_open.html (accessed 6 December 2015).
100. Zamani, P., Deen, R., and Alexander, D. (2005) Seeing on a budget: Mars rover tactical imaging product generation. *6th International Symposium on Reducing the Costs of Spacecraft Ground Systems and Operations (RCSGSO), ESA/ESOC, Darmstadt, ESA SP-601, European Space Agency*.
101. Paar, G., Griffiths, A.D., Bauer, A., Nunner, T., Schmitz, N., Barnes, D., and Riegler, E. (2009) 3D vision ground processing workflow for the panoramic camera on ESA's ExoMars mission 2016. *Conference Proceedings of Optical 3D 2009*, pp. 1–9.
102. NASA/JPL (2015) PDS; The Planetary Data System, https://pds.nasa.gov/ (accessed 6 December 2015).
103. Bajpai, A., Burroughes, G., Shaukat, A., and Gao, Y. (2015) Planetary monocular simultaneous localization and mapping. *Journal of Field Robotics*, **33**, (2), 229–242, doi: 10.1002/rob.21608.
104. Shaukat, A., Spiteri, C.C., Gao, Y., Al-Milli, S., and Bajpai, A. (2013) Quasi-thematic features detection and tracking for future rover long-distance autonomous navigation. 12th Symposium on Advanced Space Technologies in Robotics and Automation, ESA/ESTEC, Noordwijk, The Netherlands.
105. Villefranche, P., Evans, J., and Faye, F. (1997) Rosetta: the {ESA} comet rendezvous mission. *Acta Astronautica*, **40** (12), 871–877.
106. Kolbe, D. and Best, R. (1997) The {ROSETTA} mission. *Acta Astronautica*, **41** (4-10), 569–577. Developing Business.
107. Keller, H.U., Barbieri, C., Lamy, P., Rickman, H., Rodrigo, R., Wenzel, K.-P., Sierks, H., A'Hearn, M.F., Angrilli, F., Angulo, M., Bailey, M.E., Barthol, P., Barucci, M.A., Bertaux, J.-L., Bianchini, G., Boit, J.-L., Brown, V., Burns, J.A., Buettner, I., Castro, J.M., Cremonese, G., Curdt, W., Da Deppo, V., Debei, S., De Cecco, M., Dohlen, K., Fornasier, S., Fulle, M., Germerott, D., Gliem, F., Guizzo, G.P., Hviid, S.F., Ip, W.-H., Jorda, L., Koschny, D., Kramm, J.R., Kuehrt, E., Kueppers, M., Lara, L.M., Llebaria, A., López, A., López-Jimenez, A., López-Moreno, J., Meller, R., Michalik, H., Michelena, M.D., Mueller, R., Naletto, G., Origné, A., Parzianello, G., Pertile, M., Quintana, C., Ragazzoni, R., Ramous, P., Reiche, K.-U., Reina, M., Rodríguez, J., Rousset, G., Sabau, L., Sanz, A., Sivan, J.-P., Stoeckner, K., Tabero, J., Telljohann, U., Thomas, N., Timon, V., Tomasch, G., Wittrock, T., and Zaccariotto, M. (2007) OSIRIS – the scientific camera system onboard rosetta. *Space Science Reviews*, **128** (1-4), 433–506.

108. Liu, C., Yuen, P.C., and Qiu, G. (2009) Object motion detection using information theoretic spatio-temporal saliency. *Pattern Recognition*, **42** (11), 2897–2906.
109. Yan, J., Zhu, M., Liu, H., and Liu, Y. (2010) Visual saliency detection via sparsity pursuit. *IEEE Signal Processing Letters*, **17** (8), 739–742.
110. Siagian, C. and Itti, L. (2009) Biologically inspired mobile robot vision localization. *IEEE Transactions on Robotics*, **25** (4), 861–873.
111. Borji, A., Sihite, D.N., and Itti, L. (2013) Quantitative analysis of human-model agreement in visual saliency modeling: a comparative study. *IEEE Transactions on Image Processing*, **22** (1), 55–69.
112. Itti, L., Koch, C., and Niebur, E. (1998) A model of saliency-based visual attention for rapid scene analysis. *IEEE Transactions on Pattern Analysis and Machine Intelligence*, **20** (11), 1254–1259.
113. Walther, D. and Koch, C. (2006) Modeling attention to salient proto-objects. *Neural Networks*, **19** (9), 1395–1407.
114. Rensink, R.A. (2000) Seeing, sensing, and scrutinizing. *Vision Research*, **40** (10-12), 1469–1487.
115. Rensink, R.A. (2000) The dynamic representation of scenes. *Visual Cognition*, **7** (1-3), 17–42.
116. Harel, J., Koch, C., and Perona, P. (2007) Graph-based visual saliency, in *Advances in Neural Information Processing Systems*, vol. 19, MIT Press, Cambridge, MA, pp. 545–552.
117. Hou, X. and Zhang, L. (2007) Saliency detection: a spectral residual approach. Computer Vision and Pattern Recognition, 2007. CVPR '07. IEEE Conference on, pp. 1–8.
118. Seo, H.J. and Milanfar, P. (2009) Static and space-time visual saliency detection by self-resemblance. *Journal of vision*, **9** (12), 15.
119. Guo, C. and Zhang, L. (2010) A novel multiresolution spatiotemporal saliency detection model and its applications in image and video compression. *IEEE Transactions on Image Processing*, **19** (1), 185–198.
120. Hou, X., Harel, J., and Koch, C. (2012) Image signature: highlighting sparse salient regions. *IEEE Transactions on Pattern Analysis and Machine Intelligence*, **34** (1), 194–201.
121. Nascimento, J.C. and Marques, J.S. (2006) Performance evaluation of object detection algorithms for video surveillance. *IEEE Transactions on Multimedia*, **8** (4), 761–774.
122. Kasturi, R., Goldgof, D., Soundararajan, P., Manohar, V., Garofolo, J., Bowers, R., Boonstra, M., Korzhova, V., and Zhang, J. (2009) Framework for performance evaluation of face, text, and vehicle detection and tracking in video: data, metrics, and protocol. *IEEE Transactions on Pattern Analysis and Machine Intelligence*, **31** (2), 319–336.
123. Woods, M., Shaw, A., Tidey, E., Pham, B.V., Simon, L., Mukherji, R., Maddison, B., Cross, G., Kisdi, A., Tubby, W. et al., (2014) Seeker-autonomous long-range rover navigation for remote exploration. *Journal of Field Robotics*, **31** (6), 940–968.
124. Matson, J. (2010) Unfree Spirit: NASA's Mars rover appears stuck for good. *Scientific American*, **302** (4), 16.
125. Nevatia, Y.H., Gancet, J., Bulens, F., Voegele, T., Sonsalla, R., Saaj, C.M., Lewinger, W.A., Matthews, M., Yeomans, B., Gao, Y., Allouis, E., Imhof, B., Ransom, S., Richter, L., and Skocki, K. (2013) Improved traversal for planetary rovers through forward acquisition of terrain trafficability. Proceedings of International Conference on Robotics and Automation. o.A., 5.
126. Nevatia, Y., Bulens, F., Gancet, J., Gao, Y., Al-Milli, S., Sonsalla, R.U., Kaupisch, T.P., Fritsche, M., Vogele, T., Allouis, E., Skocki, K., Ransom, S., Saaj, C., Matthews, M., Yeomans, B., and Richter, L. (2013) Safe long-rangetravel for planetary rovers through forward sensing. 12th Symposium on Advanced Space Technologies in Robotics and Automation, European Space Agency, ESTEC, Noordwijk, The Netherlands.

127. Spiteri, C., Al-Milli, S., Gao, Y., and de Leon, A.S. (2015) Real-time visual sinkage detection for planetary rovers. *Robotics and Autonomous Systems*, **72**, 307–317.
128. Al-Milli, S., Spiteri, C., Comin, F., and Gao, Y. (2013) Real-time vision based dynamic sinkage detection for exploration rovers. *Intelligent Robots and Systems (IROS), 2013 IEEE/RSJ International Conference on*, pp. 4675–4680.
129. Sonsalla, R.U., Fritsche, M., Vogele, T., and Kirchner, F. (2013) Concept study for the faster micro scout rover. *12th Symposium on Advanced Space Technologies in Robotics and Automation, European Space Agency, ESTEC, Noordwijk, The Netherlands*.
130. Suzuki, S. and be, K.A. (1985) Topological structural analysis of digitized binary images by border following. *Computer Vision, Graphics, and Image Processing*, **30** (1), 32–46.
131. Zhang, R., Tsai, P.-S., Cryer, J.E., and Shah, M. (1999) Shape-from-shading: a survey. *IEEE Transactions on Pattern Analysis and Machine Intelligence*, **21** (8), 690–706.
132. Meirion-Griffith, G. and Spenko, M. (2011) A modified pressure-sinkage model for small, rigid wheels on deformable terrains. *Journal of Terramechanics*, **48** (2), 149–155.
133. Bekker, M.G. (1956) *Theory of Land Locomotion: The Mechanics of Vehicle Mobility*, University of Michigan Press, Ann Arbor, MI.
134. Richter, L.O. (2005) Inferences of strength of soil deposits along MER rover traverses. AGU Fall Meeting Abstracts.
135. Squyres, S. (2004) Mission to Mars: Risky Business. Astrobiology Magazine.
136. Estlin, T., Gaines, D., Chouinard, C., Castano, R., Bornstein, B., Judd, M., Nesnas, I., and Anderson, R. (2007) Increased Mars rover autonomy using AI planning, scheduling and execution. Robotics and Automation, 2007 IEEE International Conference on, IEEE, pp. 4911–4918.
137. Francis, R. (2015) Autonomous science for exploration missions. Presentation, 21st Improving Space Operations Workshop, Pasadena, https://info.aiaa.org/tac/SMG/SOSTC/Workshop%20Documents/2015/Track%203%20-%20Exploring%20Space%20Using%20Game%20Changing%20Approaches/5_1500_RFrancis%20Improving%20Space%20Ops%20Workshop%20v2.pdf (accessed 6 December 2015).
138. Paar, G., Tyler, L., Barnes, D., Woods, M., Shaw, A., Kapellos, K., Pajdla, T., Medina, A., Pullan, D., Griffiths, A. et al., (2013) The proviscout field trials tenerife 2012–integrated testing of aerobot mapping, rover navigation and science assessment. *12th Symposium on Advanced Space Technologies in Robotics and Automation (ASTRA 2013)*.
139. Paar, G., Hesina, G., Traxler, C., Ciarletti, V., Plettemeier, D., Statz, C., Sander, K., and Nauschnegg, B. (2015) Embedding sensor visualization in Martian Terrain reconstructions. *Proceedings of ASTRA Conference, ESA, ESTEC, The Netherlands*.
140. Pugh, S., Barnes, D., Tyler, L. et al., (2012) AUPE–A PanCam emulator for the ExoMars 2018 mission. *International Symposium on Artificial Intelligence, Robotics and Automation in Space (iSAIRAS)*.
141. Gunes-Lasnet, S., Kisidi, A., van Winnendael, M., Josset, J.-L., Ciarletti, V., Barnes, D., Griffiths, G., Paar, A., Schwenzer, S., Pullan, D., Allouis, E., Waugh, L., Woods, M., Shaw, A., and Diaz, G.C. (2014) SAFER: the promising results of the Mars mission simulation campaign in Atacama, Chile. *I-SAIRAS 2014: 12th International Symposium on Artificial Intelligence, Robotics and Automation in Space, 17-19 June, Montreal, Canada*.
142. Bishop, C.M. (2006) *Pattern Recognition and Machine Learning*, Information Science and Statistics, Springer-Verlag, New York and Secaucus, NJ.
143. Granlund, G. (2006) A cognitive vision architecture integrating neural networks with symbolic processing. *Kuenstliche Intelligenz*, **2**, 18–24.

144. Granlund, G. (1999) The complexity of vision. *Signal Processing*, **74** (1), 101–126.
145. Windridge, D., Felsberg, M., and Shaukat, A. (2013) A framework for hierarchical perception-action learning utilizing fuzzy reasoning. *IEEE Transactions on Cybernetics*, **43** (1), 155–169.
146. Granlund, G.H. (2006) Organization of architectures for cognitive vision systems, in *Cognitive Vision Systems*, Lecture Notes in Computer Science, vol. 3948 (eds H.I. Christensen and H.-H. Nagel), Springer, Berlin Heidelberg, pp. 37–55.

4
Surface Navigation

Peter Iles, Matthias Winter, Nuno Silva, Abhinav Bajpai, Yang Gao, Jan-Peter Muller, and Frank Kirchner

4.1
Introduction

What role does the rover navigation system play in a successful surface mission? A navigation system allows for a rover to explore its surroundings safely and accurately – and it is a rover's ability to perform *in situ* scientific observations in a variety of locations, which is tied to mission success.

This chapter describes rover navigation systems. First, context is provided by describing the challenges of navigating on different extraterrestrial bodies, and describing past and present flight rovers, from the Russian Lunokhods to Curiosity. Next, the design process of a navigation system is introduced through a discussion of requirements and major design concepts. This is followed by a thorough description of navigation technologies including relative and absolute localization as well as autonomous navigation. Finally, the future of planetary rovers is discussed, with a review of planned flight rovers and missions as well as enabling future technologies.

Planetary surface navigation has been the exclusive domain of wheeled rovers, and hence these are focused on in this chapter. Nevertheless, there are other types of mobility platforms that may be used for future missions, such as aerobots, hoppers, and legged robots, which would require navigation capabilities. Many principles and techniques discussed in this chapter are transferable among different platforms, though some important differences are discussed in Section 4.6.

4.2
Context

This section provides context for rover navigation systems. First, important definitions relating to navigation are defined, followed by a discussion of the different challenges of navigation on various planetary bodies. Finally, navigation systems on previous and current flight rovers are described.

Contemporary Planetary Robotics: An Approach Toward Autonomous Systems, First Edition.
Edited by Yang Gao.
© 2016 Wiley-VCH Verlag GmbH & Co. KGaA. Published 2016 by Wiley-VCH Verlag GmbH & Co. KGaA.

4.2.1
Definitions

This section defines key terms relevant to rover navigation systems. These terms are used throughout this chapter.

Pose: The rover position and orientation in a chosen frame of reference. For example, the position can be defined by latitude, longitude, altitude, and the orientation by its heading, roll, and pitch.

Absolute localization: The process of determining the pose of the rover in a global frame of reference (e.g., latitude, longitude, altitude), or in another fixed frame of reference (e.g., with respect to a stationary lander).

Relative localization: The process of determining the pose of the rover with respect to some initial rover pose (e.g., at the start of a traverse).

Autonomy: The degree to which a rover plans and implements paths without human intervention.

Teleoperation: The "direct" control of a rover by a human operator, that is, by sending specific speed and steering commands, as opposed to goal locations.

Guidance: As a component of guidance, navigation, and control (GNC), Guidance is the path-planning competency.

Navigation: As a component of GNC, this is the same as localization.

Control: As a component of GNC, control is the competency of commanding the rover mobility system.

4.2.2
Navigating on Extraterrestrial Worlds

The challenge of navigating a rover on the Moon is different from that on Mars, as it would be on Europa or other bodies. This section discusses the different constraints imposed on a navigation system, in particular, the effects of communication delays, the environment (radiation, thermal, lighting, dust, gravity, magnetic field), the appearance of the sky, and the existence of support assets.

A major consideration for navigation is communication delay. The Moon is approximately 1.3 light-seconds from Earth, meaning that there would be this amount of delay between sending a motion command from Earth and having the rover actually move, and an additional delay of this length before any telemetry from the move is received back on Earth. The delay could also be larger depending on overhead involved with the communication system (e.g., Earth's Deep Space Network, which is used for communication with the Mars rovers). With only a few seconds' control delay, teleoperation of a lunar rover from Earth is possible. This is how the Soviet Lunokhod rovers were controlled (discussed in the following section). Mars is anywhere from approximately 3 to 20 light-*minutes* from Earth, depending on the relative orbital positions of the two planets. This large delay rules out teleoperation and necessitates much more navigation autonomy, as is the case for the Mars Exploration Rovers (MERs) and Curiosity rovers (also described more in the next section).

The rover's environment is the next major navigational consideration. Gravity differences are important not only for mobility but also for navigation: gravity differences can affect wheel slip and, therefore, wheel odometry accuracy. Gravity can also affect obstacle types and sizes: lower gravity on the Moon for example, can result in more impassible cliffs and steeper slopes, due to the high angle of repose of the regolith. Lighting is also very important. Proposed lunar polar missions such as NASA's planned Resource Prospector (RP) mission [1] will have to deal with very low Sun angles and therefore long shadows, flat lighting (when the Sun is behind the rover), glare (the Sun in front of the rover), and large shadowed areas, all of which make path planning and obstacle detection much more challenging. The atmosphere (or lack of it) also affects the navigation system: a dusty atmosphere (such as on Mars) makes celestial navigation difficult, and can change the appearance and apparent size of the Sun, causing issues for Sun-sensing techniques. However, a dusty atmosphere can also benefit navigation in that it would create ambient lighting to illuminate areas shadowed from direct sunlight. The radiation, thermal, and dust environments also affect the navigation system by imposing requirements on hardware such as cameras and processors.

A third major consideration for navigating on different extraterrestrial bodies is the availability of support assets (e.g., orbiters) or data products (e.g., maps). For the Moon, the Lunar Reconnaissance Orbiter (LRO) has been gathering imagery (with the LRO Camera or LROC) and topographic information (via a laser altimeter) since 2009. The imagery and topographic data resolution is finest at the poles where 0.5 m/pixel and 5 m/pixel resolutions are available, respectively [2, 3]. On Mars, the high-resolution imaging science experiment (HiRISE) camera on the Mars Reconnaissance Orbiter (MRO) is gathering imagery down to 25 cm resolution, and this is being processed to create topographic information (from stereo images), down to 1 m resolution [4]. Navigation of the Curiosity rover is made easier due to HiRISE periodically imaging it and its tracks. An orbiter can also act as a step in the communication chain and can increase the overall data bandwidth compared with a "direct-to-Earth" communication system on the rover.

4.2.3
Navigation Systems on Current and Past Flight Rovers

This section describes and compares the navigation systems on past and current flight rovers: Lunokhod I and II, the Apollo Lunar Roving Vehicle (LRV), Sojourner, the Mars Exploration Rovers (MERs), Curiosity, and Yutu.

4.2.3.1 Lunokhod I and II
The Soviet Lunokhod rovers explored the lunar surface in the early 1970s. These eight-wheeled rovers were teleoperated from Earth with feedback from the rover cameras, making them unique among all extraterrestrial rovers. Drivers could command the rover to drive forward and backward (at two speeds), and to rotate [5]. Distance traveled was measured using wheel odometry, with a free-rolling ninth wheel used to measure slip [6]. A gyroscope measured relative changes in

orientation, and a technique involving imaging the Sun and Earth was used periodically to determine absolute orientation [6]. Processing of the navigation telemetry to measure the rover's position and orientation was done on the ground.

Teleoperating Lunokhod I was challenging. Camera views were refreshed only every 20 s, the stereo navigation cameras (NavCam) were found to be too low to the ground and pitched up too high, and the cameras provided poor dynamic range in the harsh lunar lighting [7]. The latter point caused operations to halt for 2 days each lunar noon because the lack of shadows made the scene almost featureless. Driving Lunokhod was a stressful activity. Average speed of the rover was approximately 4.8 m/h toward the end of the mission [5].

Lunokhod II improved on the teleoperation experience over Lunokhod I. The stereo NavCam were moved higher, and the images were refreshed every 3.2 s [8]. Average speed of the rover was 28 m/h, an improvement of about a factor of 6 [8]. Total distances covered by Lunokhod I and II were 10 and 39 km, respectively [9].

4.2.3.2 Apollo Lunar Roving Vehicle

NASA's goal for the Apollo LRV navigation system was for it to be simple, reliable, low-weight, and low-power [10]. The navigation system also had to be easy to use, as it would be used directly by astronauts that would have many other things vying for their attention. The system was required to enable navigation to a planned location, and to allow for navigation back to the lander. Components of the system included a gyroscope for orientation tracking, odometers on each wheel to track distance, and a simple Sun shadow device to initialize the heading measurement [10]. The most distance covered by the LRV was during Apollo 17, at almost 36 km over the course of three extra-vehicular activities (EVAs) [9].

4.2.3.3 Sojourner Microrover

Sojourner was NASA's first unmanned vehicle on another planet, and the first rover to employ autonomous navigation and obstacle-avoidance techniques. The navigation system is described in detail here [11]. In summary, the system was implemented as follows.

The rover had a modest localization system. Wheel odometers in each wheel were averaged to measure the distance traveled and a gyroscope measured the orientation. Accelerometers measured tilt. Position error was approximately 5–10% of distance traveled. A stereo camera on the Pathfinder lander provided information to the rover operators about the true pose of the rover.

Autonomous navigation was done by sending "Go to Waypoint" commands to the rover. The rover operators commanded the rover once per Martian sol, and would chart a path (series of waypoints) to a viewable goal location that would avoid any visible obstacles. The rover detected obstacles in its path by using a laser to project stripes on the terrain; based on the appearance of the stripes, rocks, drop-offs, or steep slopes could be detected (see Figure 4.1). There were also contact sensors on the rover bumpers. A sensing cycle occurred every 7 cm of travel.

Figure 4.1 Image from Sojourner showing a laser stripe used for obstacle detection. (Courtesy NASA/JPL-Caltech.)

The rover traveled approximately 2–3 m/sol, and 100 m over the course of the mission.

4.2.3.4 Mars Exploration Rovers

NASA's MERs, Spirit and Opportunity, dramatically raised the bar for autonomous navigation capabilities of an extraterrestrial robot. This was necessitated by the requirement to navigate 100 m in a single day (Martian sol), with only one control input from Earth [12]: this distance being much further on the terrain than could properly be assessed via MER imagery from the rover starting point each sol.

Localization on the rovers is performed using wheel odometry, an Inertial Measurement Unit (IMU), and optionally visual odometry (VO) that is applied between stereo images taken at the start and end of short traverses [12]. Visual odometry was included in the system to address the requirement that the rovers could localize themselves to within 10% of the path length over a 100 m traverse [13]. Visual odometry is also used to detect slip (e.g., in soft soil), which would not be measured by wheel odometry, and could result in the rover getting stuck. The MERs' visual odometry capability is described in more detail in Section 4.4.2.2. Absolute heading is measured by imaging the Sun using the narrow-field-of-view Panoramic Camera (PanCam) stereo camera outfitted with a solar filter [14].

Offline processing of the MERs' imagery improves the onboard localization estimate. Two techniques are used: an incremental bundle adjustment (IBA) algorithm for relative localization and a rover-to-orbiter photo matching for absolute localization. These techniques are described here [15], and are summarized in the two following paragraphs. These techniques are normally done once per sol [16].

The IBA algorithm links all Navigation Camera (NavCam) and PanCam imagery gathered by the rover to estimate the rover positions along the traverse. Panoramic images with these two cameras are typically gathered twice per sol, with a forward- and backward-looking set often captured in the middle of a sol (see Figure 4.2). Features ("tie points," e.g., rocks) common to neighboring stereo pairs (i.e., those pairs that are separated by a mast yaw rotation) are automatically linked. Tie points

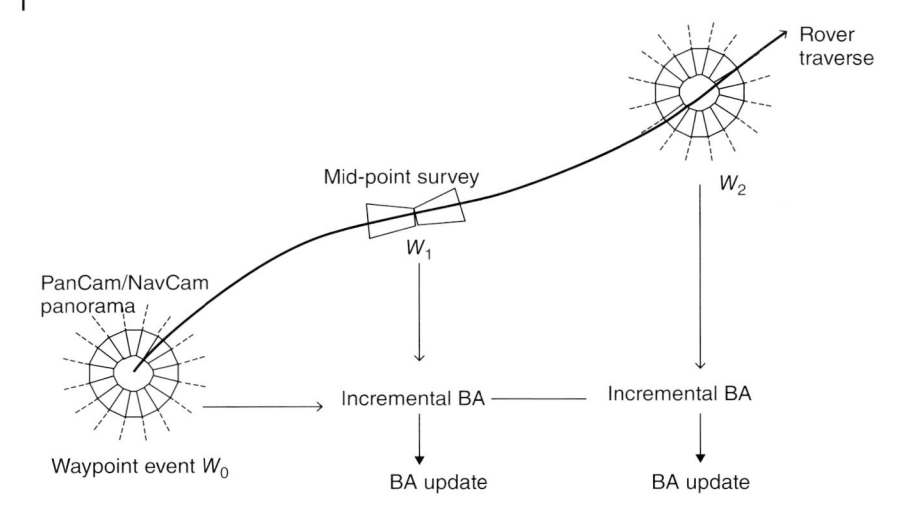

Figure 4.2 Construction of network of rover images for iterative bundle adjustment algorithm on the MERs [15]. (Courtesy American Geophysical Union.)

between images gathered from different sites are selected manually or with an automated algorithm (with a 70% success rate). The automated method has been used in MER operations since 2007. A solution for rover positions along the traverse is found by finding a best agreement for the tie-point locations among all the images. A preflight field trial of this technique on Earth demonstrated a localization accuracy of 0.2% of distance traveled (measured against global positioning system (GPS)).

A rover-to-orbiter image-matching algorithm provides absolute localization of the rovers. The orbiter data are orthorectified HiRISE imagery at approximately 30 cm resolution, which, to improve horizontal accuracy, is integrated with a digital elevation model (DEM) generated from Mars Orbiter Laser Altimeter (MOLA) data. The rover panoramic images described earlier are also converted into "orthoimages," to give the same from-above viewpoint as the orbital images. Features in the rover imagery (e.g., ridges, rocks, craters) are manually matched to the same features in the orbital imagery (see Figure 4.7). Positional accuracy of this method can be checked whenever rover tracks are visible in the orbital images. The average distance from the estimated position to the tracks is 0.75 m.

A direct comparison of these two offline algorithms yields approximately a 1.5% difference as a function of traveled distance [15], an error that is likely dominated by errors inherent in creating the orbital orthoimages. Performing an alignment between the two paths yields a roughly constant error of 3.8 m, on average.

Path-planning is mostly a manual process (though autonomy aids in obstacle avoidance, as described in following paragraphs). Short-term paths, out to approximately 20 m, are selected from NavCam stereo imagery, whereas medium- and long-term paths are selected from HiRISE imagery from the MRO. HiRISE imagery only became available about 3 years into the MER mission, before which time rover images were used for all path-planning activities [17].

The rovers can also autonomously drive to a specified waypoint. The autonomy system is described in detail in Ref. [12], but is summarized here. A 3D representation of the terrain is gathered using one or multiple image pairs from the stereo 45° NavCams or 120° Hazard Cameras (HazCams). The Terrain Assessment module analyzes this data, looking for hazards such as excessive slopes, steps, or obstacles. The output is a grid-based traversability map, in which each cell of the grid contains a traversability score for the rover if it were centered at that location. This local map is then merged into an existing map, and cropped to only keep data within 5 m of the rover. With this map, the Path Selection module then plans a path to the goal, weighting candidate paths (arcs) by how well each path makes progress to the goal and by the traversability values on the path. Path selection uses a greedy algorithm and only plans one or two steps ahead [18] (although this has been upgraded, as described later). This procedure is repeated at each motion step (up to 2 m). Safeguarding the rover during these traverses are real-time checks of vehicle tilt, motor current, and obstacles (using the HazCams) to detect dangerous situations. Also, human operators can specify "keep-out zones" to keep the rover in a specified area. A visualization of the terrain assessment is shown in Figure 4.3.

The MERs can actually be commanded using one of many control modes, thus allowing the most appropriate mode to be used given the current circumstance.

Figure 4.3 Visualization of MER terrain evaluation. Darker squares indicate decreasing traversability. (Courtesy NASA/JPL-Caltech)

Also described here [12], these include "Directed Driving" (i.e., move 5 m forward) or "Blind Goto Waypoint" (i.e., go to a particular location, but have the rover blindly select the path) in benign terrains, as well as "Autonav" (i.e., go to a particular waypoint, but have the rover select the path based on terrain assessment) in unknown terrains. Often the daily command sequence includes a "Directed Driving" command to put the rover through the terrain that can be assessed, followed by an "Autonav" command to extend the driving of the rover into unknown terrain. This control framework has worked very well on the MERs, allowing them to drive in excess of their required 100 m in a single day (the maximum being 124 m). Visual odometry can be turned on or off independently, adding to the localization accuracy for any of these modes.

In-mission flight software upgrades have added new capabilities to the MERs' navigation system, as described in Ref. [18]. The first of these upgrades was visual target tracking, by which a goal location (e.g., a rock) can be specified by its appearance, rather than its location. The rationale is that the location of the target has potentially a large uncertainty due to the increase of stereo errors with range. The second enhancement was addition of the "go-and-touch" mode, allowing the rover to make a short drive and place an instrument on an object of interest. Third and finally was a global path selection ability (to be used past the initial ground-planned route), which augments the Path Selection module described earlier. This global planner is based on the Field D* algorithm [19] and uses a much larger map of 50 m × 50 m size.

4.2.3.5 Mars Science Laboratory/Curiosity

The Curiosity rover navigation system inherited much navigation hardware and many navigation capabilities from the MERs but also has incorporated evolutionary improvements to the same. From the MERs are Curiosity's engineering cameras (HazCams and NavCams), its IMU and orientation estimation framework, its hazard detection system, and its autonomous software [20]. Same as the MERs, communication constraints from the Deep Space Network limit communication to once a day "up" and once or twice a day "down" [21], which necessitates significant autonomy on the rover. This section describes the Curiosity navigation system in the context of the MER system described in the previous section.

There are however, important differences between the MERs' and Curiosity's navigation-related hardware. Curiosity's NavCams, while build-to-print versions of the MERs NavCams, are much higher above the ground: nearly 7 ft, compared with 5 ft. Curiosity's NavCams also have a much larger stereo baseline: 42 cm compared with 20 cm [22, 23]. The higher vantage point allows Curiosity to see further and over larger obstacles, and the larger baseline allows Curiosity to produce stereo data out to almost 100 m, compared with around 25 m for the MERs [22, 23]. Curiosity's processor is also approximately 10 times the speed of the MERs' processors [20] allowing processes such as visual odometry to run in much less time. The locations of Curiosity's NavCams and HazCams (as well as other cameras) are shown in Figure 4.4.

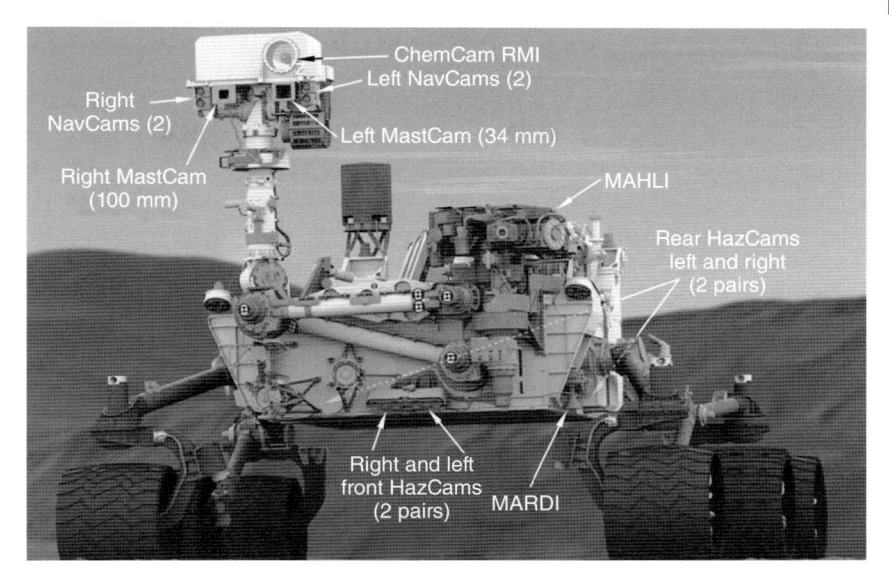

Figure 4.4 Cameras on the Curiosity rover. The NavCams and HazCams are primarily engineering (navigation) sensors. (Courtesy NASA/JPL-Caltech.)

Same as for the MERs, Curiosity depends on wheel odometry, its IMU, visual odometry, and an offline IBA technique to relatively localize itself. Visual odometry is used much more frequently than for the MERs, thanks to the reduced processing time. More detail of Curiosity's VO capability is given in Section 4.4.2.2. Slip detection is also achieved in a new way (in addition to using VO): the rover wheel treads leave a pattern of soil that can be manually inspected in imagery to check for slipping [24].

Curiosity uses the same technique for absolute localization as the MERs: panoramas gathered from rover images are manually aligned to images and DEMs gathered by the HiRISE cameras on the MRO [16]. The rover images come from the NavCams or Mast Cameras (MastCams), and are in the format of panoramas or DEMs (from stereo) that are converted to overhead projections. NavCam DEMs are useful to a range of approximately 50 m, a limit that is due to the shallow viewing angle [16]. Initial absolute localization (of the landing site) was achieved via the same technique, but using the MARDI (descent imager) images.

Path planning for Curiosity works in much the same way as for the MERs: the short-term path is selected manually out to the end of the NavCam stereo mesh (typically 20–40 m, but sometimes out to 80–90 m), whereas long-term paths are chosen using HiRISE imagery [21]. An exception to this occurred in May 2015 when mission planners spotted an interesting feature from 200 m away using the high-zoom MastCam; this caused them to change the planned path to go and investigate [17].

Curiosity's autonomy software also operates very similarly to the MERs, with differences largely due to Curiosity's hardware differences. As described earlier, paths are manually planned to the end of the NavCam mesh, but a global path planner (Field D*) can be used to efficiently reach targets beyond this range [20, 21]. Each autonomous drive step is approximately 0.5–1.5 m; before each drive, the rover assesses the terrain by panning its NavCams to capture four image pairs [25, 26]. The terrain is evaluated out to 7 m away, and checks are done for dangerous slopes, steps, and roughness [20]. The cruising speed is about 5 cm/s, and each pause takes around 2 min [27]. Benign terrain thus far in Curiosity's travels has resulted in autonomous mode only being demonstrated, not used routinely [20].

The set of control modes for Curiosity are enhanced from the MERs, a result of the increased processing capability. This set of modes was described in this presentation [21], and is summarized here. "Directed Driving" is still available, which has the rover blindly follow a specified path. This is also the fastest mode for the rover, with the only safety checks being on the motor suspensions. Curiosity also has multiple visual odometry modes. "Full VO" performs VO every 1 m or so in order to maximize relative localization accuracy and protect against slip. This was the second-most popular mode as of June 2014. The "Slip Check" is just used to detect slip, and does this by performing VO on a short movement every 20 m or so. The last VO mode, "Auto VO" is the same as "Slip Check" mode, but switches into "Full VO" mode if slip is detected. This mode has worked very well to only use VO when necessary, and as of June 2014, is the most popular method. "Autonomous" mode is used to extend driving distance past stereo range, and detects obstacles using the HazCams and NavCams. Visual Target Tracking is also available, as is multiday plans, for example, blind on day 1, auto on day 2 [21]. The control mode breakdown as of sol 740 (September 5, 2014) was "VO Full" 40%, "Auto VO": 40%, "Directed Drive": 10%, "Autonomous": 8%, and others 2% [25].

As of writing, the Curiosity rover has been operating on Mars for just over 1040 sols (Martian days) and has covered approximately 11.5 km of terrain.

4.2.3.6 Yutu/Jade Rabbit

China's Yutu ("Jade Rabbit") rover landed on the Moon in 2013. The rover suffered a mechanical malfunction before traveling away from the vicinity of the lander. The rover had a stereo camera intended for use for path planning and obstacle avoidance [28].

4.3
Designing a Navigation System

This section discusses key factors to consider in designing a rover navigation system, including requirements and important design considerations. Both of these topics are further discussed to explore the navigation system design space.

4.3.1 Requirements

This section discusses the types of requirements that can be imposed on a rover navigation system, and explains why they are important. These include performance requirements, environmental requirements, resource requirements, and others such as redundancy and interfaces.

4.3.1.1 Performance Requirements

Performance requirements for navigation determine how well a rover knows its location, how well it can reach a particular location, and whether the rover can do these tasks autonomously. This section discusses relative and global localization, control, and autonomous performance.

Relative localization is a critical capability of a rover as this determines how well the rover pose is known with respect to the terrain around it. Relative localization performance affects the rover's ability to accurately carry out requested movements (e.g., turns or straight drives), autonomously get to goal locations, avoid obstacles, and detect slip. Relative localization performance is primarily defined as a percentage of distance traveled. Also important is the update rate of the relative pose value: if pose information is refreshed at too low a rate, than control of the rover motion becomes more difficult, and off-nominal situations such as slip are harder to detect. A related requirement is processing time or complexity: this affects whether the rover can move continuously while estimating its pose, or if it must pause to perform localization calculations. Largely due to limitations of the onboard processor, Curiosity, for example, typically drives for 10–20 s before stopping and processing for 2 min [27].

Absolute localization – the knowledge of the pose of a rover with respect to a fixed (e.g. global) frame of reference – is important for path planning. Absolute localization estimates are used to update drift-prone localization estimates from the relative localization system. This performance is often specified in meters. Update rate and processing time are also important for absolute localization, though this capability is employed with a much lower frequency than a rover's relative localization system, and the processing is often done on the ground where processing power is effectively unlimited.

Tightly coupled with localization is driving control performance. This is the ability of a rover to minimize the difference between planned and estimated pose while it is traversing. This capability aids the autonomous system to implement motion faithful to the desired motion, which increases a rover's ability to navigate through challenging terrain. Coupled with localization performance, accurate control is important to enable science capabilities of the rover, for example, to get the rover sufficiently close to an interesting rock to allow for different types of close-proximity analysis. A control performance requirement can be specified in meters.

Autonomous performance enables a rover to move with little human intervention; this is critically important for the Mars rovers that are only sent control scripts once per day, but not necessary for teleoperated rovers such as Lunokhod I and II. Requirements for autonomous performance include the ability to plan a global path to a waypoint, the ability to plan a local path (e.g., to avoid an obstacle immediately in front of the rover), the ability to guide the rover on the selected path, the minimum size of obstacles that can be detected, and of course, processing complexity. Also in this category are requirements for types of control: for example, the ability of the rover to completely autonomously travel to a waypoint, or the ability of a rover to blindly follow a designated path selected by the operators.

4.3.1.2 Environmental Requirements

Environmental requirements must be considered by all rover subsystems, but the environment affects the navigation system in unique ways. The environmental requirements discussed here are thermal/vacuum, vibration, radiation, dust, and lighting.

Thermal/vacuum requirements will be imposed on navigation hardware and also affect the navigation system at a performance level. For hardware, these requirements can be challenging to meet, particularly for components such as mast-mounted stereo cameras that are far away from any temperature-controlled areas (e.g., in the rover chassis). In addition, these cameras must not only survive and operate in these large temperature ranges but must also maintain proper calibration. Other hardware such as an inertial measurement unit (IMU) may have much more relaxed temperature requirements as this unit can typically be placed in the rover chassis within which there may be compartments that are thermally maintained to a more narrow temperature range. There may also be two sets of temperature requirements: one for in-transit and one for on the surface. At a system level, the navigation system will have requirements defining the thermal envelope in which its performance requirements must be met.

Vibration requirements will be imposed not only on navigation hardware but also on navigation system performance. Typically, the most severe vibration occurs during launch (or during aerobraking or landing), but vibration requirements may also be imposed on the navigation system that requires specific performance considering vibration due to rover driving.

Navigation hardware must also meet radiation requirements. These requirements typically fall into two categories: total-dose and single-event upsets. Long-term exposure to cosmic radiation can degrade the performance of optical components (e.g., camera detectors) and processors, and hence components must be required to perform at a certain level given a specific total radiation dose. Single-event upsets are instantaneous radiation effects – for example, a component can "latch up" or data can be corrupted. Strategies to manage the effects of radiation include selecting radiation-tolerant components, providing shielding for sensitive components, designing software to detect corrupt data (e.g., by duplicating processing on different processors), and selecting components that can restart to clear the effects of single-event upsets. Thanks to the thin Martian atmosphere, the radiation environment on Mars is much more benign than that on the Moon, though rovers destined for either world must survive the harsh radiation environment experienced during transit.

Dust is also of particular importance to rover navigation. Dust creates wear on parts and can occlude optical components. On the Moon, the dust is extremely sharp due to the lack of erosion processes. On Mars, the dust hangs in the atmosphere and can be blown around in severe storms. Navigation hardware will be required to survive and perform in the presence of dust (e.g., requiring sealing, careful material selection, or other mitigating techniques), but the system must also meet its performance requirements in these conditions. For example, a derived requirement may be imposed on a visual odometry process, requiring a certain level of performance in certain optical densities due to dust, or with certain obscuration of the camera lenses. In addition, a Sun-finding algorithm may be required to detect the location of the Sun within a specified accuracy given an increase in the apparent size of the Sun due to dust in the atmosphere.

Finally, a navigation system will be required to perform in the lighting conditions found at its target mission site. On the Moon, due to the lack of atmosphere, lighting is very high contrast and the shadows are black. Shadows will be long near the lunar poles, causing challenges for navigation processes that depend on imagery such as visual odometry or teleoperation. Permanently shadowed regions - of particular interest because of the likelihood that they contain deposits of water – ice - would require powerful illumination to support camera-based navigation. On Mars, the lighting conditions are softer due to the atmosphere, but the Sun intensity is lower than on Earth or on the Moon. Requirements for performance with the Sun directly in front of the rover (i.e., likely in the field of view (FOV) of rover cameras), and in the presence of strong reflections are also typical.

4.3.1.3 Resource Requirements

Resource requirements include cost, size, mass, power, and processing. The navigation system must balance these requirements against the performance (and other) requirements. For example, increasing the number of sensors could increase localization performance, but would also increase resource usage. Typically, each subsystem (e.g., navigation, communication, power) has a budget for each resource, but these resources can be traded among subsystems if necessary.

4.3.1.4 Other Requirements

There are many other requirement types that can be imposed on a navigation system, depending on the nature of the rover or mission. Interface requirements are typical, such as mechanical, electrical, thermal, and software interfaces with other subsystems. Software may be required to have a certain type of architecture, modularity, error reporting, code reusability, and so on. The navigation system as a whole may be required to have built-in redundancy, graceful degradation to certain failures, availability (e.g., geographically, or as a function of time), usability, and so on. Another important requirement could be uplink/downlink bandwidth limitations. This could determine, for example, if performing visual odometry on the ground is possible given how many images could be downlinked to the ground.

4.3.2
Design Considerations

This section discusses design decisions that a rover navigation system design team is likely to consider given the imposed requirements on the system. This includes the selection of sensors, software algorithms, computational platform, hardware design philosophy (i.e., commercial off-the-shelf (COTS) vs space-qualified), operator control strategy, communication strategy, location of processing (i.e., on or off the rover), and level of robustness or redundancy.

4.3.2.1 Functional Components

Given the requirements imposed on the navigation system, one can start planning which functional components should be included. For example, it is typical for a rover navigation system to include a means of measuring distance traveled, of measuring change in orientation, and of determining its absolute orientation. If the rover must support autonomous control, then obstacle avoidance, path planning, and pose (driving) control functions are necessary.

Next, depending on the relative severity among the requirements, one can begin to select specific technical solutions to perform each necessary function. For example, if the localization performance requirements are challenging, for example, relative localization accuracy of 1% of distance traveled, then multiple complementary solutions for measuring distance and orientation may be necessary. If the resource requirements are strict, this might bar certain sensor or processor types from inclusion in the system.

4.3.2.2 Sensors

Choosing sensor types and particular sensor models will occur in parallel with selecting technical solutions for the navigation system. Navigation sensors for mobile robots can be broken into the following categories: vision, tactile, wheel-/motor, heading, beacons, active ranging, and motion/speed [29]. Each of these types is briefly described later – a more detailed discussion of specific sensors is given along with the corresponding technology in Section 4.4.

Vision sensors – cameras – are the most flexible of rover sensors, with many uses for navigation (notwithstanding their use for science, public outreach, etc.). Cameras can be used to provide a visual data feed to a rover teleoperator. A stereo camera pair can be used to provide depth information of a scene, which can in turn be used for obstacle detection, path-planning, or visual odometry. Every space rover has had cameras for at least one of these uses. Drawbacks of stereo imagery are that this technique requires sufficient texture in the scene, adequate lighting, and requires intensive image-disparity calculations to generate 3D points. Cameras can be used in conjunction with illuminators to improve their usefulness in poor lighting conditions.

Tactile sensors include pressure or contact sensors. These sensors could be used, for example, to measure pressure on each drive wheel (helpful information to detect slip), or to detect contact with an obstacle. The Sojourner rover had a contact sensor for obstacle detection.

Wheel or motor sensors are primarily for measuring odometry – that is, by measuring the number of wheel or motor revolutions. These sensors are standard issue on extraterrestrial rovers.

Orientation estimation is important because small errors can yield large position errors as a rover moves. Heading (or orientation) sensors can be divided into those measuring absolute heading and those measuring relative heading changes. The two types are complementary. Relative heading sensors such as IMUs have a high update rate, and are very precise and accurate over short periods of time, but their orientation estimate is subject to drift. IMUs are a fundamental part of a rover navigation system and have been used in all extraterrestrial rovers to date. Absolute heading sensors such as star trackers or Sun sensors typically provide lower-accuracy and less-frequent estimates of orientation, but their estimates do not drift.

The beacons category includes any sensor on the rover that is used to localize man-made beacons for the purpose of localizing the rover. These beacons would have known locations, and could be anything from passive, reflective targets to active RF emitters (e.g., GPS satellites) [29]. This category of sensor has yet to be exploited for extraterrestrial rovers as this would involve constructing a network of beacons, which would be difficult and costly. However, sensors of this type could localize the rover very accurately (e.g., centimeter-level) in the vicinity of the beacons. This type of sensor strategy could find a role in areas that are traversed often by a rover, for example, at a future lunar base site by rovers excavating regolith and operating within a well-defined area.

Active ranging sensors include laser triangulation, laser time-of-flight (TOF) (i.e., light detection and ranging or LiDAR, also known as LIDAR), and structured light sensors [29]. LiDARs in particular are very popular in the terrestrial autonomous robotics world. This is because of their many benefits over a competing technology, stereo cameras: LiDARs can produce 3D data that is high density, long-range, lighting-insensitive, and scene texture insensitive, all without the need for any stereo calculations. Despite this, only one flown rover, Sojourner, has had an active ranging sensor: its custom laser-line projection system to detect obstacles. The critical drawbacks of LiDARs for space applications are that they are much larger, heavier, and consume significantly more power than a stereo camera pair. The size, weight, and power (SWAP) of LiDARs will need to be reduced before rovers can accommodate them.

Speed sensors make use of the Doppler effect to measure relative speed. In space, Doppler radar could be used, although this technique has not been employed on any space rovers to date.

Finally, the sensor design and procurement philosophy must be discussed. The main decision is whether to build a new sensor that is designed and space-qualified specifically for a mission, procure an existing fully space-qualified sensor, or procure an existing COTS part (e.g., an industrial camera, which may need some customizations to change a particular material, or make it meet tougher temperature requirements). As one would expect, cost is highest for the first option, and lowest for the third. Cameras on the MERs and on Curiosity are custom, space-qualified cameras, and it is NASA's strategy to use a fully space-qualified component if the component in question is mission-critical [30]. Recently, however, there has been a drive in national space agencies to employ COTS cameras where possible in space applications in an effort to reduce costs: for example, the Japan Aerospace Exploration Agency (JAXA) and the Italian Space Agency (ASI) carried out radiation testing on COTS cameras with good results [31], and NASA has successfully space-qualified a COTS camera for use outside the International Space Station [32] (ISS, although the radiation environment is much less harsh than that of the Moon or Mars).

4.3.2.3 Software

Design of the navigation software, including the type of computational unit on which the software will run, is the next big design decision. This section discusses the navigation software system design, software techniques and algorithms, and selection of computational platform(s).

Good architectural design of the navigation system software is important to ensure that software is easily testable, maintainable, and reusable. Making the software modular and keeping related navigation functions (e.g., those related to obstacle avoidance) in the same module is a good place to start: this allows for unit testing, for individual modules to be modified and replaced easily, and for the design to be organized intuitively [33]. An overall software architecture that groups navigation functions into the major categories of localization, cognition, perception, and motion control is typical [33, 34]. For examples, the MER navigation software has a similar design though the module names are different [35], so does the ExoMars rover navigation and mobility system architecture diagram as illustrated in Figure 4.5.

The report [37] on a rover concept for the NASA Resource Prospector mission discusses other important software design considerations. These include rover states, telemetry, power on self-test (POST)/built in test (BIT), internal communications protocol, image formats, ground communications, and ground control software. Proper selection of rover states ensures that the rover responds correctly to commands and sensor inputs. The goal of telemetry design is to ensure the efficient use of limited bandwidth – for navigation, only giving the operator only what is necessary to safely command the rover. POST and BIT should be designed to ensure the rover is in a proper state to perform upcoming actions, and to detect any errors early. An internal communications protocol should be selected to ensure reliable and timely communication among software modules. Image (compression) format selection is an important decision: lossy compression (e.g., JPEG2000) reduces bandwidth requirements and may be acceptable for human interpretation, but a lossless format may need to be used in the rover to support, for example, visual odometry. Ground communications software must manage bandwidth (buffering data as necessary), and account for communication delays and drop-outs. Ground control software (discussed in Section 4.3.2.5)

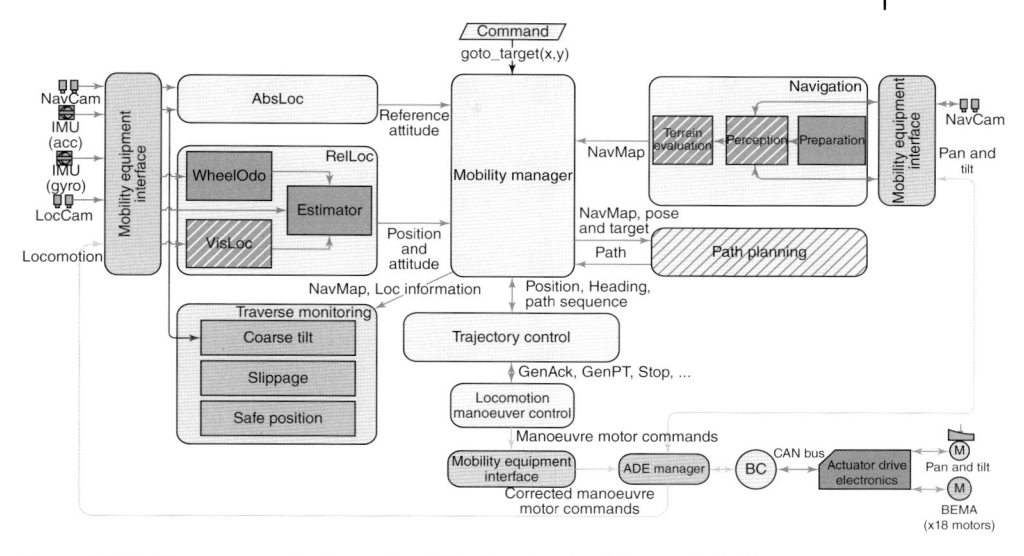

Figure 4.5 ExoMars rover navigation and mobility functional architecture [36]. (Courtesy Airbus DS)

must provide the operator with appropriate data for teleoperation or autonomous navigation, as the case may be.

The decision of which algorithms are employed is tightly coupled with the selected functional components and sensors, discussed earlier. For example, visual odometry requires stereo cameras as well as processing software to track image features in 3D space and convert this to a motion estimate. Within visual odometry, there are many algorithmic options to choose from (e.g., disparity matching for stereo, feature detection, and tracking) that can offer trade-offs between accuracy, reliability, and computational intensity. Specific localization and autonomy algorithms are described more in Sections 4.4 and 4.5, respectively.

Processing complexity is also a key consideration for selecting algorithms. As discussed in the following section, space-qualified processors have limited capabilities, and therefore care must be taken to limit the computational burdens imposed by the navigation algorithms. Processing-intensive navigation capabilities such as visual odometry, obstacle-avoidance, and path planning must be implemented efficiently, and should include only the most necessary functionality to meet the performance requirements. R&D efforts had been taken to streamline algorithm footprints for the MER navigation system [38] and the ExoMars perception system [39, 40].

Another decision at the algorithmic level is whether to code everything from scratch, or to make use of third-party libraries. The former allows for complete design and quality control over the algorithm (including ability to optimize the implementation), but comes at a cost of a high effort. Using third-party libraries for algorithms has become the option of choice for developers of terrestrial robot navigation systems because these libraries have become very mature, efficient, and well-supported. Examples are 3D point-cloud libraries (e.g., PCL [41]), computer vision libraries (e.g., OpenCV [42]), geographic libraries (e.g., GeographicLib [43]), and math/physics libraries (e.g., Geometric Tools [44]).

4.3.2.4 Computational Resources

The selection of computational resources to host the navigation software is an important decision, as power is often at a premium in rovers and current space processors are very modest in terms of their capabilities. The performance of fully radiation-hardened processors, in terms of processing speed and memory capabilities, is limited by long design cycles, and the inclusion of layers of redundancy to protect against radiation-induced errors. For example, the "state of the art" is the RAD750 processor, which is on Curiosity; this processor was released in 2001 and launched first in 2005, and is capable of performing 400 millions of instructions per second (MIPS) [45]. A contemporary quad-core i7 processor can perform at roughly 100 times this rate.

A result of this modest capability is that processing-intensive algorithms require significant time to run on their computation platforms: Curiosity, as discussed earlier, typically drives for 10–20 s and processes for 2 min [27]. This processing time cuts into important mission time and can limit the usability of these navigation capabilities.

General-purpose processors are not the only option for navigation algorithm processing, but do have the advantage of significant heritage on flight rovers. Other options are field-programmable gate arrays (FPGAs), digital signal processors (DSPs), and graphical processing units (GPUs). General-purpose processors support any type of processing and employ conventional software development techniques; however, their performance is limited by their serial computational logic, they consume a lot of power per computation, and significant investment must be made to support radiation tolerance (e.g., implementing redundant computation paths). FPGAs are on the opposite end of the spectrum: their computational blocks can be configured to perform computations in parallel, can be radiation-hardened "by design" [46], and consume very little power per calculation. FPGAs, therefore, have great potential for implementing parallel-processing-heavy vision algorithms such as visual odometry. FPGAs do have drawbacks that limit their use in many applications: they have a more complex development cycle than processors, and updating their firmware is more challenging. While FPGAs have not been used for navigation software on flight rovers, there is a strong lineup of radiation-hardened FPGAs (see here again [46]). DSPs and GPUs occupy the middle of the spectrum: DSPs are processors that are parallel by nature and are optimized for performing signal processing (e.g., speech or image processing). GPUs are optimized to perform calculations on a video frame (e.g., ray-tracing) and are found on COTS video cards. Radiation-hardened DSPs are available but have not been employed for hosting rover navigation algorithms. Radiation-hardened GPUs are not currently on the market. As described, there are alternatives to general-purpose processors for running navigation software, but processors have the key advantage of significant flight heritage.

NASA and the European Space Agency (ESA) are currently working on space-processor-related projects that are relevant to this discussion. NASA-JPL intends to incorporate an FPGA implementation of stereo vision and visual odometry on the Mars 2020 rover to enable fast rover traverses [27]. The ESA is also working on porting visual odometry algorithms to an FPGA as part of the SPARTAN/SEXTANT/COMPASS projects [47]. A joint project between NASA and the Air Force Research Laboratory (AFRL) is currently underway to design and build a "next-generation space processor" [48]. The goal of this project, currently at the risk-retirement stage, is to produce a processor that has 100 times the computational performance of the RAD750 (discussed earlier) by 2025. This project evaluated all the main computational options discussed earlier. The conclusions are that a multi-core processor will be the most flexible option and best able to support a wide array of space computation applications studied in the project. For navigation, these applications include "fast traverse" for surface robotics, automated GNC, and autonomous/telerobotic construction. Progress is, therefore, being made to increase the performance of space processors, and make better use of the processing options available.

A final processing consideration is: can processing be done off the rover? Processing power is cheap on the ground, but this approach comes with drawbacks, namely increasing bandwidth requirements (e.g., to transfer images) and adding delays due to communication time. This consideration is discussed more in the following section.

4.3.2.5 Rover Control Strategy

The rover control strategy is another pillar of rover navigation system design. In general, this is the level and type of human interaction with the rover and varies from direct teleoperation (e.g., discrete motion commands are uploaded from ground controllers) to full autonomy (e.g., one command sequence per "day" is uploaded). The strategy employed depends greatly on communication delay and bandwidth, and obviously affects the design of the rest of the navigation software (e.g., a teleoperated rover will require less autonomy). This section discusses the spectrum of control strategies, and provides examples of their application on flight rovers, or work currently being done in this area.

Teleoperation is the simplest control case from a system design perspective. Direct driving commands (e.g., forward and rotational velocity) are uploaded to the rover and telemetry (e.g., pose) and imagery feedback is sent back to the controllers. Humans are excellent problem-solvers and decision-makers, so teleoperation is, therefore, a desired control method as it keeps human controllers directly in the loop. The operators can perform real-time path-planning and obstacle avoidance, which removes the need for these capabilities to be built into the navigation software. However, teleoperation begins to break down as control delays increase and/or bandwidth decreases. In the case of control delays, the operator must upload a single (or limited) command sequence and then wait for the rover to implement the commands, and for the telemetry and images to be downloaded. This command/telemetry cycle pattern will significantly slow down the progress of the rover. Decreasing bandwidth can manifest itself as a control delay, as an operator must wait for telemetry and imagery to download. This effect can be traded off against image quality (i.e., heavily compressed images require less bandwidth), but image compression also has its issues: operator situational awareness will decrease, and obstacles or safe paths will be harder to recognize. This is an argument for adding decision-making capability on the rover: the rover has instant access to the top-quality imagery and pose information, no communication bandwidth required. As one can see, teleoperation can be a powerful control method, but it has drawbacks.

The Moon is naturally a location where teleoperation will be employed, due to its "proximity" to Earth. The Lunokhod rovers were teleoperated, although this was a challenging task for the operators, as described earlier. Also, despite a theoretical round-trip communication delay of approximately 2.6 s (the Moon being 1.3 light-seconds from Earth), the actual control delay is greater. The Canadian Space Agency (CSA) has recently been investigating lunar teleoperation strategies, specifically targeting the NASA Resource Prospector mission. To this end, the CSA recently completed a field trial in which they tested a prototype teleoperation interface and camera suite on a rover with a number of different users [49]. They estimate control delays of around 10 s due to constraints of using the Deep Space Network (DSN) communication system. To compensate for this delay, the CSA implemented tools such as predictive displays, which were incorporated into the field trial testing. NASA has also been testing teleoperation strategies, though with a different mission type in mind: they tested ISS crew teleoperating an Earth-bound rover to simulate an astronaut in a lunar orbit controlling a rover on the lunar far side [50]. The ESA is also involved in this area, for example, their Rover Autonomy Testbed (RAT) is used to test teleoperation and autonomy strategies for varying communication constraints [51]. Teleoperation, therefore, is the domain of the low-delay control scenario.

"Supervised" or "guarded" teleoperation are general terms for a teleoperation system augmented to either increase speed or safety. Safety can be increased by giving a rover the ability to detect off-nominal situations and override the teleoperation commands if necessary. Most commonly this involves having the rover monitor its roll and pitch values and its motor currents (e.g., as is the case for the MERs [12]). If any of these values enter a predefined danger zone (e.g., rover is approaching its rollover angle, or high motor currents indicate rover is pressing against an obstacle), then the rover will automatically be commanded to stop, and the operator notified. Another example would be to have a more sophisticated system that could use, for example, stereo cameras to detect an obstacle or a dangerous slope (also implemented on the MERs, but rarely used [12]). To increase rover speed, another augmentation is to command the rover at a higher level: for example, commanding it to follow a two-segment 5 m path rather than commanding its instantaneous forward and rotational speeds. This has also been tested by the CSA [49].

Full autonomy generally refers to a rover that is safely able to guide itself to a goal location defined by operators. This involves planning a global path to the goal, guiding the rover along that path, and updating the path along the way to avoid obstacles. Autonomy is necessary when control delay is large (e.g., on Mars) and the rover is required to navigate into terrain that cannot be properly assessed from the current position of the rover. This was exactly the case for the MERs, which were required to be able to navigate 100 m in a single day, with only one control input from Earth [12]. The MERs (as well as Curiosity), therefore, possess the ability to autonomously navigate to a predefined goal location. Drawbacks of incorporating autonomy into navigation mainly relate to the additional complexity: the software takes longer to develop and test, there are additional error conditions to manage, and the processing (and power, etc.) requirements on the rover increase. Of course, this must be traded off with the benefits: autonomy empowers a rover to make driving decisions on its own, thus giving it the ability to safely navigate large distances without human input.

As described earlier, the MERs incorporate a set of control modes, from full autonomy down to teleoperation-like short paths. This flexibility has served the MERs very well in meeting their odometry goals.

The control strategy employed on a rover, therefore, has far-reaching effects on navigation system design and navigation performance. There are many flavors of control to choose from, and one must consider the performance requirements and the communication limitations when selecting the strategy to be used.

4.4
Localization Technologies and Systems

This section covers the breadth of navigation and localization technologies, and references particularly rovers where these technologies have been applied. The pros and cons of each technology are discussed as well as software and hardware options for each, as applicable. The technology categories are classified into orientation estimation, relative localization, absolute localization, and combining localization sources.

4.4.1
Orientation Estimation

This section discusses techniques to estimate a rover's relative and absolute orientation. A rover will typically require both types of techniques: relative heading measurements are normally more precise and higher rate, whereas absolute heading measurements are used to constrain drift of the relative measurement. The techniques discussed in this section include Sun and star trackers, inclinometers, IMUs, and a method using a high-gain antenna.

4.4.1.1 Sun Finding

Locating the Sun in the sky yields one's absolute orientation, given knowledge of the true location of the Sun in the planetary body frame of reference (a function of approximate rover location and time), and the gravity vector (measured by accelerometers) as shown in Figure 4.6. Sun finding can be done using a purpose-built Sun-sensor, which are typically wide FOV sensors that output a vector to the Sun. Alternatively, Sun finding can be done using a pan-tilt camera on the rover, outfitted with an appropriate filter to prevent saturation of the detector. The MERs use the camera-based technique: they point the PanCams in the expected direction of the Sun and then process the image to locate the actual Sun location; a process that had been done 100 times on the rovers by January 2007 [18]. The Lunokhod rovers, too, used a technique to image the Sun and Earth in order to measure their absolute orientation. The ExoMars rover will also use the camera method; however, this will be done without applying a filter on the NavCam, so processing must deal with the resulting saturation artifacts. The Sun has an approximate angular diameter of 0.5° from Earth and 0.34° from Mars, so a heading estimation with subdegree accuracy is possible (although this depends on camera properties and pointing accuracy, and accelerometer accuracy, etc.). The Sun can also appear much larger on Mars at times due to varying atmospheric conditions. Finally, this technique works best when there is a sufficiently large angle between the Sun and the gravity vector. At low latitudes, there can be issues during large parts of the day (i.e., when the Sun is well up in the sky).

Research is active in incorporating Sun direction measurements into rover navigation software. For example, Lambert *et al.* [52] discuss incorporating regular

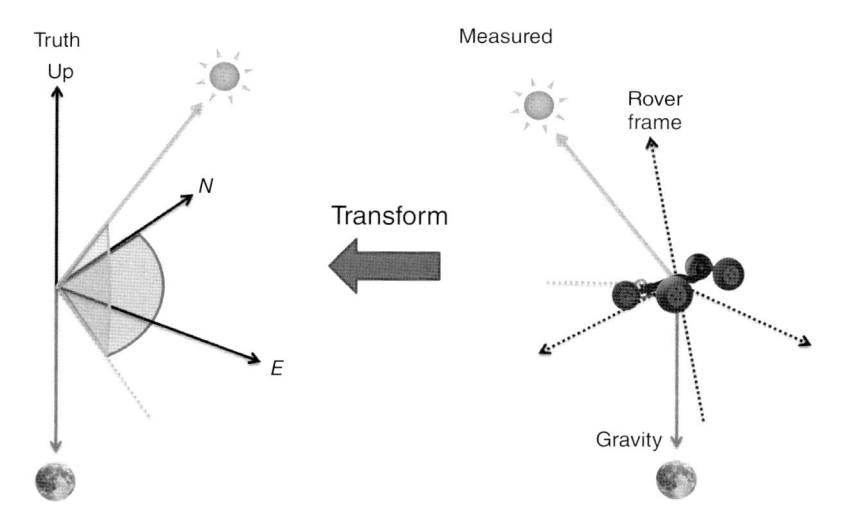

Figure 4.6 Orientation estimation via measuring local gravity and Sun vectors and aligning these to corresponding vectors in a world frame of reference. (Courtesy Neptec Design Group.)

measurements from a Sun sensor directly into a localization system, and demonstrate a resulting increase in localization accuracy.

4.4.1.2 Star Trackers

A similar method of finding absolute orientation is to determine one's orientation in the celestial reference frame by recognizing star patterns. This could again be done with a camera, but it is more typical to use a purpose-built star tracker, which is a common sensor on satellites. A star tracker can recognize star patterns and output an orientation estimate typically at a rate greater than 1 Hz. Accuracy is typically in the dozens of arc seconds.

Key specifications of a star tracker include the following. First is angular accuracy, typically measured in arc seconds. Next is tracking rate (deg/s), which defines how fast a rotation the star tracker can follow before becoming "lost in space." Update rate (Hz) is the frequency of generated solutions. Finally, acquisition time (s) defines how long a star tracker requires to generate a solution either initially or from a "lost in space" state.

Star trackers have not yet been incorporated on a flight rover, but with their high measurement accuracy, and usefulness in areas where the Sun is not visible (e.g., in craters at the lunar poles), they are an attractive option. One issue to date has been mass of the sensors. However, with small satellites becoming more popular, smaller and lighter star trackers have been developed, cutting mass from more than 2 kg (typically) to less than 100 g (e.g., the Sinclair ST-16 star tracker). Another issue is atmosphere: a star tracker would work well on the Moon, but the hazy Martian atmosphere *may* preclude use of a star tracker for navigation. Research is underway to test star trackers as part of rover localization systems (in particular, see [53]).

Star trackers can also be used to estimate rover position. This is discussed in Section 4.4.3.

4.4.1.3 Inertial Measurement Units

IMUs have been used in every flight rover to date. IMUs are primarily used to measure angular rates (and position) at a high-rate, though this measurement will drift and must be constrained by an absolute orientation measurement periodically. IMUs also measure linear accelerations.

IMUs for space applications are typically made up of three gyroscopes and three accelerometers. The gyroscopes are aligned orthogonally and measure rotational rates about each of the three axes. These rates are integrated to find angular position, and as such, small errors in the rates will cause the angular position measurement to accumulate errors, or "drift." IMUs also normally include accelerometers in the three axes. When stationary, these accelerometers can measure the gravity vector, which constrains the drift in the pitch and roll measurements. Measured accelerations can also be doubly integrated to find position, but these two integrations make the position error grow quickly (particularly because the rover experiences quickly changing accelerations as it moves through rocky terrain). Terrestrial IMUs often include a magnetometer to measure bearing, though the lack of a strong enough magnetic field on the Moon or Mars would render this useless.

IMUs are differentiated by a few key specifications. In addition to SWAP, these include *angular random walk* (ARW) and *gyro bias*. ARW defines the angular error ("walk") as a function of time, which is due to random noise in the angular rate measurements. As a point of reference, the MER IMUs drifted less than 3 deg/h of operation [13]. Gyro bias is the error contribution due to a bias of the rate measurement. Any fixed bias will typically be calibrated out at the factory, but the bias too will drift over time due to factors such as temperature.

ITAR designation is another important differentiator among IMUs. Since IMUs are used in weapon systems such as missiles, IMUs produced in the United States and having a certain level of performance will be classified as ITAR-controlled. This must be considered because an ITAR designation increases the difficulty of procuring an IMU and can affect the ITAR designation of the entire rover.

The two main types of IMU technologies are *fiber-optic gyros* (FOG) and micro electro-mechanical systems (MEMS). FOG IMUs have significant flight heritage (e.g., the LN200S IMUs on the MERs and Curiosity) and are highly accurate, but are larger, heavier, and more power-intensive than the MEMS IMUs. MEMS IMUs have limited heritage and are less accurate, but offer significant reductions in SWAP. As discussed here [37], recent developments have brought MEMS IMUs closer to space applications. One such example is a prototype being developed for the ESA [54]. This project targets an IMU of <200 g and <1 W, whereas the LN200S is 750 g and consumes 12 W [55].

4.4.1.4 Vision Techniques

The rover's vision sensors can be employed to estimate orientation as well. These techniques are described in more detail in Sections 4.4.2 and 4.4.3. First, by observing the local terrain and correlating this with a map or DEM, absolute orientation can also be estimated either manually or automatically. Also, by comparing successive images collected by the rover, relative orientation can be estimated.

4.4.1.5 Antenna Null-Signal Technique

The high-gain antenna on a rover could also be used to determine orientation, as discussed here [37]. This technique, which has not been employed on a rover, would be based on the mature "automatic direction finding" (ADF) technique used for airplanes on Earth. The technique would involve slewing the rover antenna on the rover to find the "null" (minimum) point, which would be directly away from the communication target (e.g., Earth, or an orbiting asset). With knowledge of the direction of the communication target, this technique would measure absolute orientation. A technical risk of this technique is the unknown effect of signal reflections on the surface. Also, this technique would take valuable time away from the communication window.

4.4.2
Relative Localization

This section discusses both hardware- and software-focused relative localization strategies. This includes wheel odometry, visual odometry (both 2D and 3D), other vision-based techniques, and speed sensing.

4.4.2.1 Wheel Odometry

Wheel odometry is the most mature method of localizing a rover, and has been used in all flight rovers. Distance traveled is measured by counting wheel or motor revolutions. Typically, the revolutions of all wheels are averaged to increase accuracy. Orientation could also be measured by comparing left- and right-side revolutions, though using an IMU (as described earlier) is much more accurate for this measurement.

The accuracy of wheel odometry depends greatly on the type of terrain. Accuracy of <1% of distance traveled is possible on hard, even surfaces, but slip due to steep inclines or loose soil can significantly degrade this performance. For example, the MERs encountered numerous challenges due to slip in getting to targets during their first 2 years of operation, before visual odometry was enabled [13].

Avoiding slip errors is possible by measuring odometry using a free-wheeling, unpowered wheel. This was used in the Lunokhod rovers [6].

4.4.2.2 Visual Odometry

Visual odometry is a mature technique to measure a relative change of rover pose, by comparing scene contents between successive camera pair acquisitions. This technique is well established on terrestrial robots as a means to increase localization accuracy (<1% of distance traveled is possible) and account for slip errors from wheel odometry. In addition, visual odometry has proven its worth on the MERs and Curiosity to not only improve localization accuracy, but to increase driving efficiency and keep the rover safe.

Visual odometry involves a pipeline of processing steps applied to successive stereo image pairs. The steps are described in detail in Chapter 3, but are summarized here. These steps, with options for each step, are also described in this localization report on the Resource Prospector mission rover, prepared for the CSA [56]. The first step is feature detection, in which interesting features are found in one image from each camera pair (e.g., the two left images). Second is stereo matching, in which 3D locations of each of the features is calculated using the stereo pairs (i.e., comparing each right to each left image). Third is feature tracking, or matching, in which features from the first image (i.e., the first left image) are linked to features in the second image (i.e., the second left image). Next is often a filtering step in which outlier features are removed. The final step is determining the average 3D motion of all tracked features. The final result is a 6-DOF relative pose change between the first and second stereo pairs. Table 4.1 gives a

Table 4.1 Comparison of visual odometry pipelines of MERs and ExoMars rover.

Algorithm step	MERs	ExoMars
Feature detection	Forstner or Harris [58] corner detector. Image is binned and strongest features from each bin are taken to ensure spread of features across image	FAST corner detector [59] at different image scales. Quadtree used to ensure spread of features across image
Stereo matching	1D search along epipolar line (see Section 4.5.1) using pseudo-normalized correlation to determine the best match	1D search along epipolar line using sum of absolute differences (SAD) to compute matching score
Feature tracking	Features projected into second image using motion provided by wheel odometry. A correlation-based search relocates features	BRIEF feature descriptor [60] used to calculate integer pixel location of matched features in new image pair, which is then refined to subpixel using a minimization scheme
Filtering	RANSAC outlier removal [61] used	RANSAC outlier removal used
Motion estimation	Maximum likelihood estimator	Least-squares minimization

comparison of how the MERs and the ExoMars rover perform each of these steps [13, 57].

The performance of visual odometry depends greatly on the quality and contents of the source imagery. The cameras must have accurate stereo calibrations, and the scene must be in good focus. Certain lighting conditions can degrade performance: the scene can be washed out if the Sun is in the FOV, the scene can have little contrast if the Sun is high in the sky (minimizing shadows), or the scene can be blurry if the ambient lighting is too low (requiring higher exposure times). The scene must also contain enough interesting texture or contrast to provide features for the algorithm to track. The stereo pairs must have sufficient overlap, too, to allow for features to be matched and tracked. Finally, any motion in the scene can yield erroneous displacement measurements; motion such as a moving dust cloud, or even the rover shadow (the latter actually causing an issue at times on the MERs [13]). Despite these constraints, however, visual odometry can provide a powerful source of rover localization information: accurate and independent from the measurements made from the IMU or the wheels.

The MERs have had a rich and well-documented visual odometry experience. Maimone *et al.* [13] provide a detailed description of both the algorithm and the experience on Mars. Key points are described here. Visual odometry has proven its worth on the MERs not only to improve localization performance but also to avoid obstacles (by keeping the rover on its desired path), to reach science targets more efficiently (in fewer sols), and to detect dangerous driving conditions (soft soil in particular). Visual odometry was necessary for the MERs because they were required to have the ability to localize themselves to better than 10% of distance traveled over a 100 m drive, a performance that could not be met with wheel odometry alone. In ground-based testing, accuracies of approximately 2% were found. The MERs use their mast-mounted NavCams for visual odometry, and these cameras are manually pointed at "interesting" features (sometimes their own tracks) to ensure the algorithm has sufficient image features that it can track. Wheel odometry estimates are used to seed visual odometry. Distance is *maximized* between image pairs (which increases error and is opposite to a typical visual odometry implementation) in order to reduce the number of times the computationally expensive process must run. This distance, however, is no more than 75 cm. Visual odometry is used judiciously as each iteration requires approximately 3 min to run, and thus reduces average driving speed by an order of magnitude. The performance of the algorithm has been impressive on both rovers: as of February 2006, the algorithm has converged to a solution in 95% of the cases.

On Curiosity, which inherited its visual odometry software from the MERs, the results are also impressive. Convergence rates as of sol 650 are 99.66% [25]. In one instance, visual odometry actually detected an error in the IMU-measurement logic [25]. A more powerful processor on Curiosity also enables the software to run in less than 40 s/iteration (compared with 3 min for the MERs), allowing VO to be used much more frequently (34 of the first 40 sols [20]). Curiosity also has a mode to best use visual odometry in a way to maximize efficient traverse efficiency: this mode, "Auto VO," performs slip-checks every 20 m or so, and based on the result, automatically decides whether visual odometry should be run all the time (every 1 m or so) [21]. This mode allows visual odometry to be employed when necessary, thus maximizing the average speed of the rover while still keeping it safe.

The upcoming ExoMars rover will incorporate visual odometry even more tightly into its navigation system [36]. VO updates will be done every 10 s (while the rover is in motion) and will be incorporated directly into the localization solution in a closed-loop fashion, distinct from the MERs and Curiosity, which use VO to update the position after each path segment. This allows for continuous monitoring of the rover slippage levels. The estimated average wheel slip is also used to scale the wheel odometry measurements in order to improve localization accuracy between VO updates. The VO algorithm, in its current implementation, is described here [57]. This algorithm has undergone field testing as part of the SEEKER [62] field trials.

A 10-s implementation time for ExoMars is a significant speed-up compared with the Curiosity time (approximately 40 s). This is likely due to a number of factors, but likely contributors are that ExoMars has a dedicated processor for VO, and the feature-matching solution space for ExoMars is much smaller (approximately 10 cm compared with 75 cm – 1 m).

Other future flight rovers will also include visual odometry – often with a focus of reducing its computational footprint. NASA's Mars 2020 rover in particular is planned to include an FPGA implementation of visual odometry that will dramatically reduce computation times [27, 63]. The ESA as well is working on an FPGA implementation of visual odometry as part of the SEXTANT activity [64]. A rover proposed for NASA's Resource Prospector mission incorporates both a traditional visual odometry system using mast-mounted cameras [65, 66], as well one using downward-looking cameras [67] in an effort to maximize localization accuracy.

4.4.2.3 Other Vision-based Techniques

Two other vision-based techniques can be used to perform relative localization. First is offline IBA of stereo images, as is currently used in the MERs and Curiosity, and described in Section 4.2.3.4. Next is Simultaneous Localization and Mapping (SLAM), which is described in Section 4.4.4.3

4.4.2.4 3D Visual Odometry

3D visual odometry – using active, 3D sensors such as scanning LiDARs as input – aims to improve visual odometry by capitalizing on the advantages of these sensors over traditional stereo cameras. As discussed earlier, LiDARs can produce dense point clouds out to long ranges regardless of scene lighting or texture, and without stereo calculations. Despite this promise, neither 3D visual odometry nor active 3D sensors have found themselves on a flight rover to date (the laser-striping sensor on Sojourner notwithstanding). The SWAP consumption of active 3D sensors is currently limiting the deployment of this technology for flight rovers.

Active 3D sensors on mobile terrestrial robots (e.g., SICK, Hokoyu, Velodyne), however, have become very popular in the last decade, contributing to a surge of research activity targeting 3D visual odometry. A popular approach in the literature is a variant of the well-known iterative closest point (ICP) algorithm, called generalized ICP [68], which is a robust and efficient 3D point-cloud alignment technique. The CSA has also prototyped an algorithm based on ICP [69]. Another approach, which was built to be robust to the often feature-poor terrain of a typical extraterrestrial environment, uses "curvelets" to detect features for alignment [70]. Other approaches apply the 2D visual odometry pipeline to LiDAR intensity images [71, 72], which takes direct advantage of the lighting-invariant quality of LiDARs. Additional field results from LiDAR intensity-based algorithms are given here [53].

4.4.2.5 Speed Sensing

Radio- or light-based sensors can use the Doppler effect to measure speed with respect to a target object. This has not been deployed on a flight rover, but work is being done to apply this concept to space applications, and many relevant systems exist on Earth. VORAD, a mature radar-based technology used in the trucking industry can measure speed to a resolution of 1 km/h [73]. Using light allows for increasing this resolution: NASA Langley Research Center is developing a LiDAR-based landing sensor that aims to have a velocity accuracy of 1 mm/s [74], though the prototype sensor has a mass that is currently too great for rover use.

4.4.3
Absolute Localization

This section discusses both hardware- and software-focused absolute localization strategies. This includes rover-to-orbiter data matching, Earth-based and other asset-based ranging, rover/orbiter imagery matching, skyline matching, and a star-tracker solution.

4.4.3.1 Rover-to-Orbiter Imagery Matching

As introduced in Section 4.2.3.4, the MERs use a manual rover-to-orbiter imagery matching technique to achieve accurate global localization, as shown in Figure 4.7. This process uses HiRISE camera images, which can achieve 25 cm resolution and, therefore, can resolve (some) objects that can also be seen with the rover-based cameras. A similar process can be conceived for a future lunar rover by using LROC imagery, which exists at resolutions up to 0.5 m [2, 75].

This section discusses three aspects of rover-to-orbiter imagery matching in the Martian context. First, side benefits of this matching process to tasks such as path planning are discussed. Second, the challenge of achieving a desired global localization accuracy is discussed. Finally, a discussion is provided for possible approaches of automating the matching process.

The HiRISE imagery can be used to plan rover traverses or to update existing rover traverses established using the matching method discussed earlier. The advantage is that all the rover images are within the same geospatial context and can be manipulated using geographic information system (GIS) methods including those involving WebGIS (a tool used for the MERs). This allows the scientific user of rover imagery as well as the engineering team access to a seamless 3D visual

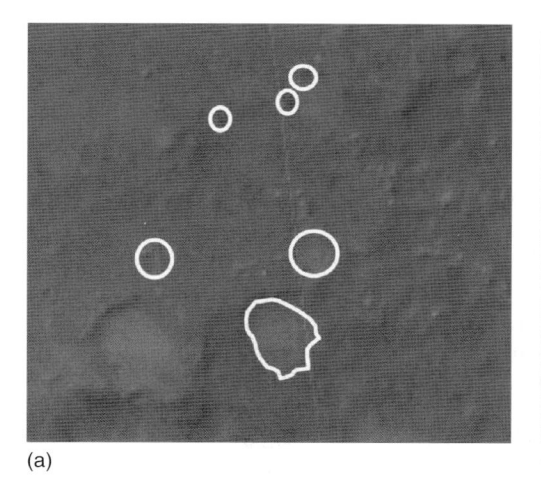

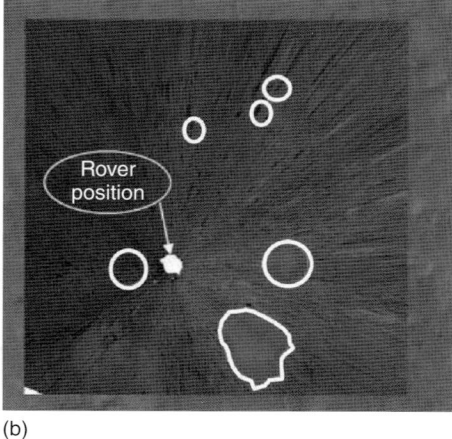

(a) (b)

Figure 4.7 Orbiter (HiRISE) imagery (a) and rover imagery (b) used for matching algorithm for localization on the MERs [15]. (Courtesy American Geophysical Union.)

experience in which the focus is on examining the entire landscape and not just the tiny strip of land around the rover traverse. The internal planning systems for MERs and MSL use rover traverses, which are now primarily obtained from visual and manual optical navigation, on such orbital images (T. Parker, personal communication) [15].

However, such very high-resolution imagery from assets such as HiRISE images are poorly georeferenced at present. Therefore, in order to be able to ensure that the geographic positions for the rover can be converted into global coordinates, they need to be coregistered to a lower resolution image source whose georeferencing error is low (on the order of tens of meters rather than hundreds of meters). One such possible source is NASA MRO Context Camera (CTX) 6 m imagery. However, CTX images are often themselves also poorly georeferenced. The 30–150 m DTMs and 12 m orthorectified images (ORIs) from the ESA Mars Express High/Super-Resolution Stereo Color (HRSC) imagery are the only high-resolution images available which are globally georeferenced to the MOLA-based global coordinate system [76]. By employing the 12 m HRSC ORIs, a source of landmarks for CTX can be established to ensure that the CTX are in a similarly good georeferenced global coordinate system and then a further coregistration stage can ensure that the HiRISE DTMs and ORIs are coregistered to the CTX and thence to HRSC and MOLA. Such an orthoimage coregistered cascade can be generated fully automatically as described here [77] and more recently refined here [78].

When HiRISE images are sufficiently accurately georeferenced to Mars global coordinates, one can manually find corresponding landmarks between HiRISE ORIs and ORIs generated from rover-based imagery whether from NavCam or PanCam. By extracting a series of such landmarks from different rover camera stations, it is feasible to find a set of common features between the rover viewpoint and the orbital viewpoint. For rover imagery, this means using primarily far-range (>10 m) imagery of similar resolution to that from HiRISE. Dr Tim Parker (of the NASA Jet Propulsion Laboratory) has developed an interactive rover traverse generator using manipulation of rover ORIs and successive locking-down of these ORIs to a HiRISE ORI raster backdrop. Tao and Muller [78] have recently developed a fully automated system for rover traverse construction and applied this to MER and MSL rover traverse generation.

Given sufficient overlapping images from as many full panoramas as possible and/or wide baseline imagery from multiple stations, rover-based imagery can be employed to generate orbital landmarks for construction of a fully automated rover traverse construction (i.e., automated matching of features). The end result is a rover traverse, which can be displayed in the context of an orbital ORI in order to find new paths through the landscape as well as assess the general landscape and select future science targets.

The ability to automatically match features between the orbiter and rover images could allow the rover to perform accurate localization on board in the future. An investigation into next-generation GNC architectures [79], based upon a model Mars Precision Lander (sample-fetching rover, SFR) mission, included suggestions of new technologies that would be required for a successful mission with increased autonomous long-distance traverses. These technologies include a localization system termed Constellation Matching, based on the principle of autonomously matching features from orbiter images (the coarse map) with features extracted from images captured by the rover's onboard camera (the local map), in order to aid with localizing the rover.

In the proposed design, groups of detected features in the local map (rocks) are coregistered with features visible in larger orbital maps, which will have been generated offline and preloaded onto the rover before launch. At regular intervals, the rover would stop to take a series of images, to construct a 360° panorama. A feature detection system attempts to identify large boulders in the scene, filtering out those boulders too small to be observed in orbital imagery. The centroids of these features can be mapped to real-world coordinates in order to produce a top-down map of local features. Using the rover's estimate of its pose together with the confidence associated with that estimate, an ellipse can be drawn on the global map, that is, the preloaded map with features derived from orbital imagery. This ellipse acts as a space that can be searched for patterns of large boulders that match the arrangement of the boulders in the local map. The two maps are compared using the iterative closest point (ICP) algorithm, which will output the rotation and translation between the two maps, providing an absolute localization for the rover in a global map.

Two specific implementations of Constellation Matching have recently been developed [80, 81] to implement absolute localization of a planetary rover by using a global map derived from orbital imagery and a local map generated from images taken by the rover during its traverse on the surface.

4.4.3.2 Rover-to-Orbiter Horizon Matching

Given a DEM of the terrain, which would have been gathered from an orbiting platform, one can estimate the appearance of the horizon from various rover locations, which can be used to match against the actual horizon seen from the rover. The best match would yield a location estimate for the rover. This concept has never been applied on a flight rover, but numerous variations exist in the literature. Examples include an approach that targets terrestrial mountainous terrain [82], and other approaches that specifically target the rover case and have variously tested the algorithms on simulated lunar data or real data from analog test sites [83–85].

4.4.3.3 Rover-to-Orbiter Digital Elevation Model Matching

A related approach to this involves performing a more direct comparison of the rover-sensed DEM to the orbiter-sensed DEM. This type of approach has also not been employed by a flight rover, but numerous approaches have been developed and tested on simulated data and real data from analog sites. This algorithm [86], created for a prototype Resource Prospector mission rover, compares rover LiDAR data to the orbital DEM using correlation and feature-based techniques and combines the results with a voting scheme. The approaches in Refs. [83] and [87] propose similar approaches, the latter using a particle filter (see Section 4.4.4.2) to estimate rover position. Finally, the approach in Ref. [88] proposes an algorithm, which compares topographic peak features in orbital DEMs to peak features detected by a LiDAR-equipped rover. This was field tested at an analog site in the Canadian High Arctic.

4.4.3.4 Orbiting Asset- or Earth-based Localization

Localization using orbiting assets is very mature on Earth: using the GPS, submeter localization accuracy is possible over much of the Earth's surface. This, however, is only possible given the massive investment of building and maintaining the multisatellite network, something that is not likely over other planetary bodies. That said, reduced systems can be conceptualized, and existing satellites can be used in some cases to localize a rover, as is discussed in this section. Also, measurements taken directly from Earth can also help to localize a rover.

Shortly after the MERs landed, they were localized using range and Doppler (speed) measurements from two DSN antennas, as well as using Doppler measurements from two Martian orbiters: Mars Global Surveyor and Mars Odyssey [89]. With periodic inputs from these sources (1–3 times/day), localization of 1 km after 5 days was possible by only using the DSN measurements, and localization to within 10 m was possible after 2 days by adding in the Doppler measurements from the orbiters.

A purpose-built Martian or Lunar "GPS" can also be conceptualized. The NASA Jet Propulsion Laboratory performed a design exercise of a Martian GPS, and estimated that a six-satellite system could yield a sub-10 m localization accuracy with a 1.5-h settling time anywhere on the Martian surface [90]. For the Lunar case, Chelmins et al. [91] from NASA discuss the possibility of using two satellites plus a local station to compute the location of an astronaut/vehicle to within 1 m in 5 min (given an initial uncertainty of 100 m). However, this scenario was designed while the Constellation program was still active.

Absolute localization can also be done using an Earth-based ranging and triangulation technique that relies on detecting reflected light from a target, as discussed here [37]. The Lunokhod 1 rover has recently been localized to within centimeters using this technique [92]. This technique involves triangulating multiple range measurements to the rover, which are achieved by beaming a laser (using a large telescope) in the general area of the rover and detecting reflected light from a corner cube. There are a number of considerations to use this technique with a future rover. First, the rover would have to be outfitted with a corner cube and then have a means of directing the cube at the Earth. A discussion with the lead researcher for this work, Prof. Tom Murphy, yielded a possible use case for the Moon scenario: take two ranging measurements of a stationary rover approximately an hour apart, which would yield a localization accuracy of about 5 m. Constraints are that it must be night time at the telescope location (New Mexico) and the sky must be clear.

4.4.3.5 Fixed Assets/Beacons Localization

If a rover is to remain relatively close to a fixed asset or assets, then this infrastructure can be used to help localize the rover. One possibility is for this asset to be on the lander, which would simplify deploying and powering such an asset. Another possibility is for a network of assets to be used, possibly deployed by the rover. Many terrestrial prototypes and commercial systems could provide such functionalities. To date, only the Sojourner rover has used this type of localization – its position was monitored using a stereo camera on the Pathfinder lander. This section discusses these types of asset-aided localization.

The first scenario is that of a single fixed asset (e.g., on the lander) providing localization information to the rover. A spectrum of functionality and complexity is possible. For example, this asset could measure range to the rover using a radio pulse, or could determine the direction to the rover from the lander, or could determine the relative 3D position of the rover using a laser rangefinder on a pan-tilt unit. Each measurement could be incorporated into the localization solution. One particular example is proposed in Ref. [93], in which a localization framework is introduced for a rover that is tracked by a single beacon that measures range. Another proposed lander-based system is discussed here [94]. This prototype system uses a total station (laser-based) tracking system at the lander to track the 3D location of the rover. The prototype demonstrated centimeter-level accuracy out to approximately 1 km from the lander. A limitation of this laser-based method is that it requires line-of-sight to localize the rover, but if the line-of-sight outages are only temporary, this method can provide regular absolute location updates to contain the drift of relative measurements.

If the rover is to spend some time in a particular area, such as in Moon-base-construction scenarios, then the "asset" could actually be a network of assets deployed by the rover to use for localization. These assets can be active or passive: active options may provide increased accuracy or reliability, but with the cost of increased complexity. In the active category, one example is "pseudolites": ground-based GPS transceivers that would provide GPS-like location accuracy. For example, the work [95] done at NASA's Ames Research Center, demonstrated that 40 cm localization accuracy was possible over a 174 m traverse that took a rover far outside the three pseudolites, which were arranged in a triangle 10 m to a side. Commercial pseudolites are now available from a company called Locata. Other technologies for active assets include Wi-Fi (e.g., see a commercial system from the company Ekahau), WLAN (see Ref. [96], and ultra waveband (a NASA effort [97]). Passive assets can be any sort of recognizable object, such as the "barcode-like" targets of the ARToolKit system [98]. A hybrid active/passive approach is a simple lighted beacon, which can be easily recognized from its surroundings at long ranges.

There are many things to consider with a fixed-asset network localization system. First is how these assets are to be deployed, and whether the system can be easily calibrated (a self-calibration concept is presented in the pseudolite reference [95]). For active beacons, a major question is how they will be powered. Issues with multipath returns (reflection) of the signal are also a concern with active beacon systems. For passive beacons, major questions are: what is the maximum range at which the beacons can be recognized, and are there any angular viewing constraints? In general, fixed-asset network localization systems can provide very high accuracy in a specific area, but with a high overhead cost of setup of the system.

4.4.3.6 Celestial Localization

It is possible to localize a rover by determining very accurately the orientation of the local gravity vector in the celestial frame of reference. The gravity vector would be measured by an inclinometer, and the celestial reference frame would be determined by a star tracker. On the Moon, an accuracy of hundreds of meters with a single measurement is reported as possible here [99], and 60 m with careful filtering of multiple measurements here [100]. Field trials of a star tracker and inclinometer localization system on Earth [101] demonstrate accuracy of better than 800 m.

Related techniques use the observed location of other celestial objects such as the Sun or the Moon of Mars to estimate position. Ning and Fang [100] summarize some of these other approaches.

4.4.4
Combining Localization Sources

As discussed in Section 4.3.2.1, a rover localization system typically requires a means of measuring distance traveled, relative orientation, and absolute orientation. In flight rovers, these functions are normally provided by wheel odometers, an IMU, and a Sun-sensing technique, respectively. Visual odometry has also been employed to increase localization accuracy and ensure rover safety in situations of wheel slip. Some means of periodically determining absolute location is also typical: whether from range measurements from orbiters or by comparing surface features to orbital imagery features.

This section discusses the problem of combining the measurements from these different localization sources, to provide a unified estimation of rover pose. Rover pose can be determined directly from the pose sources (e.g., distance calculated from wheel odometry and orientation from an IMU), but there are shortcomings with this approach. First, if a localization source is noisy, then that noise is passed directly to the rover pose. Second, if more than one localization source provides input to the rover state, for example, wheel odometry and visual odometry both measure distance traveled, then it is not straightforward how these measurements should be combined. Combining the pose information via some type of filter is a common approach to this problem.

There are two commonly used classes of filters that can be used for rover localization, Gaussian filters and particle filters. Gaussian filters, such as the extended Kalman filter (EKF) and extended information filter (EIF) represent the rover state as a multivariate Gaussian distribution. Given that the vehicle model and the measurement models are unlikely to be Gaussian in nature, these algorithms use a linear approximation of these models with an assumption of zero mean Gaussian noise in order to make use of Gaussian estimation methods.

Another option for rover localization is to use a technique called SLAM. This is a probabilistic technique for estimating the position of a mobile agent in an unknown environment while concurrently creating a map of local features. SLAM techniques are typically special implementations of Gaussian filters or particle filters.

Gaussian filters, particle filters, and SLAM techniques are discussed further in the later sections.

4.4.4.1 Gaussian Filters

Two types of Gaussian filters are the EKF and EIF. Gaussian filtering is used to combine disparate, noisy, measurements to estimate parameters of interest. Gaussian filtering can also be used to predict the future values of parameters in the absence of information. In the case of planetary and terrestrial rovers, Gaussian filtering is a means of combining the various localization sources, each with their own noise characteristics, in an optimal way [102].

This section is not meant to be a detailed description of these filters, as the material exists in many places elsewhere. A detailed discussion of Kalman filtering is given here [103] and a helpful interactive tutorial of an EKF is here [104]. The following text gives a brief overview of these filters.

Gaussian filtering for rover localization includes the following concepts. First is the rover state, x_t, which is a vector that typically includes rover position, orientation, and optionally velocity or acceleration at time t. Second is the measurement of the state z_t, which, for example, can include data from an IMU, a visual odometry process, and star and Sun sensors. All of these data sources will also have an associated margin of error (covariance), as well as a measurement model relating the measurements to the rover's state. Third is the concept of control inputs, u_t, which usually takes the form of high-level control delivered to locomotion system. Wheel odometry is typically taken as the measure of this control.

The rover state is estimated from a probability distribution conditioned upon the state at all previous times and all the previous control data and measurements:

$$P\left(x_t | x_{(0:t-1)}, z_{(0:t-1)}, u_{(0:t)}\right) \tag{4.1}$$

A Markov assumption is made, meaning that the current state of the rover is only dependent on the current inputs, and all knowledge of past inputs is contained within the state from the previous step, that is, the state is complete. The algorithm is, therefore, recursive and the state probability distribution can be written as

$$P(x_t | x_{(t-1)}, u_t) \tag{4.2}$$

This probability is called the state transition probability. One can also make the assumption that if the state is complete, the probability of observing features in the environment can be defined as

$$P\left(z_t | x_{(0:t)}, z_{(0:t-1)}, u_{(0:t)}\right) = P(z_t | x_t) \qquad (4.3)$$

This distribution is defined as the measurement probability and allows one to predict the measurements given the complete state of the rover. These two probabilities allow one to generate a belief state for the rover state.

The belief state is the robot's posterior estimate of its current state. Given the control signal inputs $u_{0:t}$ and the environmental observations $z_{0:t}$ for each step, the current state x_t is represented by the belief function:

$$bel(x_t) = P\left(x_t | z_{(0:t)}, u_{(0:t)}\right) \qquad (4.4)$$

The belief state is built using two main steps: a predict step and an update step.

1) Using the control signal, the hypothesized state is moved. This prediction is the *a priori* belief of the rover state $bel(x_t) = P(x_t | x(t-1), u_t)$. This prediction will be noisy as it is based only on the control signal and the previous rover state.
2) Sensors make measurements of the rover's true state. The rover uses the *a priori* state to predict the values of the measurements. The difference between the predicted and actual measurements (the innovation) is used to update the rover's belief state giving a posterior estimate of the rover state, which is closer to the true state.

The EKF estimates the mean and covariance of the rover state while the EIF (the dual of the EKF) instead uses the information vector and information matrix. This difference causes a reduction in complexity for the EIF of some of the steps in that there is a reduction in number of large matrix inversions, which are computationally expensive.

4.4.4.2 Particle Filters

Particle features share the concepts of belief state, measurements, and control inputs with Gaussian filters. However, they have important differences, namely they do not assume a Gaussian distribution for the vehicle and measurement model but instead attempt to approximate the distribution directly. Particle filters are a type of nonparametric filtering technique that models the posterior through a set of samples randomly drawn from the posterior. These random samples are called particles, and those with lower errors are preserved and used to perform iterative resampling.

4.4.4.3 Simultaneous Localization and Mapping

SLAM is an umbrella name for techniques that simultaneously estimate the position of a vehicle in an unknown environment while at the same time creating a map of local features. SLAM techniques are typically special implementations of Gaussian filters or particle filters, as is described later. To date, a SLAM system has not yet been implemented on a flight rover.

SLAM implementations widen the definition of the rover state, x_t, to include a "map," that is, features in the local environment. Feature detection is the identification of features within sensor data.

The sensors are typically vision sensors such as monocular or stereo cameras, or LiDARs. The most common form of sensor data in SLAM implementations is from LiDAR; however, LiDAR technology has not been extensively used in space missions, and specifically not yet on rovers. As discussed, all planetary rovers to date have included cameras, so vision-based SLAM is well suited to planetary applications.

Feature detection uses algorithms such as Speeded Up Robust Features (SURF) [105] to identify points of interest within the image. If these points can be identified in consecutive images then their position relative to the camera can be tracked and hence it is possible to retrieve information about the location of the rover.

SLAM implementations also add an "augment" step to the "predict" and "update" steps. In this step, new detected features are added to the map (and thus the state). By building this map of features as the rover moves, it is able to improve its localization estimate and reduce its error upon re-encountering features. This mechanism is termed loop closure.

Two popular particle-filter implementations of SLAM are FastSLAM [106] and FastSLAM 2.0 [107]. In these techniques, the rover state and map are separated. Each particle contains a rover position estimate, as well as an EKF for each of the features in the map. The sum covariances of each of the EKFs are used to determine the weight of each particle, which is then used to resample the distribution.

SLAM implementations can produce high localization accuracy, which can help to improve the autonomous navigation of planetary rovers, but implementing SLAM for planetary exploration will not be trivial. The environment and rover platform present significant challenges. Terrestrial SLAM techniques have tended to be used either in structured indoor environments or structured outdoor environments where the structured environment makes feature tracking a simpler task. Planetary exploration provides a much more challenging scenario, as the terrain is unstructured and fairly homogeneous. As well, the computational power of planetary rovers is also extremely limited, and therefore the application of any SLAM system would need to be adapted to run on more restrictive, space-qualified hardware. SLAM can be very computation-intensive, as its complexity is a function of the size of its map. A comparison of three algorithms (FastSLAM, an EKF-based SLAM, and an EIF-based SLAM) in the context of planetary environments is given here [108]. Another example of a SLAM approach for planetary rovers, this time for a worksite context, is given here [109].

One approach to reduce the map size is to use "salient" features in conjunction with point-based features (e.g., SURF). Visual saliency feature detection algorithms (e.g., Ref. [110]) are biologically inspired methods for identifying regions of interests (ROIs) in an image, such as rocks on a planetary surface. Point-based features can then be detected inside these ROIs, rather than in the entire image. Tracking across successive images can use the point-based features, but only the salient features need be stored in the map. Planetary Monocular SLAM [111] is an example of such an algorithm. The algorithm also uses monocular images as an input to the system to further reduce the system complexity.

While SLAM techniques have not yet found themselves on a planetary rover, they have a bright future. SLAM can provide for accurate localization, owing to the fact it considers all measurements taken by the rover at once in order to produce an optimal estimate of the rover position and map contents. The concept of loop closure can also make SLAM ideal for missions that involve returning to locations (such as sample retrieval missions). The major challenge is implementing SLAM with the constraints of a rover processor.

4.4.5
Example Systems

This section provides examples of rover localization systems: proposed, prototyped, and real. The sensors used and the method for combining measurements is given for each case. These systems give a taste of the breadth of possible localization implementations, which is driven by the unique constraints of each targeted mission.

The navigation (and localization) systems for current and past flight rovers were presented in Section 4.2.3, while the planned flight rovers and rover concepts for future missions are described in Sections 4.6.1 and 4.6.2, respectively. In large part, these systems are based on IMUs for relative orientation, some type of Sun finding technique for absolute orientation, wheel odometry for distance traveled, and visual odometry for improved distance traveled and slip detection.

A lunar absolute-localization-only system that was predicated on the implementation of the Constellation program (and its associated lunar satellites) is described here [91]. This system uses an EKF to combine measurements from the local surface station and the two satellites.

Improving the absolute heading performance of rovers is the focus this system [52]. Lambert *et al.* describe how incorporating *regular* measurements from a dedicated Sun sensor and inclinometer directly into a visual odometry pipeline can contain orientation drift. The field results show a dramatic increase in localization accuracy to below 1% of distance traveled (measured over a 1 km traverse in a planetary analog terrain). A bundle-adjustment approach (similar to an EKF) is used to combine the measurements. This is different from the MERs, which update absolute orientation only occasionally.

Other work by the same group proposes an alternative to the Sun-sensor system, by using only sensors that are appropriate for a full-dark scenario, as would often be found near the lunar poles [53]. These sensors include wheel odometers and an IMU "as usual," but also a star tracker. Field testing of this system demonstrated 0.85% localization accuracy over a 7.5 km traverse, with star tracker measurements taken at 1 Hz while the rover was in motion.

The localization system of a rover prototype designed to support NASA's Resource Prospector mission is described here [112]. This rover, Artemis Jr,, incorporates the measurements from the following systems into its EKF: an IMU, two complementary visual-odometry systems (one using a downward looking stereo camera pair), wheel odometry, a Sun-sensor, a map-based absolute localizer, and a lander-based absolute localizer.

An "infrastructure free" localization system is described here [83]. This system, implemented on a prototype rover, incorporates measurements from wheel odometry, an IMU, visual odometry, a rover-sensed 3D terrain to orbital terrain map technique, and a rover-sensed horizon to orbital terrain map technique. The latter two are absolute-localization measurements, which are added to contain the drift of the relative measurements.

This thesis [113] describes the localization of the Zoe rover, which was used in Carnegie Mellon's Life in the Atacama project. This rover with articulating axles used wheel encoders, an IMU, inclinometers, and a Sun tracker. A Kalman implementation was developed to combine the measurements, but in the end, problems with the implementation required creation of a simple technique to combine the measurements; and this simple technique allowed the localization error goal of 5% to be met. The final error was 3.3% over 3.9 km.

A simple system developed for a prototype, skid-steered, micro-rover is described here [114]. This system combines measurements from an inclinometer, an azimuth-only gyroscope, and right- and left-drive motor odometers. Two unique aspects of the system are that zero updates are performed (i.e., the pose of the rover is fixed when it is stationary and the gyroscope bias is estimated), and that wheel slippage is estimated by taking advantage of redundant azimuth measurements (from the differential odometry measurement and the gyroscope).

Batista *et al.* [93] proposes a localization system in which IMU measurements are combined with range measurements from a single beacon on a lander using a Kalman filter. This system would be appropriate for a rover that does not venture far from its lander.

Another category of localization systems are SLAM systems, discussed in Section 4.4.4. Example systems are provided in that section as well.

As demonstrated here, there are many ways to implement a localization system. The performance requirements will drive the type of sensors and technologies required.

4.5
Autonomous Navigation

This section discusses various aspects of autonomous navigation: sensing, mapping, terrain assessment, path planning (both local and global), and control.

4.5.1
Sensing

The first step in autonomous navigation is sensing the 3D geometry of the local terrain. On all autonomous flight rovers, this has mainly been done with stereo cameras, though active 3D sensors (e.g., LiDARs) are another candidate for this purpose (see Section 4.6.4.4). This section introduces a typical stereo processing pipeline to generate 3D data. Active 3D sensors would generate native 3D data and would thus not require the following computations.

The main steps in generating stereo data are image rectification, image filtering, and stereo (or correspondence) matching. The image rectification process applies a calibration-derived transformation to the images to remove lens distortion. The result is "epipolar alignment" between the cameras, which means that the ray of light defined by each pixel in one rectified image will appear horizontal in the other image. This is important to simplify the correspondence matching. Image filtering is the application of a "difference operator" such as a Sobel mask to the images to highlight edge content and zero-out constant-intensity areas, important to make correspondence-matching more robust to differences between intensities in the source images. Finally, correspondence matching is done pixel by pixel in one image: the corresponding row in the other image is searched to find the best match. The difference in column number is called the "disparity," and range decreases with increasing disparity. A window-based correlation method, such as SAD, is typically used for this task.

The MERs, Curiosity, and ExoMars rovers all use variants of this pipeline. The MERs and Curiosity reduce the computational burden by subsampling the Nav-Cam images from 1024×1024 to 256×256 [38], whereas ExoMars segments the image into the upper part of the image where the full resolution (1024×1024) is kept (to improve long-range data), and the lower part where only half resolution is kept [39]. The MERs and Curiosity also have a fixed maximum disparity whereas ExoMars has a row-dependent maximum disparity.

4.5.2
Mapping

Mapping is the next step in autonomous navigation. A point cloud or mesh is a typical output of this process, which can then be passed to a terrain assessment step, described in the next section. This section discusses a typical processing pipeline for performing mapping, and also discusses differences in implementations among flight rovers and some terrestrial systems.

It is worth mentioning here that many types of data products, other than a simple point cloud or mesh generated by the rover, fall into the category of a "map." For example, a "global map" can be a 2D image, assembled from orbital imagery, such as the HiRISE image maps used for path planning for the MERs and Curiosity (see Section 4.2.3.4). A global map may also be a DEM, which is also derived from orbital imagery (and orbital ranging), and can be used for such things as absolute localization (see Section 4.4.3.3). A "traversability" or "cost" map, which a rover normally plans its path on, is discussed in the next section. Finally, a "topological" map is an abstraction in which a map is composed of features (e.g., SURF features, or sets of images) connected by graph edges to allow for searchability. This last type of map is used for SLAM systems (Section 4.4.4.3) or bundle-adjustment localization systems such as is used for the MERs (Section 4.2.3.4). The term "map" clearly has a wide definition.

Returning to a point-cloud or mesh map, a mapping process typically builds (and updates) a local terrain map. The rover senses the local terrain (see previous section) and generates a raw point cloud of the terrain. This can be built from data from a single stereo collection, or from multiple collections gathered by panning the cameras to different directions. The pose of the rover is used to place these data into a local frame of reference. "Outlier" points, such as those that are below ground, are removed. Next, a data-reduction process is typically applied to reduce the size of the point cloud while preserving features of interest (e.g., obstacles). This can be done in different ways: selective decimation of the point cloud (see Ref. [115] for a comparison of these methods), or voxelization/gridding (assigning points to a regular grid). At this point, the data product can be considered a map, and terrain assessment can proceed.

A variant of the above-mentioned approach is to triangulate (or "mesh") the data before or after the data-reduction process. Meshes can be a useful format for path planning because they provide connections between points through which a path can pass. A simple Delaunay triangulation can be used to produce the mesh. An alternative currently popular in the Open Source community is "greedy triangulation" [116], which is a fast triangulation method that also allows for fast updating of the mesh with new information. An example of a mesh-simplification technique is the "QSlim" method introduced here [117].

With either the point-cloud or mesh-based approaches, a map generated from a single rover position can be merged into an existing map. This allows the rover to remember obstacle locations that may be shadowed (or too far) from its current perspective. The major challenge here is that uncertainty in the rover pose between terrain sensing instances can lead to artifacts in the map. Point-cloud alignment techniques, such as those introduced in Section 4.4.2.4, can be applied to minimize these artifacts.

Each of the MERs, Curiosity, and ExoMars rovers creates its map from multiple stereo images collected from a single rover position. For the MERs and Curiosity, the entire set of range data from the stereo images is preserved at this stage [118]. Merging of these data with an existing map is done at the terrain assessment stage of the autonomy process (see Section 4.5.3).

One relevant terrestrial mapping example is the Reliable Autonomous Surface Mobility (RASM) system, created by Carnegie Mellon University and implemented on various terrestrial rover prototypes [119]. RASM is designed to autonomously guide a rover, in motion, at speeds up to 5 km/h. RASM's mapping pipeline, as presented in the cited paper, is as follows. The point cloud (which can come from any 3D source) is decimated using a covariance-based technique that preserves points in geometrically interesting areas and removes points in relatively featureless areas. The point cloud is then meshed using the Delaunay method and the mesh is then further reduced to a target size by selectively removing vertices that have minimal effect on the overall mesh shape. RASM then merges the current mesh with the existing mesh in memory (gathered during the traverse) using the ICP method to smooth the interface between the old and new meshes. RASM also "optimistically" meshes across occluded areas (e.g., behind obstacles), meaning that the rover will consider that area traversable. This is based on the field observation that this typically works in a rover's favor for navigating in rock-strewn landscapes: more paths are available for the rover, and it will likely get a better view of the occluded areas as the rover approaches them. Figure 4.8 shows a sample map generated by RASM.

A second terrestrial rover example is a system implemented by the CSA [120], which includes a mapping pipeline implemented on a prototype rover where the map is generated from a single point of view. A point cloud is gathered using a 360° LiDAR, meshed using the Delaunay method, and simplified using the QSlim algorithm.

4.5.3
Terrain Assessment

Terrain assessment follows mapping. The map is analyzed for traversability by considering the rover mobility capabilities such as ground clearance and maximum safe slope. A "traversability map" is typically the output of the terrain assessment process, which is then passed to an onboard autonomous path planner.

Figure 4.8 Sample mesh-based map generated by RASM autonomy software, also showing candidate motion arcs for rover. (Courtesy Neptec Design Group, Carnegie Mellon University.)

The MERs, Curiosity, and the ExoMars rover each analyze the point cloud map to generate a local, grid-based traversability map which is then merged into a larger map. Specifically for the MERs and Curiosity, the group of 3D points within a rover diameter of the centers of each grid cell are analyzed by fitting a plane to them to find the normal, residual, and maximum/minimum elevations [118]. This information is used to check this "rover-sized patch" of terrain for dangerous slopes, steps, and roughness. The resulting map is merged into a larger map (built from measurements taken from other rover positions), and new data overwrites old data. The ExoMars rover uses a similar process. For this rover, the local map is merged into a larger map by considering the degradation of rover positional accuracy over time – a wider berth is given around "old" obstacles.

Two prototype systems assess terrain in unique ways. For the RASM system [119] introduced in the previous section, terrain assessment is not done on the entire map. Computational load is reduced by detecting only obstacles on the set of valid short-range (i.e., <5 m) path candidates. The CSA prototype system ([120], referenced earlier) detects obstacles in a mesh-based map by simply removing triangles that have very steep slopes, and then removing triangles that are then left unconnected to the main map.

The preceding description and examples of terrain assessment capabilities can all be grouped into "geometric" terrain assessment, but terrain assessment can be done in other, less-common, ways. Sancho-Pradel and Gao [121] provide a detailed survey of terrain assessment techniques and define five categories, discussed here: geometric, appearance, soil, scientific interest, and semantic mapping. Appearance-based terrain assessment can complement geometric approaches by using visual cues from cameras (e.g., color, texture) to estimate the traversability of the terrain. This can be of particular use at long ranges for which geometric information may not be available. Defining and/or learning classifiers for appearance-based approaches, however, can be challenging due to significant intraclass variability. Next, soil-based techniques can be used to estimate soil characteristics and, therefore, traversability. These techniques can use measurable features such as motor current, output from visual odometry, and wheel encoder ticks to measure features such as sinkage, torque, and wheel loads. The "scientific interest" category involves the automatic assessment of a scene to determine targets for exploration, such as interesting rocks or outcrops. The resulting metrics can then be combined with other metrics to indicate areas of the terrain that optimize rover safety as well as scientific interest. The final category of semantic mapping involves extracting features of interest from sensor data (e.g., objects, planes) and linking them to the terrain map. This is similar to the concept of a "topological" map discussed in Section 4.5.2.

4.5.4
Path Planning

Given a traversability map, the current rover pose, and a goal location, an autonomous path planner will determine a safe and efficient path to this goal.

Typically, the goal is set by human operators who pick targets from orbital imagery or rover zoom-imagery (as is the case for the MERs and Curiosity). In addition, human operators often will plan a path manually through terrain that is visible from the rover position, and will then leave the autonomous path planner to plan a path into the unknown territory (also the case for the MERs and Curiosity). This section discusses important considerations of path planners, followed by different types of path planners, and then the specific configurations for flight rovers.

There are a number of important considerations when selecting or developing a path planner. First, the planner must take into account vehicle constraints. For example, a path that is a sequence of straight lines and right-angle turns (e.g., on a grid) is not usable by a vehicle that cannot point turn. Next, a planner should be able to efficiently replan (update) the path when given new information. In the extraterrestrial context, we do not expect the terrain to change, but we do expect new information about the terrain to come in regularly from the sensors. Next, the planner should be able to guarantee some degree of optimality of the path; and this optimality may include not only distance traveled and safety but also power used. For example, straight-line driving is more power-efficient for skid-steer vehicles than arc driving. Another consideration is the time to calculate the path (i.e., computational burden). Some planners, denoted as "anytime" (and discussed later), can produce a path in any length of time, but the quality of path increases with time.

This section divides path planners into "local" and "global" planners. Local planners generate a short path segment for the rover to follow typically using only current sensor information. Global path planners, however, generate an entire path from the current rover location to the goal using all information that is available about the terrain. Local path planners are naturally simpler to implement but have drawbacks such as being susceptible to dead ends.

4.5.4.1 Local Path Planning and Obstacle Avoidance

This section discusses local path planning and obstacle avoidance: two concepts that are treated as functionally identical here. Local path planners typically use only current sensor data as they choose short-term paths to avoid obstacles and make progress toward the goal. Obstacles can come in many forms, whether they are large rocks, trenches, or high-slope areas.

There are many types of local path planners. The simplest example is the "bug" algorithm [122], which assumes only contact sensors for obstacle detection and circumnavigates each obstacle in the rover's path so as to find the best place to depart from each obstacle on the way to the goal. There are numerous extensions, including Bug2, which avoids complete circumnavigation of each obstacle, and TangentBug, which incorporates a range sensor capability to make better decisions about which direction to take. A comparison of many types of Bug algorithms is presented here [123]. A more sophisticated category of local planner is the potential fields approach. Here, the goal is modeled with a "negative" particle and the obstacles with "positive" particles. The vehicle is, therefore, modeled as a positive particle: attracted to the goal while being pushed away from obstacles. This is a simple approach, but it is difficult to model vehicle behavior, and this approach has an issue with getting trapped in local minima. An example of this implementation is on a CSA protoype vehicle [124]. Many other types of local planners are described here [125].

The MERs' path-planner, GESTALT (before the 2006 software upgrade), was also a type of local planner. As described here [126], this planner used a voting structure to plan the next one or two short arc steps for the rover. The three votes were: avoid hazards, minimize steering time, and progress toward goal. The hazard-avoidance vote is based on the traversability map described earlier, though only "local" data (out to 6 m) is considered. The steering time vote is greater the less turning is necessary to begin the arc. The progress-to-goal vote is based on how close each arc gets the rover to the goal. The arc (pair) with the highest overall vote is selected.

Local path planners can also be used in tandem with a global planner (described in the next section): for example, the local planner can plan a small deviation off the global path, such as around an isolated obstacle that was not seen when the global path was planned. This avoids the need to regenerate a global path (often a computationally expensive process) each time such an obstacle is encountered.

Obstacle detection can be done in a number of ways. The MERs and Curiosity use body-mounted stereo cameras to detect a variety of obstacle types. LiDARs, popular in terrestrial rovers, may be used for future space rovers. Contact sensors were used in Sojourner to detect contact with an obstacle. Accelerometers (tilt sensors) can detect whether the rover is at a dangerous slope. Once an obstacle is detected, the rover can either stop or replan a path using any of the local or global methods discussed in this and the following sections.

4.5.4.2 Global Path Planning

Global path planners will plan a path to a goal using current and recent sensor information – data that is typically in the form of a traversability map, as described previously in Section 4.5.2. Also distinct from local planners, global planners will plan the entire path from the rover to the goal and can, therefore, estimate the time required to reach the goal. This section describes graph-search methods that can be used to plan a path on the traversability map, and also describes how the traversability map can be further manipulated to improve algorithm robustness or run-time. Finally, example systems are described.

A grid- or mesh-based traversability map can be used directly to plan a path using one of the graph-search methods described in this section, but the traversability map can also be further processed to improve results. For example, graph-search methods often require significant computation time to complete if a goal is inaccessible from the current rover position. Flood-fill algorithms can be used to remove unreachable areas of the map, thus determining quickly if a goal is reachable (this is done for the MERs and Curiosity [126]). Dead-end paths (or "fingers") can also be removed from the map to improve performance of graph-search methods.

A traversability map can also be transformed into something other than a regular grid or mesh – the goal being reducing the number of "nodes" on the map to improve graph-search algorithm performance. Two categories of transformations are described briefly here: cell decomposition and road maps (also described in more detail here [127]). Cell decomposition involves finding a cell-based representation of the traversability map, different from regular square (or triangular) cells. For example, the cells can be amalgamated using a quadtree approach in which large passable areas are represented by a few large cells, and more congested areas are represented in more detail by smaller cells. A road map, on the other hand, abandons the concept of cells and instead represents the traversability map as a network of passable paths (or roads). One example of this is a "visibility graph" in which a road segment is created between each obstacle corner and all other obstacle corners that are visible from that location (which of course would result in the rover passing very close to obstacles as it traverses the map). Conversely, a Voronoi diagram of the traversability map would create a network of roads that maximizes distance from all obstacles. A final example is a probabilistic roadmap for which nodes are randomly sampled on a map and are connected with each other wherever the resulting road segment is passable.

All of these approaches will significantly reduce the number of nodes in a traversability map, but these techniques have not been implemented in any flight rover autonomy system, nor are they that common on terrestrial rovers. This is likely due to the facts that many of the popular graph-search methods described later have been developed with a regular grid in mind, and that using a traversability map as-is reduces overall implementation complexity (and, therefore, testing and performance characterization complexity) of a path planner.

Employing a graph-search algorithm on the traversability map will yield a path from the rover location to the goal. This is an active area of research, and hence there are many options to choose from. This report [128] describes a number of graph-based path planners, which are summarized here.

The A* method, first introduced in 1968 [129], begins at an initial node (i.e., the rover location) and employs a best-first search to expand outward and toward the goal. For each evaluated node, a cost to reach that node and the estimated cost to reach the goal location from that node are stored. This information is used to determine preferred connections between nodes, which are saved as the algorithm progresses. When the algorithm reaches the goal location, the optimal path can be directly extracted from these connections. Variable terrain costs (i.e., traversability costs) can be incorporated into the cost function along with distance costs, making this algorithm very appropriate to apply to grid-based traversability maps. Drawbacks of A* are that replanning a path requires rerunning the entire algorithm, and that when given a regular grid, it will produce a path limited to horizontal, vertical, and diagonal moves. Simple examples of A* paths on a grid and on a graph are shown in Figure 4.9.

Numerous other algorithms have been developed to improve on the performance of A*. The D* (or "Dynamic A*") algorithm [131] improves on A* by supporting fast replanning of the path; in the rover case, this translates to when the

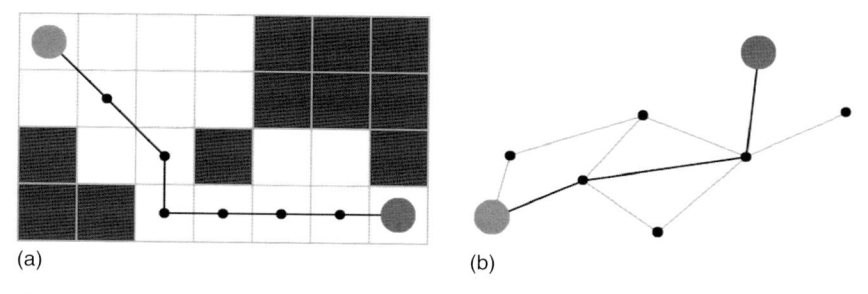

Figure 4.9 A* paths on a grid (a) and on a connected graph (b) [130].

rover moves and the traversability map is updated. D*-Lite [132] matches the performance of D*, but is algorithmically simpler. The Field D* [19] is a variant of D* and D*-Lite that allows for interpolation between grid squares, resulting in movements that can be at any angle (i.e., not just at 45° or 90°). The Incremental Phi* algorithm [133] improves on the Field D* path quality while maintaining the fast-replanning capability. Finally, the Anytime D* [134] is a variant of D*-Lite that produces a path in "anytime," that is, a path is always available, but its quality increases with time.

The final graph-search path, however, does not consider vehicle constraints such as turning radius. A workable solution to this issue, one used by the MERs, Curiosity and the ExoMars rover, is to first consider a set of valid short path (e.g., arcs) that can be achieved by the rover followed by a graph-search path from the end of each short path to the goal. The MERs and Curiosity consider a set of arc options (one or two arcs in length) and then use the Field D* algorithm as one of the three "votes" in their planner (replacing the "progress towards goal" vote) [126]. The ExoMars rover, on the other hand, will use the A* algorithm, but the search space will start with an (optional) point-turn, followed by a set of three curves, then a standard path on a regular grid. The terrestrial RASM system, discussed previously, also picks the short-term path (from a set of arcs) using a metric based on obstacles and progress to the goal. This progress measurement is done with A*.

This section has described how global path planners can produce a path on a traversability map when given the rover location and a goal location. The specific implementation on the MERs, Curiosity, and the ExoMars rover has also been described.

4.5.5
Control

The final piece of the autonomy puzzle is control: how to issue drive commands to guide the rover along its chosen path. This section discusses important factors in control and the control strategy used for flight rovers.

Control modes can be broken down into "closed loop" and "open loop" control. With closed-loop control, the rover keeps a real-time estimation of its location and orientation as it moves so as to follow the path as closely as possible. This localization can be done with gyroscopes, wheel odometry, visual odometry, and so on. With open-loop control, however, the rover will drive a short segment "blind": it will command the wheels to turn a prescribed amount that the rover estimates will get it to follow the planned path. Only between path segments does the rover update its position and orientation with its localizations sensors. Closed-loop control will naturally yield better performance, particularly in cases of slippery terrain or slopes, for which it is difficult to estimate the effect of the terrain on the traverse. The ExoMars rover will implement closed-loop control to keep it very close to its planned path (including using visual odometry every 10 s during motion), which was a priority given the expected difficulty of the terrain it will encounter. The MERs and Curiosity, however, operate in open loop: an arc is commanded, and they try to follow the arc as close as possible without real-time feedback on their progress. These rovers can detect wheel slip (as discussed previously in Section 4.2.3), but not during motion.

The control algorithm used is a second consideration for rover control. For example, the Astolfi control algorithm [135] keeps running values for the distance to goal (or next waypoint), delta orientation to goal, and delta orientation to goal orientation (a goal is assigned an orientation). These values (scaled by tuning parameters) get incorporated into a control law that outputs values for forward and rotational velocities. A potential field local planner (described earlier) will also output drive directions and speeds that can be used to control a vehicle. The ExoMars has separate controllers to keep the lateral error from the path and the orientation error within limits [36]. Control of a rover is made somewhat easier because rover dynamics will likely need not be considered given the slow speed of the vehicles.

4.6
Future of Planetary Surface Navigation

This section discusses the future of planetary rover navigation. The navigation systems of the ExoMars and Mars 2020 rovers are described, along with planned future missions that will require new types of rovers. Next, recent field trials as proving grounds for future rover missions are described. Finally, capabilities that may be a part of future rovers are described, from new sensor classes to alternate mobility concepts.

4.6.1
Planned Flight Rovers

This section describes the navigation systems of two currently planned flight rovers: the ESA ExoMars rover and the NASA Mars 2020 rover. The ExoMars rover navigation system is described in detail, to provide the reader with an in-depth view into this state-of-the-art system.

4.6.1.1 ExoMars Rover

ESA will land the ExoMars rover on Mars in 2018, with a goal of determining whether life ever existed on the planet [36]. The ExoMars rover will have to achieve a level of autonomy not yet tried on the surface of another planet.

Environmental, autonomy, and performance requirements drive the design of the rover navigation system. The rover must survive and operate under extreme temperatures and in the presence of dust. Equipment must survive temperatures ranging from -120 to $+40\,°C$, and must operate between -50 and $+40\,°C$. The rover must be able to continue performing its mission without ground controllers in the loop for 2 sols. This may take the rover outside the terrain which an operator can safely assess (up to 20 m range); therefore, there is a need to autonomously ensure the rover safety while driving. Performance requirements (driven by science needs) include the rover being able to drive 70 m/sol with maximum autonomy active. After this distance, the rover must also be placed within 7 m of its intended target, and with a heading within $5°$ of the command.

The umbrella of sensor and processor hardware includes a number of components. The stereo NavCams support the Perception, Navigation, and Path Planning functions. However, their images are also used to determine the Sun direction. This measurement, together with the gravity measured by the accelerometer (part of the IMU, which also includes a three-axis gyroscope), is used to initialize the rover's absolute attitude. The NavCam is on top of a mast 2 m above ground and is pointed using the pan and tilt mechanism. The Localization Cameras (LocCam) are of the same design as the NavCam. They provide 1024×1024 (or 512×512) 8-bit panchromatic stereo images. The LocCam provides images as the rover moves such that Visual Localization provides an update of the rover pose every 10 s. These sensors are shown in the architecture diagram of Figure 4.5. The processor used by the navigation algorithms is a LEON2 processor at 96 MHz with 512 MB of RAM, and benchmarked at 75 MIPS.

The rover supports a sliding scale of driving modes, from full autonomy to full manual control. The top level corresponds to maximum autonomy, where all functions are used. This level corresponds to the nominal operational mode where operators provide a target in Mars Local Geodetic (MLG) frame for the rover to autonomously reach. This mode corresponds to the lowest rover net speed. One step lower is a mode in which the rover follows a predefined path, but still must check that the path is safe. Going down the levels of autonomy, the rover can also follow a path without the checks mentioned earlier, but still performs closed-loop trajectory control. This is followed by ground directly commanding open loop maneuvers such as Ackerman geometries to follow (duration or distance limited - the latter still requires localization) or point-turn maneuvers. If one drops localization, only duration-limited maneuvers are then possible; this level also allows for direct commanding of each actuator. Finally, in a much reduced mode, it is possible to bypass most of the functionality and directly command at bus-level each actuator.

For absolute localization, the process will be similar to that used by the MERs and Curiosity. Landmarks seen in orbital imagery (e.g., from HiRISE) can be compared with those seen in rover panoramas. This will be used to help guide the long-term path of the rover (e.g., to avoid large craters).

Absolute orientation of the rover is estimated using the IMU and detecting the Sun location in the sky. The roll and pitch angles are updated each time the rover stops using inputs from the IMU [36]. At the beginning and/or end of each day of operations, the entire absolute attitude is reinitialized when the rover is at rest. A method based on the "triad" algorithm is used taking as inputs the gravity and Sun direction vectors. A low Sun angle maximizes the separation of these vectors [36]. In order to obtain the Sun direction, the NavCam image(s) must be processed such that the direction is extracted. The NavCam does not have any specific filter to image the Sun, so there are artifacts in the image the algorithm has to cope with.

Relative localization is achieved via wheel odometry, the IMU, and visual odometry. Visual odometry (VO) will be used much more heavily than for the MERs and Curiosity: it will run on images collected at regular 10-s intervals while the rover is driving, meaning that the rover will often be in motion when the image pairs are captured. The VO result will overwrite the between-frame estimation of position and orientation calculated by the wheel odometry and gyroscope. This frequent and regular source of position information is a key safety feature since it allows for continuous monitoring of the rover slippage levels. The estimated average wheel slip is also used to scale the wheel odometry measurements in order to improve localization accuracy between VO updates [36]. Additional discussion of the ExoMars VO functionality is given in Section 4.4.2.2.

Target locations, to be reached via autonomous traverses, will be selected manually by ground-based operators using either rover or orbital imagery, depending on operational circumstances. At the end of one sol, if the rover does not reach a target, it will continue to travel toward that target on the next sol. By requirement, in order to detect whether something went wrong in communications with ground (safety), the rover only drives two consecutive sols without any ground communication.

The terrain is traversed via the rover repeatedly planning and following a path sequence toward the target location. Each path sequence is an up to 2.3-m long series of straights, curves, and point turns. On average, a path sequence must be planned by the navigation processor in 135 s or less, a derived requirement from the 70 m/sol requirement. The ability to plan these path sequences is split into three modules: Perception, Navigation, and Path Planning, which are discussed here. The entire process is also shown in Figure 4.10.

The Perception module generates disparity maps from NavCam stereo images, and is not allowed to take longer than 15 s per stereo pair in average processing time. A multi-resolution approach is used to improve the disparity accuracy in the far field while at the same time not violating the processing time requirement [39].

At each "navigation stop" between path sequences, the Navigation module creates the maps used by the Path Planning module [36]. This must be done in an

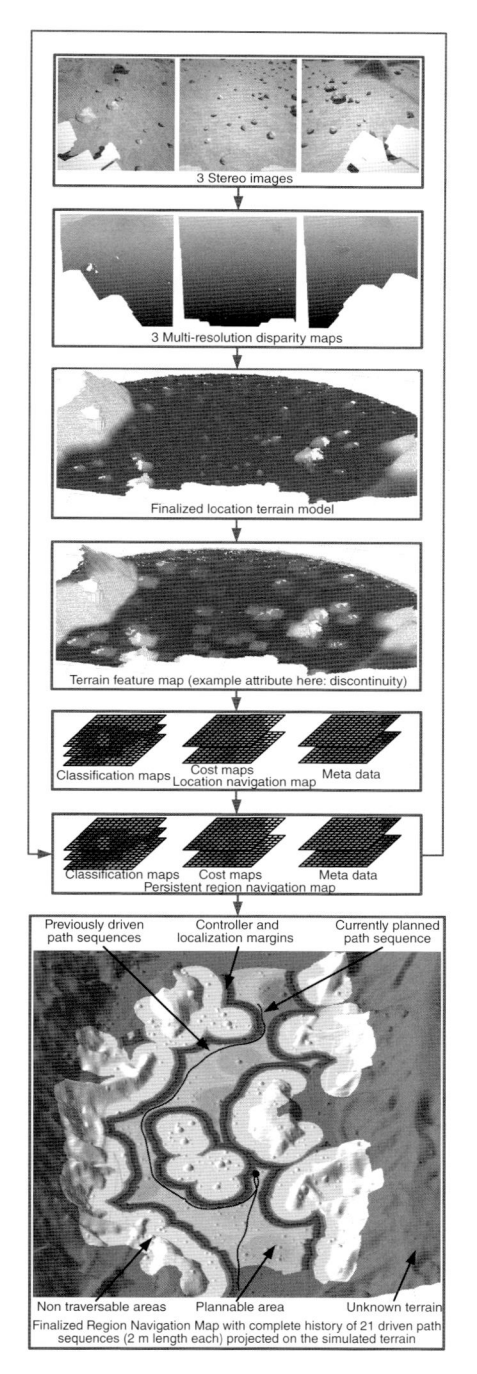

Figure 4.10 Navigation processing overview for the ExoMars rover [36]. (Courtesy Airbus DS).

average of 78 s or less. A DEM-based 3D terrain model is first created from all disparity maps generated at the same navigation stop (three pairs of images, acquired by panning the NavCam). Next, this model is analyzed with respect to the capability of the locomotion system and is converted into a 2D navigation (traversability) map, which is in a Cartesian grid format. The map specifies traversable, nontraversable, and unknown areas. Moreover, cost values are defined for traversable areas based on how difficult and risky traversing them will be. The content of this map is then merged into a region navigation map containing the data acquired at previous locations. It is important to note that this map must take into account that with every navigation stop the uncertainty in the relative position of previously mapped areas increases – for example, if the rover detects an obstacle 2 m to the left of it and then drives an estimated 30 m forward, then the obstacle might not be 30 m behind it and 2 m to the left because of the uncertainty in the estimate of the localization system. Therefore, the rover's mapping system does not know the exact coordinates of the obstacle any longer, but utilizing uncertainty estimates from the localization system, it can define a larger area that contains the obstacle.

There are three outputs generated at each stop: a "finalized region navigation map" in which the Path Planning module will plan the next path sequence, the "escape boundary," which is a map specifying possible intermediate end points for the long-term path planning, and the "traverse monitoring map," which is a map specifying the areas the rover position estimate is allowed to enter.

The outputs take into account the expected performance of the relative localization system and the trajectory control system when driving the next path sequence. For example, an area of unknown or nontraversable terrain in the persistent region navigation map will lead to a slightly larger area the rover position estimate is not allowed to enter in the traverse monitoring map. This is needed to guarantee rover safety considering the uncertainty in the position estimate while driving the next path sequence. The same area will lead to a yet larger area that no path sequence is allowed to be planned though in the finalized region navigation map. This is needed to avoid triggering a safety failure when the motion controller – due to a disturbance such as a rock or a slope – is temporarily not able to keep the position estimate perfectly on the planned path.

The Path Planning module plans a safe path sequence toward the target on the region navigation map, which it must execute in an average time of 12 s or less [36]. At each navigation stop, the Path Planning module plans the next path sequence to be driven. Most of the time, the majority of the terrain from the rover to the target is not known because the target is too far away from the current rover position. Therefore, the Path Planning module selects a point on the escape boundary as temporary target. This point is called the escape point. Next, the Path Planning module plans the route to the escape point, which consists of the following segments in order: an optional point turn, three smooth curves, and a nonsmooth path. The path selection process utilizes an A* algorithm operated across a hybrid search graph. This consists of lattice edges and nodes representing the initial point turn and the smooth curves, and classical Cartesian grid edges and nodes for the search between the initial smooth curves and the escape point. Only the

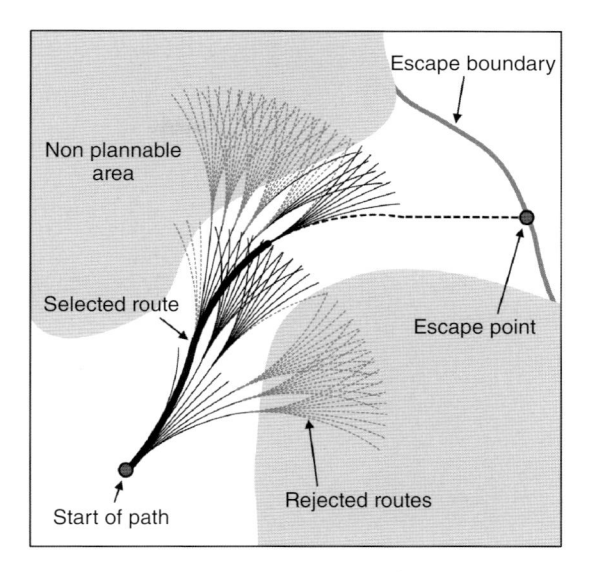

Figure 4.11 Path planning visualization for the ExoMars rover [36]. (Courtesy Airbus DS.)

optional initial point turn and the first two smooth curves are output as current path sequence to trajectory control. The rest of the path is only created to take into account the cost and dynamics of follow-up path sequences. Figure 4.11 provides a visualization of the path planning functionality.

The ExoMars rover will follow a path using a closed-loop control strategy. Separate controllers keep the lateral error from the path and the heading error within limits [36], and the full localization estimate (including VO) is used in real-time on board the rover. This decision was driven by the expectation that the ExoMars rover will drive over challenging terrains and will thus require more accurate trajectory control. During driving, the rover is also monitored for entering a dangerous configuration: like excessive tilt, large bogie angles, excessive slip, or entering a forbidden area.

4.6.1.2 Mars 2020 Rover

This mission will include an essentially build-to-print copy of the Curiosity rover, though with an enhanced payload package. The rover will also "inherit MSL's C++ code base" according to NASA [25]. This means that the navigation system will be mostly identical to Curiosity's, although with a couple of enhancements. The first enhancement is that the engineering cameras (NavCams and HazCams) will be color and will be of higher resolution than their Curiosity versions [136], which may lead to enhanced visual odometry and 3D perception capabilities. Second, as discussed previously, NASA plans to include a "fast traverse capability" in the rover enabled by visual odometry calculations being done on an FPGA to reduce the algorithm run-time [27]. This will mean shorter stops, and more roving time. There may be other enhancements to the navigation system that are incorporated before launch, so the reader is encouraged to follow the pre-mission updates.

4.6.2 Future Rover Missions

This section discusses rover navigation requirements and capabilities for missions currently on the drawing board. Described first is the ESA Mars Precision Lander mission, which will include a sample-fetching rover (SFR). Next, the proposed NASA Resource Prospector mission is described as well as a concept for a rover.

4.6.2.1 Mars Precision Lander

The planned Mars Precision Lander ESA mission is part of the international Mars sample-return initiative. This mission will land a small (approximately 85 kg) SFR and an ascent vehicle to a location near where a cached sample is located [137] (possibly cached by the Mars 2020 rover). The rover will be up to 10 km distant from the cache when it lands. Distinct from other rovers, the SFR must return to its starting location – an intriguing problem with many possible solutions (e.g., SLAM-based loop-closure, leaving a trail of beacons that can be tracked, vision-based path following).

A concept for the SFR, and a detailed analysis of the navigation requirements is described in Ref. [138]. This is summarized here. The estimated average navigation speed is 55 m/h and distance per sol is 210 m. The baseline navigation system includes continuous navigation with perception based on a single stereo pair (for navigation and localization), and an IMU. The small rover size limits the number and size of sensors that can be included. Absolute localization would be performed on the ground via a bundle-adjustment technique that compares the positions of natural features in orbital and descent imagery with those in the rover imagery. This paper also highlights priority areas for European technology development, including porting vision-based algorithms onto FPGAs (e.g., SLAM), further developing the imagery-matching technique, and developing a lower-mass and lower-power IMU. Lower-priority technology developments include active sensors (e.g., LiDAR) for improved rover perception, and the detection of dangerous soil characteristics (e.g., soft soil).

The SFR, therefore, embodies an exciting new navigation challenge. It has a high-speed requirement, low mass and power requirements, must be able to find and acquire a cached sample, and perform the novel task of returning to an ascent vehicle.

4.6.2.2 Resource Prospector Mission

NASA's planned Resource Prospector mission seeks to "demonstrate prospecting for volatiles and extraction of oxygen directly from lunar regolith" [1]. Planned for a 2019 launch, and targeting one of the lunar poles, this mission includes a rover to enable the prospecting activity. The mission duration is extremely short – 1 week or less – which is a result of the harsh lighting and thermal environment at the poles. The roving will be within a 1 km radius of the lander, and will include multiple stops to perform drilling operations in the regolith. This mission is also a "Class D" mission, which is the most risk-tolerant, and low-cost of the NASA mission types. This unique mission will require a unique navigation system on the rover: one that must not only be low-cost but also be able to perform in the challenging polar environment.

A rover concept for the Resource Prospector mission is described in this report for the CSA [37]. Additional requirements and details were specified by the CSA for this project, most notably: a requirement for 5% relative localization accuracy (target: 2%), and an assumption of a 10-s communication delay. The navigation system concept is as follows. Wide-FOV mast-mounted stereo cameras support multiple purposes: teleoperation, visual odometry, terrain imaging, and obstacle-detection. A star tracker is used for orientation estimation with a low-cost IMU as backup. Rover-to-orbiter image mapping is used for absolute localization. Visual odometry is performed by sending images to the ground for processing (after each short traverse), and wheel odometry rounds out the localization methods. The rover is controlled via a supervised teleoperation in which the rover is sent motion commands from the ground, which can be overridden by the obstacle detection functionality on the rover. The rover can follow prescribed paths and even avoid small detected obstacles. This navigation system strives to minimize cost and complexity (e.g., limited autonomy, low sensor count) but still provides the necessary mission functionality.

4.6.3
Field Trials as Proving Grounds for Future Navigation Technologies

This section discusses field trials and mission simulations that have been carried out to test future rover navigation technologies. To put these activities in context, a review of the types of testing locations is first presented.

Broadly speaking, test sites can be broken down into four categories, from low to high testing fidelity: indoor test "yards," outdoor test "yards," local field sites, and remote field sites. Indoor test yards such as the Airbus Defence and Space test yard (built to test the ExoMars rover) normally feature representative surface material (e.g., sand), an assortment of obstacles, controlled lighting, and significant supporting infrastructure (e.g., for measurement, communication, power). Outdoor test yards, such as the CSA Analog Terrain (see Figure 4.12), are normally similar in appearance, but are larger, and perhaps with reduced infrastructure, and obviously no control over lighting or other environmental factors. Local field sites are typically representative terrains that are much larger than the outdoor test yards, such as gravel pits or quarries, but have little in the way of supporting infrastructure. Finally, remote field sites are those sites that are selected for their similarity with the target mission site, such as the Atacama Desert as a Mars analog. These

Figure 4.12 CSA Analog Terrain.

sites are typically very large (multiple kilometers of traversable terrain), but naturally can require significant travel and logistics effort, and have little in the way of supporting infrastructure.

Clearly there are benefits to each of these different types of test sites. For example, indoor test yards are valuable for early technology validation because they provide for a slightly representative test environment, and are very time-efficient for developers to use (little travel necessary, equipment can be left set up day-to-day, uninterrupted testing, etc.). Moving through the categories to the more distant sites, the main trade-off becomes time and effort efficiency versus realism and testing fidelity. The remote field sites are appropriate for testing late in the technology development cycle. They are invaluable for verifying robustness of technologies in realistic scenarios (such as long-duration, challenging lighting, new terrain). A comparison table of the different test site categories, with examples for each category, is given in Table 4.2.

Recently, international space agencies have increasingly found value in performing hi-fidelity field trials (and "mission simulations") at remote field sites. These field trials have been used not only to test navigation (and other) technologies but also to test operational mission concepts such as mission planning and operator control strategies. The following sections summarize the objectives and results of a number of recent navigation-related field trials conducted by the ESA, NASA, and CSA.

4.6.3.1 RESOLVE/Resource Prospector Mission (NASA/CSA)

This mission simulation was a joint NASA and CSA effort and was conducted on the slopes of Mauna Kea, Hawaii, in July 2012 [139]. The objective was to perform a hi-fidelity simulation of the possible NASA Regolith and Environment Science and Oxygen and Lunar Volatiles Extraction (RESOLVE) mission, now called the RP

Table 4.2 Comparison of types of field testing sites.

Testing site type	Example sites	Environmental control	Time efficiency	Support infrastructure	Realism/testing fidelity
Indoor test "yards"	Airbus Defence & Space Mars Yard (30 × 13 m, Stevenage, UK), University of Toronto Mars Dome (40 m diameter, Toronto, Canada)	Yes	High	High	Low
Outdoor test "yards"	CNES Mars Yard (80 × 50 m, Tolouse, France), JPL Mars Yard III (66 × 36 m, California, USA), CSA Analog Terrain (120 × 60 m, Montreal, Canada), RAL Mars Yard (Harwell, UK)	No	Medium/high	Medium/high	Medium
Local field sites	Beaches, sandpits, quarries, and so on	No	Medium	Low	Medium/high
Remote field sites	Atacama Desert (Chile), Devon Island (Canadian High Arctic), Tenerife (Spain), Mauna Kea (Hawaii, USA), Mars Desert Research Station (Utah, USA)	No	Low	Low	High

mission (see Section 4.6.2.2). This simulation involved six mission days in which the rover was teleoperated or commanded to drive autonomously from a remote "mission control" location to prospect for volatiles in the soil using sensors in the NASA payload.

NASA specified mandatory and desirable objectives for the mission simulation. The ones that were relevant to the rover navigation system included a mandatory requirement to map volatiles in the terrain over a point-to-point distance of at least 100 m, and desirable objectives to map the distribution of volatiles over a distance of 500 m, and to travel a minimum of 3 km in total. Only the 3 km goal was not met, as the rover traveled a total distance of 1.1 km.

The CSA rover Artemis Jr was used for the mission simulation (as previously introduced in Section 4.4.5) with a localization system detailed in Ref. [112]. Figure 4.13 shows the rover (and the "lander" in the distance) at the field site.

A summary of the performance of the rover at the mission simulation is described in Ref. [140], and discussed in this paragraph. Rover localization was primarily provided by a lander-based system which tracked a reflective target on the rover and communicated the calculated Cartesian coordinates to the rover. This kept the localization error below 1 m for the mission duration. On two occasions, however, line-of-sight contact with the lander was lost due to self-occlusion of the rover. During these times, the rover operated for 60–70 m relying only on its relative localization system, which consisted of wheel odometry, an IMU, and stereo visual odometry. The rover communicated its estimated position to the lander system, which allowed the lander to reacquire the reflective target later and begin tracking the rover again. During these two occasions, relative localization

Figure 4.13 The CSA Artemis Jr rover at the RESOLVE mission simulation on Mauna Kea, Hawaii.

error was measured at about 2%. The rover was primarily teleoperated using a mast camera as input, but it did perform two autonomous traverses for a total distance of 200 m. The average speed of the rover was approximately 1.5 m/min, which included stoppages for navigation and science decision-making.

4.6.3.2 SEEKER and SAFER (ESA)

Two linked ESA field trials in the Atacama Desert were the SEEKER field trial of June 2012 and the SAFER field trial of October 2013. The same navigation and autonomy system was used for both field trials.

The SEEKER field trial was an ESA effort which brought a prototype Mars rover to the Atacama Desert in Chile for long-distance autonomous navigation testing [62]. The goal of the field trial was to have a rover autonomously traverse 2 km each day for a period of 3 days. In doing so, this would demonstrate whether state-of-the-art vision and navigation techniques were up to this task, or whether further development was necessary. The rationale for this goal was that longer traverses (over the approximately 200 m/day currently possible) would be necessary to satisfy the challenging requirements of the next generation of Moon and Mars exploration missions.

This field trial was much further afield than previous ESA field trials to the CNES Mars Yard or to the island of Tenerife in Spain, and featured a much more representative environment. The test area was approximately 2 km × 3 km.

The rover featured relative and absolute localization systems that were exercised over the course of the field trial. The rover uses wheel odometry, an IMU, and two different stereo camera visual odometry techniques, all combined with a Kalman filter, to perform relative localization. Initial results in UK field testing demonstrated that 0.5% relative localization error (a percentage of distance traveled) was possible. The stereo cameras were also used to generate coarse (far-field) and fine (near-field) DEMs. Absolute localization was achieved by comparing the coarse DEM to a pre-existing map using a particle-filter approach. This was performed occasionally by the team when there were strong local features and after long distances traveled. The pre-existing map was generated using ortho images from unmanned aerial vehicle (UAV) flights. The map product was reduced to 1 m/pixel to be representative of a real mission.

The path-planning function used all available information to guide the rover. At the highest level, an operator created a path using the DEM, which avoided obstacles in the aerial imagery. This path was converted to local goals every 2 m, which were passed to the rover-based autonomy system. This system used the rover stereo cameras to generate local DEMs to navigate to the goals. Using a terrain-assessment approach, which took into account ground clearance, max slope, and roughness, the DEM was converted into a traversability grid with 15 cm cells. The autonomy system would try to get to each of the 2 m-spaced goals, but had the capability to miss goals to avoid newly detected obstacles, as long as it stayed close enough to the global path. There was no real-time obstacle detection implemented (e.g., using "HazCams"), but this functionality was avoided by re-running the autonomous path-planning algorithm every 2 m of travel. The D* search algorithm was used for path planning.

Results from the field test were positive, if they did not quite reach their stated goals. During prep days, the rover covered a total of 10 km. Over the three formal test days, the rover covered an additional 5 km, including one completely autonomous run of 1.2 km. The absolute localization algorithm was successfully demonstrated. The two visual odometry algorithms were well-exercised in the challenging lighting and terrain appearance. One of the algorithms proved much more accurate with a 1.9% localization error compared with 11.2% localization error for the other one. Perhaps most importantly, the team recognized the necessity of performing field trials in representative environments:

> …Although the team had carried out extensive and progressive testing in the United Kingdom in Mars Yards, quarries and beach locations there is no substitute for the diversity and combination of both adverse terrain and lighting conditions when testing a vision-based system…. The main challenges in the analog site included featureless, saturated terrain, un-predictable slip and variable slope, varying size boulder fields.

Performing part of the ExoMars reference surface mission (specifically the sample acquisition procedure) was the primary objective of the SAFER field trial, which took place in the Atacama Desert in October 2013 [141]. Secondary goals were to test control of the rover from an operations center at ESA Harwell (therefore, the project team was split between Chile and Harwell), and to continue to improve ESA's experience with field trials. This field trial used the Airbus Defence and Space Bridget rover, which used an autonomy system derived from the SEEKER system.

The rover traveled a total of 300 m over 5 days (six simulated sols), with all traverses performed autonomously. A 134-m traverse was performed on the final mission day. As the mission days passed, it was shown that the remote operations team began to trust the autonomous system more. This resulted in longer traverses and greater scientific output. The rover localization system was also essential in allowing the remote operations team to precisely locate a ground-penetrating radar and point a narrow field-of-view camera in order to capture the desired scientific data [142].

There was another unique lesson-learned from the SAFER field trial, a result of there being a local and a remote team. The on-site team noted occasions in which the remote team missed identifying interesting features in the rover surroundings – a finding that demonstrated that sensing and bandwidth limitations can affect science returns. This provided insight into the design of future sensor package designs, and for future concepts such as autonomous science.

4.6.3.3 Teleoperation Robotic Testbed (CSA)

The CSA performed a multi-week field trial in a local sand quarry in the fall of 2013. This field trial, introduced in Section 4.3.2.5 [49], focused on testing teleoperation strategies in a lunar scenario with a multi-second control delay. The NASA Resource Prospector mission was targeted as the reference mission.

In particular, the CSA sought the following: to study how the operators used a set of wide-field-of-view cameras on the rover for situational awareness, and to test some augmented control strategies and tools. The cameras proved helpful as they provided complete visual coverage around the rover, but the testing revealed that some type of 3D sensor would be very helpful in resolving ambiguities in the 2D images and, therefore, helping to ensure rover safety. As well, the augmented control strategies, which included a predictive display (to lessen the impact of the control delay) as well as a means to command short paths (e.g., go straight 5 m), also proved useful.

A unique aspect to this trial was the way CSA conducted the tests. The testing was quite exhaustive: a total of 16 operator pairs with varying backgrounds each had 3 h to perform the same "mission." As well, none of the operators had seen the test site. In the end, the rover had covered 5.8 km of terrain with an average speed of 2–4 m/min.

4.6.3.4 Other Field Trials

Some other field trials have been funded through research projects, for example, the EU FP7 PRoViScout field trial, or the UKSA Chameleon field trial.

The Planetary Robotics Vision Scout (PRoViScout) field trial was held on the island of Tenerife in September 2012 [143]. Its primary goal was autonomous sample identification, and a secondary goal was to demonstrate vision-based navigation. The prototype rover Idris has a laser-scanner-based obstacle-detection system and localized using visual odometry (with stereo cameras), wheel odometry, and an IMU. The stereo cameras were also used to generate DEMs, which were used to create hazard and slope maps. A tethered aerobot provided aerial imagery, which was used to create a map for the field trial. Autonomous target selection was demonstrated, although mostly in a stand-alone fashion. A number of challenges such as lighting and scene clutter were noted. The rover was driven mostly by teleoperation using cameras on the rover and on the terrain for situational awareness, although two autonomous drives were also accomplished.

Part of the SEEKER and SAFER team returned to the Atacama in the fall of 2014 as part of the UKSA Chameleon project. This project had a specific navigational goal to test an adaptive system that can select its autonomous mode based on visual characteristics of the terrain [144]. These visual characteristics, such as ground texture or 3D morphology, could be analyzed by the system, which could in turn decide to modify how it will function, such as by turning off unnecessary sensors. The assumption is that this autonomy may yield energy savings and could, therefore, increase distance traveled by the rover, something that will be important for future rovers such as the SFR described earlier. The equipment at the field trial included a rover called SOLO as well as a UAV, which provided simulated orbital data. In addition to the typical navigation sensors, the rover sensor package also included a laser range finder and an infrared (IR) camera. The field trial was essentially a data-gathering exercise to support offline analysis [144]. One definite result the team noted was that the vision-based algorithms were clearly the most power-intensive, so reducing the resolution of images when possible yielded significant power savings.

4.6.4
Future Capabilities

Future planetary exploration missions will require rover traverses that are longer than are achieved currently and that take the rover through more challenging environments, such as in shadows or in rough terrain. Advances in autonomy, sensing, onboard processing, and even in mobility platforms will all help achieve the necessary performance. This section discusses specific technologies that can help address these challenges, in particular, SLAM systems, cooperative robotics, new mobility concepts, enhanced processors, new sensors, and new approaches to using orbital imagery.

4.6.4.1 SLAM Systems

Future Mars missions, such as the Mars Precision Lander mission, are likely to include a requirement for sample return. The collection of distant samples and the return journey to a sample capsule are ideally suited to SLAM algorithms (see Section 4.4.4.3), taking advantage of loop closure. Such a mission would likely involve long traverses, through differing types of terrain. The rover would have to be able to pinpoint its position with a high degree of accuracy in order to locate useful samples and also to find the cache and deposit them. Other applicable uses are construction scenarios, for example in berm construction around a lunar landing pad, for which a rover often returns to the same location.

Utilizing SLAM can have ancillary benefits as well. 3D or feature maps generated by the SLAM algorithm are a useful product for various applications. SLAM could also reduce rover hardware costs, by allowing for increased localization performance with lower-end sensors (e.g., IMUs).

4.6.4.2 Cooperative Robotics and New Mobility Concepts

Exploration surface robotics of the future may not always look like the solitary six-wheeled rovers that we are used to seeing over the past two decades. The rationale for different mobility concepts for future exploration missions will be the type of environments that these robotics must operate in. For example, extremely rough terrain may preclude the use of wheels (or even tracks), and ultra-low gravity targets, such as asteroids, may also require nonwheeled robotics. Why target such environments? They can represent high-profit areas, where profit refers to high impact with respect to the data or samples one can collect.

Different surface mobility concepts may take any of a number of forms: walking, hopping, different wheel concepts, and even fixed- or rotary-wing concepts for low-elevation exploration. With these new mobility concepts will come new navigation challenges and opportunities: measuring odometry for a walking robot will be challenging, but obstacle detection and avoidance may take a much lower priority owing to the increased mobility capabilities of such a robot. New mobility concepts, in the navigation context, are discussed further in this section.

The motivation for legged systems is that they can operate in extremely rough terrains. Considering this scenario, however, reveals that the "issue" of the rough terrain can actually be a plus: all of the structure in the environment (rocks, hills, depressions) will provide a lot of statistically significant features for visual odometry that are not found in smoother environments. As well, legged systems will provide a rich telemetry source in order to evaluate motion, such as current consumption, frequency and amplitude of swing and stance phases and even walking speed and patterns. Therefore, odometry in legged systems may not be such a challenge, but rather just requires one to reconsider how to compute it, exploiting both the environment and the rover system.

There have also been tremendous developments in legged systems in the past decade. First, actuators of today are high-force and light-weight compared with those of a decade ago. Microcontrollers also have increased in performance and decreased in weight and power consumption. Batteries too have a higher energy density, which for the same mass provides a legged system with a meaningful mobility lifetime between charges. All of these things combine to allow legged systems to carry more sophisticated sensor packages (such as LiDARs) than has been possible before.

Cooperative robotics is another approach to addressing the needs of challenging missions of the future. One example is given here, which is the Forward Acquisition of Soil and Terrain data for Exploration Rover (FASTER) mission concept [145]. This envisions a highly mobile scout rover evaluating terrain in front of a primary rover in order to inform future path-planning decisions of the primary rover and to speed up the overall rate of traverse. This concept specifically targets the SFR for Mars sample return (see Section 4.6.2.1), which has challenging speed requirements for the rover traverse.

Similarly, aerial robotics may also play a role in future surface exploration missions. Paired with a primary rover, an aerial robot can also act as a scout to produce map products and inform path planning. One such example is the helicopter that may accompany the Mars 2020 rover [27].

Finally, hopping "rovers" can be appropriate to maneuver across low-gravity objects such as comets or asteroids. The Minerva hopping rover was part of the Japanese Hayabusa mission from a decade ago [146] although it never landed on the target asteroid. The Minerva-II hopping rover is part of the Hayabusa-2 mission currently on its way to a near-Earth asteroid [147]. For navigation, this rover has a MEMS three-axis gyro, an accelerometer, and eight photodiodes. The gyro is used for real-time attitude estimation and is used to discriminate between when the rover is hopping or at rest. The photodiodes are used to detect the Sun for the purpose of absolute attitude estimation. The rover also has a stereo camera pair for close-up imaging the comet's surface – with a tiny baseline of 3 cm. This sensor suite is impressive for a tiny vehicle that weighs only 1.1 kg [148]. One can picture future hoppers or microrovers that build on this precedent of very-low-weight robotics.

4.6.4.3 Enhanced Processing Capabilities

Onboard processing resources are limiting factors for rover speed. Contemporary autonomous rovers must sit and wait while their processors sense the terrain (and perform stereo processing), model the terrain, and plan a path. As mission requirements become more demanding, rover processors must become more capable to increase a rover's motion duty cycle.

As discussed in Section 4.3.2.4, efforts are underway to address this issue: from porting algorithms to an onboard FPGA (e.g., Mars 2020, ESA efforts) to developing a more powerful general space processor (NASA/AFRL effort [48]). Another possibility not yet on the horizon would be a space-qualified graphics processing unit, which would very efficiently process imagery.

Increasing computational resources will allow for increased rover speed, but can also have side benefits, such as opening up more processing room for science purposes (e.g., automated onboard processing of images to send back only an ROI).

4.6.4.4 New Sensors

Visible-band stereo cameras, wheel odometers, and IMUs are currently the sensors of choice for rover navigation. This section discusses other sensors that could be employed to improve navigation performance, including active 3D sensors (e.g., LiDARs), infrared cameras, and star trackers.

Scanning LiDARs, such as the Velodyne HDL32 [149], have become the navigation and mapping sensors of choice on terrestrial autonomous robotics. This is owing to their lighting immunity and ability to generate wide-FOV 3D data that is dense, long-range, and centimeter-accurate. LiDARs also have space heritage, with a LiDAR being used in the Space Shuttle and the Orbital Cygnus vehicle for rendezvous and docking with the International Space Station. Use of these sensors on rovers, however, has not occurred, a fact that is primarily due to LiDARs having larger SWAP footprint when compared with stereo cameras, their primary alternative.

Demanding planned missions, however, may push for the inclusion of these sensors in future rover designs. First, the lunar-pole Resource Prospector mission discussed earlier will feature severe lighting conditions (due to the low Sun angle) and a need to travel long distances in a short period of time, which make a LiDAR an attractive sensor for navigation. Also, the SFR, part of the Mars Precision Lander mission discussed earlier, must also travel long distances to achieve the mission requirements. This supports the need for a LiDAR-based navigation system to allow for long-range path planning and terrain assessment.

As well, efforts are underway to address the SWAP limitations of LiDARs. The CSA has recently completed an initial study of a Compact Active Sensor Technology, which focused on developing a miniaturized LiDAR concept. The ESA as well has a miniaturized LiDAR initiative, "MILS." Compact LiDAR prototypes that target rover applications include [150, 151]. NASA, too, has recently released a Request For Information about a rover LiDAR that would weight less than 5 kg and require less than 25 W of power, to be used as early as 2019 for a lunar mission.

Other active vision technologies have emerged in the last decade, although none appear to have the promise of scanning LiDARs for rover navigation due to limitations imposed by their operating principles. These technologies include TOF flash LiDARs (such as [152, 153]) and structured light sensing (such as the Microsoft Kinect). TOF flash LiDARs are different from scanning LiDARs primarily in that they gather an entire "image" at once, spreading their laser light over the entire FOV of the sensor. This imposes limits on their performance as range is traded off against FOV. These sensors also typically have poor dynamic range, which is an issue for rover applications since terrain must be imaged both near and far from a rover. Commercial structured light sensors have become popular recently because of their low cost; however, their range is limited and they cannot perform outdoors in ambient lighting.

IR cameras are appealing for use in space given their affordability, low SWAP footprint, and immunity to ambient lighting conditions. For rover navigation, these cameras could be used in shadowed areas to discriminate between rocks versus sandy areas (therefore, between obstacles and clear terrain) and could detect recently disturbed earth (e.g., rover tracks, to return to safety). This could be used to increase the operational envelope of planetary rovers in harsh lighting conditions. Also, a thermal imager could be used to track a scout rover, or locate a sample cache exhibiting a particular heat signature.

There have been relevant uses of IR cameras in space. The MERs' mini-thermal emission spectrometer (TES) is a narrow-FOV science instrument to investigate rock mineralogy, but it has also collected data that shows its potential for navigation. In a collage of individual measurements that map out a typical camera FOV, the Mini-TES has shown that the thermal signature of rocks on Mars shows up differently (cool, blue) than that of the surrounding loose terrain (warm, red) [154]. The THEMIS infrared camera on the Mars Odyssey orbiter has also demonstrated this effect [155]. Finally, the MastCam on the Curiosity rover also has an IR filter for science applications, although to date this has not been used for navigation.

Star trackers, discussed in Section 4.4.1.2, are also candidates for inclusion in navigation systems of the future. Their ability to detect absolute orientation at a relatively high rate may allow them to steal "market" away from IMUs for calculating rover orientation.

This section has described the applicability of new types of sensors for rover navigation. These include active 3D sensors (LiDARs), infrared cameras, and star trackers. Demanding rover missions of the future may result in some or all of these sensor types being included in the next wave of rovers.

4.6.4.5 New Applications of Orbital Imagery

This section discusses future uses of orbital imagery in aid of rover navigation. "Super-resolution restoration" (SRR) images and their possible uses are discussed first, followed by a summary of the benefits of being able to autonomously match features between rover and orbiter images.

The generation of orthorectified HiRISE imagery for Mars (and similar imagery such as LROC on the Moon) means that there now exists an image base for a variety of different tasks (introduced in Section 4.4.3.1). However, 25 cm imagery is still coarse for many day-to-day operations tasks (equivalent to NavCam imagery at 10 m range) such as producing obstacle maps.

An alternative to the use of HiRISE at its native resolution has recently been investigated, which was first reported here [156]. If five or more repeat HiRISE 25 cm images are available, which have sufficient clarity in individual images, then each image can be automatically coregistered to a HiRISE ORI and using a model of the distortion of each and every pixel in the image can be used to generate an SRR image of up to 5 cm resolution. An example of this is shown in Figure 4.14 which displays the original image and an SRR version of the so-called "Homeplate area explored by the Spirit MER. Note the presence of the rover tracks in the SRR image.

Such orbital SRR images have a number of possible applications, which will be discussed here. Firstly, they can be employed to assist with path planning by generating very high resolution rock obstacle maps and "no-go" areas due to steep inclines, high density of surface rocks or surface discontinuities. Secondly, they can be used to assist in the selection of future landing sites, by eliminating those where the rock or other navigation obstacles are too severe. Thirdly, they can be employed to better select camera stations to try to balance the very limited return bandwidth with the need to maximize the exploration of an area.

A limiting factor to the capabilities of Curiosity and the MERs, is that human operator input is required to find the most accurate estimate of position (either

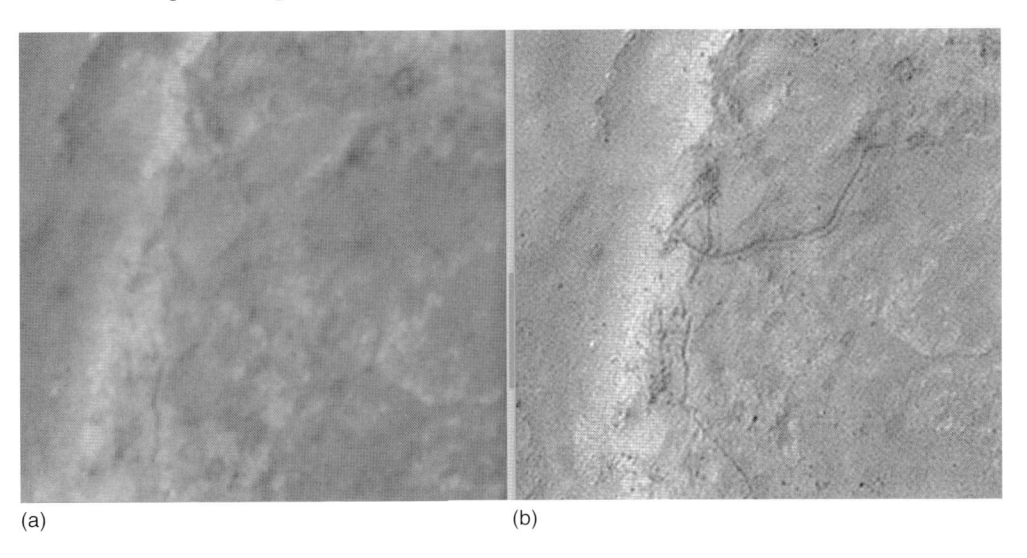

Figure 4.14 Example of a 25 cm HiRISE orthorectified image created from a HiRISE stereo-pair (a), with an example of a 5 cm super-resolution restoration image derived from a stack of 8 HiRISE 25 cm images (b) taken over 7 years. (Left image: Courtesy NASA/JPL/University of Arizona. Right image: Courtesy Planetary Space Science.)

IBA, or rover-to-orbiter image matching). This means that when the rover has completed a long path, the rover will have drifted somewhat from its intended location – something that can only be corrected through communication with Earth. This can slow the deploying of instruments for making scientific observations at the site.

Reference [79], referenced in Section 4.4.3.1, discusses automatically matching features between rover and orbital images. Being able to localize autonomously could allow the rover to perform mission objectives at multiple sites without the need for human input. Accurate map building would also allow the rover to revisit sites it may have passed previously with relative ease, rather than planning an entirely new route. Reducing the amount of human intervention required to move the rover could significantly reduce the amount of time taken to achieve its exploration goals.

References

1. Andrews, D. (2014) Resource prospector mission to the moon. *International Astronautical Congress*.
2. Arizona State University (2015) Lunar Reconnaissance Orbiter Camera Data Node, http://lroc.sese.asu.edu/ (accessed February 2015).
3. Massachusetts Institute of Technology (2015) LOLA Data Archive, http://imbrium.mit.edu/LOLA.html (accessed February 2015).
4. The University of Arizona (2015) High Resolution Imaging Science Experiment, https://hirise.lpl.arizona.edu/ (accessed February 2015).
5. Harvey, B. (2007) *Soviet and Russian Lunar Exploration*, Chapter 7, Springer-Verlag, Berlin, p. 255.
6. Kassel, S. (1971) Lunokhod-1 Soviet Lunar Surface Vehicle. Technical Report R-802-ARPA, RAND Corporation. Prepared for the Advanced Research Projects Agency.
7. Harvey, B. (2007) *Soviet and Russian Lunar Exploration*, Chapter 7, Springer-Verlag, p. 257.
8. Harvey, B. (2007) *Soviet and Russian Lunar Exploration*, Chapter 7, Springer-Verlag, p. 267.
9. NASA Jet Propulsion Laboratory (2015) Driving Distances on Mars and the Moon, http://www.jpl.nasa.gov/images/mer/2014-07-28//odometry140728.jpg (accessed February 2015).
10. Young, A. (2007) Navigation subsystem, *Lunar and Planetary Rovers*, Springer-Verlag, p. 42.
11. Matijevic, J. (1998) Autonomous navigation and the sojourner microrover. *Science*, **276** (592), 454–455.
12. Maimone, M. and Biesiadecki, J. (2006) The Mars exploration rover surface mobility flight software: driving ambition. IEEE Aerospace Conference.
13. Maimone, M., Cheng, Y., and Matthies, L. (2007) Two years of visual odometry on the Mars exploration rovers. *Journal of Field Robotics, Special Issue on Space Robotics*, **24** (3), 169–186.
14. NASA Jet Propulsion Laboratory (2015) Mars Exploration Rovers - Mission Page, http://mars.nasa.gov/mer/mission/spacecraft_rover_eyes.html (accessed February 2015).
15. Li, R., He, S., Chen, Y., Tang, M., Tang, P., Di, K., Matthies, L., Arvidson, R., Squyres, S., Crumpler, L., Parker, T., and Sims, M. (2011) MER Spirit rover localization: comparison of ground image- and orbital image-based methods and science applications. *Journal of Geophysical Research*, **116**, E00F16, doi: 10.1029/2010JE003773.
16. Parker, T.J., Malin, M.C., Calef, F.J., Deen, R.G., Gengl, H.E., Golombek, M.P., Hall, J.R., Pariser, O., Powell, M., and Sletten, R.S. (2013) Localization and 'Contextualization' of curiosity in gale crater, and other landed Mars

missions. *International Astronautical Congress*.

17. Lakdawalla, E. (2015) Curiosity update, sols 949-976: Scenic road trip and a diversion to Logan's Run, http://www.planetary.org/blogs/emily-lakdawalla/2015/20150506-curiosity-update-sols-949-976.html (accessed May 2015).

18. Maimone, M., Leger, C., and Biesiadecki, J. (2007) Overview of the Mars Exploration Rovers' autonomous mobility and vision capabilities. *IEEE International Conference on Robotics and Automation*.

19. Ferguson, D. and Stentz, A. (2005) The Field D* Algorithm for Improved Path Planning and Replanning in Uniform and Non-uniform Cost Environments. Technical report CMU-RI-TR-05-19, Robotics Institute.

20. Maimone, M. (2013) Curiouser and Curiouser: surface robotic technology driving Mars rover curiosity's exploration of gale crater. *International Conference on Robotics and Automation*.

21. Maimone, M. (2014) What drives curiosity? Robotic technologies on the Mars science laboratory (plenary session). *International Symposium on Artificial Intelligence, Robotics and Automation in Space*.

22. Maki, J., Thiessen, D., Pourangi, A., Kobzeff, P., Litwin, T., Scherr, L., Elliott, S., Dingizian, A., and Maimone, M. (2012) The Mars Science Laboratory Engineering cameras. *Space Science Reviews*, **170** (1-4), 77–93.

23. Maki, J.N.et al., (2003) Mars exploration rover engineering cameras. *Journal of Geophysical Research*, **108** (E12), 1–23.

24. Lakdawalla, E. (2014) Curiosity update, sols 671-696: out of the landing ellipse, into ripples and pointy rocks, http://www.planetary.org/blogs/emily-lakdawalla/2014/07241401-curiosityupdate-sols-671-696.html (accessed May 2015).

25. Maimone, M. (2014) C++ on Mars: incorporating C++ into Mars rover flight software. *The C++ Conference*.

26. Maimone, M. (2013) Leave the Driving to Autonav, Curiosity Rover Report, http://www.nasa.gov/mission_pages/msl/multimedia/curiosity20130919.html (accessed May 2015).

27. Volpe, R. (2014) 2014 robotics activities at JPL. *International Symposium on Artificial Intelligence, Robotics and Automation in Space*.

28. Liu, H. (2014) An overview of the space robotics progress in China. *International Symposium on Artificial Intelligence, Robotics and Automation in Space*.

29. Siegwart, R. and Nourbakhsh, I.R. (2004) *Introduction to Autonomous Mobile Robots*, Chapter 4, MIT Press, p. 91.

30. Yoder, G. (2004) Implementation of COTs hardware in non critical space applications: a brief tutorial. *17th Annual Microelectronics Workshop*.

31. Zampato, M., Finotello, R., Ferrario, R., Viareggio, A., and Losito, S. (2004) Radiation susceptibility trials on COTS cameras for international space station applications. *SIRAD Workshop*.

32. Gazarik, M., Johnson, D., Kist, E., Novak, F., Antill, C., Haakenson, D., Howell, P., Pandolf, J., Jenkins, R., Yates, R., Stephan, R., Hawk, D., and Amoroso, M. (2006) Development of an Extra-Vehicular (EVA) Infrared (IR) Camera Inspection System. *Defense and Security Symposium*.

33. Siegwart, R. and Nourbakhsh, I.R. (2004) *Introduction to Autonomous Mobile Robots*, Chapter 6.3, MIT Press, p. 291.

34. Moreno, S. (2013) CNES robotics activities - towards long distance on-board decision-making navigation. *Advanced Space Technologies in Robotics and Automation*.

35. Biesiadecki, J. and Maimone, M. (2006) The Mars exploration rover surface mobility flight software: driving ambition. *IEEE Aerospace Conference*.

36. Winter, M., Barclay, C., Pereira, V., Lancaster, R., Caceres, M., McManamon, K., Nye, B., Silva, N., Lachat, D., and Campana, M. (2015) ExoMars rover vehicle: detailed description of the GNC system. *Advanced Space Technologies in Robotics and Automation*.

37. Artemis Jr/DESTiN Rover Team (2013) Lunar Tele-Operated ISRU Platform Concept Study. Technical report NDG01183, Neptec Design Group, Prepared for the Canadian Space Agency.
38. Goldberg, S., Maimone, M., and Matthies, L. (2002) Stereo vision and rover navigation software for planetary exploration. *IEEE Aerospace Conference.*
39. McManamon, K., Lancaster, R., and Silva, N. (2013) ExoMars rover vehicle perception system architecture and test results. *Advanced Space Technologies in Robotics and Automation.*
40. Yuen, P.C., Gao, Y., Griffiths, A., Coates, A., Muller, J.-P., Smith, A., Walton, D., Leff, C., Hancock, B., and Shin, D. (2013) ExoMars rover PanCam: autonomy & computational intelligence. *IEEE Computational Intelligence Magazine*, **8** (4), 52–61.
41. Open Source (2015) Point Cloud Library, http://pointclouds.org/ (accessed June 2015).
42. Open Source (2015) OpenCV, http://opencv.org/ (accessed June 2015).
43. Karney, C. (2015) GeographicLib, http://sourceforge.net/projects/geographiclib/ (accessed June 2015).
44. Geometric Tools, LLC (2015) Geometric Tools, http://www.geometrictools.com/ (accessed June 2015).
45. BAE Systems (2008) RAD750 Radiation-Hardened PowerPC Microprocessor. Component Datasheet PUBS-08-B32-01, BAE Systems, Manassas, VA.
46. Violante, M., Battezzati, N., and Sterpone, L. (2011) *Reconfigurable Field Programmable Gate Arrays for Mission-Critical Applications*, Chapter 5, Springer-Verlag, p. 179.
47. Lentaris, G., Stamoulias, I., Diamantopoulos, D., Maragos, K., Siozios, K., Soudris, D., Rodrigalvarez, M.A., Lourakis, M., Zabulis, X., Kostavelis, I., Nalpantidis, L., Boukas, E., and Gasteratos, A. (2015) Spartan/-sextant/compass: advancing space rover vision via reconfigurable platforms, in *Applied Reconfigurable Computing*, Lecture Notes in Computer Science, vol. **9040** (eds K. Sano, D. Soudris, M. Hübner, and P.C. Diniz), Springer International Publishing, pp. 475–486.
48. Doyle, R., Some, R., Powell, W., Mounce, G., Goforth, M., Horan, S., and Lowry, M. (2014) High Performance Spaceflight Computing (HPSC) Next-Generation Space Processor (NGSP) a joint investment of NASA and AFRL. *International Symposium on Artificial Intelligence, Robotics and Automation in Space.*
49. Gingras, D., Allard, P., Lamarche, T., Rocheleau, S., and Gemme, S. (2014) Lunar rover remote driving using monocameras under multi-second latency and low-bandwidth: field tests and lessons learned. *International Symposium on Artificial Intelligence, Robotics and Automation in Space.*
50. Bualat, M., Fong, T., Schreckenghost, D., Kalar, D., Pacis, E., and Beutter, B. (2014) Results from testing crew-controlled surface telerobotics on the international space station. *International Symposium on Artificial Intelligence, Robotics and Automation in Space.*
51. Visentin, G. (2013) ESA robotics overview. *Advanced Space Technologies in Robotics and Automation.*
52. Lambert, A., Furgale, P., and Barfoot, T. (2012) Field testing of visual odometry aided by a Sun sensor and inclinometer. *Journal of Field Robotics*, **29** (3), 426–444.
53. Gammell, J.D., Tong, C.H., Berczi, P., Anderson, S., and Barfoot, T. (2013) Rover odometry aided by a star tracker. *IEEE Aerospace Conference.*
54. Rehrmann, F., Schwendner, J., Cornforth, J., Durrant, D., Lindegren, R., Selin, P., Carrio, J.H., Poulakis, P., and Kohler, J. (2011) A miniaturized space qualified MEMS IMU for rover navigation: requirements and testing of a proof of concept hardware demonstrator. *Advanced Space Technologies in Robotics and Automation.*
55. Grumman, N. (2013) LN200S Inertial Measurement Unit.
56. Bilodeau, V.S. and Hamel, J.-F. (2013) Lunar Tele-Operated ISRU Platform: Relative Localisation Trade-Offs.

Technical report 7_LTOIP-TN-001-NGCCA, NGC Aerospace, Prepared for the Canadian Space Agency.

57. Shaw, A., Woods, M., Churchill, W., and Newman, P. (2013) Robust visual odometry for space exploration. *Advanced Space Technologies in Robotics and Automation.*

58. Harris, C. and Stephens, M. (1988) A combined corner and edge detector. *Proceedings of 4th Alvey Vision Conference*, pp. 147–151.

59. Rosten, E. and Drummond, T. (2006) Machine learning for high speed corner detection. *9th European Conference on Computer Vision*, pp. 430–443.

60. Calonder, M., Lepetit, V., Strecha, C., and Fua, P. (2010) BRIEF: binary robust independent elementary features. *Proceedings of the 11th European Conference on Computer Vision: Part IV*, ECCV'10, Springer-Verlag, Berlin, Heidelberg, pp. 778–792.

61. Fischler, M.A. and Bolles, R.C. (1981) Random sample consensus: a paradigm for model fitting with applications to image analysis and automated cartography. *Communications of the ACM*, **24** (6), 381–395.

62. Woods, M., Shaw, A., Tidey, E., Pham, B.V., Artan, U., Maddison, B., and Cross, G. (2012) SEEKER - autonomous long range rover navigation for remote exploration. *International Symposium on Artificial Intelligence, Robotics and Automation in Space.*

63. Howard, T.M., Morfopoulos, A., Morrison, J., Kuwata, Y., Villalpando, C., Matthies, L., and McHenry, M. (2012) Enabling continuous planetary rover navigation through FPGA stereo and visual odometry. *IEEE Aerospace Conference.*

64. Lourakis, M., Chliveros, G., and Zabulis, X. (2014) Autonomous visual navigation for planetary exploration rovers. *International Symposium on Artificial Intelligence, Robotics and Automation in Space.*

65. Bilodeau, V.S., Hamel, J.-F., and Iles, P. (2013) A rover vision-based relative localization system for the RESOLVE moon exploration mission. *International Astronautical Congress.*

66. Bilodeau, V.S., Beaudette, D., Hamel, J.-F., Alger, M., Iles, P., and MacTavish, K. (2012) Vision-based pose estimation system for the lunar analogue rover 'Artemis'. *International Symposium on Artificial Intelligence, Robotics and Automation in Space.*

67. Wagner, M., Wettergreen, D., and Iles, P. (2012) Visual odometry for the lunar analogue rover 'Artemis'. *International Symposium on Artificial Intelligence, Robotics and Automation in Space.*

68. Segal, A., Haehnel, D., and Thrun, S. (2009) Generalized-ICP. *Robotics: Science and Systems.*

69. Gemme, S., Gingras, D., Salerno, A., and Dupuis, E. (2012) Pose refinement using ICP applied to 3-D LiDAR data for exploration rovers. *International Symposium on Artificial Intelligence, Robotics and Automation in Space.*

70. Ahuja, S., Iles, P., and Waslander, S. (2014) 3D scan registration using curvelet features in planetary environments. *International Symposium on Artificial Intelligence, Robotics and Automation in Space.*

71. McManus, C., Furgale, P., and Barfoot, T. (2013) Towards lighting-invariant visual navigation: an appearance-based approach using scanning laser-rangefinders. *Robotics and Autonomous Systems*, **61** (8), 836–852.

72. Tong, C.H., Anderson, S., Dong, H., and Barfoot, T. (2014) Pose interpolation for laser-based visual odometry. *Journal of Field Robotics: Special Issue on Field and Service Robotics*, **31** (5), 731–757.

73. Siegwart, R. and Nourbakhsh, I.R. (2004) *Introduction to Autonomous Mobile Robots*, Chapter 4, MIT Press, p. 117.

74. Amzajerdian, F., Pierrottet, D., Petway, L., Hines, G., and Barnes, B. (2012) Doppler LiDAR descent sensor for planetary landing. *Concepts and Approaches for Mars Exploration.*

75. Robinson, M.S.et al., (2010) Lunar Reconnaissance Orbiter Camera (LROC) instrument overview. *Space Science Review*, **150** (1-4), 81–124.

76. Gwinner, K.et al., (2010) Topography of Mars from global mapping by HRSC

high-resolution digital terrain models and orthoimages: characteristics and performance. *Earth and Planetary Science Letters*, **294** (3-4), 506–519.
77. Kim, J.R. and Muller, J.P. (2009) Multi-resolution topographic data extraction from Martian stereo imagery. *Planetary Space Science*, **57** (14–15), 2095–2112.
78. Tao, Y. and Muller, J.-P. (2014) Automated navigation of Mars rovers using HiRISE-CTX-HRSC Co-registered orthorectified images and DTMs. *EGU General Assembly Conference Abstracts*.
79. Shaukat, A., Al-Milli, S., Bajpai, A., Spiteri, C., Burroughes, G., Gao, Y., Lachat, D., and Winter, M. (2015) Next generation rover GNC architectures. *Advanced Space Technologies in Robotics and Automation*.
80. Lourakis, M. and Hourdakis, E. (2015) Planetary rover absolute localization by combining visual odometry with orbital image measurements. *Advanced Space Technologies in Robotics and Automation*.
81. Boukas, E., Gasteratos, A., and Visentin, G. (2015) Matching sparse networks of semantic ROIs among rover and orbital imagery. *Advanced Space Technologies in Robotics and Automation*.
82. Baatz, G., Saurer, O., Köser, K., and Pollefeys, M. (2012) Large scale visual geo-localization of images in mountainous terrain. *Proceedings of the 12th European Conference on Computer Vision - Volume Part II, ECCV'12*, pp. 517–530.
83. Nefian, A., Bouyssounousse, X., Edwards, L., Dille, M., Kim, T., Hand, E., Rhizor, J., Deans, M., Bebis, G., and Fong, T. (2014) Infrastructure free rover localization. *International Symposium on Artificial Intelligence, Robotics and Automation in Space*.
84. Palmer, E., Gaskell, R., Vance, L., Sykes, M., McComas, B., and Jouse, W. (2012) Location identification using horizon matching. *43rd Lunar and Planetary Science Conference*.
85. Cozman, F., Krotkov, E., and Guestrin, C. (2000) Outdoor visual position estimation for planetary rovers. *Autonomous Robots*, **9** (2), 135–150.
86. Hamel, J.-F., Langelier, M.-K., Alger, M., Iles, P., and MacTavish, K. (2012) Design and validation of an absolute localisation system for the lunar analogue rover 'Artemis'. *International Symposium on Artificial Intelligence, Robotics and Automation in Space*.
87. Van Pham, B., Maligo, A., and Lacroix, S. (2013) Absolute map-based localization for a planetary rover. *Advanced Space Technologies in Robotics and Automation*.
88. Carle, P., Furgale, P., and Barfoot, T. (2010) Long-range rover localization by matching LiDAR scans to orbital elevation maps. *Journal of Field Roboticss*, **27** (5), 534–560.
89. Guinn, J. and Ely, T. (2004) Preliminary results of Mars exploration rover in-situ radio navigation. *14th AAS/AIAA Space Flight Mechanics Meeting*.
90. Ely, T., Anderson, R., Bar-Sever, Y., Bell, D., Guinn, J., Jah, M., Kallemeyn, P., Levene, E., Romans, L., and Wu, S. (1999) Mars network constellation design drivers and strategies. *AAS/AIAA Astrodynamics Specialist Conference*.
91. Chelmins, D., Nguyen, B., Sands, S., and Welch, B. (2009) A Kalman Approach to Lunar Surface Navigation using Radiometric and Inertial Measurements. Technical report NASA/TM—2009-215593, NASA Glenn Research Center.
92. Murphy, T.W., Adelberger, E.G., Battat, J.B.R., Hoyle, C.D., Johnson, N.H., McMillan, R.J., Michelsen, E.L., Stubbs, C.W., and Swanson, H.E. (2011) Laser ranging to the lost Lunohod 1 reflector. *Icarus*, **211** (2), 1103–1108.
93. Batista, P., Silvestre, C., and Oliveira, P. (2010) Single beacon navigation: observability analysis and filter design. *American Control Conference*.
94. Molina, P., Iles, P., and MacTavish, K. (2012) Lander-based localization system of the lunar analogue rover 'Artemis'. *International Symposium on Artificial Intelligence, Robotics and Automation in Space*.
95. Matsuoka, M., Rock, S.M., and Bualat, M.G. (2004) Autonomous deployment

of a self-calibrating pseudolite array for Mars rover navigation. *Position Location and Navigation Symposium*.
96. Patmanathan, V. (2006) Area localization using WLAN. Master of Science Thesis. Royal Institute of Technology (KTH).
97. Ni, J., Arndt, D., Ngo, P., Phan, C., Dekome, K., and Dusl, J. (2010) Ultra-wideband time-difference-of-arrival high resolution 3D proximity tracking system. *Position Location and Navigation Symposium*.
98. University of Washington (2015) ARToolKit homepage, http://www.hitl.washington.edu/artoolkit/ (accessed March 2015).
99. Sigel, D.A. and Wettergreen, D. (2007) Star tracker celestial localization system for a lunar rover. *International Conference on Intelligent Robots and Systems*.
100. Ning, X. and Fang, J. (2009) A new autonomous celestial navigation method for the lunar rover. *Robotics and Autonomous Systems*, **57**, 48–54.
101. Enright, J., Barfoot, T., and Soto, M. (2012) Star tracking for planetary rovers. *IEEE Aerospace Conference*.
102. Siegwart, R. and Nourbakhsh, I.R. (2004) *Introduction to Autonomous Mobile Robots*, Chapter 5, MIT Press, p. 227.
103. Grewal, M.S. and Andrews, A.P. (2001) *Kalman Filtering - Theory and Practice Using MATLAB*, John Wiley & Sons, Inc.
104. Levy, S. and Washington and Lee University (2015) The Extended Kalman Filter: An Interactive Tutorial for Non-Experts, http://home.wlu.edu/~levys/kalman_tutorial/ (accessed July 2015).
105. Bay, H., Ess, A., Tuytelaars, T., and Van Gool, L. (2008) Speeded-up robust features (SURF). *Computer Vision and Image Understanding*, **110** (3), 346–359.
106. Montemerlo, M., Thrun, S., Koller, D., and Wegbreit, B. (2002) FastSLAM: a factored solution to the simultaneous localization and mapping problem. *Proceedings of the AAAI National Conference on Artificial Intelligence, AAAI*, pp. 593–598.
107. Montemerlo, M., Thrun, S., Koller, D., and Wegbreit, B. (2003) FastSLAM 2.0: an improved particle filtering algorithm for simultaneous localization and mapping that provably converges. *Proceedings of the International Conference on Artificial Intelligence (IJCAI)*, pp. 1151–1156.
108. Shala, K. and Gao, Y. (2010) Comparative analysis of localisation and mapping techniques for planetary rovers. *International Symposium on Artificial Intelligence, Robotics and Automation in Space*.
109. Tong, C.H., Barfoot, T.D., and Dupuis, E. (2012) Three-dimensional SLAM for mapping planetary work site environments. *Journal of Field Robotics*, **29** (3), 381–412.
110. Shaukat, A., Spiteri, C., Gao, Y., Al-Milli, S., and Bajpai, A. (2013) Quasi-thematic feature detection and tracking for future rover long-distance autonomous navigation. *Advanced Space Technologies in Robotics and Automation*.
111. Bajpai, A., Burroughes, G., Shaukat, A., and Gao, Y. (2015) Planetary monocular simultaneous localization and mapping. *Journal of Field Robotics*, **33** (2), 229–242.
112. Iles, P., Wagner, M., Hamel, J.-F., Simard-Bilodeau, V., MacTavish, K., and Molina, P. (2012) Localization system of the lunar analogue rover 'Artemis Jr.'. *Global Space Exploration Conference*.
113. Villa, D. (2005) Position estimation for a planetary, autonomous, long-distance rover. Masters of Science Thesis, Carnegie Mellon.
114. Polotski, V., Ballotta, F.J., and James, J. (2014) Terrain exploration, planning and autonomous navigation with MRPTA rover. *International Symposium on Artificial Intelligence, Robotics and Automation in Space*.
115. Pauly, M., Gross, M., and Kobbelt, L. (2002) Efficient simplification of point-sampled surfaces. *Visualization*.
116. Open Perception Foundation (2015) Fast Triangulation of Unordered Point Clouds, http://pointclouds

117. Garland, M. and Heckbert, P. (1997) Surface simplification using quadric error metrics. *International Conference on Computer Graphics and Interactive Techniques (SIGGRAPH)*.
118. Autonomous Navigation Results from the Mars Exploration Rover (MER) Mission (2004) An overview of the space robotics progress in China. *International Symposium on Experimental Robotics*.
119. Wettergreen, D. and Wagner, M. (2012) Developing a framework for reliable autonomous surface mobility. *International Symposium on Artificial Intelligence, Robotics and Automation in Space*.
120. Gingras, D., Lamarche, T., Bedwani, J.-L., and Dupuis, E. (2010) Rough terrain reconstruction for rover motion planning. *Canadian Conference on Computer and Robot Vision*.
121. Sancho-Pradel, D.L. and Gao, Y. (2010) A survey on terrain assessment techniques for autonomous operation of planetary robots. *Journal of the British Interplanetary Society*, **63** (5-6), 206–217.
122. Lumelsky, V. and Skewis, T. (1990) Incorporating range sensing in the robot navigation function. *IEEE Transactions on Systems, Man, and Cybernetics*, **20** (5), 1058–1068.
123. Ng, J. and Braunl, T. (2007) Performance comparison of bug navigation algorithms. *Journal of Intelligent and Robotic Systems*, **50**, 73–84.
124. Gingras, D., Dupuis, E., Payre, G., and Lafontaine, J. (2010) Path planning based on fluid mechanics for mobile robots using unstructured terrain models. *IEEE International Conference on Robotics and Automation*.
125. Siegwart, R. and Nourbakhsh, I.R. (2004) *Introduction to Autonomous Mobile Robots (Obstacle Avoidance)*, Chapter 6.2.2, MIT Press, pp. 272–290.
126. Carsten, J., Rankin, A., Ferguson, D., and Stentz, A. (2009) Global planning on the mars exploration rovers: software integration and surface testing. *Journal of Field Robotics*, **26** (4), 337–357.
127. Siegwart, R. and Nourbakhsh, I.R. (2004) *Introduction to Autonomous Mobile Robots (Path Planning)*, Chapter 6.2.1, MIT Press, pp. 261–267.
128. Molina, P. (2013) Surface Reconstruction Algorithm and Path Planning Integration: Final Report. Technical report. Carleton University, Neptec Design Group, Prepared for Neptec Design Group.
129. Hart, P., Nilsson, N., and Raphael, B. (1968) A formal basis for the heuristic determination of minimum cost paths. *IEEE Transactions on Systems Science and Cybernetics*, **4** (2), 100–107.
130. Imms, D. (2015) A* Pathfinding Algorithm, http://www.growingwiththeweb.com/2012/06/a-pathfinding-algorithm.html (accessed November 2015).
131. Stentz, A. (1993) Optimal and efficient path planning for unknown and dynamic environments. *International Journal of Robotics and Automation*, **10**, 89–100.
132. Koenig, S. and Likhachev, M. (2002) D*lite. *18th National Conference on Artificial Intelligence*, American Association for Artificial Intelligence, Menlo Park, CA, pp. 476–483.
133. Nash, A., Koenig, S., and Likhachev, M. (2009) Incremental Phi*: incremental any-angle path planning on grids. *Proceedings of the International Joint Conference on Artificial Intelligence (IJCAI)*, pp. 1824–1830.
134. Likhachev, M., Ferguson, D., Gordon, G., Stentz, A., and Thrun, S. (2005) Anytime dynamic A*: an anytime, replanning algorithm. *Proceedings of the International Conference on Automated Planning and Scheduling (ICAPS)*.
135. Astolfi, A. (1999) Exponential stabilization of a wheeled mobile robot via discontinuous control. *Journal of Dynamic Systems, Measurement, and Control*, **121** (1), 121.
136. Farley, K. and Williford, K. (2015) Mars 2020 mission update. *Mars Exploration Program Analysis Group*.

137. Visentin, G. (2014) ESA AI and robotics at iSAIRAS 2014. *International Symposium on Artificial Intelligence, Robotics and Automation in Space.*
138. Merlo, A., Larranaga, J., and Falkner, P. (2013) Sample Fetching Rover (SFR) for MSR. *Advanced Space Technologies in Robotics and Automation.*
139. Sanders, G.B.et al., (2012) *RESOLVE Lunar Ice/Volatile Payload Development and Field Test Status*, Lunar Exploration Analysis Group.
140. Reid, E.et al., (2014) The Artemis Jr. rover: mobility platform for lunar ISRU mission simulation. *Advances in Space Research*, **55** (10), 2472–2483.
141. Gunes-Lasnet, S., Van Winnendael, M., Chong Diaz, G., Schwenzer, S., and Pullan, D. (2014) SAFER: the promising results of the Mars mission simulation in Atacama, Chile. *International Symposium on Artificial Intelligence, Robotics and Automation in Space.*
142. Woods, M., Shaw, A., Wallace, I., Malinowski, M., and Rendell, P. (2014) Demonstrating autonomous Mars rover science operations in the Atacama Desert. *International Symposium on Artificial Intelligence, Robotics and Automation in Space.*
143. Paar, G.et al., (2013) The PRoViScout field trials tenerife 2013 - integrated testing of aerobot mapping, rover navigation and science assessment. *Advanced Space Technologies in Robotics and Automation.*
144. Woods, M., Shaw, A., Wallace, I., and Malinowski, M. (2015) The Chameleon field trial: toward efficient, terrain sensitive navigation. *Advanced Space Technologies in Robotics and Automation.*
145. Nevatia, Y., Bulens, F., Gancet, J., Gao, Y., Al-Milli, S., Kandiyil, R., Sonsalla, R., Frische, M., Vogele, T., Allouis, E., Skocki, K., Ransom, S., Saaj, C., Matthews, M., Yeomans, B., Richter, L., and Kaupisch, T. (2013) Safe long-range travel for planetary rovers through forward sensing. *Advanced Space Technologies in Robotics and Automation.*
146. Yoshimitsu, T., Nakatani, I., and Kubota, T. (1999) New mobility system for small planetary body exploration. *International Conference on Robotics and Automation (ICRA)*, pp. 1404–1409.
147. Yoshimitsu, T., Kubota, T., Tomiki, A., and Kuroda, Y. (2014) Development of hopping rovers for a new challenging asteroid. *International Symposium on Artificial Intelligence, Robotics and Automation in Space.*
148. Yoshimitsu, T., Kubota, T., and Tomiki, A. (2015) MINERVA-II rovers developed for Hayabusa-2 mission. *Low Cost Planetary Missions Conference.*
149. Velodyne (2015) HDL-32e, http://velodynelidar.com/lidar/hdlproducts/hdl32e.aspx (accessed June 2015).
150. Bakambu, A.J., Nimelman, M., Mukherji, R., and Tripp, J.W. (2012) Compact fast scanning LiDAR for planetary rover navigation. *International Symposium on Artificial Intelligence, Robotics and Automation in Space.*
151. Ishigami, G. and Mizuno, T. (2014) Towards space-hardened small-lightweight laser range imager for planetary exploration rover. *International Symposium on Artificial Intelligence, Robotics and Automation in Space.*
152. Advanced Scientific Concepts (2015) DragoneEye 3D Flash LiDAR Space Camera, http://www.advancedscientificconcepts.com/products/older-products/dragoneye.html (accessed June 2015).
153. Dissly, R., Weimar, C., Masciarelli, J., Weinberg, J., Miller, K., and Rohrschneider, R. (2012) Flash LiDAR for planetary missions. *International Workshop on Instrumentation for Planetary Missions.*
154. Arizona State University (2015) Mini-TES Project Page, http://themis.asu.edu/projects/minites (accessed June 2015).
155. Arizona State University (2014) THEMIS Project Page, http://themis.asu.edu/node/5402 (accessed June 2015).
156. Muller, J.P.et al., (2014) European geospatial image understanding tools for Mars exploration. *8th International Conference on Mars.*

5
Manipulation and Control

José de Gea Fernández, Elie Allouis, Karol Seweryn, Frank Kirchner, and Yang Gao

5.1
Introduction

The use of robotic systems for planetary exploration has been successfully shown in recent years with outstanding examples such as the Mars rovers Opportunity, Spirit, and Curiosity. While the first phase of planetary exploration tends to deal with "passive" *in situ* exploration of unknown areas by using cameras, measuring values of the atmosphere by using onboard instruments, and so on, the relevant scientific experiments can be performed as soon as samples (such as soil or rocks) can be analyzed either by bringing them back to Earth or by analyzing them *in situ*. In either case, the rover is required to have the ability to gather samples, drill into the rocks, or bring the samples to the rover's onboard scientific instruments in order to get them analyzed; hence, robotic manipulation is an importance functionality in these missions. In case of advanced missions such as to establish extraterrestrial outposts on the Moon, the robots need to be equipped with one or several manipulators to be able to grasp, transport, and assemble infrastructure.

This first section reviews current and existing robotic manipulation systems for planetary exploration. The rest of the chapter discusses relevant design requirements, specifications, and procedures, describes underlining technologies such as dynamical and motion control of robotic arms, and presents various future needs and directions in this area.

5.1.1
Review of Planetary Robotic Arms

As explained in Chapter 1, planetary robotic systems belong to a class of their own where robustness and accuracy need to be achieved with limited mass, power, and processing capabilities. Hence, the design and operation of a robotic arm for space should be challenging and in this case rather unique too. Up to date, a fewer number of planetary robotic arms have been designed, flown, or operated successfully compared with the total number of successful missions that have landed on the extraterrestrial bodies. This section reviews these existing developments

Contemporary Planetary Robotics: An Approach Toward Autonomous Systems, First Edition.
Edited by Yang Gao.
© 2016 Wiley-VCH Verlag GmbH & Co. KGaA. Published 2016 by Wiley-VCH Verlag GmbH & Co. KGaA.

5 Manipulation and Control

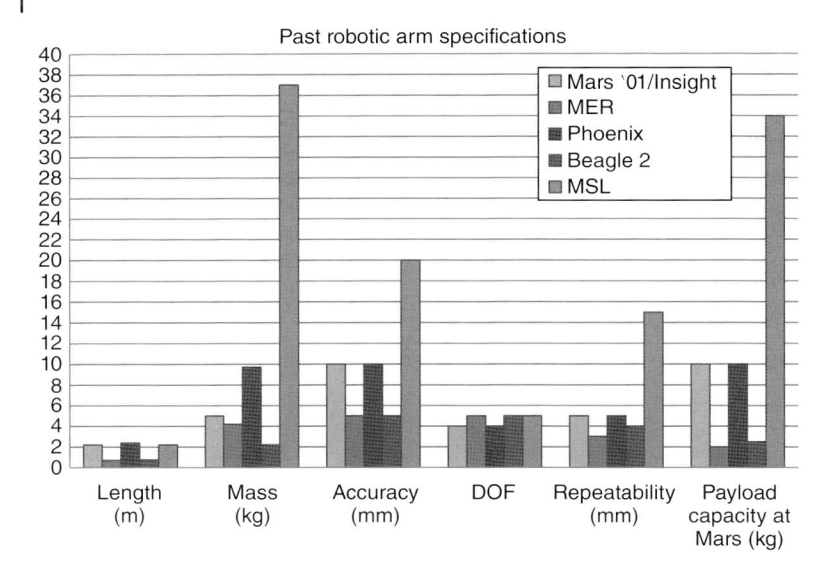

Figure 5.1 Specifications of existing planetary robotic arm systems.

Table 5.1 Overview of existing planetary arms.

Arms	Length (m)	Mass (kg)	Accuracy (mm)	DOF (–)	Repeatability (mm)	Payload (kg)	Materials capacity	Max speed (deg/s)
Beagle 2	0.75	2.2	5	5	4	2.5	Ti joints-CFRP	0.5
Mars'01	2.2	4.1	10	4	5	10	Graphite epoxy	2–6
MER	0.7	4.2	5	5	3	2		
Phoenix	2.4	9.7	10	4	5	≈10	Alu/Ti	
MSL	2.2	37	20	5	10	34		3.2–5.2

in real-world missions and the state-of-the-art planetary robotics they represent. Figure 5.1 and Table 5.1 provide a summary and comparison of properties of these robotic arm systems.

5.1.1.1 Mars Surveyor '98/'01

The Mars Surveyor 2001 robotic arm as shown in Figure 5.2 is a low-mass 4 degree-of-freedom (DOF) manipulator with a backhoe design inherited from the Mars Surveyor '98 robotic arm [1, 2]. The end-effector consists of a scoop for digging and soil sample acquisition, secondary blades for scraping, an electrometer for measuring tribo-electric charge and atmospheric ionization, and a crowfoot for deploying the rover from the lander to the surface. Control of the arm was achieved by a combination of software executing on the lander computer and

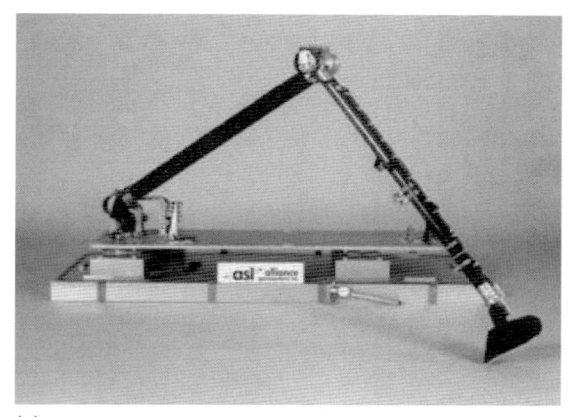

(a)

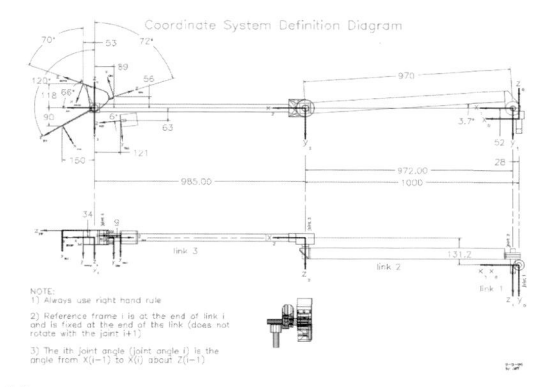

(b)

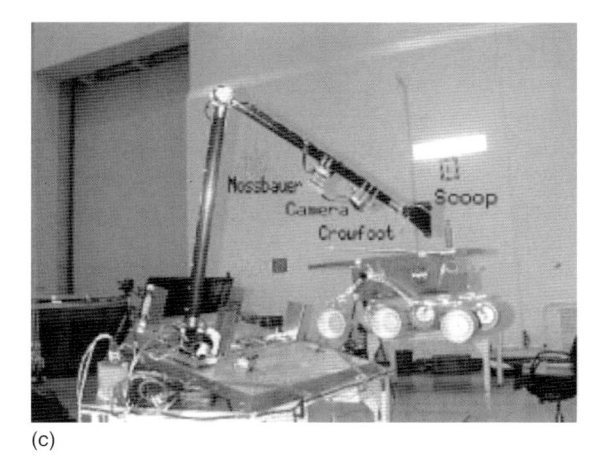

(c)

Figure 5.2 MARS'01 arm [2, 3]. (a) Photo, (b) kinematics, (c) in operation for rover deployment. (Courtesy NASA/JPL-Caltech.)

firmware resident in the robotic arm's electronics. The arm was an essential instrument in achieving the scientific goals of the Mars Surveyor '01 Mission by providing support to the other Mars Surveyor 2001 science instruments as well as conducting arm-specific soil mechanics experiments. While the arm was not flown, it was space qualified to be launch on the canceled Mars'01 mission. In addition to the acquisition of samples, it was designed to deploy a small 10 kg rover the size of the Sojourner rover. At 2.2 m long its graphite/epoxy construction made it a lightweight system with good accuracy for its purpose.

5.1.1.2 Phoenix

Building on the Mars'01 design, the Phoenix Mars Lander robotic arm as shown in Figure 5.3 is a 2.4 m arm with a aluminum/titanium construction, pushing its mass to a recorded mass of 9.7 kg [4]. It operated for 149 sols after landing on May 25, 2008. During its mission, it dug numerous trenches in the Martian regolith and acquired samples of Martian dry and icy soil.

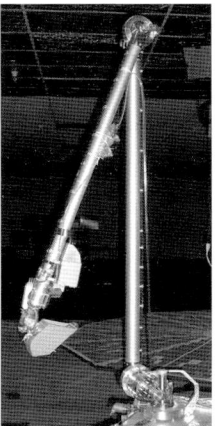

Figure 5.3 Phoenix arm [4]. (a) Calibration and testing, (b) kinematics. (Courtesy NASA/JPL-Caltech.)

5.1.1.3 MARS Exploration Rovers (MERs)

The successful instrument deployment device (IDD) as shown in Figure 5.4 has been used in the two MERs and has accommodated 5 DOFs with an overall length of 0.7 m and a mass of 4.2 kg [5]. It accommodated a range of payload and was tasked to deploy them directly to target locations onto rocks and the surface. The driving system requirements for the IDD are primarily concerned with the absolute and relative positioning performance associated with the placement of the instruments on targets of interest including rock and soil targets as well as rover-mounted targets. The absolute positioning requirement stated that each *in situ* instrument should be positioned to within 10 mm in position and 10° with respect

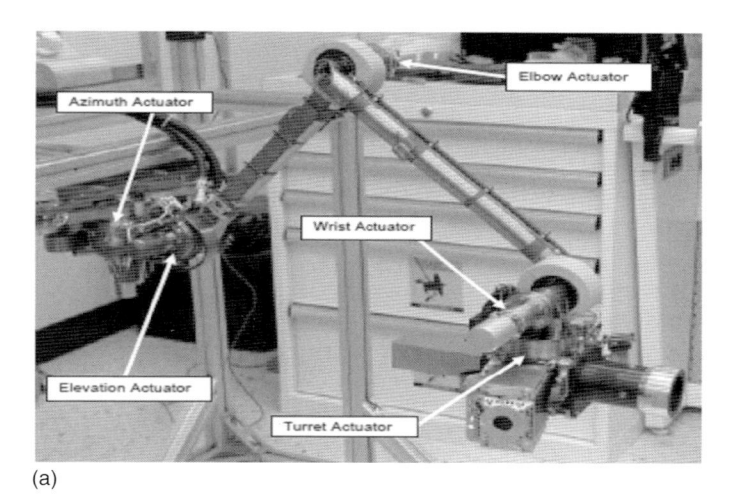

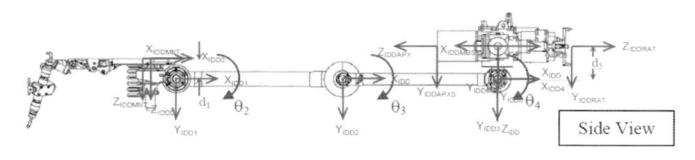

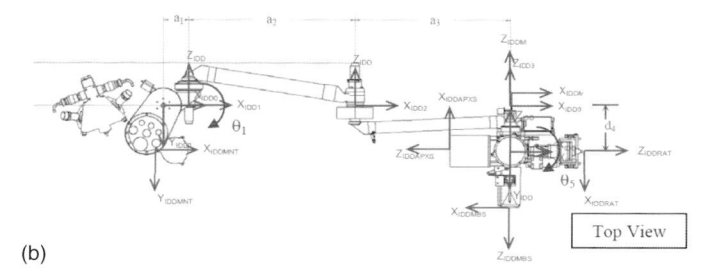

Figure 5.4 MER IDD [5]. (a) Breadboard, (b) kinematics. (NASA/JPL-Caltech.)

to the surface normal of a science target that has not been previously contacted by another *in situ* instrument. This requirement was then broken down into two error budgets associated with the ability of the IDD to achieve a certain instrument position and orientation and the ability of the front HazCam stereo camera pair to resolve the 3D position and surface normal of a science target. Therefore, the overall absolute positioning and orientation error requirements were split equally into two error budgets. The IDD was required to be capable of achieving a position accuracy of 5 mm and an angular accuracy of 5° in free space within the dexterous workspace of the IDD. Factors that affect the ability of the IDD to meet this requirement include knowledge of the IDD kinematics (link lengths, link offsets, etc.), knowledge of the location of actuator hardstops used to home the actuators, actuator backlash effects, closed-loop motion controller resolution, and knowledge of IDD stiffness parameters. A calibration procedure was utilized to experimentally determine the parameters that affect the IDD positioning performance. The remaining half of the error budget was assigned to the front HazCam stereo pair such that the vision system was required to determine the location of the science target with a position accuracy of 5 mm and the angular accuracy was 5° with respect to the target's surface normal. Factors that affect the ability of the stereo camera pair to meet this requirement include camera calibration errors, stereo correlation errors, and image resolution issues. For RAT grinding operations, the IDD is required to place and hold the RAT on the rock target with a specified preload. The IDD is required to provide the RAT with a preload of at least 10 N within 90% of the reachable science target workspace. As mentioned previously, each instrument carried proximity sensors to detect contact between the instrument and the target surface.

5.1.1.4 Beagle 2

The Beagle 2 arm [6] as shown in Figure 5.5 was designed, built, and space qualified for the Beagle 2 mission and is to date the only European planetary robotic arm to be designed up to flight level, including planetary protection qualification. It had a length of 0.75 m and 5 DOFs to deploy an instrument workbench of 2.5 kg. The arm was particularly compact and at 2.5 kg and the lightest arm sent to Mars with the highest mass to payload ratio of about 1 : 1. Each joint comprises a Maxon DC brushed motor and high-ratio planetary gearbox, driving through a 100 : 1 harmonic drive gearbox. Joint position is detected by a potentiometer mounted directly to the output shaft. All structural items were manufactured from titanium in order to closely match the thermal expansion of the bearings, while minimizing the mass. It is also the best choice where the carbon fiber arm tubes are bonded to the end fittings. The performance of the arm was driven by the accuracy of the stereo camera pair on the position adjustable workbench (PAW) of ±2 mm and the resolution of the potentiometer used to measure joint position (±0.2 mm). A total positional accuracy of ±5.34 mm was achieved. All the instruments on the PAW used to examine rocks required a contact force to be applied. The arm was, therefore, capable of generating the 5 N required (Table 5.2).

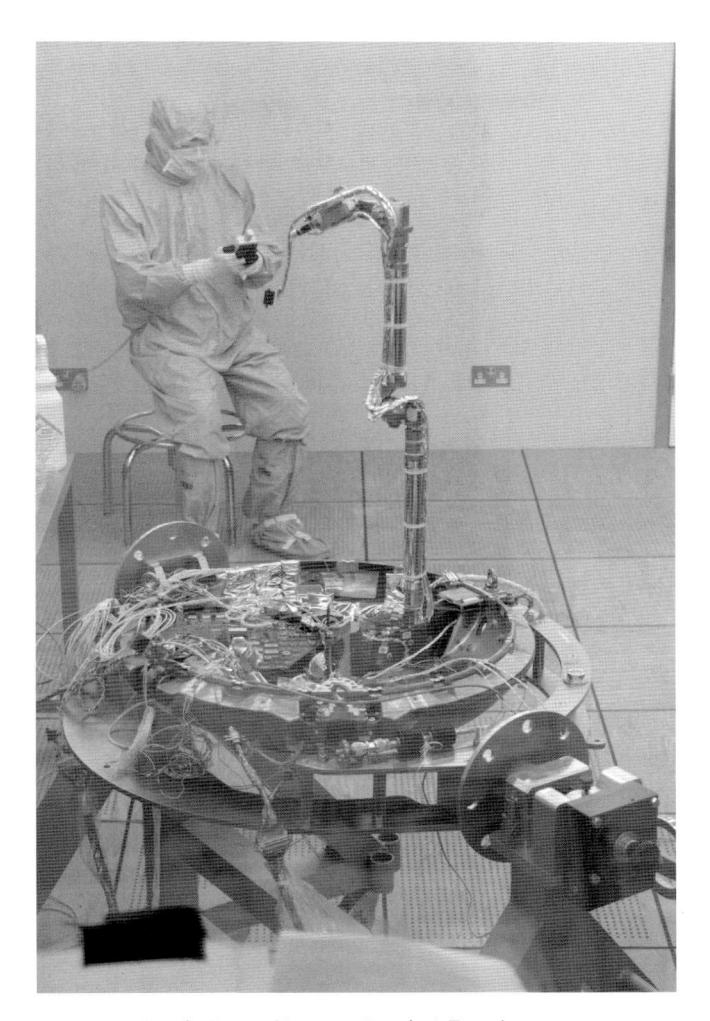

Figure 5.5 Beagle 2 arm. (Courtesy Beagle 2 Team.)

5.1.1.5 Mars Science Laboratory

The Mars Science Laboratory robotic arm as shown in Figure 5.6 is a critical, single fault-tolerant mechanism in the MSL science mission that must deliver 5 out of the rover's 12 science instruments to the Martian surface [7].

The main attributes of the MSL robotic arm are

- 5 DOFs;
- 2.2 m outstretched length from base to center of instrument turret;
- 67 kg mass without turret instruments;
- 5 turret instruments with mass of 34 kg;
- electrical cabling system with 920 signals traversing the length of the arm;

Table 5.2 Beagle 2 arm performance.

Characteristics	Performance achieved
Mass (kg)	2.2
Maximum reach (m)	0.709
Max output torque (N m)	25
Back-driving torque (N m)	21
Max rotational speed (deg/s)	0.5
Supply voltage (V)	12
Drive current (mA)	100 (Max)
Position feedback	10 kΩ, 0.1% Linearity potentiometer
Operating temperature range (°C)	−40 to +30
Nonoperating temperature range (°C)	−100 to +125

- two dual-use caging mechanisms capable of surviving landing loads of over 20 g, passively re-stowing the RA after deployment, and surviving rover driving loads of 8 g;
- capable of surviving temperature range of −128 to +50 °C and operating within a temperature range of −110 and +50 °C.

The primary function of the arm is to position the turret-mounted instruments and tools with respect to Mars surface or rover-mounted targets [8]. The key requirement levied against absolute positioning accuracy is the arm shall be placed with an accuracy of 20 mm in position and 10° in orientation relative to the surface normal of a target selected in stereo imagery. Positioning accuracy is to be 15 mm for instruments that can sense contact with the target (lateral accuracy). The arm is required to have a repeatability of 10 mm. A key contact science requirement is for the system to be capable of deploying and placing an instrument on a surface target determined from stereo imagery, retracting the instrument and placing another instrument or tool within a single command cycle. All requirements related to contact science are applicable to rover tilts up to 30°. The requirements regarding science target operations are specified with respect to the arm's primary workspace, which is a 1 m tall vertically oriented cylinder with a radius of 800 mm as shown in Figure 5.6. It is located 1.1 m in front of the rovers front panel and extends 200 mm below the level of the rovers front wheels. The system is required to be capable of sample acquisition at 90% of the reachable targets within the primary workspace. In addition, the system is required to be capable of acquiring, processing, and delivering samples to science instruments when the rover has a tilt up to 20°. Each of the five arm joints is driven by an actuator consisting of a brushless DC motor integrated with an encoder, a planetary gearhead, a resolver, a brake, and hard stop hardware. Motor angular position is measured by the incremental encoder. Actuator output angle (joint angle) is inferred from motor position through the gearhead reduction ratio. The joint angle is also measured directly by the resolver. After joint motions are complete, the brakes are engaged to hold the

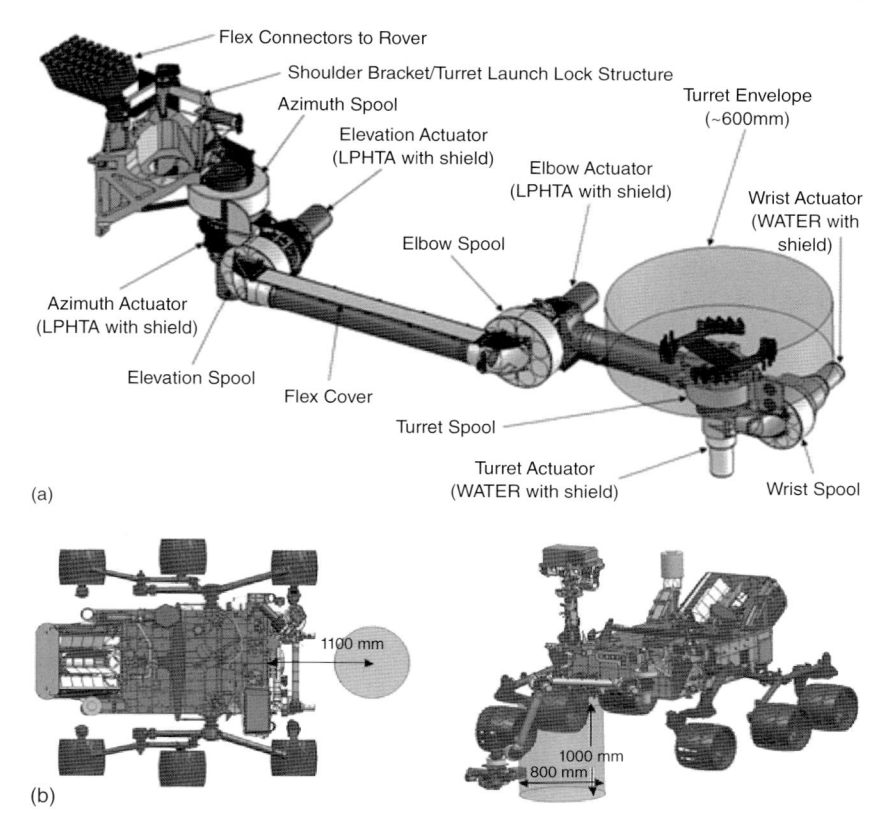

Figure 5.6 MSL robotic arm [8]. (a) Drawing, (b) primary workspace. (Courtesy NASA/JPL-Caltech.)

motor position without servoing (and without power). The two shoulder joints and the elbow use low-power high-torque actuators (LPHTAs). The remaining joints use wrist and turret actuators (WATERs). A summary of the capabilities of these actuators is shown in Table 5.3.

5.2
Robotic Arm System Design

5.2.1
Specifications and Requirements

The design of a planetary robotic arm system needs to first investigate the performance requirements (such as accuracy and repeatability), environmental requirements (such as lighting, dust, and thermal), as well as the design requirements

Table 5.3 Curiosity arm parameters.

Parameters	Units	LPHTA	WATER
Gear ratios	None	7520	4624
Max output torque	N m	1143	259
Max current limit	A	5	3
Max speed	RPM	0.532	0.865
Backlash	mrad	3.64	4.36
Brake holding torque	N m	1313	517
Mass	kg	7.8	4.24

(such as size, mass, power, speed, payload, reach, number of DOFs, redundancy, workspace or working envelope, and autonomous capabilities).

The requirements of a planetary manipulator design are translated from the tasks that the manipulator is expected to perform. For instance, for a simple pick-and-place operation in the manufacture industry, a robotic arm with 3 DOFs is deemed sufficient. Similarly, payload of the robotic arm depends on the anticipated target objects to be handled. Ideally, the manipulator should be designed with extra flexibility so that it can potentially perform a variety of tasks. But the space system design constraints (typically with regard to the mass) are expected to impose restrictions to the design (e.g., end up limiting the DOF to reduce the mass).

5.2.1.1 Performance Requirements

The performance of a manipulator is primarily determined by the chosen kinematic structure, by the actuator technology used and by the control algorithms in place. Those decisions can influence a number of performance measures but basically related to the following:

- Accuracy: This measure indicates how good a manipulator can position itself in a certain location in space and is an important figure especially if the manipulator is used for nonrepetitive movements. The accuracy of the manipulator is in turn influenced by the accuracy of the arm kinematic model available, which usually needs to be enhanced with experimental calibration procedures. Typical required accuracy for the industrial manipulator is around tenths of millimeter.
- Repeatability: This measure indicates the ability of the manipulator to return repeatedly to the exact same location when given the same target position and is an important figure for a reliable manipulator, in this case, especially for repetitive tasks with preprogrammed positions. The magnitude of the required repeatability depends on the task. For example, typical industrial robots are below 1 or 2 mm, that is, an order of magnitude higher than the accuracy.

5.2.1.2 Design Specifications

- Workspace: The workspace is the spatial volume in which the manipulator can work; in other words, the position and orientations in space that the robot can reach. It is defined by three characteristics of the mechanical design, namely the length of each link, the range of movement of each joint, and the type of joint (revolute or prismatic) used. The design should try to maximize the so-called "dexterous workspace" (set of positions in space that can be reached with multiple orientations of the robot's end-effector). The "dexterous workspace" is not only interesting in the phase of the mechanical design, but can also be computed and used during the execution of a manipulation movement in order to relocate the rover or the arm to manipulate in the most convenient area.

- Reach: The reach can be seen as the maximum extent of the "reachable" workspace ("reachable workspace" is the workspace defined by the positions, which can be reached at least with one orientation). Its value depends on the tasks that the robot needs to fulfill and the usual location of the objects to grasp (in front of the rover, on the sides, behind) with respect to the location of the robotic system on the rover or lander.
- Size: The size (length) of the manipulator is determined primarily by the requirements on the robot's workspace: that is, how far should it reach. The size, in terms of how "bulky" the manipulator is, is primarily given by the requirements on the payload (the more payload, the larger motors are required, and consequently, the more weight of the structure) and the required precision (the more precision, the stiffer the robot, that is, usually thicker and thus heavier links).
- Payload: The payload is the maximum load that the manipulator is able to lift (excluding its own weight). Obviously, payload is related to the acceleration or speed at which the object being held needs to be moved. The greater the acceleration, the lower the payload that can be moved for the same robot. In extraterrestrial bodies, the fact of usually dealing with low gravity and requiring very low motion speeds allows for relative large payloads without requiring large torques (thus not requiring heavy motors).
- Mass: The mass of the robotic system depends on the previous points related to the size of the system. Its maximum value is likely given as a mission constraint on the robot's maximum weight, which in turn influences size and structure of the robot.
- Power: Similarly, the power required to drive the robotic manipulator depends on the size, payload, and mass of the manipulator. In addition, the environmental conditions (duration of the day and the nights) of the planetary body influence the size of the energy system. The robot needs to be able to work during the day and keep enough energy to withstand the cold night during which probably some energy is used to heat up some subsystems.
- Number of DOFs: The number of degrees of freedom (which equals to the number of joints in a usual serial open chain manipulator) is determined by the demand of the task. Six DOFs would be required to place the end-effector at any arbitrary position and orientation within the "dexterous workspace" of the robot. However, many tasks can be accomplished with fewer DOFs as they might not need arbitrary orientations. For instance, a simple pick-and-place

operation could be achieved with only 3 DOFs if the object can be always picked up vertically from above.
- Redundancy: Redundancy in terms of the manipulator kinematics refers to the fact of possessing more DOFs than the task requires. For instance, if the task requires 3 DOFs (to position the robot at a certain 3D position regardless of its orientation), a robot with 4 or more DOFs is called redundant. However, a classical redundant robot is the one having 7 or more DOFs, because generally a robot manipulator needs to be able to reach a given 3D position and orientation (thus, requiring 6 DOFs). One benefit of having the redundancy is to allow additional criteria for the motion of the robot without modifying the end-effector position and orientation. A typical example is for a manipulator holding an object in a 3D space, keeping fixed position and orientation of the end-effector, while at the same time being able to move its "shoulder" (or any other) joint to avoid obstacles or to reconfigure the arm posture, and so on. On the other hand, the control algorithms are reasonably more complex, as they need to use some type of redundancy resolution method.
- Autonomous capabilities: An additional requirement can be given by the level of autonomy (LoA) that the robot needs to possess, ranging from a manually teleoperated system to a fully autonomous system (as introduced in Section 1.2 and explained in Section 2.6.2 from system design point of view). The chosen LoA drives various hardware systems design; for example, in order to achieve full autonomy, the robot needs to be equipped with powerful onboard computational resources and dedicated sensor devices. A thorough discussion on autonomy for planetary robots is detailed in Chapter 6.
- On- or off-board processing: Depending on the required LoA, more or less onboard processing power need to be integrated within the robot. For a teleoperated system, the robot requires minimum onboard computational power to implement low-level controllers, which move the robot joints according to the commands received via a remote connection. But if the robot needs to be able to plan collision-free trajectories for the arm, the requirements for the onboard processing increase substantially.
- Software control algorithms: Low-level (joint level) control algorithms is the minimum requirement for the arm in order to drive the joints, which refers to feedback controllers that send commands to the joints (positions or velocities) and use sensors to monitor the accomplishment of the commands. In addition, feedforward signals might be used to compensate for dynamics effects. In case of using lightweight materials, which might introduce vibrations or inaccuracies on the positioning due to bending of the mechanical arm links, vibration suppression algorithms are also required onboard. High-level control algorithms (i.e., output references to be followed by the low-level controllers) are additional demands, which can be implemented either on board the robot at the cost of computational power or on ground at the cost of uplink communication latency.

One can take the specifications of the robotic arm mounted on the Mars Science Laboratory rover (Curiosity) as an example [8]. The Curiosity rover was required to gather soil samples and drill rocks in order to provide samples to the instruments on board the rover. Moreover, the manipulator had to be able to place science instruments over interesting surface targets to study their chemical and mineral composition. As such, the manipulator had four primary requirements to achieve those tasks [8]: (i) to be able to place instruments over surface targets (rocks and regolith), which requires stereo camera systems to extract the 3D location of the targets, (ii) to be able to place the arm to acquire soil samples or position the drill over rock targets, (iii) to be able to support the processing of samples by shaking the onboard instruments, and (iv) to be able to drop off samples to onboard instruments. Basically, the manipulator had to be able to position its tools relative to desired targets over the Mars surface or over onboard instruments carried by the rover. The required accuracy for successfully achieving those tasks was defined as to be lower than 20 mm in position and 10° in orientation and a repeatability lower than 10 mm. The workspace was defined as a vertically placed cylinder of 1 m in height, with radius 800 m and located 1.1 m in front of the rover. The final design selected a configuration with 5 DOFs and a 30-kg payload with a reach of around 2 m. The joints are driven by brushless DC motors equipped with incremental encoders, planetary gears, and brakes. What is called the low-level arm control includes forward and inverse kinematics computations, trajectory generation, and deflection compensation among others. The latter is required to compensate for the deflection of the arm links due to the long reach of the arm (over 2 m) and the heavy payload (30 kg). The forward and inverse kinematics use a rigid-body assumption, which is compensated on board by using a model of the stiffness of the arm to compute its deflection depending on the current pose. In this way, the positioning error can achieve the required specifications. High-level behaviors are a series of sequences of low-level behaviors to ease the use of the rover for recurring operations. For example, the behavior ARM_PLACE_TOOL would place the tool in the specified target position by sequentially calling a series of low-level behaviors.

5.2.1.3 Environmental Design Considerations

For a planetary robot hence the manipulator, environmental conditions of the specific extraterrestrial body in many respects drive the robot design and selection of components. A thorough study on environment-driven design considerations is presented in Chapter 2.4. Here, some of the design factors are re-enforced with regard to the robotic manipulator system.

- Light: One of the most interesting areas to yet be explored on extraterrestrial bodies are craters, where the use of a manipulator to gather samples from inside the crater can be of high scientific interests. Within the craters, the areas with "eternal darkness" are most interesting for science, for example, the famous Shackleton crater at the South Pole on the Moon where measurements from the Lunar Reconnaissance Orbiter/LCROSS indicate the presence of water ice. However, the eternal darkness needs to be considered while designing the camera systems, which would be used by a manipulator to recognize the targets.

- Dust: Dust storms have been a major problem for missions on Mars. In 2007, dust storms covered the solar panels of the rovers Spirit and Opportunity preventing sunlight hence power generation. In addition, dust on the lens of a microscopic imager mounted on Spirit reduced the quality of the images delivered by that camera. The abrasive nature of the dust allows it to seep into mechanisms through seals and reduce the efficiency of the actuators, leading to jamming and ultimately failure. For a planetary manipulator, the dust problems need to be taken into consideration, either by having some way of self-cleaning or by designing the system such that it can work with reduced performance, energy, or image quality.
- Gravity: In a low-gravity environment compared with Earth, the weight of the manipulator will be lower, which in turn requires lower torques to drive its joints, and vice versa. For example, the working Canadarm in space can have trouble to work on Earth as the actuators of the manipulator are likely unable to lift its own weight under Earth surface gravity. Gravity (lower or higher) also affects the dynamic properties of the robot. For example, the manipulator is designed to use the computed-torque control (as described in Section 5.3.1.3) that requires identification of the dynamic model of the robot manipulator, then the dynamic model needs to be reidentified (or at least adapted) if the control needs to be used on another planetary body.
- Temperature: It is known that the temperatures on other planetary bodies can vary abruptly. On Mars, for instance, temperatures might reach $20\,°C$ at noon on the equator, whereas at the poles the temperatures can be as low as $-150\,°C$. As mentioned earlier, the polar regions (and their craters) are usually most scientifically interesting for sample collection and manipulation. Consideration needs to be taken for the design of the electronics and mechanical parts to withstand such temperature range and extremes.

5.2.2
Design Trade-Offs

This section presents design trade-offs for developing the planetary robotic manipulator systems given the requirements discussed in the previous section, including selection of the kinematics (i.e., singularities, manipulability, etc.), the structure and material, and the required sensors.

5.2.2.1 Arm Kinematics
- Singularities: The selection of a certain mechanical manipulator structure determines a certain number of configurations in which the robot enters a "singularity," that is, a configuration in which the robot is not able to generate end-effector velocities (or forces) in a certain direction. Near a singularity, the joint velocity required to achieve a certain Cartesian velocity can be extremely large. In other words, the robot "loses" DOF: mathematically stated, the Jacobian of the robot manipulator loses rank, that is, two or more columns of the Jacobian matrix become linearly dependent. Given the fact that the number

of degrees of a manipulator is usually carefully selected in order to achieve a certain task, the fact of entering singularities and thus losing DOF is a situation needs to be carefully studied. On the one side, the mechanical design can try to minimize the number and type of singularites (by choosing the mechanical structure) and on the other side, the control algorithms can and need to be aware of singularities in order to avoid approaching or working close to them.
- Manipulability: In order to assess how close (or how far) is the manipulator from a singularity, the concept of "manipulability" has been defined. It describes the ability of the robot to move freely in all directions within its workspace (thus, it is closely related to the robot singularities). The manipulability can refer to (i) the ability to reach a certain position (related to the workspace of the robot, thus a global measure) or (ii) to the ability of changing the position or orientation given a certain robot configuration (a small movement around the current configuration, thus, it is a local measure). To study manipulability, one studies the Jacobian of the manipulator, to relate infinitesimal joint movements to infinitesimal Cartesian movements. To this aim, many different manipulability measures have been proposed in the literature, introduced initially by Yoshikawa [9]. By using this information, the manipulator can adjust its configuration (if available, by making use of the robot redundancy) to maximize the manipulability and thus maximize the probability of succeeding on the execution of the manipulation task. The original work from Ref. [9] describes the manipulability index as a distance to singularities computed as

$$w = \sqrt{det(J(q) \cdot J^T(q))} \tag{5.1}$$

where J is the Jacobian of the robot manipulator and q the joint positions. Higher values of manipulability w indicates higher ability to move in all directions around the current configuration of the robot.

5.2.2.2 Structure and Material

Usually, robot manipulators are being designed as rigid as possible so as to minimize the deflection of the links while moving its own weight or a payload. The stiffer the robot, the more accurate the positioning can be achieved without requiring complex control algorithms. However, stiff rigid links generally mean more weight of the mechanical structure, which in space applications is of main concern. For this reason, lighter materials (i.e., usually more flexible materials such as carbon and glass-fiber composites) can be used to build the manipulator at the price of possibly reducing the robot positioning accuracy, increasing the risk of vibrations during the movement, and thus, requiring more complex algorithms to deal with those issues.

5.2.2.3 Sensors

Apparently from the previous section, a decision needs to be made about the choice of sensors on board the manipulator, and the need to consider saving the mass by using the minimum set of sensors required to successfully accomplish the assigned tasks.

- Position and velocity sensors: Joint positions (and velocities) are one of the basic manipulator sensors and of crucial importance. There are different technologies: from simple and low-resolution potentiometers to optical or magnetic encoders providing high resolution. Similarly, they can be found as incremental or as absolute encoders. The first type starts counting revolutions from the current position of the motor after power-up; thus, before having passed through a reference point, there would be no way of knowing in which absolution position the motor is in. This usually requires some sort of homing procedure (joints move until finding its "zero" position) at every start-up. On the other hand, as the name already indicates, absolute encoders signal the absolute position of the motor within a rotation.
- Force or torque sensors: The typical force sensor is attached to the wrist of the manipulator and measures forces in the three Cartesian directions and torques in the three axes of rotation of the end-effector. These measurements might be interesting for accurately controlling the forces exerted by the robot. Torque sensors on each joint are very rarely used, although they might be interesting for several purposes: for implementing computed torque control or vibration control algorithms, especially in the case of joints presenting some flexibility.
- Cameras: The use of cameras is one of the basic exteroceptive sensors that a planetary arm uses. On the one side, there is the need of recognizing the location of the objects to be grasped, which is usually accomplished by using a pair of cameras mounted as a stereo vision system. On the other side, the manipulator is probably designed using lightweight materials due to the constraints on the mission's payload, thus making the manipulator less accurate than a rigid one. In this case, a precise calibration can be performed to improve the accuracy of the arm but usually combined with the use of the camera information to drive the motion of the manipulator toward the object (what is known as "visual servoing").

5.3
Robotic Arm Control

Manipulation refers to the process of moving or rearranging objects in the environment. In order to control a robotic manipulator system to perform a desired motion, several control components are required. Those components can be grouped into three functions as shown in Figure 5.7.

1) High-level control: The highest manipulator control layer, planning the path for the end-effector and implementing the LoA for the robot, which can range from the teleoperation to the fully autonomous operation with varying LoA in between.
2) Trajectory generation: Components that provide references for the joint controllers, basically acting as interface between the high-level commands and the low-level control commands.

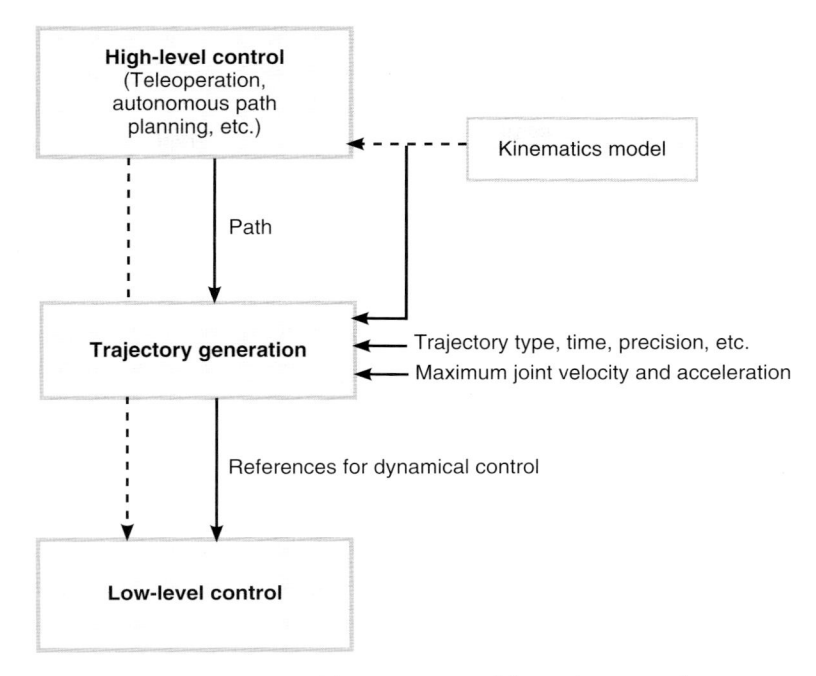

Figure 5.7 Layered structure of the motion control for a robot manipulator.

3) Low-level control: The control strategies dealing with the commands at joint level to achieve the desired motion references dictated by the high-level control (usually given in Cartesian space).

In addition to the three-level control, collision avoidance is often regarded explicitly where the control strategies need to avoid collisions of the robot with the environment and with itself (self-collision), particularly true for complex and redundant robots. In the following sections, the basic concepts and state-of-the-art design examples for each of these functions are described.

5.3.1
Low-Level Control Strategies

5.3.1.1 Position Control
Position (or velocity) trajectories, which ideally the robot manipulator should be able to follow according to its kinematic constraints, are received by the robot joints, which will try to follow the trajectories as close as possible. This step is usually known as kinematic control.

In practice though, those trajectories are not always achieved due to the dynamical characteristics of the robot (such as inertia, frictions). The dynamical model of a robot is highly nonlinear, multivariate, coupled, and of time-varying parameters, which makes the dynamical control usually a hard problem. For that reason, in practice usually some assumptions are taken, which simplify the control scheme. Factors such as the gear ratio or the viscous friction can actually help to simplify some dynamics. For instance, a high gear ratio as commonly found in industrial robotics allows to consider the links of the robot as decoupled. In those conditions, a set independent proportional and derivative (PD) controllers for each joint might be a more than suitable solution. Note that, however, a high gear ratio does not come for free: it also adds undesired viscous friction, joint play, and joint elasticity.

This section shows the two typical independent joint feedback control schemes for stiff robots (high-gear ratios), which assume the manipulator as a decoupled system (a joint motion has no influence on the other joints' motion). In the rest of the section, the following symbols are used:

- The end-effector position and orientation is denoted as X. This 6×1 vector is defined as $X = (p^T \varphi^T)^T$, where the vector p describes the end-effector position and φ is a set of Euler angles from the rotation matrix describing the orientation of the end-effector.
- The three-dimensional force f as well as the three-dimensional moment m exerted by the robot's end-effector are the components of the wrench $h = (f^T m^T)^T$.
- The joint positions are denoted by q.
- The commanded torques are denoted by τ.
- The subscript d denotes a "desired" reference value for the specific magnitude (either force or position/orientation in Cartesian or joint space).

Feedback Position Control with Position Reference The simplest and most common form of robot joint control is independent PD control, described as

$$e_p = q_d - q \tag{5.2}$$
$$\tau = K_P \cdot e_p + K_D \cdot \dot{e}_p \tag{5.3}$$

The error signal is formed by comparing the joint position reference with the current joint position measurement. This error is then used by the proportional part (P) and the derivative part (D) of the controller to eliminate (or minimize) the error signal. The selection of the controller parameters K_P and K_D determines the behavior of the controller. The controller scheme can be seen in Figure 5.8.

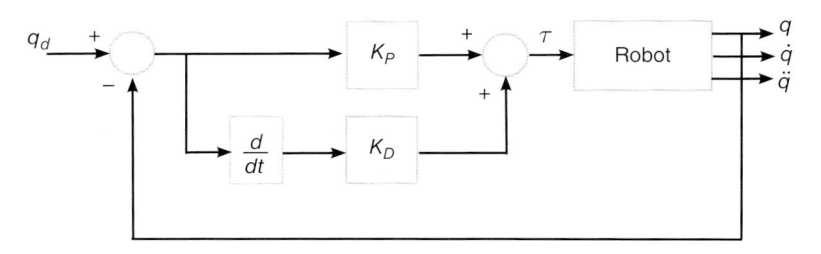

Figure 5.8 Feedback controller with position reference.

Feedback Position Control with Position and Velocity Reference A slightly better performance is obtained by using also a velocity reference in addition to the position reference, as described by

$$e_p = q_d - q \tag{5.4}$$
$$e_d = \dot{q}_d - \dot{q} \tag{5.5}$$
$$\tau = K_P \cdot e_p + K_D \cdot e_d \tag{5.6}$$

The controller scheme can be seen in Figure 5.9.

5.3.1.2 Force Control

Position control is the best solution as long as no (or little) contact with the environment is necessary. However, in case of unstructured environments such a planetary body, where the knowledge of the environment is imprecise, or in complex motion-constrained tasks, some kind of force control is required in order for the robot to attain its goal. More specifically, robots need to sense contacts with the environment and, by sensing, comply with the current environment, regardless of its unknown nature.

One of the required components for a robust and adaptive robot is the ability to adapt the contact interaction forces when manipulating an object or contacting a surface (in this area, any object/surface with which the robot makes contact is generally defined as "environment"). This problem domain is generally known as compliance control and tries to guarantee that the robot accommodates the interaction forces rather than resisting to the constraints posed by the contact with the environment.

Active compliance control methods use measurements from the contact forces (and moments) and robot's motion, which are fed back to the robot's controller in order to generate appropriate motion commands according to a desired robot's behavior. The research on active methods to provide robust force control in any of its flavors has been gaining importance in the last three decades and a great deal of research papers is available. A first description of the state of the art in the 1980s can be found at [10]. Similarly, [11] reviews the state of the art in the 1990s. Some books also appeared at that time that focused exclusively on force control [12]. More recently, Lefebvre et al. [13] surveyed the state-of-the-art of the required subcomponents for active compliant systems. The recent "Handbook of

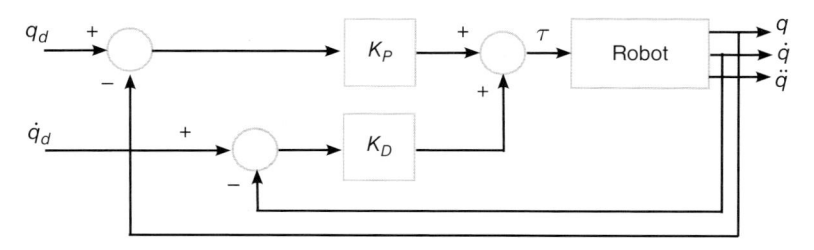

Figure 5.9 Feedback controller with position and velocity reference.

Robotics" [14] includes also a chapter that reviews the current state of robot force control. In the area of planetary rovers, the robotic arm mounted on the Mars Science Laboratory rover (Curiosity) includes the capability to autonomous drill a rock, being its main component a force feedback loop that controls the forces during the execution of the drill. The algorithms to achieve those operations are described in Ref. [15]. The drill on the rover Curiosity is the first autonomous extraterrestrial drill into a rock.

In general, any active control method for compliance control tries to somehow add or combine motion and force errors, and use a controller or set of controllers to send the most proper commands to the robot's joint actuators. The way these errors are combined is what creates the basic distinction between direct and indirect compliance control methods.

- Direct force control methods are those where the controller directly regulates the contact force to a desired reference. A classical force control strategy or a hybrid force/motion control belongs to this category.
- Indirect force control methods are those where the force is controlled indirectly via motion control. Impedance control in its different "flavors" belongs to this category.

Both approaches differ in the way to specify the interaction task. For instance, in the hybrid force/motion control (or direct method), the task is specified in the geometric space, as can be seen later that the user defines which directions are controlled on force and which ones are controlled on position. In the case of impedance control (indirect method), the designer can define a dynamic relationship between force and motion. In the end, the set of impedance parameters defined will determine the robot's behavior. When comparing the design methodologies for compliance control in a practical sense, both are quite similar: where a designer would identify a constrained direction in hybrid force/motion control, he would probably define a more compliant behavior when using an impedance controller; where a designer would identify an unconstrained direction in hybrid force/motion control, he would probably define a stiffer behavior when using an impedance controller. Interaction control methods might also be classified according to the static or dynamic performance of the control method:

- Dynamic model-based control methods are those that are concerned with the transient, that is, with the dynamical response of the system. To this category belong impedance and admittance control schemes, as well as hybrid and parallel force/motion strategies. For any of them, a complete dynamic model of the robot is required, thus being more complex, both to be designed and implemented. Furthermore, force measurements are necessary in order to obtain a decoupled and linear interaction model that is easily tractable.
- Static model-based control methods are those that are only concerned with the steady-state response of the system. In the case of impedance control, the static case is called stiffness control. In the case of admittance control, the static case is known as compliance control. That is, compliance and stiffness control are subsets of admittance and impedance control, respectively. These methods are thus

easier to implement, as they do not require dynamic models but only knowledge about the gravity terms.

A review of several interaction control methods, divided into dynamic and static model-based methods, is given in Ref. [16], which also includes experimental evaluation.

Hybrid Force/Motion Control This method [17] treats the contact interaction as a geometric problem, where there are a set of geometric constraints to be taken into account. After careful examination of the interaction task, a number of robot's DOFs are regarded as "force-controlled," whereas the rest are considered to be "motion-controlled." That means that the hybrid controller controls exclusively motion along unconstrained, "motion-controlled," directions, and force/moment along constrained, "force-controlled," directions. This approach is based on the assumption that for most of the usual constrained robotic tasks, it is possible to split the task into two mutually independent subspaces, one controlling contact forces and one controlling robot's motion. Thus, making use of different subspaces, motion, and force can be controlled simultaneously. Figure 5.10a shows a diagram of a hybrid force/motion controller.

Parallel Force/Motion Control The parallel approach [18] as shown in Figure 5.10b is classified as a direct control method as it starts with the geometric constraints also used for hybrid force/motion control.

The difference with a hybrid force/motion approach is that parallel force/motion control does not use different control subspaces for force and motion, but combines and weights the contributions of motion and force controllers into one controller using a single matrix. In this case, the controller gives priority to force errors, which dominate the controller's response. Thus, a position error would be "tolerated" along a constrained direction in order to ensure proper force tracking.

Impedance Control The general term impedance control is commonly used indistinctly when referring to either impedance or admittance control. Both pursue the same goal – active modification of the mechanical impedance of the robot – but they do it from different perspectives. Impedance control basically works by measuring position and outputting force, whereas admittance control works by measuring force and outputting position. Due to the different ways of solving the control problem, the accuracy of the compliance method lies on different factors. The impedance control depends on the accuracy of the position sensors and the bandwidth and accuracy of the force-controlled actuators, whereas in admittance control the accuracy depends on the force sensors used and the bandwidth and accuracy of the position-controlled actuators.

Active impedance control as shown in Figure 5.11a indirectly regulates the contact forces by generating an appropriate motion that ends up in a desired dynamic relationship between the robot and the environment. In contrast to hybrid force/control methods, impedance control uses a single control law to regulate simultaneously both position and force by specifying a target dynamic relationship

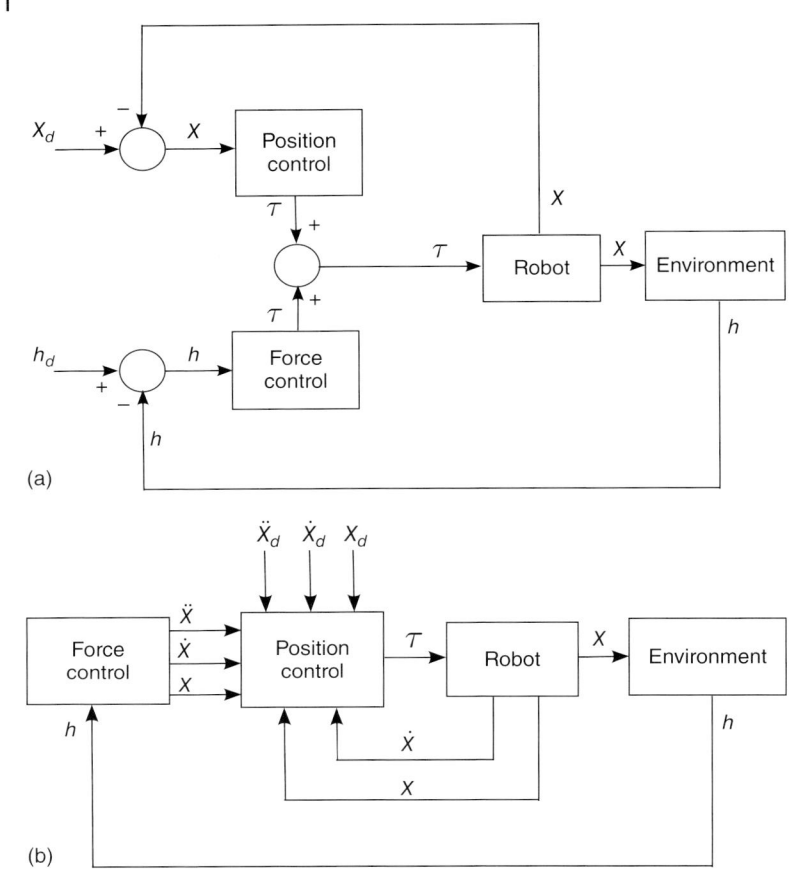

Figure 5.10 Force/motion control. (a) Hybrid, (b) parallel.

between them. In other words, it weighs the contributions of the force and motion controllers using a set of weighing matrices. In contrast to hybrid force/control where there are also weighing matrices, the matrices used in active impedance control have physical dimensions of impedance: that is, stiffness, damping, and inertia parameters. These parameters thus shape the dynamical behavior of the robot as if mechanical springs, dampers, and extra inertia were included into the robot's end-effector. The design of the target impedance that ensures a proper behavior is though not an easy task. It is clear to see that the behavior of the robot when contacting an environment needs to be different than when moving freely, especially when a good position tracking is desired in free space (remember that a stiff robot is a good position tracker, whereas by definition a stiff robot is not compliant when contacting the environment). Moreover, even a well-defined target impedance depends finally on the dynamics of the environment, which needs to be estimated as good as necessary to ensure not only stability on a first level but also the necessary robot's performance.

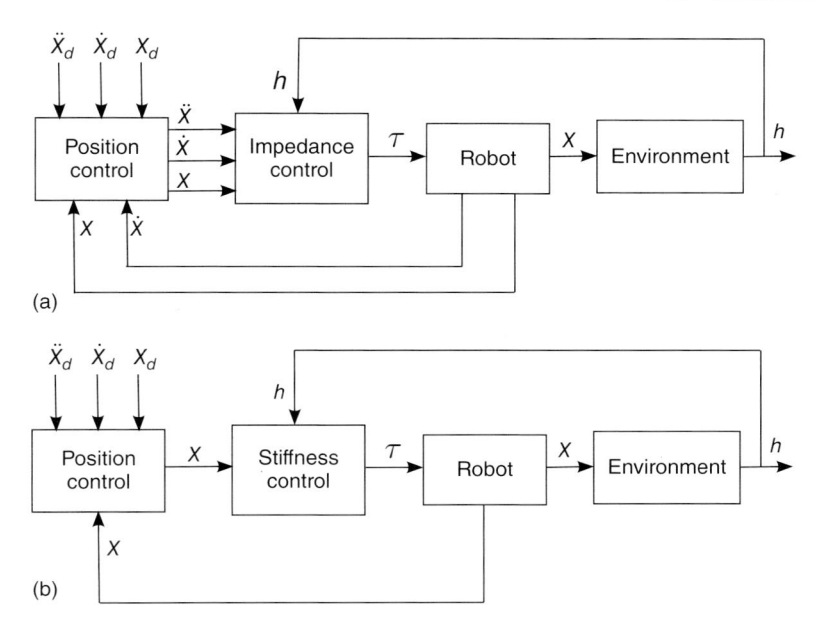

Figure 5.11 Control strategies. (a) Impedance control, (b) stiffness control.

The work from Hogan [19] is considered the benchmark for impedance control as it is the first to describe the concept of a virtual mechanical impedance for controlling robot-environment interaction forces. However, it was not the first idea on the use of virtual mechanical elements to control forces: in Ref. [20], a generalized spring and damper system for force control is presented that is later implemented in Ref. [21]. One of the best current examples of the use of active impedance control are the LWR lightweight arms from DLR [22]. Since then a vast literature has been published around the topic in order to cope with pitfalls of the original description or deal with implementation issues. For instance, Hogan itself [23] and later other authors [24] analyzed the stability of the impedance control law or the instability originated after contacting stiff environments [25]. In order to tackle the problem of the uncertainty on the parameters of the environment model as well as of the robot, some works propose adaptive [26–29] or robust [30, 31] impedance control strategies that would deal with uncertainties. Moreover, as the original description lacks of force tracking capabilities, some works propose methods for enhancing the controller to track forces [32, 33]. The impedance control area has seen as well the use of neural network implementations of the controller [34, 35], in most cases used as a method to minimize the problems arising from the model's uncertainties. Learning algorithms have also been proposed [36, 37], as well as impedance control methods have been not only applied to single-arm systems but also to multiple-arm systems [38–40].

Stiffness Control Stiffness control is only concerned with the steady state of the end-effector, thus this is a simplification of a complete impedance control. Only a proportional action (a single matrix) is necessary to define the behavior of the controller, which controls the behavior of the robot in such a way that behaves as a 6-DOF spring with respect to the forces (and moments) applied to the robot's end-effector. Since this method focuses specifically on the steady-state response, it does not require knowledge about the complete robot's dynamics. In this case, only the gravity terms of the Lagrangian equation representing the dynamics of the system is necessary. In brief, the stiffness control regulates the static relationship between the forces exerted on the environment and the deviation (if and as much as necessary) of the position and orientation from the desired values. Figure 5.11b shows a block diagram on the stiffness control that only receives information about position, since it does not deal with higher derivatives of this variable.

Admittance Control   Admittance control holds basically the same working principle as impedance control. However, it separates explicitly the position control from the impedance control. The position controller is an inner loop, as in industrial robots, designed to be stiff to reject position disturbances robustly. The trick is that the position controller does not receive directly the desired motion but the output of the "impedance controller." Figure 5.12a shows a diagram sketching the admittance control. As it can be seen, the "admittance controller" to be more correct, since the input is force (wrench) and the output is position, receives two inputs: the desired position and the end-effector wrench (force and moment). A proper target impedance can generate an output that is a suitable position/orientation for the robot that maintains the desired dynamical relationship between force and motion.

Note that the admittance control diagram shows similarity to the previous parallel force/motion control, which requires knowledge of the interaction task in order to define the geometric constraints. The difference with admittance control is the criteria to define the force controller (physical impedance parameters vs force-controlled task directions).

Compliance Control   As stiffness control is for impedance control, compliance control is a subclass of admittance control, where the controller is only concerned with the static relationship between the forces exerted by the robot and the deviation of the position/orientation from the desired values. Figure 5.12b shows the block diagram of the compliance control, where the controller only generates a reference position (no higher derivatives) for the position controller.

5.3.1.3   Dynamic Control

Dynamics studies the relations between forces acting on a body and its resulting motion. For that reason, the purpose of a robot dynamical model is to obtain the relation between the movement of the robot and the forces acting on it. The dynamical model can be used to achieve several objectives:

- simulation of the robot motion,
- design and evaluation of the mechanical structure of the robot,

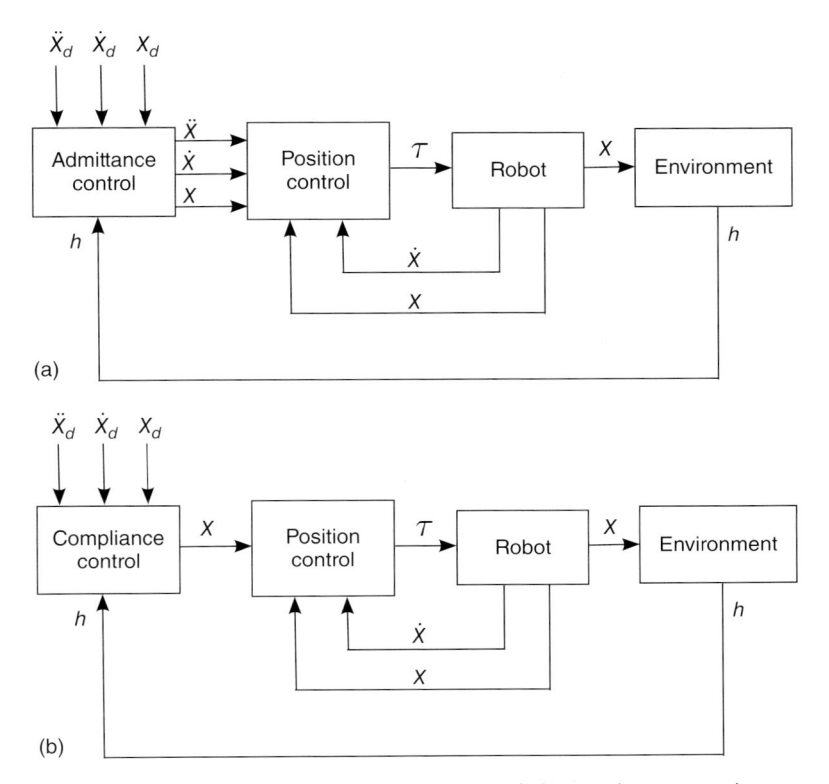

Figure 5.12 Control strategies. (a) Admittance control, (b) compliance control.

- selection of the required actuators for the robot,
- design and evaluation of the dynamical control of the robot.

This last point is of great significance, since the quality of the dynamical control determines the robot's precision and speed of movement. In other words, the use of dynamic control strategies are used to ensure that the trajectories followed by the robot are as close as possible to the ones proposed by the kinematic control.

However, the high complexity associated with the formulation of a dynamical model obliges in most cases to assume simplifications on the model. Needless to say, a complete dynamical model of a robot not only includes the dynamics of its links but also of the transmission system, the actuators, and the power/control electronics. Those elements include new inertia, friction, and saturation effects, which even make the dynamical model more complex.

The purpose of a dynamic control is to compute the torques τ, which make possible to obtain the desired joint positions q (possibly also joint velocities and accelerations). Obviously, torques are not directly sent to the motors, but the computer generates some torque-related "quantities," which in turn are transformed into analog voltages by the PWM generator. The motor then produces torques

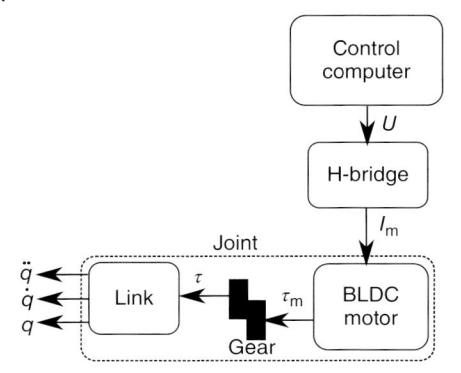

Figure 5.13 Overview of a torque control scheme of a brushless DC motor-based joint.

that provide the actuating torques τ at the joint through the gear ratios as seen in Figure 5.13.

The dynamic equation relating τ and q is well-known:

$$\tau = M\ddot{q} + C\dot{q} + G + \tau_f = M\ddot{q} + H \tag{5.7}$$

where M is the inertia matrix ($n \times n$) of the link, C are of Coriolis forces/torques ($n \times 1$), G is the vector of gravity forces/torques ($n \times 1$), and τ_f is the vector of friction forces/torques ($n \times 1$).

Mechanical Motor Model The simplified (without regard to Coriolis forces) dynamic equation of the mechanical part of the motor can be described as

$$\tau_m = J_m \ddot{q}_m + f_m \dot{q}_m + C_m \tag{5.8}$$

where τ_m is the motor torques, J_m is the inertia matrix of the motor, $\ddot{q}_m$ is the motor acceleration, $\dot{q}_m$ is the motor velocity, f_m is the motor viscous friction, and C_m is the resisting torque at the motor. We can include the gear ratio (N) to relate motor and link velocities/angles as $\dot{q}_m = N\dot{q}$. Thus, the resisting torque at the motor side is related with the link torque as

$$C_m = N^{-1}\tau \tag{5.9}$$

By using the gear ratio (N) and Eq. (5.9) in Eq. (5.8) and replacing τ by the expression in Eq. (5.7), we obtain

$$\tau_m = M'\ddot{q} + H' \tag{5.10}$$

where

$$M' = J_m N + N^{-1} M \tag{5.11}$$

and

$$H' = f_m N\dot{q} + N^{-1} H \tag{5.12}$$

Electrical Motor Model Regarding the electrical part of the model, a DC motor model is used as an example.[1] The voltage vector at the motor input can be defined as

$$U_m = r_m I_m + L_m \frac{dI_m}{dt} + E_m \quad (5.13)$$

where r_m is the motor armature resistance, I_m is the motor's armature current, L_m is the inductance of the motor, and E_m is the motor back-emf (counter-electromotive) voltage. For DC motors, we additionally have that

$$E_m = K_e \dot{q}_m \quad (5.14)$$

and

$$\tau_m = K_c I_m \quad (5.15)$$

What it is finally missing is the power electronics (H-bridge) model, which can be assumed as a proportional relation between input voltages U and I_m motor currents as

$$U = K_v I_m \quad (5.16)$$

Feedforward Plus Feedback Control The simplest dynamic controller is implemented making use of the inverse model as feedforward signal. This allows its computation directly (and exclusively) from reference signals of position/velocity/acceleration, thus it can be computed offline. The controller is described by

$$e_p = q_d - q \quad (5.17)$$
$$e_d = \dot{q}_d - \dot{q} \quad (5.18)$$
$$\tau_{FB} = K_P \cdot e_p + K_D \cdot e_d \quad (5.19)$$
$$\tau_{FF} = \hat{M}(q_d)\ddot{q}_d + \hat{C}(\dot{q}_d, q_d)\dot{q}_d + \hat{G}(q_d) \quad (5.20)$$
$$\tau = \tau_{FB} + \tau_{FF} \quad (5.21)$$

The controller scheme can be seen in Figure 5.14. In this scheme, it might be necessary to include a zero-order delay after the computation of the inverse model since we are feeding forward a "future" signal into the inner feedback loop.

Computed Torque Control. In this control strategy, an inverse model is used in the feedback loop of the controller, that is, it needs to be computed online. In addition, there is a PD controller, which works against a simple linear model for the acceleration. Given the knowledge about Eq. (5.7), a nonlinear decoupling controller can be written as $\tau = \alpha \cdot \tau' + \beta$, and choosing $\alpha = M$ and $\beta = H$. This can decouple the system and gives $\tau' = \ddot{q}$, which means that the PD controller needs to deal only

---

[1] Assumption is valid for a BLDC motor using a vector control strategy, which mathematically transforms the BLDC motor into a "pure" DC motor.

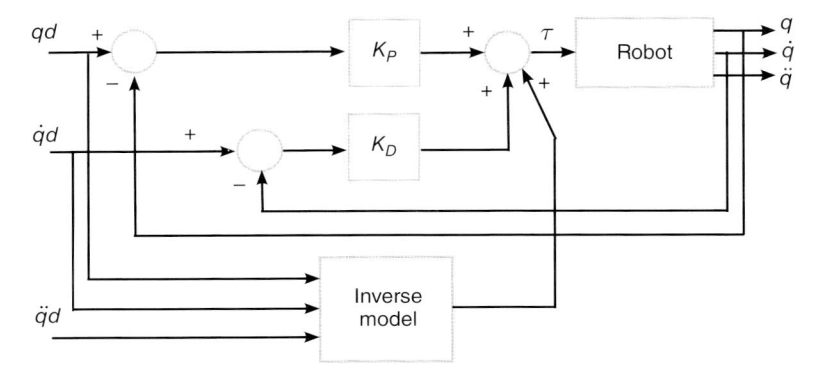

Figure 5.14 Feedforward (inverse model) plus feedback controller.

with a linear system. In summary, the controller is described by

$$e_p = q_d - q \tag{5.22}$$
$$e_d = \dot{q}_d - \dot{q} \tag{5.23}$$
$$\tau' = K_P \cdot e_p + K_D \cdot e_d + \ddot{q}_d \tag{5.24}$$
$$\tau = \hat{M} \cdot \tau' + \hat{H} \tag{5.25}$$

The schematic use of the inverse model (and its components) within the computed torque control scheme can be seen in Figure 5.15. The complete controller scheme can be seen in Figure 5.16.

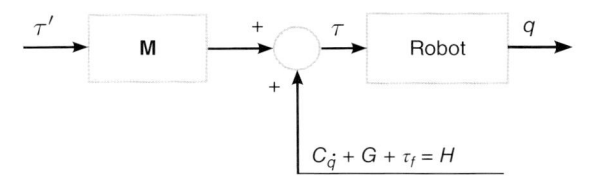

Figure 5.15 Inverse model for decoupling robot model.

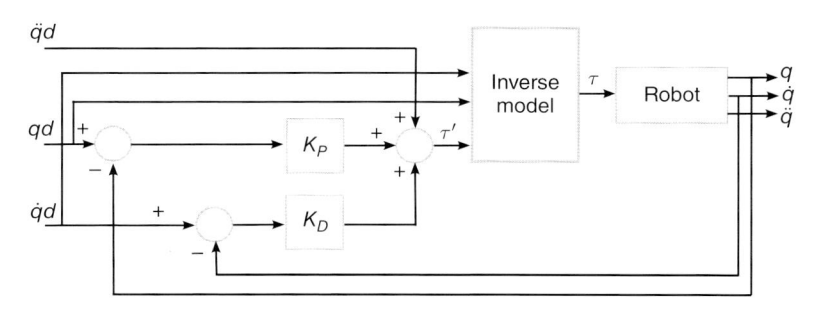

Figure 5.16 Computed torque control.

5.3.1.4 Visual Servoing

When a robot operates in an unstructured environment, the use of sensor-based control strategies for ensuring the correct execution of the manipulation tasks are unavoidable. Position or velocity control is used in most robot systems, as we just saw in the previous sections. Force control and other compliance control methods are also of interest when operating in nondeterministic scenarios as previously seen. Another source of information is vision; the use of the information from a camera as feedback for controlling the motion of a manipulator is in general known as *Visual Servoing*. The method allows to deal with real-time changes of the relative pose of the object with respect to the robot and has the advantage of possessing great accuracy and robustness. In contrast to the methods described earlier, the visual servoing does not need a explicit reference signal but always tries to reactively move toward the object by minimizing the error between the pose of the object and the pose of the gripper.

Hill and Park [41] introduced the term *Visual Servoing* in order to differentiate with previous works. In this approach, visual features from the camera image are used in order to orient, for instance, the robot gripper with the object to be grasped. The camera can be placed on a fixed position with respect to the manipulator (the so-called *eye-to-hand* approach) or can be directly be attached to the manipulator's end-effector (the so-called *eye-in-hand* approach). The general advantage of the eye-to-hand approach is that the camera is not moving and usually has a better overview over the workspace and the object/gripper. On the other side, in this case, the robot needs to track both the object and the robot's end-effector. In the eye-in-hand approach, since the camera is mounted on the gripper, the pose of the gripper can be extracted directly from the pose of the camera.

One of the first eye-to-hand systems was shown in 1973 by Shirai and Inoue [42]. In Reference [43], a static camera was used to grasp an object that is moving on a plane. Kragic [44] used a motion tracking system in connection to a grasp simulator in order to plan the grasping motion. However, the eye-in-hand configuration is used more frequently, see Refs [45, 46]. By using two cameras, complete 3D information of the object can be retrieved, even when the object is unknown. One of the first stereo visual servoing systems was implemented using a eye-to-hand configuration [47].

Apart from the location and number of cameras, visual servoing can be also categorized according to the control strategy used, namely imaged-based, position-based, or hybrid strategies [48]. In the image-based visual servoing (see Figure 5.17), the position of the gripper is extracted directly from features of the image, with a so-called feature sensitivity matrix is computed [49, 50]. By using this matrix, the motion required to move the manipulator toward the object to be grasped can be estimated. In the position-based visual servoing (see Figure 5.18), the complete 3D position and orientation (pose) of the target object are extracted from the camera information and the error between the pose of the end-effector and the object pose is brought to zero [51, 52]. The advantage of image-based method is that they do not require stereo camera system although it has to deal with the problems related to tracking features are lost from the image. When

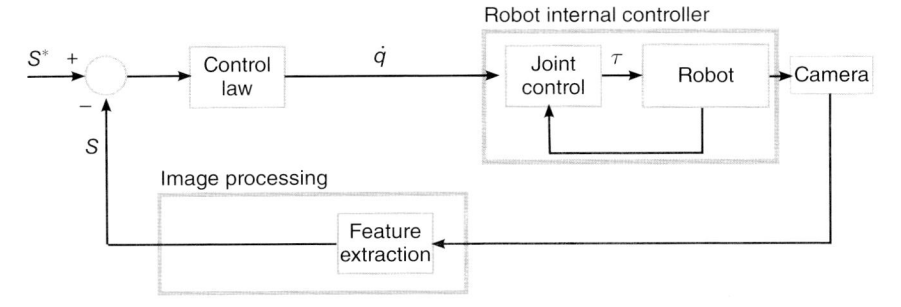

Figure 5.17 Image-based visual servoing control scheme.

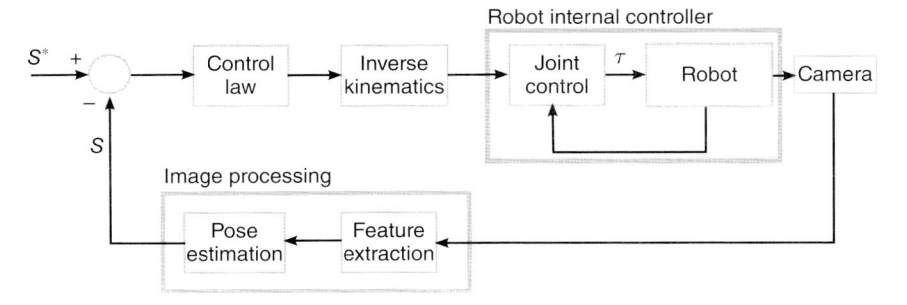

Figure 5.18 Position-based visual servoing control scheme.

using position-based methods, and since the control is done in Cartesian space, possible constraints on the path can be taken into account, for instance, to avoid collisions. However, those methods require a higher computational effort and 3D information about the object.

5.3.2
Manipulator Trajectory Generation

This section focuses on how to allow a robot manipulator moves from some initial Cartesian position X_0 to some desired final position X_f. Usually, the path to follow is given by the high-level control (for instance, by an automatic path planning module). The path consists of the initial, final, and via points that the robot is going through. A trajectory also includes the time information at which each joint reaches a certain position, velocity, and acceleration. In other words, the trajectory generation involves time-parameterizing the path given by a higher level controller. Thus, the trajectory generation module establishes the trajectories to be followed by each joint with respect to time and in order to meet the user's requirements, for example, target position, type of trajectory, and time used for the movement, and so on.

5.3.2.1 Trajectory Interpolation

In most cases, paths are given and formulated in Cartesian (task) space in which it is easier to define tasks and the motion of the manipulator. Furthermore, the generation of a collision-free trajectory is also simplified when working in task space. On the other side, working on joint space can avoid problems with singularities and requires less computation. Either case, before sending the commands to the manipulator joints, the Cartesian trajectory needs to be converted to joint space.

In order to translate a certain Cartesian trajectory into a joint space trajectory, the inverse kinematics of the manipulator is used. This topic is out of the scope of this chapter, but two important things need to be kept in mind: the mapping between a Cartesian trajectory and the joint space trajectory is not unique and depends on the number of DOFs of the manipulator. That is, several joint space trajectories might yield the same Cartesian trajectory. In addition, there might be no solution to the inverse kinematics problem. That might happen because of singularities of the manipulator or because the target positions are out of its reachable workspace.

Given either a Cartesian trajectory or a joint space trajectory, it is not possible to store all the trajectory points in memory and it might not be possible to find an analytical expression to describe it. The usual procedure is then to store the initial, final, and intermediate via points of the trajectory through which the manipulator needs to pass. The conversion between this set of discrete points and the continuous trajectory points to be sent to the robot is what is called *trajectory interpolation*.

The simplest case is the use of a linear interpolator. Given two known points of the trajectory (and its associated time stamps), a linear interpolator draws a straight line between them. That function can help find intermediate points between the two initial known points. This is a very simple method to compute and ensure continuity in the position. However, it does not come for free: the velocity is kept constant between two successive points, which might create abrupt changes on the velocities between sets of adjacent points and, in theory, it might require infinite values for the acceleration on those changes. This is one of the common reasons for jerky robot movements.

In order to avoid those problems, the solution can be to use higher-order polynomial functions to represent the trajectory. A usual choice is a cubic interpolator, which uses a third-order polynomial; that is, four coefficients are available to describe two adjacent points. In other words, by using a cubic interpolator, there is the possibility to specify four constraints that determine the cubic trajectory which are usually the initial and final positions (as the linear interpolator) plus the initial and final velocities. Equation (5.26) describes a cubic polynomial:

$$\theta(t) = a_0 + a_1 \cdot t + a_2 \cdot t^2 + a_3 \cdot t^3 \tag{5.26}$$

where $a_0, a_1, a_2,$ and a_3 are the four coefficients that specify the cubic trajectory. As an example, given boundary conditions set to zero, that is, initial position and velocity ($\theta(0)$ and $\dot{\theta}(0)$) as well as final position and velocity ($\theta(t_f)$ and $\dot{\theta}(t_f)(0)$) defined as zero, and solving Eq. (5.26) with the previous boundary conditions, we obtain the values for the four unknown coefficients as

$$a_0 = \theta_0 \qquad\qquad a_1 = 0$$
$$a_2 = \frac{3}{t_f^2}(\theta_f - \theta_0) \qquad\qquad a_3 = \frac{2}{t_f^3}(\theta_f - \theta_0)$$

Thus, the cubic trajectory at time t is defined as

$$\theta(t) = \theta_0 + \frac{3}{t_f^2}(\theta_f - \theta_0) \cdot t^2 + \frac{2}{t_f^3}(\theta_f - \theta_0) \cdot t^3 \qquad (5.27)$$

Figure 5.19 shows an example of using Eq. (5.27) with the input parameters: initial position $\theta(0) = 0$, initial velocity $\dot{\theta}(0) = 0$, final position $\theta(t_f) = 10$ (degrees), and final velocity $\dot{\theta}(t_f) = 0$ in a total time of 5 s. Notice that the slope at the beginning and end of the curve is zero, corresponding to the given initial and final velocities as zero.

By using such a cubic trajectory, the acceleration between sets of two adjacent points is kept linear. This might cause problems in some cases, as real robots have upper limits on the maximum accelerations (i.e., torques) that can achieved. In that case, other interpolators can be used that keep the acceleration constant or under a certain limit. For instance, a linear interpolator with parabolic blends can be used. This interpolator is divided in three parts: a linear part (middle part of the trajectory) and quadratic functions that defined the beginning and the end of the trajectory. The interpolator thus *blends* from quadratic to linear and from linear to quadratic to keep the acceleration constant. Similarly, a minimum-time trajectory interpolator can be used, which limits the maximum acceleration to a desired value and in addition create a trajectory from one point to the next in the minimum time possible [53].

5.3.2.2 On-Line Trajectory Generation

The aforementioned interpolators are *offline* computed trajectories, that is, they cannot be easily modified during its execution. Sometimes though, it is interesting to have trajectories that can be slightly (or majorly) adapted during the robot's

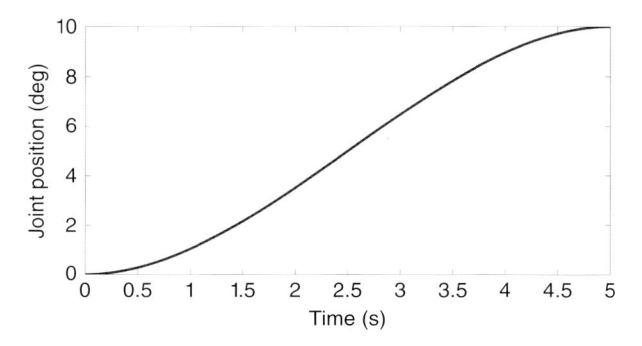

Figure 5.19 Example of a trajectory interpolation between two points using a cubic polynomial.

motion to deal with a dynamic environment or react to a certain sensor event [54, 55].

One recent example of such online trajectory generation algorithms is the *Reflexxes* software library [56]. This library can compute jerk-limited motions within one control cycle so that robots can almost instantaneously react to (unexpected) sensor events. Since the computations are performed very fast (at around 1 ms), a new time-optimal state of motion can be calculated for the same control cycle given an arbitrary initial state of motion and the kinematic constraints in order to reach a desired target state of motion. The output values can be directly sent to the low-level controllers thus achieving very reactive motions.

5.3.3
Collision Avoidance

Collision avoidance has been traditionally associated with high-level control, especially in the area of path planning, which aims at finding collision-free trajectories (i.e., guarantee that the robot manipulator reach the target pose without colliding with the environment). The low-level control described in Section 5.3.4.1 executes those collision-free trajectories.

However, it is obvious that, in dynamic environments, objects might change their position or obstacles might come into the path of the manipulator between the time the robot is planning a trajectory and the moment that trajectory is finally being executed. Ideally then, the low-level control layer should also be able to react in real-time to possible collisions. One option has been already seen in the previous Section 5.3.2.2: the connection between sensor feedback and low-level controllers allows to react instantaneously to sensor events. In this case, that could be, for example, a certain unexpected force measured on the end-effector.

The seminal work in this area was presented in Ref. [57]. In this article, real-time obstacle avoidance at low-level control is described based on the concept of artificial potential fields. The idea behind the potential field forces is that the target position is an attractive pole for the end-effector and the obstacles on its way act as repulsive poles. In this case, the low-level control is able to avoid in real-time known obstacles coming into the path of the moving manipulator. However, as a local method, it might fall into a local minima, and it did not consider collisions with the robot itself, apart of the possibility of including joint limits. In Reference [58], a real-time collision avoidance method was also presented which used a spring-damper model to generate virtual forces when the arm entered an non-allowed area (i.e., obstacle). An outer force–control loop can keep those forces at zero by modifying the Cartesian position of the end-effector, thus *escaping* the collision. In Reference [59], a method for detecting and reacting in real-time to collisions was shown, which used the total energy and the generalized momentum of the robot to detect anomalies (i.e., collisions) and react fast by moving the robot away from the collision area. In Reference [60], a fast method to detect collisions between a robot and moving obstacles was presented. The method evaluates distances between the robot and obstacle directly from depth data of a Microsoft

Kinect sensor to generate repulsive vectors that disturb the trajectory being executed. The work emphasizes on fast perception of the distances to the obstacle to achieve a reactive system at the risk of being not necessarily accurate, because the rationale is control accuracy can be compensated by a fast reaction to an incoming collision.

5.3.3.1 Self-Collision Avoidance

The previous methods mostly aim at avoiding collisions between a robot and an obstacle, either static or moving. However, and especially for manipulators with many DOFs, mounted on an articulated torso, it is also necessary to provide with some methods that avoid that one or more of the links of the robot collide with parts of itself. One of the first works on this area was presented in Ref. [61], in which a geometrical approach using minimum distance measurements for convex hulls is used to detect interference between links. In Reference [62], the focus is on self-collision method for robots cooperating with humans. In this case, the self-collision method is based on representing the robot as elastic elements and adding two priority functions to also consider both task and environmental constraints during the self-collision avoidance motion. The detection of the contact between the elastic elements generates a virtual reaction force, which is used to control the robot motion. An interesting method is shown in Ref. [63], the so-called *skeleton* method for real-time self-collision avoidance. In this approach, a model of the whole robot (the skeleton) is used for analytical computation of the closest points to a collision along the skeleton, generating repulsing forces via potential fields, and computing the required torque commands via the transpose Jacobian which are added to the driving joint torques. In Reference [64], a new repulsion method based on potential fields for self-collision avoidance is presented, which extends the previous work [63]. In this case, a damping is added to avoid critical situations that caused instabilities and oscillations in the previous method. Moreover, a mechanism is also included that determines whether the collision is avoidable via the repulsion method or, instead, if the braking should be activated. The collision model is based on the concept of swept volumes as referenced later in this paragraph [65]. In Reference [66], a self-collision velocity-based controller is proposed with a new distance computation making use of patches of spheres and toruses, which aim at ensuring a continuous gradient, that is, a smooth commanded trajectory. Another interesting method is also presented in Ref. [65]. The algorithm checks pairwise for collisions making use of swept volumes of all body parts. These swept volumes are represented as usual as convex hulls but, in this case, being extended by a buffer radius for an efficient and numerically stable computation, which could run in hard real-time conditions. The algorithm takes joint angles and velocities as input and computes whether (and when) a braking action should start in order to have time to stop the robot before a collision occurs. The algorithm requires two models: a kinematic model of the robot and a geometrical model with the robot rigid bodies (collision model). In References [5, 67], the self-collision mechanism used at the Mars Rovers Opportunity and Spirit to avoid collisions between the manipulator and the rover is described.

Those robot arms receive a sequence of planned commands from Earth and prior to the execution of the movement, each via point both in Cartesian and joint space are checked for collisions. The algorithm determines whether there are intersections between the geometric models of the arm and the rover (and possible payload). The collision checking is also used as part of the ground validation of the commands that are sent to the rover. In addition, the ground validation also includes the detection of collisions with the terrain.

5.3.4 High-Level Control Strategies

The high-level control is to plan the path for the end-effector of the manipulator, whereby different strategies refer to the LoA of the manipulator operation. As introduced in Section 1.2, the LoA for any spacecraft or space system's onboard operation has been defined by the European Cooperation for Space Standardization (ECSS) in four levels (E1–E4). A planetary robotic manipulator (being a space system) can, therefore, be designed to operate in all four levels, for example, under complete user control (or teleoperation) known as the E1 LoA, under semi-autonomy known as the E2 or E3 LoA, and full autonomy known as the E4 LoA.

5.3.4.1 Path Planning

In general, path planning determines a path, connecting an initial and final desired configuration of the manipulator, while avoiding the possible obstacles on the way. Depending on the LoA being used, the robotic manipulator has corresponding level of onboard capability to find and choose which trajectories to use in order to reach the target end-effector positions.

The planning of a motion for manipulation has been historically divided into a phase to reason about the manipulator path at a global scale (or called *gross motion planning*), and a phase to deal with uncertainties, forces, friction, and alike that appear when the robot contacts the environment (or called *fine planning*) [14]. Automatic motion planning is realized by exploring the configuration space using probabilistic algorithms to search for a collision-free path that would connect start and goal configurations [68]. Within that area, sampling-based planners are a group of widely used methods, which use collision avoidance algorithms to search for robot configurations that are collision-free. Such planners are not complete as they cannot guarantee a solution in a finite amount of time, but provide a weaker form of completeness: given enough time and if a solution exists, the planner will eventually find it. A typical classification sorts the algorithms into multi-query approaches and single-query approaches. The algorithms in the first group construct a roadmap once that can subsequently accept multiple queries. The best examples are probabilistic roadmap methods (PRM) [69]. The algorithms of the second group are built online each time a query is made. The best proponents of this group are rapidly exploring random tree (RRT) algorithms [70]. Alternatively, the *potential field*-based techniques have been proposed to solve specific problems in an efficient way [71]. This methodology is derived from methods used in obstacle avoidance in which there is no explicit roadmap to be constructed but a real-valued function represented by potential fields that can guide the robot to the goal configuration. The potential field is constructed based on the *attraction primitive field* that attracts the robot to its goal position and *repulsive primitive field* that pushes the robot away from obstacles.

There are also methods that do not use automatic motion planning. One example is the central pattern generators (CPGs), which are used for generating rhythmic motion. These rhythmic patterns have been used to perform manipulation tasks such as hammering, sawing, or playing drums [72, 73] or for multifingered grasping [74]. Learning approaches have also been developed for manipulation and grasping [75]. In this area, learning by demonstration [76, 77] and imitation learning [78] are regarded as appropriate methods for learning, as humanoid robots have similar kinematics to humans and are primarily designed to solve human tasks.

The MERs (Spirit and Opportunity) combined ground-based motion planning sequences (i.e., LoA E1) with autonomous onboard motion planning (i.e., LoA E4) [5, 79, 80]. The E4 motion planning is only used for the rover navigation on board, while the robotic arm motion planning is performed on ground at E1 and the onboard software only checks for collisions between arm and rover before executing the trajectories. For the Mars Science Laboratory rover (Curiosity)'s first 7 months' operation of the robotic, the creation and validation of the arm motion control sequences are responsibility of the rover planners after a plan has been approved by the ground control station [81], which corresponds to the LoA E2 or E3.

5.3.4.2 Telemanipulation

The case of controlling a manipulator through teleoperation by human operators is often called *telemanipulation*. In telemanipulation, the human operator controls all the movements of the manipulator remotely. In the case that the manipulator is programmable and able to adapt to the environment by using its own sensors so that the human operator only needs to communicate intermittently with the manipulator to supervise it, the operation is also called *telerobotics*.

In teleoperation, one key element is feeding back the sensor information from the remote system back to the human operator. The simplest way of achieving this is to deploy a number of cameras in the remote area. In fine manipulation tasks, the availability of force feedback from the remote manipulator is useful to simplify the job of the human operator. In terms of the control, the manipulator in telemanipulation is known as the "slave" system and the device used by the human operator is known as the "master" system.

Unilateral Control In the simplest teleoperation, the control strategy used is called unilateral control whereby there is no feedback from the "slave" back to the "master" and thus the "master" does not require to be equipped with motor units but just with encoders to measure the joint positions. Such a "master" is usually called a "passive" master. The name unilateral control comes from the fact that control is only performed in one direction, that is, from the master to the slave. The master generates the reference signals (position or velocity), which are sent to the low-level controllers of the "slave." The "slave" or the manipulator receives the reference signals from a remote location instead of from a local control computer. The control of a teleoperated slave manipulator can be much simpler than the control of a typical industrial manipulator since there is no need for computation of inverse kinematics or dynamics.

Bilateral Control In order to increase the performance of the teleoperated system and ease the job of the human operator, it is beneficial to send force sensor information from the remote manipulator back to the operator. Such systems allow bilateral control since the reference signals flow in two directions, that is, the "master" sends position or velocity reference signals to the "slave," and the "slave" transmits forces to the operator that are applied to his/her hand or the arm. To achieve the bilateral control, the "master" needs to be equipped with actuators.

Basically, there are two bilateral control schemes based on the variables used for the manipulator control:

- Position–position control: This control scheme was first proposed and applied by Goertz and Thompson [82] and has been extensively used since then. The idea behind is to control the master system in a local position control loop whose position reference is the current position of the slave system as shown in Figure 5.20a. The slave system is also in position control mode, with its reference signal being the current position of the master system. Thus, this scheme is completely symmetric in terms of the control.
- Force–position control: This is probably the most popular control scheme in telemanipulation. Similar to the position–position scheme, the slave is controlled in position by taking as reference signal the current position of the master. However, the master receives a sensory feedback from the force sensors on the slave, which allows the operator to feel the external forces acting on the slave as shown in Figure 5.20b.

Performance Parameters There are a number of parameters that allow to assess the performance of the teleoperation system, as described earlier:

- Stability: probably the most important criterion to take into account. Bilateral control schemes tend to be unstable, especially when the slave is in contact with stiff environments; for example, a small change in position might translate into large contact forces. Another major source of instability is due to the delays in the communication between the master and slave.
- Error: in position and force between the master and slave.
- Force feedback ratio: defined as the relation between the reaction force between the slave and the environment and the force generated by the master in order to reflect that force to the operator. The ratio needs to be selected so that it matches the capabilities of the human operator to avoid fatigue.

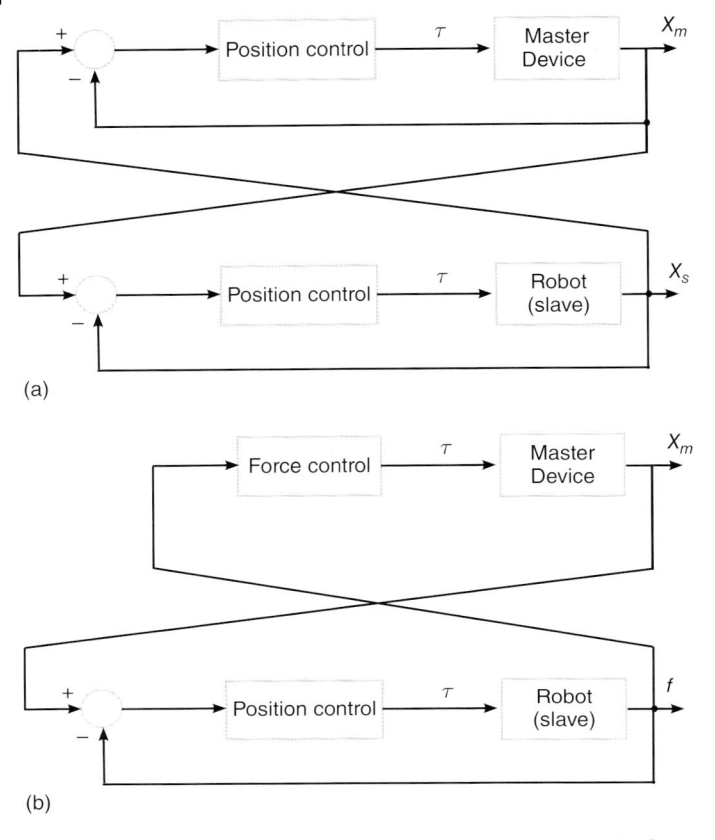

Figure 5.20 Bilateral teleoperation. (a) Position–position control, (b) force–position control.

- Transparency: to achieve the goal of transmitting position and forces between master and slave in such a way that the operator feels perfectly coupled with the environment. In other words, the teleoperation system (master, control, and slave) behaves "transparently" between the operator and the remote area.

The first teleoperated system in space was the experiment ROTEX (RObot Technology EXperiment) [83], developed by the German Aerospace Center (DLR) and performed on board the International Space Station (ISS) in 1993. The robotic "slave" system was a small 6-DOF robot with a length of around 1 m equipped with 6-DOF force/torque wrist sensor, tactile sensors, and stereo cameras, among others. The robot was controlled in several modes: teleoperated from the ISS by the astronauts using a TV monitor (i.e., E1), teleoperated from the ground (i.e., E1), or follow preprogrammed control commands given by the ground (i.e., E2). The highlight of the control was a multi-local sensory feedback allowing higher LoA for the robot, as well as the use of predictive "displays" or simulations to cope with the transmission delays of up to 7 s when using the teleoperation from the ground [84]. In a subsequent mission called ROKVISS [85], a two-joint robotic manipulator was installed outside the ISS in January 2005, which could be teleoperated from the ground via a direct communication link. Because of the direct link, the communication delay was only around 20 ms, which allowed the use of a bilateral control strategy using force feedback from Earth. Due to the fact that the communication window with the robotic arm was limited to a maximum of 7 min, the system also had an E2 operation mode. In January 2015, a NASA astronaut teleoperated from the ISS a device located at the ESA Telerobotics Laboratory on Earth by using a body-mounted joystick with force feedback capabilities. The focus of the experiment was to understand the human capabilities for manipulation and to apply forces in a weightless environment [86].

5.3.4.3 Higher Autonomy (E2–E4)

In case of planetary missions such as on Mars, the typical time delay of communication often impedes the usage of a suitable direct teleoperation of the robot. This requires the robots to have higher LoA beyond E1 to complete tasks on their own. In the literature of manipulator high-level control, various strategies (particularly representing LoA E2 and E3) have been developed, named as, for example, the supervisory, shared, or traded control. In all cases, the mission is completed by combining teleoperation and onboard sensory feedback of the manipulator hence involving operations of different LoA.

Supervisory Control or E2 LoA In the supervisory control [87], the operator specifies the high-level plans and initiates them, whereas the robot carries out the tasks in an automated fashion. As aforementioned, in various Mars missions (containing the Spirit, Opportunity, and Curiosity rovers) the motion planning for the onboard robotic arm was performed on ground, and validated and executed by the arm on itself [81, 88, 89].

Shared Control or E3 LoA In the shared control [90, 91], the robotic system and the human operator work on the same task together. In this scheme, the human operator generates gross motions, which then get rectified by the robot using local sensory feedback. In case of no delay, this feedback gets sent to the operator as well. In case of major delays, the operator and robot work simultaneously but on different parts of the tasks. An example is the concept of virtual fixtures, that is, virtual surfaces or objects that are superimposed to the visual and force feedback of the operator. They act as "active constraints" in the sense that they limit the motion of the robot in order to ease the task and avoid false or dangerous movements. Imagine the analogy of controlling the mouse pointer on the computer screen: the remote system (pointer) is controlled by the master (mouse) anywhere inside the computer screen but cannot leave the limits imposed by the screen frame (virtual fixture). Thus, the user can focus on the task without worrying about the pointer going outside the limits or into restricted areas.

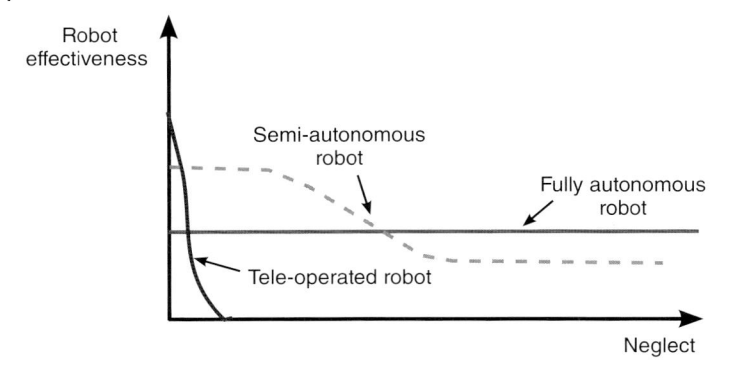

Figure 5.21 The neglect curve. (Reproduced from Ref. [95].)

Traded Control or E4 LoA In the traded control [92, 93], the human operator and the robotic system also work on the same task but at different times. In this control scheme, the robot mostly performs the task autonomously but at certain times the human can intervene to complete a specific part of the task, often requested by the robot itself (e.g., after the robot identifying a unknown situation or task).

Designing LoA It is clear that a robotic manipulator can be operated in a range or a combination of LoA based on mission constraints and task requirements. The following system-level design of LoA has been investigated:

- adaptive autonomy, if the robotic system has exclusive control over the determination of the LoA;
- adjustable autonomy, if the user can adjust the LoA;
- mixed initiative, where both user and robot have equal responsibility on the decision of choosing the LoA [94].

There exists a certain relation between the robot's effectiveness of performing a task and the level of attention or stress of the operator (indirectly related to the LoA) [95], as shown by the so-called neglect curve in Figure 5.21. While teleoperated systems achieve higher level of effectiveness if the user is entirely focused on the task, the effectiveness abruptly drops if the operator loses attention (i.e., "neglects" the system). On the other hand, a fully autonomous robotic system might have lower effectiveness but that level can remain constant independent of the level of attention of the operator. A semi-autonomous system is expected to perform somewhere in the midway, with relative high level of effectiveness even when the level of attention from the operator is relatively low.

5.4
Testing and Validation

Testing and validation are critical to the design and implementation of a planetary robotic arm. They represent important steps in the design process to finalize the control scheme, to test the operation of the arm in the representative environmental conditions and to validate the design performance. The arm design and testing activities are inherently linked and form part of the underlying development and testing strategy established at the beginning of the project, balancing the need for a realistic yet implementable test scenario.

5.4.1 Testing Strategies

As detailed in Section 2.4, the robotic systems for planetary exploration are expected to face a variety of environmental conditions, for example, air or airless extraterrestrial bodies, extreme cold or hot temperature. The robotic systems can also experience a range of gravity, from low-gravity found on asteroids or small moons to major gravity fields around planets such as Mars and Venus. The robotic manipulator system considers gravity as a key environmental factor that influences its design and testing philosophy. Three testing strategies are available to count for the gravity effect, each presenting its own engineering and system challenges:

- Design to target gravity: This strategy involves design of the arm to sustain in the gravity field of the target planet, which is likely to provide the smallest mass for the arm. However, this approach makes testing more challenging due to the need to offload the arm during the validation process. The offloading setup can use a range of configurations such as a mass pulley system or balloons that offset the weight of the arm to mimic the target gravity. However, such setups inherently constrain some aspects of the testing, mainly because the arm goes through a range of poses that may not be compatible with the offloading mechanism.
- Self-supporting arm with scaled mass payload: This strategy involves the design of an arm that can be tested on normal Earth conditions (or at 1 g) but carries a payload mass scaled to the target gravity. For example, an arm designed to work on Mars can be tested on Earth with a payload mock-up a third of its mass. This approach provides a higher mass solution compared with the previous strategy, but allows a wider range of testing solutions, thanks to the elimination of the gravity offloading rig. The use of typical design margins can to some extent bridge the gap between the two gravity environments. For example, a Martian arm could be designed with a motorization actor of 2 (i.e., a joint torque capability is doubled to account for various inefficiencies and contingencies in the system) can be allowed to work on Earth without the margin, limiting the impact on the arm design (especially the resulting mass).
- Full arm mass and payload: This strategy sees the design of the arm to be fully compatible with terrestrial testing with its actual full mass payload. This approach produces the highest mass by sizing the arm and its payload for the Earth

gravity field (not for the target gravity), resulting in significantly oversized joints. The operation of the actual payload would provide the mission operators a powerful tool to validate the operation and enable realistic rehearsal of the payload activities. However, the tuning of the arm control loop will be significantly affected by the dynamics of the arm. The higher gravity on Earth does not only have a damping influence on arm and joint dynamics but also has an effect on the flexure of the arm and joints, unlike what is expected to be experienced in the actual extraterrestrial environment.

Each strategy presented earlier has its pros and cons, affecting mainly the overall system mass as well as the realism of the testing. The "self-supporting arm with scaled payload" is one that provides a good middle ground, providing enough scope or testing on Earth as well as minimizing the extra mass required. Most planetary missions that used a robotic arm had selected this strategy, from the small arm on Beagle 2 lander to the long arms of the Phoenix and InSight lander (see Figure 5.22) where the benefits are even greater.

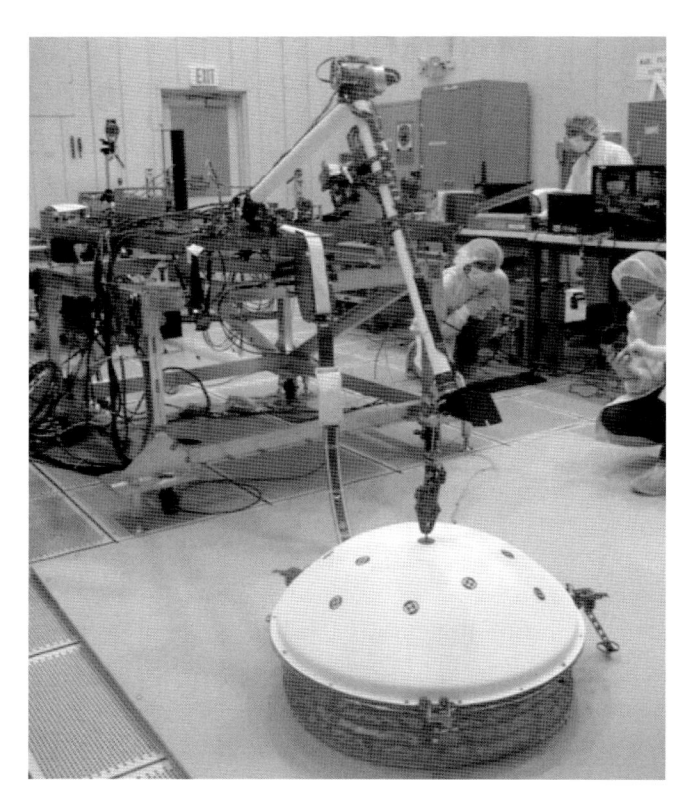

Figure 5.22 InSight testing with scaled-mass payload.

5.4.2
Scope of Testing Activities

The purpose of a planetary arm can vary from one mission to the next; however, it will invariably involve the precise deployment of the end-effector at specific locations. Whether it needs to place instruments or tools to the target locations or retrieve samples from the surface before delivering them precisely to the onboard instruments for processing, the robot must be able to reliably and constantly perform its tasks.

The purpose of the testing is twofold: (i) to physically characterize the manipulator to perform a calibration of its control and (ii) to assess the performance to execute the task with the appropriate resolution, repeatability, and accuracy which are defined as follows:

- *Resolution* is the smallest incremental move the manipulator can produce.
- *Repeatability* is the measure of the ability that the robot can move back to a pose (position and orientation).
- *Accuracy* is the ability of the manipulator to precisely reach a pose in 3D space.
- In addition, with the motion of the arm comes the concept of *dynamic accuracy and repeatability* that capture the ability of the manipulator to follow a prescribed trajectory.

As discussed by Roth and Mooring [96, 97], three levels of calibration for robotic arms are used, namely the joint level, kinematic model level, and dynamic model level.

- The joint calibration focuses on the determination of the correct relationship between the signal of the joint sensors and the actual joint displacement.
- The kinematic calibration focuses on the physical characterization of the arm to minimize the impact of geometric errors. These are the errors in the parameters that define the geometric relationship between the axes of motion. These stem from the manufacturing and machining errors resulting from machining tolerances, link length, and joint orientation. Due to the machining techniques, these can be minimized, but are intrinsically unavoidable. The testing will, therefore, result in a basic kinematic model on the arm with the correct joint-angle relationships.
- The dynamic calibration finally focuses on the remaining nongeometric (nonkinematic) errors introduced by the inherent joint compliance, gear backlash, friction, and link flexion.

While the joint calibration can be performed with a minimal setup, the kinematic and dynamic calibrations require more extensive testing scheme. The calibration of manipulators is a complex operation and a topic investigated extensively for industrial manipulators. References such as [97] provide thorough descriptions of the process, both functionally and mathematically. Given the significance of kinematic calibration to the planetary robotic arm, it is discussed further in this section.

5.4.2.1 Kinematic Calibration

Once the joint calibration has been performed, the kinematic calibration is followed and concerned with the minimization of the error between the actual pose and its intended target pose. Once built the arm physical implementation is often fixed and cannot be changed, but these errors can be mitigated through software. The process for a kinematic model-based calibration is implemented in four discrete steps [97]:

1) Modeling: A kinematic model is generated to provide a mathematical representation of a perfect manipulator. The Denavit–Hartenberg or other methods can be used to describe the properties and relationship between the arm links and joints.
2) Measurement: Moving onto the real hardware, this activity focuses on the gathering of the necessary data that will be used to measure errors and ultimately help with their compensation. To this end, the manipulator is exercised across its workspace and the actual position of the tool is compared with the position predicted by the theoretical model. To perform accurate measurement of the end-effector position, a number of metrology methods can be used including single-point or complete pose measurements. Single-point measurement, as the term indicates, measures precisely the location of a single point and can implement theodolites, laser interferometers, acoustic sensors, or coordinate measurement machines (CMMs). The measurement process is lengthy and time consuming to change the pose of the arm and measure its outcome. In recent years, complete pose measurements have been made easier though the use of optical systems such as a marker/camera setup that can resolve the position of the markers precisely. These systems can be implemented in a number of ways including, the use of typical cameras and fiducial markers, IR cameras and active IR illumination of reflective markers, or IR cameras and active IR markers. These systems have seen significant development and improvement in recent years both in terms of frequency and accuracy driven by the needs of the film and entertainment industries as a motion capture (MoCap) system. By capturing the position of a wide range of points accurately at an appropriate frequency, these systems provide not only valuable data for the kinematic calibration but also to the dynamic calibration, allowing the recording of dynamic events such as trajectories, vibration, and the deflections of limbs.
3) Parameter identification: To bridge the gap between an idealized model and the test data, parameter identification focuses on the quantification of the kinematic parameters that cause the computed pose to match as closely as possible the test data. It is in essence an optimization problem that can involve a number of methods, pending on the nature of the errors or the model (e.g., deterministic/stochastic, linear/nonlinear), and typically involve the use of least mean square techniques. (See Ref. [97] for more in-depth discussion.)
4) Implementation of compensation: Once the kinematic parameters have been identified, the accuracy of the manipulator is improved by the implementation of the new model into the controller. The previous steps of modeling,

measurement and parameters identification address the propagation of errors from joint to end-effector pose in the forward model, addressing therefore the forward calibration of the arm. The implementation phase is, therefore, critical to improve the calibration of the inverse model and, therefore, provide inverse calibration.

5.4.2.2 Beyond Calibration

Calibration in a laboratory environment can provide a good model for a newly built robotic arm as it is then fit onto its deployment platform. Nevertheless, a range of environmental and mechanical aspects need to be taken into consideration throughout the lifetime of the robot.

- Launch and landing: During launch and landing, the system is subjected to a range of violent mechanical loads including vibration shocks through pyro actuation, parachute opening, and landing, and so on. These may affect the mechanical alignment of the structure or elements of the joints.
- Operation: Operation on a robotic platform (such as a rover) may see a range of shocks due to the platform overcoming rocks and falling from them. Across its lifetime, wear of the gears and bearing, and possibly the seepage of dust into the mechanisms can affect the properties of each joint in a different way, affecting the internal friction or the accuracy of the positioning.
- Thermal environment: The operation of the arm across a wide temperature range needs to take into account the coefficient of thermal expansion (CTE) of its constituting elements. As materials get warmed up and cooled down, they contract and expand, which potentially affect the accuracy of the end-effector placement through displacement or twisting. To mitigate these aspects, the CTE of various key elements would be matched to minimize local stresses and deformations across the temperature range.

To monitor and compensate for these errors over the mission lifetime, manipulators can be dispatched to dedicated known locations on the body of the lander or rover to assess the errors in the system. Through the recording of the final end-effector pose, an *in situ* measurement step, the kinematic parameters can be updated to correct these errors and improve the overall accuracy of the manipulator. The future missions are envisaged to implement some level of visual servoing, making the precise calibration of the arm not as critical as before because the arm is controlled either by a camera overseeing the operation or placed on the end-effector (i.e., eye in hand).

5.4.3
Validation Methods

Lack of detailed measurement data and operational experience directly from planetary environment stays behind the difficulties in defining the key parameters of robots that should be evaluated. Moreover, it is even in some cases difficult to answer the question of where these parameters should be tested: here on Earth or rather at the destination? Having in mind these uncertainties and based on the current knowledge, the following possible phenomena need to be measured before the operations of the robotic arm are defined:

1) Due to reduced gravity, the lander or rover is weakly connected to the ground. The manipulator arm motion can generate reactions that are too high to be transferred to the ground. This effect might be destructive to the lander or rover's stability.
2) Due to relatively low stiffness on both the structure and the joints of the arm, the frequency characteristics might be a good starting point to evaluate the potential influence of robotic arms on other subsystems (e.g., solar panels or lander gear).
3) In low-gravity environment, the predominant load of manipulator joints comes from the dynamics of motion. Therefore a situation, where the torque crosses zero point and backlash significantly affects the arm operations, may often occur. This effect can be enhanced by specific tribological behavior of mechanisms in vacuum.
4) Exploration of planets often includes regolith sampling processes, which is typically a combined task of manipulator arm and sampling mechanism. Due to unknown regolith parameters *a priori*, the interactions between the robotic arm and the regolith should be tested.

Various methods can be used to simulate some of the effects indicated earlier:

- One possible approach to simulate near-zero gravity conditions is to perform experiments during *parabolic flights*. During such a flight, an aircraft flies a trajectory that provides about 25 s of free fall.
- The other possibility is to perform tests using *drop towers* where the maximal time of experiment is much shorter (in most facilities free fall lasts only up to 10 s, and residual gravity acceleration is around $10^{-3}-10^{-6}$.
- In case of reduced but nonzero gravity conditions, the use of planar *air-bearing tables (ABTs)* with movable platform might be a possible choice. In such an approach, the test objects (e.g., a manipulator intended for sampling operations) are mounted on a planar ABT that allows for almost frictionless motion on a flat surface that is inclined to simulate a certain level of gravity. In effect, the reduced gravity conditions act properly on the mechanisms in two transnational and one rotational DOF, and the residual gravity acceleration is around $10^{-3}-10^{-5}$ depending on the type of table's surface and other parameters. Therefore, disturbances experienced using ABT are smaller compared with using parabolic flights and most drop towers. Depending on the size of the air container, duration of experiments can last at least several minutes and even up to many tens of minutes, thus significantly longer than the previous two options. The main disadvantage of using ABT is the limited number of DOFs. As indicated, the motion can be performed without disturbances in one plane only with two translational DOFs and one rotational DOF. Special designs allow additional rotational DOF by taking advantage of spherical air-bearings, but the residual

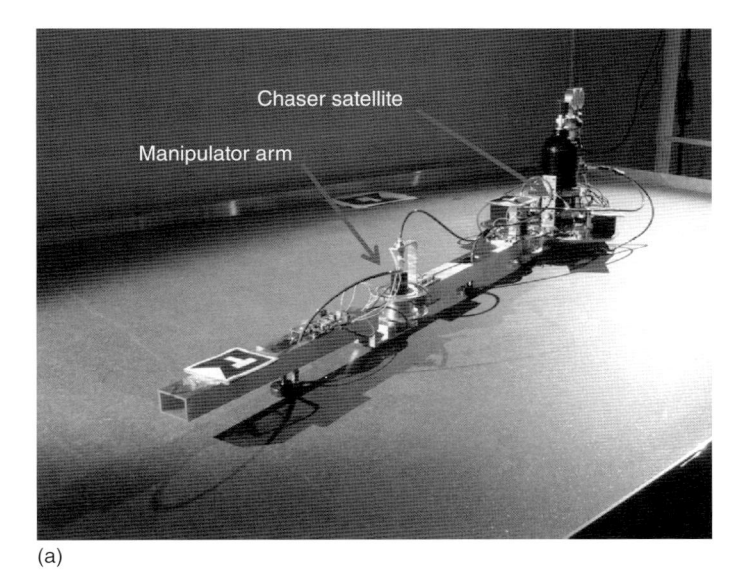

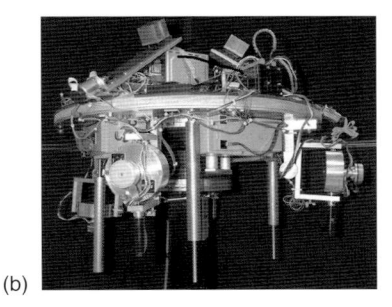

Figure 5.23 Two types of ABT configurations. (a) Flat ABT testbed simulating the chaser spacecraft during capture maneuver with a robotic arm. (Courtesy CBK PAN.) (b) Round ABT testbed simulating frictionless motion around the roll, yaw, and pitch axis. (Courtesy Surrey Space Centre.)

gravity impact is then higher and only the first-order influence can be analyzed. An example of a flat ABT, developed originally to test orbital robots working in space, is illustrated in Figure 5.23a [98, 99]. The size of table (2×3 m) allows for performing complicated maneuvers such as rendezvous maneuvers, final docking, or detumbling of a gathered target [100]. An example of a round ABT, developed typically for testing spacecraft attitude control, is illustrated in Figure 5.23b [101].

5.4.3.1 Use of ABTs
The low-gravity planetary application of the ABTs can be achieved by inclining the table by different degrees depending on the target bodies (e.g., for the Moon is 9°, and for Mars' moon Phobos is just 0.05°). To simulate and analyze the

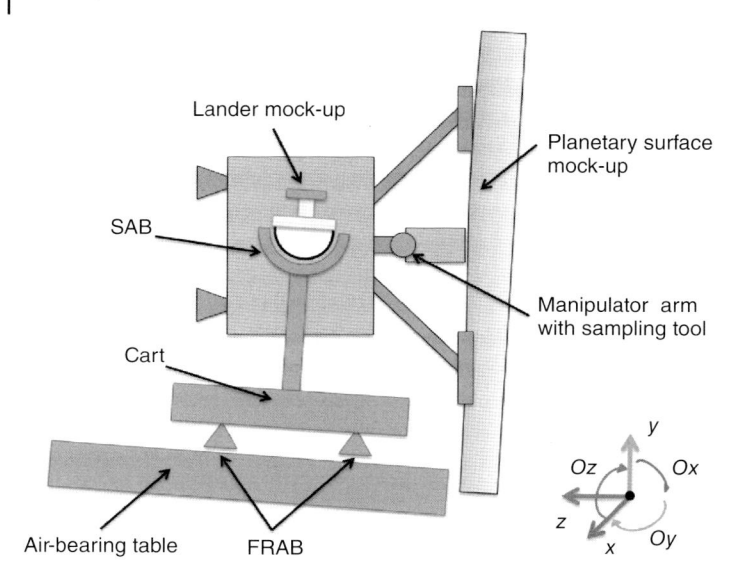

Figure 5.24 A lander mock-up installed through a spherical air bearing (SAB) on a cart moving on an inclined flat granite table with air bearings (FRAB), and the lander legs are touching a vertically oriented planetary surface mock-up. (Courtesy CBK PAN.)

manipulator's behaviors in such planetary condition, a planar ABT testbed can be set up using the granite surface as illustrated in Figure 5.24. The lander or rover mock-up can stand on the cart, which produces air against the granite table for frictionless movements. The robotic arm on board the lander or rover can be installed on the side. The arm can be tested to measure the manipulator and lander interactions including forces, momentum transfer, or induced lander motion. Given different mission scenarios, additional payloads can be added for simulation, for example, penetrometers, sampling tools or a regolith surface mock-up [102].

Analysis of similarities and differences between the ABT and real environment on a low-gravity body is crucial since not all measured parameters are realistically reproduced by the testbed. In low-gravity planetary conditions, an unanchored lander is free to move with 3 DOFs and rotate with 3 DOFs. In this description as shown in Figure 5.25, the gravity vector is along $-z$ axis and the regolith local normal to the surface is along $+z$ axis. The lander motion in $-z$ axis is limited by interactions with the planetary surface. The loads acting on the lander are generated by manipulator dynamics and may create lander's motion on the surface, and in some specific cases the lander rebounds or even escapes from planetary surface. If the manipulator arm's end-effector interacts with the regolith, the static or quasistatic reactions may appear in all directions.

It is worth mentioning that interactions between sampling tools, drills or penetrometers with the regolith can create forces and torques through the manipulator arm, which in turn leads to the lander motion with respect to the sampling

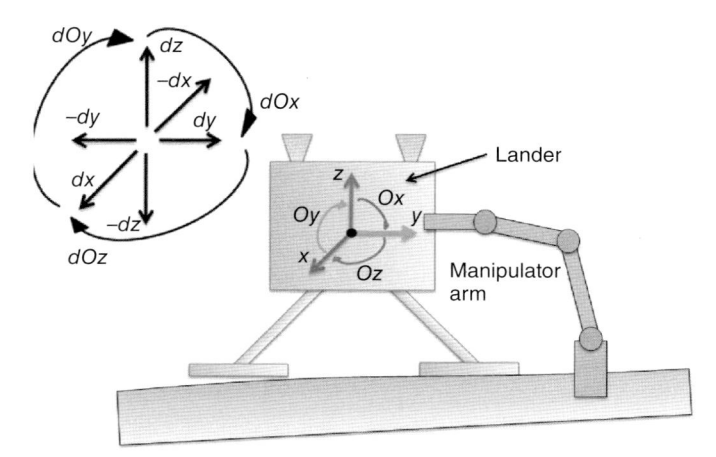

Figure 5.25 A possible configuration of a lander and a manipulator arm on planetary surface. (Courtesy CBK PAN.)

tool. This can have major impact on the manipulator itself, since its last joint will be the first one with all related consequences such as loads increase and control issues.

As aforementioned, the motion of the cart with lander mock-up shown in Figure 5.24 is limited by the flat configuration of the ABT, where both main motion of subsystems and generated forces/torques also have flat nature. For example, the test configuration of mock-up lander, manipulator arm, or other devices should be chosen in such a way that the force generated creates a motion not affected by the Earth gravity. As shown in Figure 5.26, the forces F_x and F_z generate lander's motion in x, z and around Oy direction, all of which are not affected by the Earth gravity. In another case where sampling tool generates the torque M_z and force F_z, the induced motion for the lander mock-up generates linear motion in x and z direction as well as rotation around the Oz axis, whose impact can be analyzed taking advantage of a spherical air bearing (SAB). However, that type of motion is limited to a couple of degrees hence only first order effects can be analyzed.

In summary, the major advantage of using the ABT as a testbed to validate the planetary manipulator arm is related to the natural conservation of momentum, which is helpful to study the interactions between the robotic arm and its housing platform. The limitation of the flat ABT is the two dimensions used and the fact it does not allow a full study of elasticity or vibration phenomena and joint friction, which is significantly reduced in the 2D "world."

The ABT testbed is typically used for testing spacecraft's attitude control systems or for validating dynamical models and numerical simulations. In addition to testing algorithms and concepts, flight hardware and engineering models of sensors and instruments can also be performed. Typical applications that these experiments are related to include rendezvous maneuvers, contact dynamics during

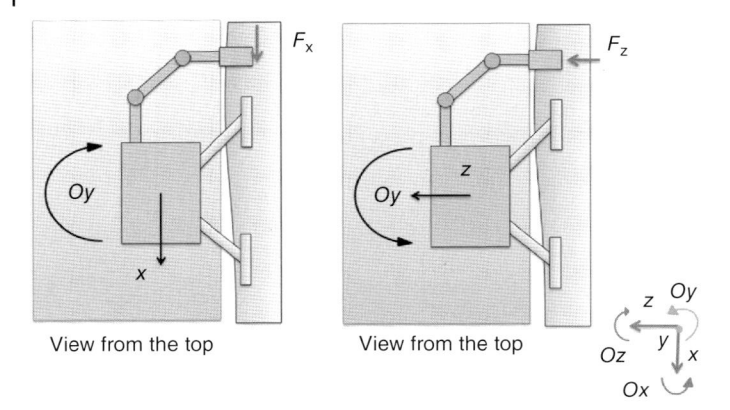

Figure 5.26 An ABT testbed allowing tests of sampling tool interactions where all major components of motion induced by forces (F_x, F_z) are flat. (Courtesy CBK PAN.)

docking, proximity navigation, guidance and control, and landing for low-gravity bodies. It is generally a good practice or advisable to perform as many tests on Earth as possible through testbeds such as the ABTs, and if possible to perform further tests on orbit in genuine space environment (e.g., at the International Space Station or ISS).

5.5
Future Trends

This section aims to shed some light on potential capabilities of planetary manipulation robots in the future, hence has chosen to present a number of example technologies in areas such as the dual-arm manipulation, whole-body control, and mobile manipulation. Some of these named systems or technologies have been investigated on Earth for some years to support different terrestrial applications; for example, the mobile manipulation has recently emerged and been recognized as a research field of its own merit.

5.5.1
Dual-Arm Manipulation

Dual-arm manipulation is an increasingly important research area where the manipulation robot is equipped with two arms that cooperate with each other to perform tasks with dexterity similar to that of a human. Driven by the interests of making more humanoid robots, the two-arm manipulators are also deemed more effective while performing complex assembly tasks. This can be extremely relevant to future planetary missions whereby building infrastructure such as the lunar or Martian outposts for human permanent presence/habitation is envisaged.

When comes to planning and executing tasks using the dual-arm manipulation, coordination between the arms is crucial. For example, holding and transporting an object with both arms requires maintaining the relative pose (position and orientation) between the arms, which is usually known as *planning with constraints*. One of the first approaches for controlling the dual-arm system used a master–slave configuration by minimizing the error between the relative poses of both arms [103]. Another approach used a similar concept [104] with one robot as a leader and another as a follower. The follower tracks the motion of the leader while taking into account the constraints imposed by the relative pose to be maintained. Work by Hayati [105] is one of the first that tackled the challenge to simultaneously control the motion and force of a multiarm system, where the arm forces are split between the arms. The study in Ref. [106] developed a closed-chain dynamical model for a dual-arm system. More recent developments in this subject include [107] where a cooperative control scheme for a dual-arm system capable of transporting rigid objects was described. The proposed control scheme integrated force and position control as well as vibration minimization. Another relevant work presented in Ref. [108] proposed a decentralized control scheme, which allows a multiarm system to move a single object in a coordinated way without using geometrical relations between the arms. There have been some well-known developments focused on dual-arm robotic platforms, to name a few, the DLR's Justin robot as shown in Figure 5.27 where a framework to parallelize the planning and execution of bimanual operations has been proposed [109], or another German institute Karlsruhe Institute of Technology (KIT)'s humanoid ARMAR-III robot that offers a redundant two-arm manipulation system [110] and a solution for efficient motion planning and especially adaptation of those plans under consideration of the redundancy of the robot [111].

The dual-arm manipulation systems can be combined with additional mobility (such as on wheels or legs), hence leading to more powerful holistic control strategies. Such whole-body control systems represent promising technologies for future planetary missions following the footsteps of their terrestrial counterparts.

Figure 5.27 Justin robot. (Courtesy DLR, Creative Commons BY 3.0.)

5.5.2
Whole-Body Motion Control

Future, complex, and highly redundant robotic systems are expected to possess mobility (with locomotion of the wheels or legs, etc.) and manipulation capabilities, whereby these different capabilities should not be treated independently from each other. Hence, the holistic approach that considers control of a complex robotic system as a whole has been proposed and most popular in recent years. This approach is also known as the *whole-body control*. To give an example: in the case of a humanoid robot walking on two legs, balance has to be guaranteed at all times especially when the arms manipulate or contact the environment; the option of simply blocking the joints of the legs (and torso) to keep a fixed posture while manipulating would be a valid solution but a rather inefficient one.

Whole-body control frameworks operate in between single-joint controllers and possible high-level planners. These frameworks can take care of multiple and simultaneous control objectives, which is especially relevant for highly redundant robots as humanoids or mobile manipulators. Moreover, they make use of real-time feedback to control the robot. As a result, robots using whole-body control approaches are more adaptive and can react promptly to unexpected sensory feedback signals, resolving at run time for an optimal use of all the available DOFs.

The origin of whole-body motion generation comes from the generation of walking for humanoid robots while trying to ensure balance of the system, not yet considering manipulation or multiple-contacts with the environment other than those at the feet of the humanoid robot. In early days, a common approach was to divide the problem into three independent stages [112]: The *first stage* is to compute a coarse movement for all DOFs. This is usually done by using a motion tracking system to observe humans walking or performing some actions [113, 114]. It is also possible to use probabilistic path planners, which can generate automatically a path according to certain constraints and goals [68, 69]. Kinematic models of humans and humanoid robots might be similar, but their dynamic models are never the same. For this reason, the *second stage* is to compute a physically feasible motion from the observed or computed coarse movement [115]. The *final stage* is then to provide online stabilization via sensory feedback during the movement execution in order to deal with unforeseen or unmodeled dynamics. The first approach for generating full-body motion was presented in Ref. [116] where the authors used a discrete set of foot steps selected by vision for the locomotion and automatic path planning for the manipulation. Work using online modification of the zero moment point (ZMP) trajectory during the execution of the motion was also documented in Ref. [117]. More recently proposed approaches have focused more on generating real-time whole-body motions.

In References [118, 119], the term "whole-body control" was used for the first time to refer to a floating-base task-oriented dynamic control and prioritization framework that enables a humanoid robot to fulfill simultaneous real-time control objectives. Prioritization and coordination of several controllers is achieved using a hierarchy that handles conflicts and selects the one with highest priority. This approach, in comparison to previous ones, was the first one that focuses not only on the walking but on the manipulation tasks while on two legs. The most recent work of this team group has been renamed to "whole-body operational space control" (WBOSC) to avoid confusion from the more broadly defined term "whole-body control" used nowadays [120]. The system prototype of WBOSC was also extended to include the possibility of internal force control and is available as an open-source software named ControlIt! [121]. Figure 5.28 shows the overall control diagram proposed by WBOSC as well as the joint torque controllers. In essence, it is a distributed control system in which the joint controllers take care of the actuator dynamics while the central control or WBOSC remains responsible for the overall robot dynamics.

A recent work in Ref. [122] presents a generalized hierarchical control, which is able to deal with both strict and nonstrict priorities in an arbitrary number of tasks. In dynamical environments, a certain nonstrict priority might become a strict one and the presented approach enables the possibility of switching task priorities to deal with those situations. Another prominent approach for whole-body control is based on controlling the robot's linear and angular momentum at

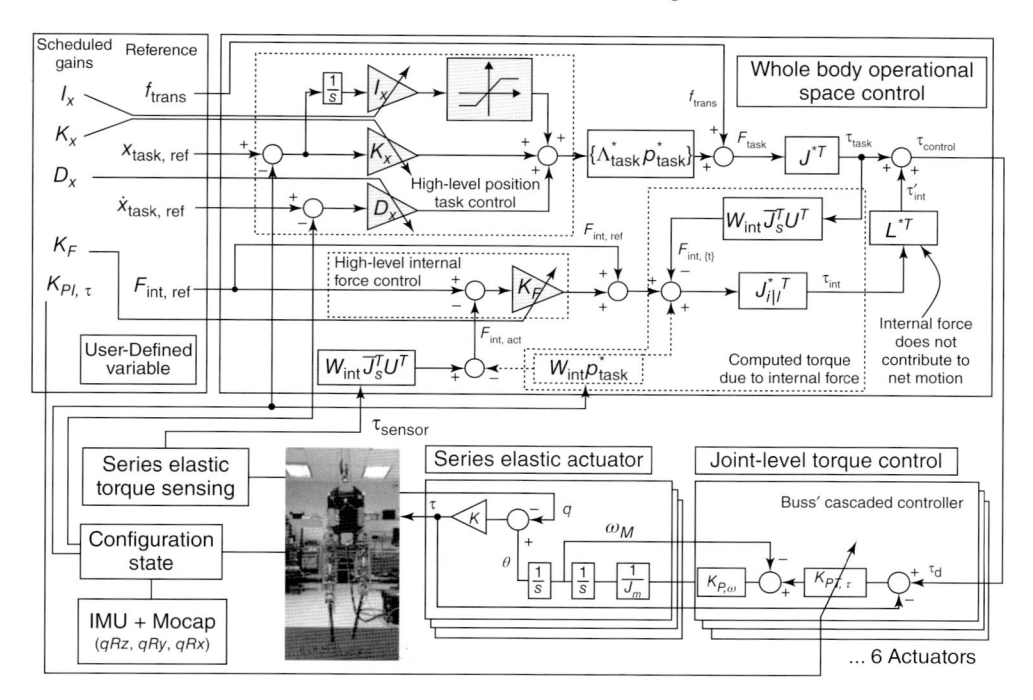

Figure 5.28 Whole-body operational space control (WBOSC) diagram. (Courtesy Luis Sentis, HCRL, University of Texas at Austin.)

each point in time [123], the so-called resolved momentum control (RMC). The approach taken in Ref. [124] is to use a set of *a priori* typical basic postures, which are combined to perform the current task so that it is dynamically balanced and natural. The software framework iTaSC can also be used to generate whole-body motions by specifying constraints between parts of the robot, and between the robot and the environment, allowing the specification and sensor-based reactive control of complex robotic tasks, which require multiple and simultaneous subtasks [125]. An example of its use can be seen in Figure 5.29 [126], in which the robot AILA makes use of a whole-body reactive control for performing a task at a mock-up ISS. Similarly in Ref. [127], the framework controls the robot by using sensor-based control tasks, which are simultaneously executed and synchronized by a so-called "stack of tasks." Another framework proposed in Ref. [128] is implemented as a teleoperation system that allows the operator to select and control only the necessary points of the robot's body for manipulation. A switching method allows the operator to select specific point among the body parts and whole-body motions are then automatically generated based on RMC to maintain stability and maximize the reaching workspace. The methodology in Ref. [129] is to observe humans with a MoCap system and generate dynamically stable and physically plausible whole-body motions from the captured data.

The recent frameworks on whole-body control are mostly used in mobile manipulation (discussed in the following section) and humanoid robots. As long as manipulation is involved, contacts with the environment are desirable and not treated as disturbances. The complex robotic systems need to deal with simultaneous multicontact forces (such as between feet or mobile base and the ground, or between manipulator(s) and the objects being manipulated), with the aim of keeping balance or an optimal posture. This requires efficient and online control

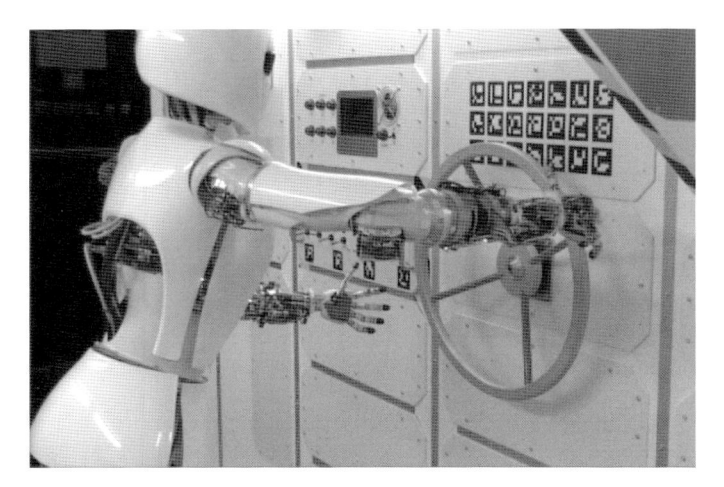

Figure 5.29 Robot AILA using whole-body control to perform some tasks within a mock-up International Space Station. (Courtesy DFKI GmbH.)

strategies based on real-time feedback that can make optimal usage of the redundancy within such robotic systems.

5.5.3 Mobile Manipulation

The field of mobile manipulation overlaps with the field of humanoid robotics both of which are closely related to the formerly described "whole-body control." These fields of robotics focus on manipulation systems mounted on mobile platforms (either on wheels or legs, with or without human morphology). Unlike whole-body control that primarily focuses on the optimal selection of the robot's DOF and the prioritization and execution of simultaneous subtasks, the mobile manipulation focuses on the mobile manipulators, which are able to solve complex tasks in unknown, unstructured, and changing environments (such as in outer space). Thus, mobile manipulation will likely use the whole-body control concepts as a building block of the mobile manipulator system.

In recent years, autonomous mobile manipulation has emerged to a new research topic in robotics and been identified as critical to future robotic applications [130]. Since the research focus of mobile manipulation lies on producing a new category of robots beyond the current state of the art in order to be able to solve complex tasks in unstructured and changing environments, the work needs to deal with the execution of complex manipulation tasks that might require mobility (such as moving from one location to another) in such challenging environments. Many robotic solutions have been deployed in well-known static environments such as factories where uncertainties and unexpected events are minimized. Future robots that benefit from having mobile manipulation are expected to be reliable and deployable in the household environment, during man-made or natural disasters, or for extraterrestrial exploration, and so on. These robots then need to be able to perform a large variety of tasks in completely or partially unknown scenarios by acquiring and reusing generic skills which can be adapted to novel or unexpected situations.

Mobile manipulation aims at maximizing task generality, that is, increasing the variety of tasks that the robots can autonomously accomplish and, at the same time, minimizing the dependency on prior information about the environments in which robots are deployed. The main challenge similar to most robotics research is about how to integrate various subsystems together (including the perception, manipulation, planning, control, cognition, artificial intelligence, and mobility, etc.) so that the mobile manipulation system is able to cope with a wide range of real-world situations.

5.5.3.1 Mobile Manipulators as Research Platforms

Some notable examples of existing terrestrial mobile manipulators include (but not limited to) the following:

- The PR-2 robot developed by Willow Garage [131]. This is a dual-arm robot with an omni-directional mobile base. It includes a variety of sensors such as tilting laser scanner on the head, laser scanner on the mobile base, two pairs of stereo cameras, and an IMU located inside the body. Currently, the PR-2 is probably the most advanced autonomous mobile manipulator capable of performing complex manipulation and navigation tasks.
- The "butler" robot HERB developed at Intel Research Labs in collaboration with Carnegie Mellon University [132]. This is an autonomous mobile manipulator that was designed to perform complex manipulation tasks in the home environment. The robot is able to search, recognize, store new objects as well as manipulate door handles and objects, and navigate in cluttered environments.
- The Rollin' Justin developed at the DLR Institute of Robotics and Mechatronics [133]. Justin is a progressive development whose two arms were initially built on the lightweight arms from DLR (LWR-III) and was later equipped with a mobile base to enhance the robot's field of work.
- The UMan from the Robotics and Biology Lab at the University of Massachusetts Amherst [134]. This robot consists of a modified Nomadic XR4000 holonomic mobile base with 3 DOFs, a WAM 7-DOF manipulator arm, and a 4-DOF hand from Barret Technologies.
- The robot AILA developed at DFKI Robotics Innovation Center [135]. This robot is a mobile dual-arm system developed as a research platform for mobile manipulation. AILA has 32 DOFs, including 7-DOF arms, 4-DOF torso, 2-DOF head, and a mobile base equipped with 6 wheels, each of which has 2 DOFs.

5.5.3.2 DARPA Robotics Challenge (DRC)

The DRC and its latest competition in 2015 set the benchmark of industrial robotic platforms for mobile manipulation [136]. The competition has been organized by the US Defense Advanced Research Projects Agency (DARPA) and aims to promote and foster development of semi-autonomous robots to perform complex tasks in realistic disaster scenarios. The involved tasks include driving a vehicle, walking among rubble, removing debris blocking the way, opening doors, climbing a ladder, using tools to drill a hole on the wall and rotate a valve. The competition was initiated in 2012, had a first virtual challenge that took place in 2013, and later on two live demonstrations including the trials in December 2013 and the finals in June 2015. For the trials, most of the teams used teleoperated systems for solving most of the tasks. At the finals, 25 robotic systems from different countries participated among which some robots applied certain LoA. In this final competition, three teams scored the maximal attainable number of points (see photos in Figure 5.30), including the following:

- The KAIST team from South Korea and their robot Hubo, which has been developed since 2002 and whose complete and more powerful version was specifically designed for the DRC competition. The biped robot has a total height of 180 cm and a mass of 80 kg.

(a) (b)

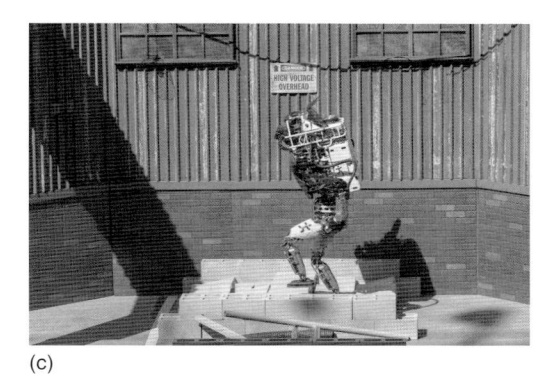

(c)

Figure 5.30 DRC robots in 2015 finals.

- The Tartan Rescue team from United States with their robot Chimp, which was developed at Carnegie Mellon University, is 150 cm in height and weighs 201 kg. Chimp combined high-level operator commands with certain low-level autonomy (e.g., autonomously plan and execute joint and limb movements or grasps). The robot rolls on four legs using rubber tracks such as a tank. In order to manipulate, the robot then stands up on two legs and manipulates with the front limbs.
- The Florida Institute of Human Machine Cognition (IHMC) Robotics team from United States with their Atlas robot Running Man, which was developed by Boston Dynamics and made available for the IHMC team participating at the DRC. The biped robot is 190 cm in height and weights 175 kg.

The final decision of the winner was judged by which robot used the least time to perform all tasks; hence, the first place went to the KAIST team.

5.5.3.3 Mobile Manipulators for Space

NASA's Johnson Space Center (JSC) has built a humanoid robot (initially named Valkyrie, latest prototypes known as "R5," see Figure 5.31). During the DRC competition, JSC collaborated with the University of Edinburgh and the IHMC on

Figure 5.31 NASA's R5 or Valkyrie robot. (Courtesy NASA.)

its control. However, the status of R5's development did not allow the robot to show its full capabilities during the competitions (in fact, the robot only participated without much success on the trials). In November 2015, NASA awarded two Valkyrie robots to two university teams (i.e., MIT in Cambridge and Northeastern University in Boston) in preparation for an upcoming NASA's Space Robotics Challenge (SRC) that aims to explore the technology readiness level of sending humanoid robots into space, specifically to Mars. As reported in a press release, NASA's interest in humanoid robots is driven from the observation that they have huge potential to efficiently operate in human-built environments and effectively cooperate with astronauts. The robots developed for the DRC demonstrate commonalities between robots designed to work in a disaster scenario and those for the extreme environments such as the planetary bodies. The tasks that the Valkyrie robots need to fulfill at the SRC are not yet released at the time of writing this book, though initial ideas have been made public. It is envisaged that in 2017 the two robots need to compete against each other by performing various tasks such as exiting a habitat and using a ladder, removing cables from one storage and attaching them to another location while traversing irregular terrain, repairing of components such as a broken valve or a tire, and collecting samples of soil or rocks as typically seen in planetary missions.

References

1. Bonitz, R.G. (1997) Mars surveyor '98 lander MVACS robotic arm control system design concepts. *Proceedings of the 1997 IEEE International Conference on Robotics and Automation, Albuquerque, New Mexico, USA, April 20-25, 1997*, pp. 2465–2470.
2. Bonitz, R.G., Nguyen, T.T., and Kim, W.S. (2000) The Mars surveyor '01 rover and robotic arm. Aerospace Conference Proceedings, 2000 IEEE, Vol. 7, pp. 235–246.
3. Barnes, D., Phillips, N., and Paar, G. (2003) Beagle 2 simulation and calibration for ground segment operations. *Proceedings of the 7th International Symposium on Artificial Intelligence, Robotics and Automation in Space*.
4. Shiraishi, L. and Bonitz, R.G. (2008) NASA Mars 2007 Phoenix lander robotic arm and icy soil acquisition device. *Journal of Geophysical Research*, **113**, pp. 1–10.
5. Baumgartner, E.T., Bonitz, R.G., Melko, J.P., Shiraishi, L.R., and Leger, P.C. (2005) The Mars exploration rover instrument positioning system. *Aerospace Conference, 2005 IEEE*, pp. 1–19.
6. Phillips, N. (2001) Mechanisms for the Beagle 2 lander. *Proceedings of the 9th European Space Mechanisms and Tribology Symposium*, pp. 25–32.
7. Fleischner, R. and Billing, R. (2011) Mars science laboratory robotic arm. *ESMATS*.
8. Robinson, M., Collins, C., Leger, P., Kim, W., Carsten, J., Tompkins, V., Trebi-Ollennu, A., and Florow, B. (2013) Test and validation of the Mars Science Laboratory Robotic Arm. *8th International Conference on System of Systems Engineering (SoSE) 2013, Maui, HI, USA, 2-6 June 2013*, pp. 184–189.
9. Yoshikawa, T. (1985) Manipulability of robotic mechanisms. *International Journal of Robotics Research*, **4** (2), 3–9.
10. Whitney, D.E. (1987) Historical perspective and the state-of-the art in robot force control. *International Journal of Robotics Research*, **6**, 3–14.
11. De Schutter, J., Bruyninckx, H., Zhu, W.-H., and Spong, M.W. (1997) Force control: a bird's eye view, in *Control Problems in Robotics and Automation: Future Directions* (ed. B. Siciliano), Springer-Verlag, pp. 1–17.
12. Siciliano, B. and Villani, L. (eds) (1999) *Robot Force Control*, Kluwer Academic Publishers, Boston, MA.
13. Lefebvre, T., Xiao, J., Bruyninckx, H., and De Gersem, G. (2005) Active compliant motion: a survey. *Advanced Robotics*, **19** (5), 479–499.
14. Siciliano, B. and Khatib, O. (eds) (2008) *Springer Handbook of Robotics*, Springer-Verlag, Berlin, Heidelberg.
15. Helmick, D.M., McCloskey, S., Okon, A., Carsten, J., Kim, W.S., and Leger, C. (2013) Mars science laboratory algorithms and flight software for autonomously drilling rocks. *Journal of Field Robotics*, **30** (6), 847–874.
16. Chiaverini, S., Siciliano, B., and Villani, L. (1999) A survey of robot interaction control schemes with experimental comparison. *IEEE/ASME Transactions on Mechatronics*, **4** (3), 273–285.
17. Raibert, M.H. and Craig, J.J. (1981) Hybrid position/force control of manipulators. *Journal of Dynamic Systems, Measurement, and Control*, **102**, 126–133.
18. Chiaverini, S. and Sciavicco, L. (1993) The parallel approach to force/position control of robotic manipulators. *IEEE Transactions on Robotics and Automation*, **9**, 361–373.
19. Hogan, N. (1985) Impedance control-an approach to manipulation, part I-

theory, part II-implementation, part III- application. *Journal of Dynamics Systems, Measurement, and Control-Transactions of the ASME*, **107** (1), 1–24.

20. Nevins, I. and Whitney, D.E. (1973) The force vector assembler concept. *Proceedings of 1st CSIM-IFToMM Symposium on Theory and Practice of Robots and Manipulators.*

21. Whitney, D.E. (1977) Force feedback control of manipulator fine motions. *ASME Journal of Dynamic Systems, Measurement, and Control*, **99**, 91–97.

22. Albu-Schäffer, A. and Hirzinger, G. (2002) Cartesian impedance control techniques for torque controlled lightweight robots. *ICRA*, pp. 657–663.

23. Hogan, N. (1988) On the stability of manipulators performing contact tasks. *IEEE Journal of Robotics and Automation*, **4**, 677–686.

24. Surdilovic, D. (1996) Contact stability issues in position based impedance control: theory and experiments. *Proceedings of IEEE ICRA96*, pp. 1675–1680.

25. Kazerooni, H. (1990) Contact instability of the direct drive robot when constrained by a rigid environment. *IEEE Transactions on Automation and Control*, **35**, 710–714.

26. Seraji, H. and Colbaugh, R. (1993) Adaptive force-based impedance control. Proceedings of the 1993 IEEE/RSJ International Conference on Intelligent Robots and Systems, pp. 1537–1544.

27. Colbaugh, R., Seraji, H., and Glass, K. (1993) Direct adaptive impedance control of robot manipulators. *Journal of Robotic Systems*, **10** (2), 217–248.

28. Singh, S.K. and Popa, D.O. (1995) An analysis of some fundamental problems in adaptive control of force and impedance behavior, theory and experiments. *IEEE Transactions on Robotics and Automation*, **11**, 223–228.

29. Matko, D., Kamnik, R., and Badj, T. (1999) Adaptive impedance force control of an industrial manipulator. *Proceedings of IEEE International Symposium on Industrial Electronics*, pp. 129–133.

30. Lu, A.A. and Goldenberg, Z. (1995) Robust impedance control and force regulation: theory and experiments. *International Journal of Robotics Research*, **14**, 225–254.

31. Jung, S. and Hsia, T.C. (2000) Robust neural force control scheme under uncertainties in robot dynamics and unknown environment. *IEEE Transactions on Industrial Electronics*, **47** (2), 403–412.

32. Seraji, H. and Colbaugh, R. (1997) Force tracking in impedance control. *International Journal of Robotics Research*, **16**, 97–117.

33. Jung, S., Hsia, T.C., and Bonitz, R.G. (2001) Force tracking impedance control for robot manipulators with an unknown environment: theory, simulation, and experiment. *International Journal of Robotics Research*, **20** (9), 765–774.

34. Jung, S. and Hsia, T.C. (1998) Neural network impedance force control of robot manipulator. IEEE Transactions on Industrial Electronics, pp. 451–461.

35. Jung, S., Yim, S.B., and Hsia, T.C. (2001) Experimental studies of neural network impedance force control for robot manipulators. *Proceedings of IEEE International Conference on Robotics and Automation (ICRA), 2001, vol. 4*, pp. 3453–3458.

36. Cheah, C.C. and Wang, D. (1995) Learning impedance control for robotic manipulators. *Proceedings of IEEE International Conference on Robotics and Automation (ICRA), vol. 2*, pp. 2150–2155.

37. Cohen, M. and Flash, T. (1991) Learning impedance parameters for robot control using an associative search network. *IEEE Transactions on Robotics and Automation*, **7** (3), 382–390.

38. Bonitz, R.G. and Hsia, T.C. (1996) Robust internal-force based impedance control for coordinating manipulators-theory and experiments. *Proceedings of IEEE International Conference on Robotics and Automation (ICRA), vol. 1*, pp. 622–628.

39. Lin, S. and Tsai, H. (1997) Impedance control with on-line neural network

compensator for dual-arm robots. *Journal of Intelligent Robotics Systems*, **18** (1), 87–104.
40. Moosavian, S.A.A. and Papadopoulos, E. (1998) Multiple impedance control for object manipulation. *Proceedings of IEEE/RSJ International Conference on Intelligent Robots and Systems (IROS)*, vol. 1, pp. 461–466.
41. Hill, J. and Park, W. (1979) Real-time control of a robot with a mobile camera. Proceedings of the 9th International Symposium on Industrial Robots, pp. 233–246.
42. Shirai, Y. and Inoue, H. (1973) Guiding a robot by visual feedback in assembling tasks. *Pattern Recognition*, **5** (2), 99–106.
43. Buttazzo, G.C., Allotta, B., and Fanizza, F.P. (1994) Mousebuster: a robot for real-time catching. *IEEE Control Systems Magazine*, **14** (1), 49–56.
44. Kragic, D. (2001) *Visual servoing for manipulation: robustness and integration issues*. PhD thesis. KTH, Numerical Analysis and Computer Science, NADA.
45. Espiau, B., Chaumette, F., and Rives, P. (1992) A new approach to visual servoing in robotics. *IEEE Transactions on Robotics and Automation*, **8**, 313–326.
46. Hashimoto, H., Ogawa, H., Umeda, T., Obama, M., and Tatsuno, K. (1995) An unilateral master-slave hand system with a force-controlled slave hand. ICRA'95, pp. 956–961.
47. Andersson, R.L. (1989) Dynamic sensing in a ping-pong playing robot. *IEEE Transactions on Robotics and Automation*, **5** (6), 728–739.
48. Vahrenkamp, N., Wieland, S., Azad, P., Gonzalez, D., Asfour, T., and Dillmann, R. (2008) Visual servoing for humanoid grasping and manipulation tasks. Humanoid Robots, 2008. Humanoids 2008. 8th IEEE-RAS International Conference on, pp. 406–412.
49. Weiss, L., Sanderson, A., and Neuman, C. (1987) Dynamic sensor-based control of robots with visual feedback. *IEEE Journal of Robotics and Automation*, **3** (5), 404–417.
50. Hosoda, K. and Asada, M. (1994) Versatile visual servoing without knowledge of true Jacobian. Intelligent Robots and Systems '94. 'Advanced Robotic Systems and the Real World', IROS '94. *Proceedings of the IEEE/RSJ/GI International Conference on*, vol. 1, pp. 186–193.
51. Wilson, W.J., Williams Hulls, C.C., and Bell, G.S. (1996) Relative end-effector control using Cartesian position based visual servoing. *IEEE Transactions on Robotics and Automation*, **12** (5), 684–696.
52. Martinet, P. and Gallice, J. (1999) Position based visual servoing using a non-linear approach. Intelligent Robots and Systems, 1999. IROS '99. *Proceedings. 1999 IEEE/RSJ International Conference on*, vol. 1, pp. 531–536.
53. Rajan, V. (1985) Minimum time trajectory planning. Robotics and Automation. Proceedings. 1985 IEEE International Conference on, vol. 2, pp. 759–764.
54. Biagiotti, L. and Melchiorri, C. (2008) *Trajectory Planning for Automatic Machines and Robots*, Springer-Verlag, Berlin Heidelberg.
55. Kröger, T. (2010) *On-Line Trajectory Generation in Robotic Systems*, Springer Tracts in Advanced Robotics, vol. 58, Springer-Verlag, Berlin, Heidelberg, Germany.
56. Kröger, T. (2011) Opening the door to new sensor-based robot applications – The reflexxes motion libraries. *Proceedings of the IEEE International Conference on Robotics and Automation, Shanghai, China*.
57. Khatib, O. (1986) Real-time obstacle avoidance for manipulators and mobile robots. *International Journal of Robotics Research*, **5** (1), 90–98.
58. Seraji, H. and Bon, B. (1999) Real-time collision avoidance for position-controlled manipulators. *IEEE Transactions on Robotics and Automation*, **15** (4), 670–677.
59. De Luca, A., Albu-Schaffer, A., Haddadin, S., and Hirzinger, G. (2006) Collision detection and safe reaction with the DLR-III lightweight manipulator arm. Intelligent Robots and Systems, 2006 IEEE/RSJ International Conference on, pp. 1623–1630.

60. Flacco, F., Kroger, T., De Luca, A., and Khatib, O. (2012) A depth space approach to human-robot collision avoidance. *Robotics and Automation (ICRA), 2012 IEEE International Conference on*, pp. 338–345.
61. Kuffner, J., Nishiwaki, K., Kagami, S., Kuniyoshi, Y., Inaba, M., and Inoue, H. (2002) Self-collision detection and prevention for humanoid robots. *Robotics and Automation, 2002. Proceedings. ICRA '02. IEEE International Conference on*, vol. 3, pp. 2265–2270.
62. Seto, F., Kosuge, K., and Hirata, Y. (2005) Self-collision avoidance motion control for human robot cooperation system using robe. *Intelligent Robots and Systems, 2005. (IROS 2005). 2005 IEEE/RSJ International Conference on*, pp. 3143–3148.
63. De Santis, A., Albu-Schaeffer, A., Ott, C., Siciliano, B., and Hirzinger, G. (2007) The skeleton algorithm for self-collision avoidance of a humanoid manipulator. *Advanced intelligent mechatronics, 2007 IEEE/ASME international conference on*, pp. 1–6.
64. Dietrich, A., Wimbock, T., Taubig, H., Albu-Schaeffer, A., and Hirzinger, G. (2011) Extensions to reactive self-collision avoidance for torque and position controlled humanoids. *Robotics and Automation (ICRA), 2011 IEEE International Conference on*, pp. 3455–3462.
65. Taubig, H., Bauml, B., and Frese, U. (2011) Real-time swept volume and distance computation for self collision detection. *Intelligent Robots and Systems (IROS), 2011 IEEE/RSJ International Conference on*, pp. 1585–1592.
66. Stasse, O., Escande, A., Mansard, N., Miossec, S., Evrard, P., and Kheddar, A. (2008) Real-time (self-)collision avoidance task on a HRP-2 humanoid robot. *Robotics and Automation, 2008. ICRA 2008. IEEE International Conference on*, pp. 3200–3205.
67. Leger, C. (2002) Efficient sensor/model based on-line collision detection for planetary manipulators. *Proceedings of the 2002 IEEE International Conference on Robotics and Automation, ICRA 2002, May 11-15, 2002, Washington, DC, USA*, pp. 1697–1703.
68. Kuffner, J., Nishiwaki, K., Kagami, S., Inaba, M., and Inoue, H. (2005) Motion planning for humanoid robots, in *Robotics Research*, Springer Tracts in Advanced Robotics, vol. 15 (eds P. Dario and R. Chatila), Springer-Verlag, Berlin / Heidelberg, pp. 365–374.
69. Kavraki, L.E., Svestka, P., Latombe, J.C., and Overmars, M.H. (1996) Probabilistic roadmaps for path planning in high-dimensional configuration spaces. *IEEE Transactions on Robotics and Automation*, **12**, 566–580.
70. Lavalle, S.M. and Kuffner, J.J. (2000) Rapidly-exploring random trees: progress and prospects. Algorithmic and Computational Robotics: New Directions, pp. 293–308.
71. Ge, S.S. and Cui, Y.J. (2002) Dynamic motion planning for mobile robots using potential field method. *Autonomous Robots*, **13**, 207–222.
72. Williamson, M.M. (1999) *Robot arm control exploiting natural dynamics*. PhD thesis, Massachusetts Institute of Technology.
73. Ijspeert, A.J., Nakanishi, J., and Schaal, S. (2002) Learning rhythmic movements by demonstration using nonlinear oscillators. *Proceedings of the IEEE/RSJ International Conference on Intelligent Robots and Systems*, pp. 958–963.
74. Kurita, Y., Lim, Y., Ueda, J., Matsumoto, Y., and Ogasawara, T. (2004) CPG-based manipulation: generation of rhythmic finger gaits from human observation. ICRA, pp. 1209–1214.
75. Saxena, A., Driemeyer, J., Kearns, J., Osondu, C., and Ng, A. (2008) Learning to grasp novel objects using vision, in *Experimental Robotics*, Springer Tracts in Advanced Robotics, vol. 39 (O. Khatib, V. Kumar, and D. Rus), Springer-Verlag, Berlin / Heidelberg, pp. 33–42.
76. Zöllner, R., Asfour, T., and Dillmann, R. (2004) Programming by demonstration: dual-arm manipulation tasks for humanoid robots. *IEEE/RSJ International Conference on Intelligent Robots and Systems (IROS)*.

77. Argall, B.D., Chernova, S., Veloso, M., and Browning, B. (2009) A survey of robot learning from demonstration. *Robotics and Autonomous Systems*, **57** (5), 469–483.
78. Billard, A., Epars, Y., Calinon, S., Schaal, S., and Cheng, G. (2004) Discovering optimal imitation strategies. *Robotics and Autonomous Systems*, **47** (2-3), 69–77.
79. Tunstel, E., Maimone, M.W., Trebi-Ollennu, A., Yen, J., Petras, R., and Willson, R.G. (2005) Mars exploration rover mobility and robotic arm operational performance. *SMC*, IEEE, pp. 1807–1814.
80. Trebi-Ollennu, A., Baumgartner, E.T., Leger, P.C., and Bonitz, R.G. (2005) Robotic arm in-situ operations for the Mars exploration rovers surface mission. *Proceedings of the IEEE International Conference on Systems, Man and Cybernetics, Waikoloa, Hawaii, USA, October 10-12, 2005*, pp. 1799–1806.
81. Robinson, M., Collins, C., Leger, P., Carsten, J., Tompkins, V., Hartman, F., and Yen, J. (2013) In-situ operations and planning for the Mars Science Laboratory Robotic Arm: the first 200 sols. *System of Systems Engineering (SoSE), 2013 8th International Conference on*, pp. 153–158.
82. Goertz, R. and Thompson, R. (1954) Electronically controlled manipulator. *Nucleonics*, **12** (11), 46–47.
83. Hirzinger, G., Brunner, B., Dietrich, J., and Heindl, J. (1993) Sensor-based space robotics-rotex and its telerobotic features. *IEEE Transactions on Robotics and Automation*, **9** (5), 649–663.
84. Hirzinger, G., Heindl, J., and Landzettel, K. (1989) Predictive and knowledge-based telerobotic control concepts. *Robotics and Automation, 1989. Proceedings., 1989 IEEE International Conference on, vol. 3*, pp. 1768–1777.
85. Preusche, C., Reintsema, D., Landzettel, K., and Hirzinger, G. (2006) Robotics component verification on ISS Rokviss - preliminary results for telepresence. *Intelligent Robots and Systems, 2006 IEEE/RSJ International Conference on*, pp. 4595–4601.
86. European Space Agency (2015) First Handshake and Force-Feedback with Space.
87. Ferrell, W.R. and Sheridan, T.B. (1967) Supervisory control of remote manipulation. *IEEE Spectrum*, **4** (10), 81–88.
88. Wright, J., Hartman, F., Cooper, B., Maxwell, S., Yen, J., and Morrison, J. (2006) Driving on Mars with rsvp. *IEEE Robotics Automation Magazine*, **13** (2), 37–45.
89. Wright, J.R., Hartman, F., Maxwell, S., Cooper, B., and Yen, J. (2013) Updates to the rover driving tools for curiosity. *System of Systems Engineering (SoSE), 2013 8th International Conference on*, pp. 147–152.
90. Sheridan, T.B. and Verplank, W.L. (1978) Human and computer control of undersea teleoperators (Man-Machine Systems Laboratory Report).
91. Crandall, J.W. and Goodrich, M.A. (2002) Characterizing efficiency of human robot interaction: a case study of shared-control teleoperation. *Intelligent Robots and Systems, 2002. IEEE/RSJ International Conference on, vol. 2*, pp. 1290–1295.
92. Hayati, S. and Venkataraman, S.T. (1989) Design and implementation of a robot control system with traded and shared control capability. Robotics and Automation, 1989. Proceedings., 1989 IEEE International Conference on, vol. 3, pp. 1310–1315.
93. Kortenkamp, D., Bonasso, P., Ryan, D., and Schreckenghost, D. (1997) Traded Control with Autonomous Robots as Mixed Initiative Interaction. Haller, S. and McRoy, S. (Eds.). *Symposium on Mixed Initiative Interaction.*
94. Dorais, G.A., Bonasso, R.P., Kortenkamp, D., Pell, B., and Schreckenghost, D. (1999) Adjustable autonomy for human-centered autonomous systems. Proceedings of the 1st International Conference of the Mars Society.
95. Crandall, J.W. and Goodrich, M.A. (2001) Experiments in adjustable autonomy. Systems, Man, and Cybernetics,

2001 IEEE International Conference on, vol. 3, pp. 1624–1629.

96. Roth, Z.S., Mooring, B.W., and Ravani, B. (1987) An overview of robot calibration. *IEEE Journal on Robotics and Automation*, **3** (5), 377–385.

97. Mooring, B., Driels, M., and Roth, Z. (1991) *Fundamentals of Manipulator Calibration*, John Wiley & Sons, Inc., New York.

98. Rybus, T. et al. (2013) New planar air-bearing microgravity simulator for verification of space robotics numerical simulations and control algorithms. *Proceedings of the 12th ESA Symposium on Advanced Space Technologies in Robotics and Automation (ASTRA 2013)*.

99. Rutkowski, K., Rybus, T., Wawrzaszek, R., Seweryn, K., and Grassmann, K. (2015) Design and development of two manipulators as a key element of a space robot testing facility. *Archive of Mechanical Engineering*, **LXII** (3), doi: 10.1515/meceng-2015-0022.

100. Banaszkiewicz, M. and Seweryn, K. (2008) Optimization of the trajectory of a general free - flying manipulator during the rendezvous maneuver. *AIAA Guidance, Navigation, and Control Conference*.

101. Wu, Y.-H., Gao, Y., Lin, J.-W., Raus, R., Zhang, S.-J., and Watt, M. (2013) Low-cost, high-performance monocular vision system for air bearing table attitude determination. *Journal of Spacecraft and Rockets*, **51** (1), 66–75.

102. Wawrzaszek, R., Banaszkiewicz, M., Rybus, T., Wisniewski, L., Seweryn, K., and Grygorczuk, J. (2014) Low velocity penetrators (LVP) driven by hammering action - definition of the principle of operation based on numerical models and experimental test. *Acta Astronautica*, **99**, 303–317.

103. Alford, C.O. and Belyeu, S.M. (1984) Coordinated control of two robot arms. *Proceedings of IEEE International Conference on Robotics & Automation*, pp. 468–473.

104. Zheng, Y.F. and Luh, J.Y.S. (1986) Joint torques for control of two coordinated moving robots. *Proceedings of the IEEE International Conference on Robotics and Automation*, pp. 1375–1380.

105. Hayati, S. (1986) Hybrid position/force control of multiarm cooperating robots. *Proceedings of IEEE International Conference on Robotics and Automation*, pp. 1375–1380.

106. Tarn, T., Bejczy, A., and Yun, X. (1987) Design of dynamic control of two co-operating robot arms: closed chain formulation. *Robotics and Automation. Proceedings. 1987 IEEE International Conference on*, vol. 4, pp. 7–13.

107. Yamano, M., Kim, J.-S., Konno, A., and Uchiyama, M. (2004) Cooperative control of a 3D dual-flexible-arm robot. *Journal of Intelligent and Robotic Systems*, **39**, 1–15.

108. Wang, Z.-D., Hirata, Y., Takano, Y., and Kosuge, K. (2004) From human to pushing leader robot: leading a decentralized multirobot system for object handling. *Robotics and Biomimetics, 2004. ROBIO 2004. IEEE International Conference on*, pp. 441–446.

109. Zacharias, F., Leidner, D. et al. (2010) Exploiting structure in two-armed manipulation tasks for humanoid robots. *The IEEE International Conference on Intelligent Robots and Systems (IROS)*.

110. Vahrenkamp, N., Do, M., Asfour, T., and Dillmann, R. (2010) Integrated Grasp and motion planning. *Robotics and Automation (ICRA), 2010 IEEE International Conference on, IEEE*, pp. 2883–2888.

111. Vahrenkamp, N., Berenson, D., Asfour, T., Kuffner, J., and Dillmann, R. (2009) Humanoid motion planning for dual-arm manipulation and re-grasping tasks. *IEEE/RSJ International Conference on Intelligent Robots and Systems (IROS '09)*.

112. Azevedo, C., Poignet, P., and Espiau, B. (2004) Artificial locomotion control: from human to robots. *Robotics and Autonomous Systems*, **47** (4), 203–223.

113. Lee, N., Rietdyk, C.S.G., and Naksuk, S. (2005) Whole-body human-to-humanoid motion transfer. *IEEE-RAS International Conference on Humanoid Robots*, pp. 104–109.

114. Mombaur, K. and Sreenivasa, M.N. (2010) HRP-2 plays the yoyo: from

human to humanoid yoyo playing using optimal control. ICRA, pp. 3369–3376.

115. Yamane, K. and Nakamura, Y. (2003) Dynamics filter - concept and implementation of online motion generator for human figures. *IEEE Transactions on Robotics and Automation*, **19**, 421–432.

116. Kuffner, J.J., Kagami, S., Nishiwaki, K., Inaba, M., and Inoue, H. (2002) Dynamically-stable motion planning for humanoid robots. *Autonomous Robots*, **12**, 105–118.

117. Nishiwaki, K., Kagami, S., Kuniyoshi, Y., Inaba, M., and Inoue, H. (2002) Online generation of humanoid walking motion based on a fast generation method of motion pattern that follows desired ZMP. *IEEE/RSJ International Conference on Intelligent Robots and Systems*, pp. 2684–2689.

118. Sentis, L. (2007) *Synthesis and control of whole-body behaviors in humanoid systems*. PhD thesis, Stanford University, Stanford, CA.

119. Sentis, L., Petersen, J., and Philippsen, R. (2013) Implementation and stability analysis of prioritized whole-body compliant controllers on a wheeled humanoid robot in uneven terrains. *Autonomous Robots*, **35**, 301–319.

120. Sentis, L., Park, J., and Khatib, O. (2010) Compliant control of multicontact and center-of-mass behaviors in humanoid robots. *IEEE Transactions on Robotics*, **26** (3), 483–501.

121. Fok, C.-L., Johnson, G., Yamokoski, J.D., Mok, A.K., and Sentis, L. (2015) Controlit! - A software framework for whole-body operational space control. CoRR, abs/1506.01075.

122. Liu, M., Tan, Y., and Padois, V. (2015) Generalized hierarchical control. *Autonomous Robots*, **XX** 1–15.

123. Kajita, S., Kanehiro, F., Kaneko, K., Fujiwara, K., Harada, K., Yokoi, K., and Hirukawa, H. (2003) Resolved momentum control: humanoid motion planning based on the linear and angular momentum. *Proceedings IEEE/RSJ International Conference on Intelligent Robots and Systems*, pp. 1644–1650.

124. Nishiwaki, K., Kuga, M., Kagami, S., Inaba, M., and Inoue, H. (2005) Whole-body cooperative balanced motion generation for reaching. *International Journal of Humanoid Robotics*, **2** (4), 437–457.

125. Smits, R., De Laet, T., Claes, K., Bruyninckx, H., and De Schutter, J. (2009) iTASC: a tool for multi-sensor integration in robot manipulation, in *Multisensor Fusion and Integration for Intelligent Systems*, Lecture Notes in Electrical Engineering, vol. **35**, Springer-Verlag, pp. 235–254.

126. de Gea Fernandez, J., Mronga, D., Wirkus, M., Bargsten, V., Asadi, B., and Kirchner, F. (2015) Towards describing and deploying whole-body generic manipulation behaviours. Space Robotics Symposium.

127. Mansard, N., Stasse, O., Chaumette, F., and Yokoi, K. (2007) Visually-guided grasping while walking on a humanoid robot. *Proceedings of 2007 IEEE International Conference on Robotics and Automation*, pp. 3042–3047.

128. Neo, E.S., Yokoi, K., Kajita, S., Kanehiro, F., and Tanie, K. (2005) A switching command-based whole-body operation method for humanoid robots. *IEEE/ASME Transactions on Mechatronics*, **10** (5), 546–559.

129. Kim, S., Kim, C.H., You, B.-J., and Oh, S.-R. (2009) Stable whole-body motion generation for humanoid robots to imitate human motions. IROS, pp. 2518–2524.

130. Brock, O. and Grupen, R. (2005) NSF/NASA workshop on autonomous mobile manipulation (Amm), Houston, USA, http://robotics.cs.umass.edu/amm.

131. Chitta, S., Cohen, B., and Likhachev, M. (2010) Planning for autonomous door opening with a mobile manipulator. *Proceedings of IEEE International Conference on Robotics and Automation (ICRA)*.

132. Srinivasa, S.S., Ferguson, D., Helfrich, C.J., Berenson, D., Collet, A., Diankov, R., Gallagher, G., Hollinger, G., Kuffner, J., and Vande Weghe, M. (2009) HERB: a home exploring robotic butler. *Autonomous Robots*, **28** (1), 5–20.

133. Fuchs, M., Borst, Ch., Giordano, P.R., Baumann, A., Kraemer, E., Langwald, J., Gruber, R., Seitz, N., Plank, G., Kunze, K., Burger, R., Schmidt, F., Wimboeck, T., and Hirzinger, G. (2009) Rollin' Justin - design considerations and realization of a mobile platform for a humanoid upper body. ICRA'09: *Proceedings of the 2009 IEEE International Conference on Robotics and Automation*, IEEE Press, Piscataway, NJ, pp. 1789–1795.

134. Katz, D., Horrell, E., Yang, O., Burns, B., Buckley, T., Grishkan, A., Zhylkovskyy, V., Brock, O., and Learned-Miller, E. (2006) The UMass mobile manipulator UMan: an experimental platform for autonomous mobile manipulation. *Workshop on Manipulation in Human Environments at Robotics: Science and Systems*.

135. Lemburg, J., de Gea Fernandez, J., Eich, M., Mronga, D., Kampmann, P., Vogt, A., Aggarwal, A., Shi, Y., and Kirchner, F. (2011) AILA - design of an autonomous mobile dual-arm robot. *Robotics and Automation (ICRA), 2011 IEEE International Conference on*, pp. 5147–5153.

136. Iagnemma, K. and Overholt, J. (2015) Special issue: DARPA robotics challenge. *Journal of Field Robotics*, **32** (2), 87–188.

6
Mission Operations and Autonomy

Yang Gao, Guy Burroughes, Jorge Ocón, Simone Fratini, Nicola Policella, and Alessandro Donati

6.1
Introduction

Automation and autonomy have become growingly important to space mission and system operations in recent years, covering applications in Earth observation, space stations, planetary, and deep space explorations. The capabilities of autonomous systems in space have improved from simple automation in the Sputnik satellite to goal-oriented driving in Mars Science Laboratory (MSL) rover. Nevertheless, autonomy has a great deal more to offer within planetary robotics, including complete goal-oriented mission operations.

Autonomy is best defined for our discussion as the ability of an intelligent system to independently compose and select among different courses of action to accomplish goals based on its knowledge and understanding of the world, itself, and the situation. In general, it may entail any function and functionality that help build perceptions, learn, and make decisions, to name a few: execution and monitoring of planned actions, detection of faults and failures and deployment of recovery strategies, planning of activities to reach specific objectives, identification of scientifically interesting features, collaboration or cooperation between multiple spacecraft, and so on.

There are three main factors that illustrate the need for autonomy in planetary robotic missions, namely communication latency, environment uncertainty, and operational cost [1]:

- **Latencies in Communication:** Due to discontinuous availability of communication links and inevitable round-trip delays, a certain level of autonomy (LoA) is mandatory for robotic missions particularly when long distance comm link is involved. For example, the distance between the Earth and Mars is around 54.6–401 million km, which results in a delay of approximately 3–22 min in one-way transmissions. Limited communication bandwidth adds further complications to the issue.
- **Uncertainties of Environment:** In general, the need for autonomy grows with the degree of uncertainties of the environment to which a system interacts.

Contemporary Planetary Robotics: An Approach Toward Autonomous Systems, First Edition.
Edited by Yang Gao.
© 2016 Wiley-VCH Verlag GmbH & Co. KGaA. Published 2016 by Wiley-VCH Verlag GmbH & Co. KGaA.

The representation of the spacecraft's environment that ground control has is only a small subset of the real environment in which the spacecraft operates, and this model is not provided in real-time due to the inherent communication limitations. This limits the ability of human operators to interact with the environment optimally.

- **Costs of Operation** In the traditional approach to spacecraft operations based in a ground control center, a team of engineers are typically in charge of a large number of functions such as planning, elaboration and execution, spacecraft internal hardware state tracking, and functional verification. Due to the increasing complexity and limited budget of the space missions, the operation team and communication costs can be constraint. By adding autonomy to a spacecraft, human operators can send high-level commands, and upload procedures that would allow the system to react to the environment. Autonomy becomes more critical in cases of error recovery, or when dealing with exceptional circumstances where the spacecraft needs to be able to understand the impact of the error on its previously planned sequence and reschedule in the light of the new information with potentially degraded capabilities.

These considerations also act to measure the LoA that is needed in a mission. The more of an issue the three considerations are, the higher the level of autonomy is necessary. The total LoA required and the nature of such autonomy depend on the characteristics of the mission itself. Using an example of a planetary rover mission, implications of the three factors can be further elaborated as follows:

- The communication limitations will be tremendous given increased distances and lack of direct-to-Earth communication during long periods. For missions on Mars, the communication window per Martian day can be as short as 2 h.
- The planetary surface is an environment with intrinsic uncertainties that the rovers need to heavily interact with, such as the energy required to travel to a target location.
- Any effort to reduce the cost of operations has an added value. Taking the Mars Exploration Rover (MER) as an example, the mission operation involved 240 human operators working 24/7 during the nominal mission and cost $4–4.5 million/day according to Ref. [2].

The remainder of the chapter is divided into seven sections. Section 6.2 introduces the basic concepts in mission operation, various processes and procedures, and typical operation modes of planetary robotic systems. Section 6.3 discusses the design issues to establish the software architecture for a given mission operation. Section 6.4 describes the core technology of planning and scheduling that enables high-level autonomy in mission operations and its representative design techniques/solutions. Section 6.5 presents the core technology that allows reconfiguration of autonomous software within mission operation. Section 6.6 covers various tools and techniques for validation and verification (V&V) of autonomous software. Section 6.7 presents a design example of mission operation software for Mars rovers. Section 6.8 outlines some over-the-horizon R&D ideas in achieving autonomous operations and systems for future planetary robotic missions.

6.2
Context

The standard operations of a planetary robotic mission involving onboard spacecraft, ground control stations, and their interface can be complicated, and filled with technical jargon and specific procedures. This section aims to clarify, elucidate, and define many of these terms and procedures.

6.2.1
Mission Operation Concepts

The operations of any space system is broken into a number of segments—some enforced by hardware, some by environment, and some we fabricated to simplify the creation and operation of subsystems. The European Committee for Space Standardization (ECSS) defines a space system to consist of two high-level segments [3]: **space (or onboard) segment** and **ground segment** (Figure 6.1). This shows an abstract and physical separation between the hardware, software, and operating procedure of the spacecraft in space and the Earth-based ground control station. This separation could be seen as a simplified view of potentially complicated space systems; nevertheless, even if a metaphorical separation can be drawn it is useful. Essentially, it can be said that human operators have full real-time access to the ground segment, but not the onboard segment.

The separation is made more apparent by the communication limitations inherent in space mission operations. The communication or interface between the ground and onboard segments is limited; thus, some specific procedures should be utilized to maximize the communication potential. Two procedures that are used in such communication are known as the uplink and the downlink [1].

- The *downlink* is the process in which information (e.g., telemetry or payload data) is sent from the space to the ground segment.
- The *uplink* is the process in which information, such as commands, is transmitted from the ground to the space segment.

The ground and space segments are further divided by hardware and software with various layers of abstraction in between as typically seen in most computer systems (see Figure 6.2). For example, the ground segment needs to have a user

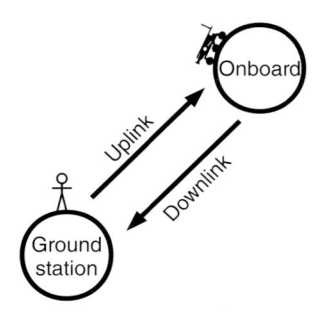

Figure 6.1 Two segments and communication links in mission operations.

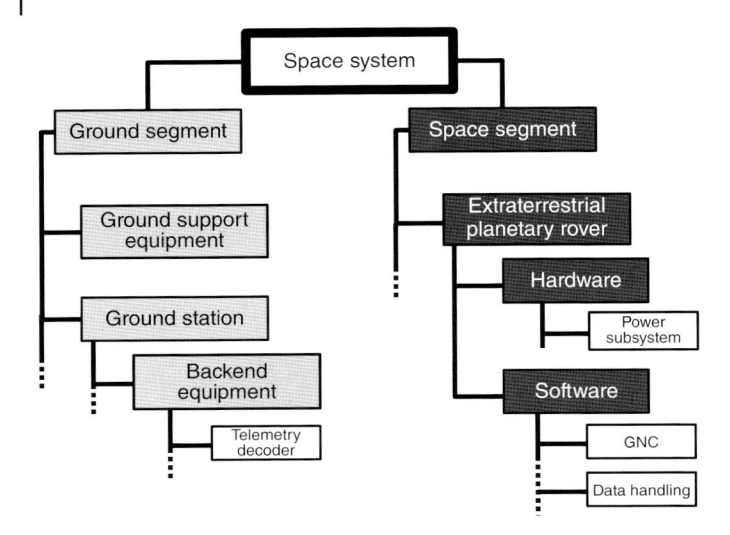

Figure 6.2 An example of ECSS space system model.

interface on top of the software layer, which lays out information and choices available for the human operators in a user friendly manner. The crew on ground often involves mission scientists, engineers, and project managers or coordinators. The different personnel groups can have different goals or priorities that are competing. The personnel interface is usually choreographed behind a well-defined process to maximize the utility of a mission. Furthermore, these interface procedures are split by their characteristic time lengths, and shall be discussed further in Section 6.2.2.

To ensure compatibility, interoperability, safety, repeatability, and quality engineering, standards are defined to regulate designs for mission operations and autonomy. For example, the aforementioned ECSS was established in 1993 as an organization that works to improve standardization within the European space sector. The Consultative Committee for Space Data Systems (CCSDS) was established in 1982 as a more intercontinental organization for national space agencies to discuss and develop standards for space data and information systems.

Several existing ECSS standards are specific to mission operations and autonomy, including the following:

- *Space Segment Operability* [4]: This standard defines sets of requirements for the design of onboard functions in unmanned spacecraft. It presents the ECSS concept for onboard operational autonomy: "the capability of the space segment to manage nominal or contingency operations without ground segment intervention for a given period of time." Although autonomy is often assimilated within space segment, it is clear that the ground segment plays a crucial role in any space system. Therefore, if performance improvements at the system level are to be assessed, it is necessary to conceive the system as a whole,

considering the implications for the ground segment when the LoA at the space segment is increased. In addition, a sufficient LoA shall also be incorporated in the ground segment part, with the aim of supporting the operations of the autonomous spacecraft.
- *Telemetry and Telecommand Packet Utilization* [5]: This standard, commonly referred to as "PUS" (from its previous ESA version, the "packet utilization standard"), defines the application-level interface between the ground and space segments. It relies on the CCSDS protocols for data transfer and describes a number of operational concepts that are satisfied by a set of services. These services comprise almost all common operations to control and monitor the space segment. In this context, the space segment is the service provider, while the ground segment is the "main" consumer; onboard applications can also consume services. The interface between the segments is defined in terms of packet requests (telecommands) and reports (telemetry). The adoption of PUS for a given mission promotes the gradual increase of autonomous behavior. Its implementation starts from minimal sets of capabilities, which only have the basics required for spacecraft operation, to extended sets of capabilities that transfer to the spacecraft advanced functions such as the conditional execution of interlocked time-tagged commands and even onboard procedures. It is up to the mission designers to define what capabilities to implement for each service.
- *Monitoring and Control Data Definition* [3]: This standard defines two main concepts for autonomy of a space mission: the Space System Model (SSM) and the Domain-Specific Views. A SSM is a "representation of the space system in terms of its decomposition into system elements, the activities that can be performed on these system elements, the reporting data that reflects the state of these system elements, and the events that can be raised and handled for the control of these system elements, activities or reporting data." The SSM is a means to capture the knowledge about a space system. It is intended to be a combined textual and graphical description of the entire mission, covering all its segments. A Domain-Specific View is a subset of the SSM that is relevant for a given application. For example, a Domain-Specific View for attitude control simulation software could have as a root of the AOCS element (refer to Figure 6.2). The SSM and the Domain-Specific View are important concepts in implementing autonomous software.

Six technical areas of CCSDS standards are specific to mission operations and autonomy [6]:

- Spacecraft Onboard Interface Services,
- Space Link Services,
- Space Internetworking Services,
- Mission Operations and Information Management Services,
- Cross Support Services,
- System Engineering.

6.2.2
Mission Operation Procedures

According to Ref. [7], the overall operation process for a planetary robotic mission can be broken into several subprocesses as shown in Figure 6.3.

1) **Nominal and extended science operation process**
   - *Period*: A few months.
   - *Teams*: Engineers, scientists, coordinators, principal investigators (PIs).
   - *Scope*: This process involves senior member of all stakeholder teams gathering and determining the high-level mission or scientific goals. These goals can take several months to complete and do not specify detailed plans of actions. They may also discuss high-level administrative issues such as staff or personnel matters. This process can generate extended mission documents, such as the MSL Extended Mission 1 Science Plan [8].

2) **Strategic operation process**
   - *Period*: A few days to a few weeks.
   - *Teams*: Engineers, scientists.
   - *Scope*: This process can be defined as the set of ground operations for planning the medium-term robotic activities to ensure the high-level scientific objectives of the mission and support the tactical operations process. During the strategic operations, scientifically important target locations for the robots can be identified. This process involves the following activities:

   robot engineering model updates,
   – tracking of the mission scientific goals,
   – onboard software management, for example, software patch,
   – postmission data product generation (i.e., archive and dissemination),
   – planning of communication windows, inter-mission coordination and mid-term planning,
   – offline engineering and science product generation and nonoperational critical data,

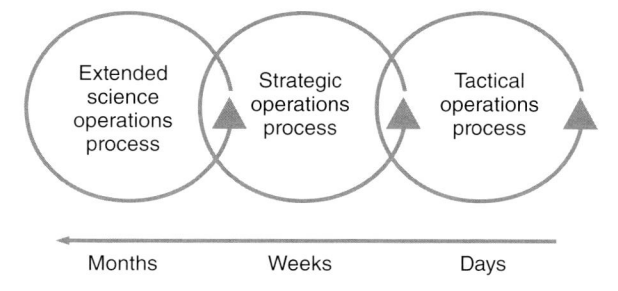

Figure 6.3 Three operation processes and respective time frame.

– specification and validation of new activities/tasks, or redesign of the existing ones.
3) **Tactical operation process**
    - *Period*: A few days.
    - *Teams*: Engineers, Scientists.
    - *Scope*: The tactical operation process is meant for the algorithmic process in order to define and execute short-term scientific experiments.

As illustrated in Figure 6.4, a typical strategic and tactical operation process interfaces with the space segment via the ground segment. Both the uplink and downlink contain different steps of the operation process carried out by different combination of personnel. Some steps can be performed autonomously, such as using database handling and plan validation techniques. There are many steps in the process that the science as well as the engineering teams need to be involved with. Given most missions have multiple objectives, various stakeholders may need to compete for resources.

Taking ESA's future ExoMars mission as an example, the envisioned strategic operation process involves individual science teams analyzing the data from the downlink and producing a potential science plan. These potential science plans are then consolidated by a Science Operations Working Group (SOWG) to output a preliminary science plan. An engineering plan that is concurrently generated by the engineering team will be integrated with the consolidated science plan. The plans are further validated in a feedback loop until a Rover Integrated Planning Team (RIPT) agrees on a final plan. The RIPT would consist of representatives from the engineering team and the SOWG. The validation process will involve various metrics and simulations of the proposed plan. Following this, the uplink to the onboard segment will be initiated driven by the decisions made by the tactical operation process. The onboard segment is then expected to carry out planned tasks in a planned LoA, ranging from human teleoperation to goal-oriented autonomy. Some of the mandatory tasks for the onboard segment include the following:

- self-reporting,
- safety monitoring,
- power management,
- database management,
- communications management,
- fault detection, isolation, and recovery (FDIR).

Automation and autonomy within different operation processes (on ground or space segment) can be very powerful. For example, housekeeping can be performed automatically; the planning, time-line creation, and validation process can be potentially controlled and performed autonomously. However, there are challenges associated with achieving autonomous operations due to tight time windows in which the operations must be performed. This has implications on

328 | *6 Mission Operations and Autonomy*

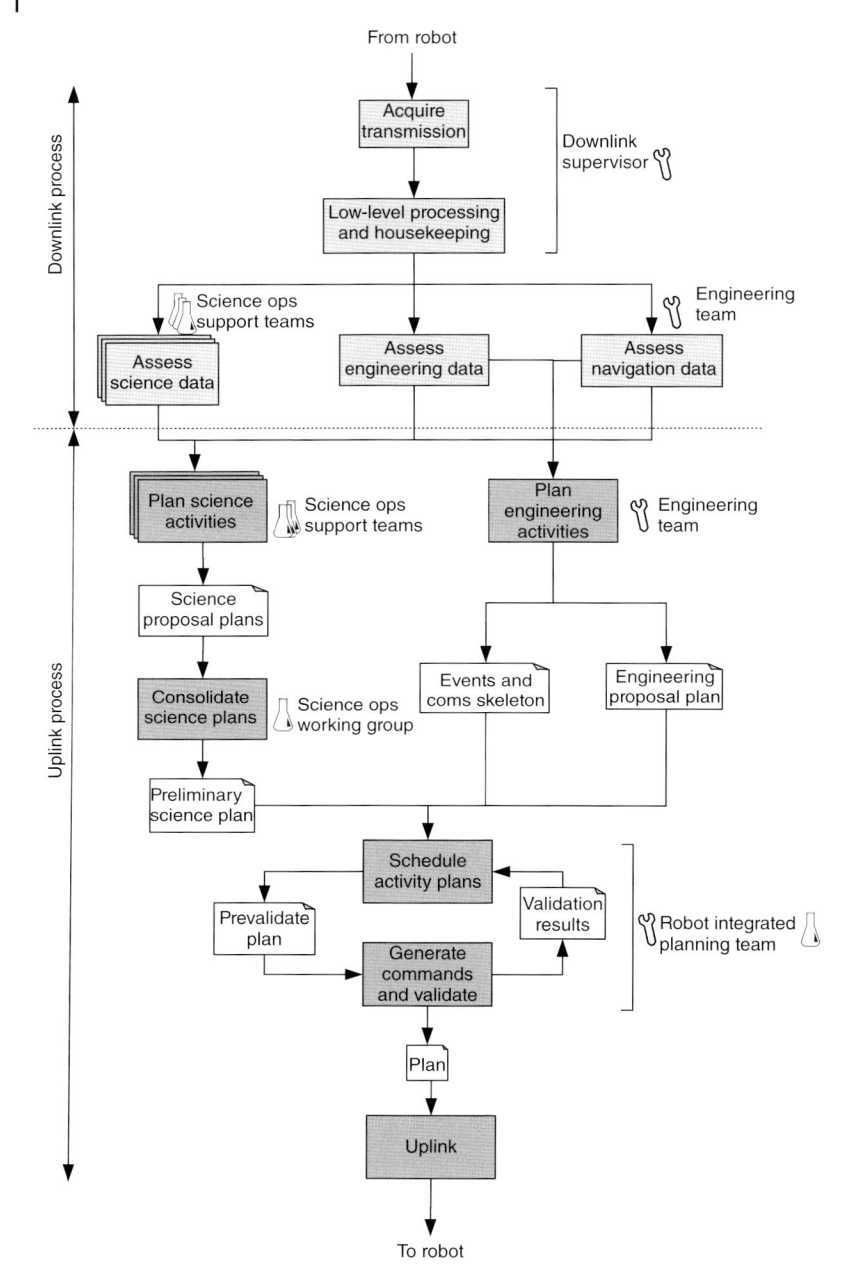

Figure 6.4 Strategic and tactical operation flow.

how fast or well the processes need to be planned and performed. Complexity of automated plan creation grows with the fidelity and length of the plan. A planetary robotic mission tends to require lengthy, complex, and detailed plans, whereas an optimal plan may not be possible to get generated within a given time window. In real-time space mission operations, a plan must be generated by a deadline even if it is suboptimal.

6.2.3
Onboard Segment Operation Modes

The ECSS has defined four LoAs for the onboard segment of a space system as shown in Table 6.1. A planetary robotic system or its onboard segment is often designed to work in different operation modes to reflect different LoAs that the mission operation aims to achieve.

The LoA from E1 to E4 represents increased autonomy and decreased human control. The LoA or corresponding operation modes for the space segment can be set by the ground segment between the tactical and strategic operation processes after assessing the risk and envisaged mission goals. For instance, a planetary rover may work in LoA E4 or autonomous, goal-oriented mode before approaching the edge of a crater; after downlink the tactical operation process can decide the rover must switch to LoA E1 or real-time control, telecommand mode for safety purposes once entering the ridge of the crater. This is a legitimate operation mode change to de-risk the rover, hence the mission.

In real-world design of planetary robotic systems, their onboard segment operation modes can be more sophisticated, driven by practical considerations and may not simply map to the LoA one-on-one. For example in the MER mission, the rover operation modes are differentiated by its maximum traveling speed as shown in Table 6.2. The Blind Drive mode in lower LoA allows the rover to drive at a faster rate but subjects to higher risk, thus must be limited to low risk traverses; whereas the VisOdom model is comparably slow but has a lowest risk factor, which makes it suitable for unknown scenarios.

Table 6.1 ECSS LoA for space segment.

LoA	Description	ESA name	NASA name
E1	Mission execution under ground control; limited onboard capability for safety issues	Real-time control	Telecommand
E2	Execution of preplanned, ground-defined, mission operations on board	Preplanned	Blind Drive
E3	Execution of adaptive mission operations on board	Adaptive	Semiautonomous
E4	Execution of goal-oriented mission operations on board	Goal-oriented operation	Autonomous

Table 6.2 MER onboard operation modes.

Mode	Description	LoA	Speed (m/h)	Risks
Blind Drive	Directed commands that perform only reactive motion and safety checks	E2	124	Increased risk of entering dangerous terrain and larger risk of missing planned trajectory
Autonav	Obstacle avoidance using visual cues	E3	36	Decreased risk of entering dangerous terrain. Same risk of slip off plan path as Blind Drive
VisOdom (benign terrain)	Following visual odometry at all steps	E3	10	Lowest risk strategy, for both terrain and slippage

There exist other operation modes in design of planetary robots such as the MER, which are not necessarily based on LoA. These modes tend to relate to the safety operation of the mission, such as

- *Safety mode*: When a major fault is detected and reported, the onboard segment can set itself in safety mode to stall on planned operation except for housekeeping activities. Recovery from safe mode involves reestablishing communication between the space and ground segments, downloading any diagnostic data and sequencing power back onto the spacecraft subsystems to resume the mission. The recovery time can be anywhere from a few hours to days or even weeks depending on difficulties of reestablishing the communication links, conditions found on the spacecraft, distance to the spacecraft, and the nature of the mission, and so on.
- *Hibernation mode*: Just before a dust storm season, the MER enters in a low-power hibernation mode. In this mode, the rover clock keeps running but communications and other activities are suspended in order to keep the available energy into heating and battery recharging. When the battery is charged adequately, the rover attempts to wake up and communicate on a schedule it is given.

6.3
Mission Operation Software

The operation software applies to both the ground and onboard segment for any space mission but with potentially different sets of design criteria and priorities. The most obvious difference between the two partitions are resource limitations, specifically the onboard segment tends to have scarce resources, whereas the ground segment has relatively unlimited resources other than time.

6.3.1
Design Considerations

There are various design considerations for the mission operation software covering the ground and space segment as well as their respective subsystems. Priorities of these considerations may vary depending on the required LoA.

For **ground segment** of the operation software, the following considerations are foreseen:

- *Data*: Supporting efficient data flows through the ground station, and how/which data is stored and archived.
- *Engineering*: Being able to reconstruct the onboard actions, to diagnose onboard faults, to generate detailed terrain and environment maps, as well as to determine new plans for the onboard segment. The ease, speed, and accuracy are important design criteria.
- *Graphical user interface (GUI)*: Offering usable, effective human interfaces.
- *Human Readability*: Having mission plans and reasoning of the plans self-explanatory to human operators.
- *Human Responsibility*: Allowing human in the loop of any decision-making steps within the operations.
- *RF Link*: Supporting automatic and efficient uplink and downlink processes, and a wide range of RF bands and missions.
- *Scalability*: Allowing expansion of the hardware or software configuration to the operation without major redesign, and allowing operations of multiple missions and/or spacecraft simultaneously.
- *Scheduling*: Scheduling of the teams, individuals, meetings, and processes that can cope with events.
- *Science*: Abilities to identify opportunistic science, to analyze payload data, to identify new objectives, and so on. The ease, speed, reprehensibility and accuracy are important design criteria.
- *Standards*: Applying or complying with standards for greater interoperability, reusability, extensibility, and good engineering practices.
- *Validation & Verification*: Providing V&V (e.g., through simulations) of the ground segment operation processes, software, and generated plans.

The **onboard segment** of operation software has its own sets of considerations, which can include the following:

- *Adaptability*: Having planning capability that adapts to changing environments and capabilities.
- *Data*: Allowing data to be saved and archived correctly and efficiently.
- *Environment*: Detecting and using environmental conditions at run time and for safety of the onboard segment.
- *FDIR*: Allowing detection, diagnosis, and mitigation of all faults that are multi-component, novel, and intermittent.

- *Guidance, navigation, and control (GNC)*: Typically containing GNC functions for the robots such as abilities to localize, map, and/or move on unknown terrain accurately and efficiently.
- *Graceful Degradation*: Being able to operate at some low-level functionality when hardware components start to fail or degrade.
- *Operation Modes*: Ability to choose appropriate operation modes (e.g., safety mode).
- *Payload*: Abilities to carry out scientific experiments and analyzes data or results.
- *RF Link*: Supporting automatic and efficient downlink and uplink processes.
- *Resources Management*: Ability to be resource efficient.
- *Standards*: Applying or complying with standards for greater interoperability, reusability, extensibility, and good engineering practices.
- *Updatability* Abilities to update, patch, and/or upgrade the software at run time without risk to continuous use.
- *Validation & Verification*: Providing V&V of the space segment operation processes, software, and generated plans.

Many design considerations presented here will lead to features within the operation software that can offer different LoA based on given design criteria. Table 6.1 has introduced the LoA defined by ECSS. In real-world design of mission operations, the LoA definitions can be more complex within relevant features of the operation software such as Information Interpretation (II), GNC, Decision Making (DM), Payload (PL) and FDIR. The LoA for these subsystems is explained in Table 6.3.

The more detailed LoA definitions are reflected in practical designs of operation software. For example, Figure 6.5 shows the LoA achieved by a number of past, present, and future planetary rover missions including Lunokhod, Sojourner, MER, MSL, and ExoMars.

6.3.2
Ground Operation Software

The ground operation software is designed to manage and support the downlink and uplink processes as outlined in Section 6.2.2, which requires the operation software to take care of data flow, basic housekeeping, archiving, managing communications windows, managing, and scheduling personnel. In addition, the software must aid mission specific tasks such as robot path reconstruction, engineering plan creation, science planning, plan merging, validation, and simulation [9].

An iconic example of the ground operation software is the **ESA's Ground Operation System (EGOS)** developed by the European Space Operations Centre (ESOC) to harmonize all the software developed and maintained within the ground segment [10]. It becomes a common platform that hosts various ESA ground software such as

6.3 Mission Operation Software | 333

Table 6.3 LoA for practical features within operation software.

Subsystem	LoA	Description
II	1	Automatic information processing
	2	Automatic model building
	3	Automatic model parametric adjustment
	4	Automatic model structural adjustment
DM	1	Following of plans and schedules of human operators
	2	Planning and scheduling of low-level actions based on high-level plans produced by human operators
	3	Adaptive decision making on low-level actions against changing information
	4	Autonomous planning and scheduling of all actions
FDIR	1	No FDIR, only hardware driven safety switches
	2	Component-based
	3	Subsystem-based
	4	System reconfiguration based on faults
GNC	1	Full teleoperation by ground control
	2	Following preplanned path
	3	Adaptive way-point following
	4	Goal-oriented autonomy for GNC
PL	1	Complete teleoperation of the payload
	2	Following a payload plan
	3	Adaptive payload plan following
	4	Goal-oriented autonomy or opportunistic science for the payload

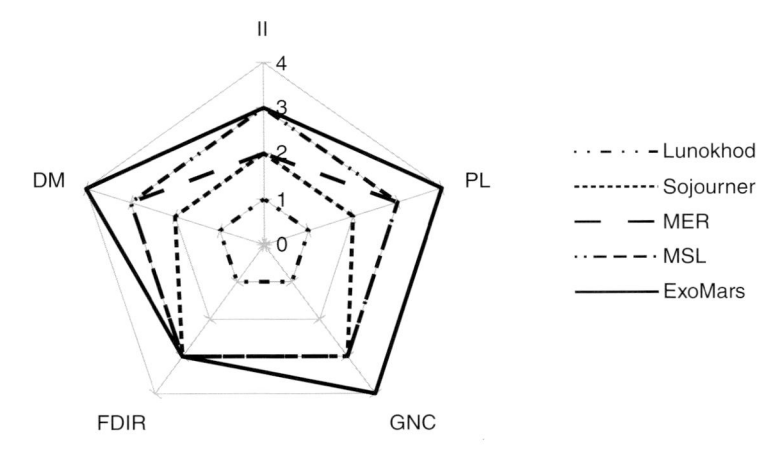

Figure 6.5 LoA of existing rover missions.

- MICONYS: Mission control system (MCS) infrastructure, including SCOS-2000 and its ancillary applications [11].
- SIMULUS: Simulation system infrastructure, including a satellite simulator, a set of models to simulate different aspects of the satellite environment, and the ground station equipment [12].
- TEVALIS: Ground test and validation system used to simulate and test the whole chain of equipment from the MCS to the spacecraft.
- NAPEOS: A portable navigation software for Earth-orbiting satellite missions [13].
- DABYS: Generic database software, an EGOS specialized framework for building database management applications [14].
- EUD: EGOS user desktop, providing a generic user interface that can be used by all EGOS applications [10].

The SCOS-2000 that stands for Satellite Control and Operation System 2000 is a generic MCS software infrastructure. A MCS provides the means for the human operators to monitor and control one or more spacecraft. It provides commanding chain, telemetry chain, and archive. The commanding chain is based on the CCSDS Frame standard. Command stacks can be loaded either manually or automatically, and are validated against predefined constraints prior to release. The telemetry chain is based on the CCSDS Frame and Packet standards. CCSDS frames are received from the ground station through a Space Link Extension interface and demultiplexed into telemetry packets. Typical functions performed upon receipt of a telemetry packet include checking whether the parameters are within range (hard and soft limit checking) and validation of sent telecommands. Finally, the software provides automated packet and parameter archives, capable of managing high data volumes required by modern space missions. The MCS does not provide science data processing, which is typically performed at a dedicated science center of the mission. The ground operation system developed for planetary robotic missions generally follow a similar schema, for example, the Rover Operation Control Centre (ROCC) for ESA's upcoming ExoMars mission [7].

Another example of integrating operation and planning tools within the ground operation software is the **NASA's Ensemble** [15] used on extended operations of three Mars missions including Phoenix, MER, and MSL. Ensemble is hence a multi-mission toolkit for building activity planning and sequencing systems. It is responsible for strategic mission planning, science observation planning, engineering activity planning, and sequence command uplink, and therefore can support both user and automated reasoning operations needed to modify activity plans. Ensemble uses a vast array of software tools during various mission phases, such as

- APGEN: part of Mixed Initiative Activity Plan GENerator (MAPGEN) with three core capabilities/components:
  - Activity plan database: Database with set of activities, each at a specific time. This database has no notion of constraints between activities, but does support context-free activity expansion.

– Resource calculations: A method for calculating, using forward simulation, resource states that range from simple Boolean states to complex numerical resources.
– Graphical user interface: An interface for viewing and editing plans and activities.
- EUROPA: An artificial intelligence (AI)-based planner (see Sections 6.4.3 and 6.4.4 for details), also part of MAPGEN.
- Constraint Editor: APGEN has no notion of variables and constraints. The Constraint Editor was developed as an enhancement of APGEN interface to deal with daily constraints to coordinate scientific observations.
- SAP: Science Activity Planner used by the science teams to facilitate science telemetry analysis and for science activity planning.
- RSVP: Rover Sequencing and Visualization Program used for advanced Martian surface and rover visualizations and to generate the actual rover command sequences.
- SEQGEN: Sequence Generator used to predict events that will occur on the spacecraft as a result of the sequence, giving warnings when the sequence violates rules or causes spacecraft subsystems to be misused.
- SLINC: Spacecraft Language Interpreter and Collector is responsible for the translation of a Spacecraft Sequence in the form of a Spacecraft Sequence File (SSF) into a Command Packet File (CPF) for radiation to the spacecraft. In addition, a binary UNIX file may be formatted into a CPF for transmission to the spacecraft.
- RSFOS: Reengineered Space Flight Operations Schedule is a program that reads user maintainable ASCII tables and an input Predicted Events File (PEF) generated by SEQGEN to generate two output files.
- CAST: Common Allocation Scheduling Tool is an integrated set of multimission software tools that assist projects negotiate Deep Space Network (DSN) coverage.
- SEQREVIEW: Sequence Review Tool used to reformat and extract information from sequence products (e.g., PEF), which allows users or external applications to analyze the data.

While there exist many designs of ground operation software by space agencies or industrial companies, the basic software architecture of these designs are understandably rather common to fulfill standard operation procedures (see Figure 6.6). The key building blocks within the architecture are further explained as follows:

- The **network interface** module manages the direct communications with the onboard segment such as packet management and verification.
- The **monitoring and control** module manages housekeeping, communication preparations, and scheduling. One example is ESA's SCOS 2000.
- The **database manager** module manages the archiving of all the information within the architecture including the incoming and outgoing communications packets.

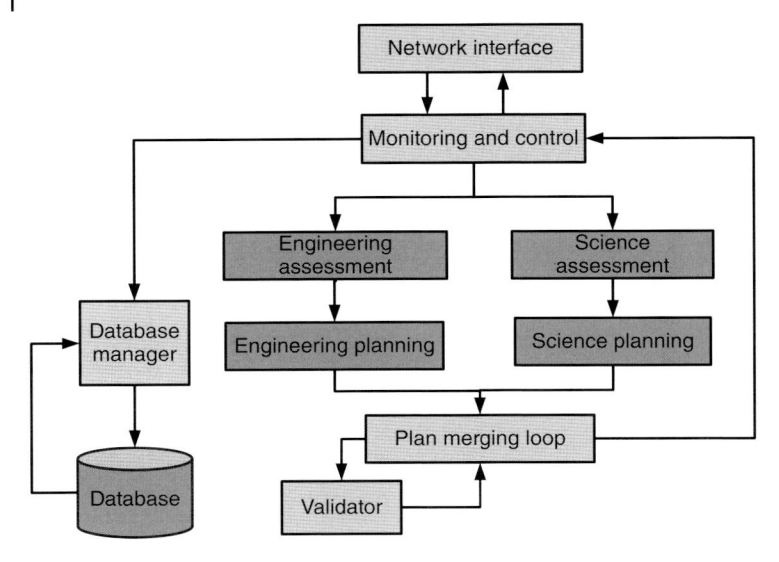

Figure 6.6 Block diagram of basic ground operation software architecture.

- The **engineering assessment** module manages analysis of incoming engineering data such as the engineering model and robot path reconstruction. Both of these tasks can be handled autonomously or manually with a user interface. The path reconstruction, in particular, has the potential to be fully autonomous.
- The **engineering planning** module manages robot path planning and engineering subsystem planning. This module also includes validation and parametric simulation functions. The path planning, validation, and simulation can be performed autonomously with minimal human supervision.
- The **science assessment** module aids the assessment procedure for the science team. It is mission and payload specific.
- The **science planning** module aids the science planning and plan consolidation for the science team, hence is mission specific. The required LoA is subjective to the science goals. One example is NASA's SAP [16] used within MER and MSL rover missions as the primary science data downlink analysis and uplink planning tool.
- The **plan merging loop** module manages planning and scheduling for completing all goals set by the science and engineering team. The module contains simulators and user interfaces to be used and reviewed by the Integrated Planning Team. The validation, simulation, planning, and scheduling functions can be potentially carried out autonomously.

6.3.3
Onboard Operation Software

Onboard operation software for the planetary robots is used to manage the spacecraft (such as power, payload, and safety) and achieve mission plans created by

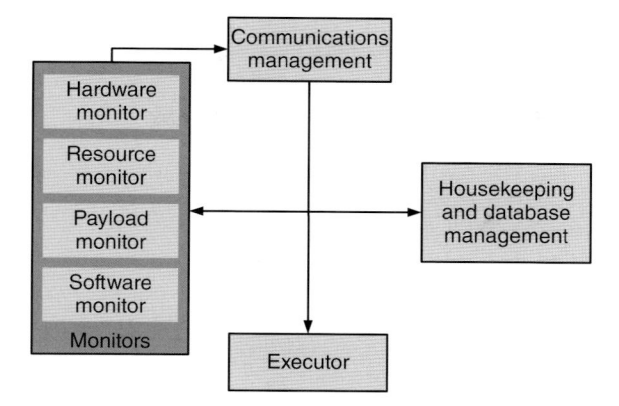

Figure 6.7 Basic architecture of onboard operation software with low LoA.

the ground segment. More advanced functions can be required by the onboard software as the required LoA increases.

A basic configuration or architecture of onboard operation software with relatively low LoA (e.g., up to E3) is illustrated in Figure 6.7 and envisaged to contain the following building blocks:

- The **communication management** module automates communication links.
- The **housekeeping and database management** module automates housekeeping and database management onboard the robots.
- The **monitor** module watches for every subsystem and can move the robot to safety mode if an anomaly is detected.
- The **executor** module executes the plan generated by the ground without much decision autonomously made onboard.

This basic architecture is seen in many existing onboard software of planetary robots, rovers in particular. Examples include the MER software architecture described in Refs [17, 18], the CREST software architecture for an ExoMars-like rover [19], and even the latest MSL rover except that it has modular executors [20] that may operate at a different LoA or adaptability though still are limited to E3.

To accommodate higher LoA such as E4, the software architecture requires advanced modules such as

- The **planner and scheduler** module receives high-level mission goals determined by the robot integrated planning team in the ground segment, and autonomously plans tasks and schedules actions to complete the goals. It can include replanning based on external stimuli as illustrated by the data flow loop in Figure 6.8, unlike the low LoA version in Figure 6.7. Techniques involved in this module is discussed in depth in Section 6.4.
- The **validator & simulator** module is responsible of validating safety and the proposed plan, and closing the plan loop.

338 | *6 Mission Operations and Autonomy*

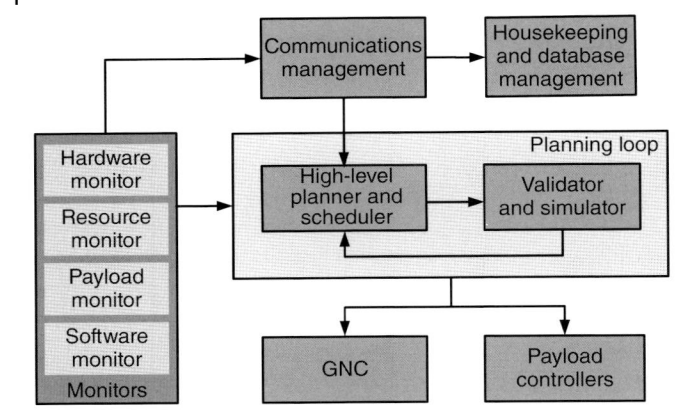

Figure 6.8 Basic architecture of onboard operation software with high LoA.

- The **GNC** module autonomously controls the navigation, localization, and locomotion subsystems of the robot.
- The **payload controller** module can autonomously operate the payload and manage its data outputs.

Figures 6.7 and 6.8 have both presented the onboard autonomy in terms of the software functional building blocks. In general robotics, such autonomous software is often deemed to follow a three-layer control architecture, which connects "sense," "plan," and "act" as shown in Figure 6.9. A brief description of the three layers are given as follows (from low to high level of hierarchy):

- The **functional layer** includes all the basic, built-in robot sensing, perception, and actuation capacities.
- The **executive layer** controls and coordinates the execution of the functions distributed in the software according to the task requirements.

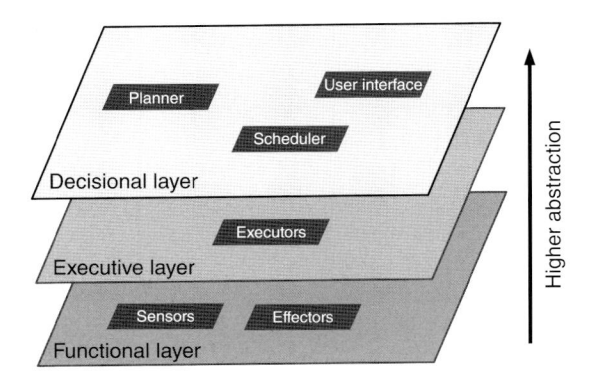

Figure 6.9 Three-layer control architecture in robotics.

- The **decision layer** includes the capacities to produce the task plan and supervise the execution, and potentially react to events from the previous level at the same time.

The three-layer architecture allows higher level of abstraction to be gained at the higher level of the hierarchy. It also helps archiving greater modularity and deliberative/reactive decision making to be integrated at different levels of abstraction. The onboard operation software with LoA E1–E3 typically includes the functional and executive layer, while leaves the decision layer in the ground segment. LoA E4 onboard software contains all three layers.

Continuous efforts have been made to employ the three-layer architecture or achieve LoA E4 within planetary robotic systems. One example being a three-tiered software architecture designed for planetary rovers by LAAS-CNRS [21]. This design uses a distributed set of nonhomogeneous processes to perform real-time event-based or task-based control loops, with three levels of abstraction. The first layer embeds the various functions that endow the rover with abilities to navigate autonomously. This layer is developed in Generator of Modules (GenoM) [22], a software development framework that allows the definition and the production of modules that encapsulate algorithms. A module is a standardized software entity that is able to offer services that are provided by a set of algorithms. Modules can start or stop the execution of these services, pass arguments to the algorithms, and export the data produced. The GenoM provides a description language and standard templates. The templates allow the developer to describe a module, the services it can offer, and for each service the list of expected parameters and algorithms that will be executed, the results along with their description, the failure messages, and so on. The next, execution layer is primarily the R^2C, which performs state checks on the functional layer as to complete a scheduled plan. It also checks for fault states [23]. The final layer is the decision layer, which is broken into OpenPRS [24] and Indexed Time Table (IxTeT) [25], where the IxTeT is a temporal planner that creates a plan for the Open Procedural Reasoning System (OpenPRS) to refine and execute, and the OpenPRS is an execution-only system and composed of a set of tools to represent and execute procedures including

- a *database* to represent the world;
- a *library of procedures* to describe a particular sequence of actions and tests that may be performed to achieve given goals or to react to certain situations. Each procedure is self-contained and describes in which conditions it is applicable and the goals it achieves;
- a *task graph* that corresponds to a dynamic set of tasks currently executing. Tasks are dynamic structures which track the state of execution of the intended procedures and the state of their posted subgoals.

NASA has developed and launched onboard autonomous software for satellite systems such as the Remote Agent Experiment (RAX) in Deep Space 1 (DS-1) mission [26] and Autonomous Sciencecraft Experiment (ASE) in Earth Observing 1 (EO-1) mission [27]. The ASE uses a three-layer equivalent architecture and

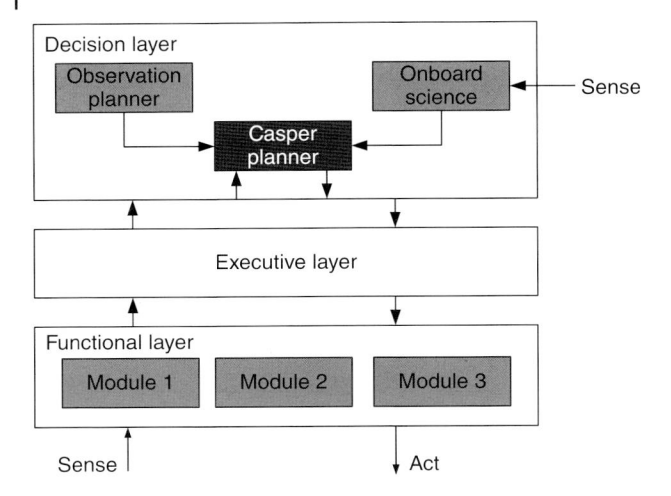

Figure 6.10 ASE's software architecture [27].

with a planner called Continuous Activity Scheduling Planning Execution and Replanning (CASPER) in the decision layer. The executive layer of ASE accepts the CASPER-derived plan as its input and expands the plan into low-level commands for the functional layer as shown in Figure 6.10. The ASE has three major software components, namely onboard science algorithms, execution software spacecraft command language (SCL), and onboard planning and scheduling software CASPER. The science algorithms can analyze the image data to detect the interesting features, which helps the satellite to search the valuable science data to reduce data volumes on board and retargeting at the interesting regions by itself. This also helps the spacecraft to capture the short-lived science phenomena for increasing the opportunistic science back. The model-based goal-oriented onboard planner CASPER takes the goals from science algorithms onboard or operators on the ground as input, and outputs the schedule of activities to execution system SCL. The robust execution system SCL accepts the CASPER-derived plan as an input and expands the plan into low-level commands.

ESA has developed onboard software for planetary rovers called Goal-Oriented Autonomous Controller (GOAC) that can achieve LoA E4 [28]. GOAC follows a divide-and-conquer approach to complexity, by splitting the deliberation problem into subproblems, thus making it more scalable and efficient. Moreover, planning and execution are intertwined. GOAC has a hybrid architecture that can be mapped with the three-layer architecture (see Figure 6.11). It consists of a set of reactors (such as deliberative reactors and a command-dispatcher reactor) and a functional layer. Each deliberative reactor uses a planner based on APSI. The APSI planner has capabilities of dynamic replanning and step-wise deliberation. The functional layer of GOAC is based on GenoM and BIP [29]. The basic self-contained design unit in GenoM is a module. Each module encapsulates a function

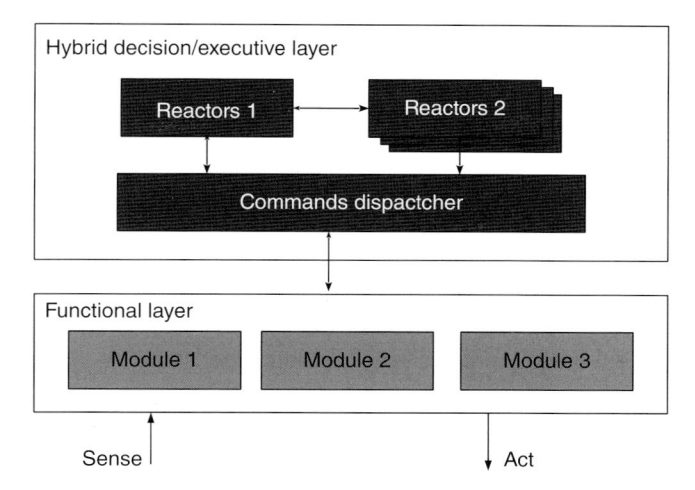

Figure 6.11 GOAC's software architecture.

of the rover. The BIP framework provides a methodology for building real-time systems consisting of heterogeneous components. It is used in order to reduce a posterior validation as much as possible by putting focus on the following challenges: composition of components, correctness-by-construction, and automated component integration. Details of APSI and GOAC are presented in Sections 6.4.4 and 6.7, respectively.

6.3.4
Performance Measures

As aforementioned in Section 6.2.2, the mission operation at the ground or space segment needs to work with priorities and interests of different user groups that potentially have conflicts. Performance measures can be used to identify design and implementation issues in such complex scenarios and maximize utility of the hardware, software, and user-interface of the space system. One measure to be considered is temporal performance given time is a critical consideration for any space mission. The temporal performance in the ground segment is not as critical as it is in the onboard segment. When comes to autonomy, the former can use offline planners with little resource constraints and no reactive behaviors, while the latter requires online planners that must react fast with scarce resources. For ground-based planner, additional performance measures can include interaction with the user where the user interface must clear and precise, and the time available for operators to generate a new plan so that a plan can be generated within a fixed time window even if it is suboptimal.

A metric containing various performance measures can be written as follows:

$$\begin{pmatrix} \text{Mission} \\ \text{Metrics} \end{pmatrix} + \begin{pmatrix} \text{GNC} \\ \text{Metrics} \end{pmatrix} + \begin{pmatrix} \text{Payload} \\ \text{Metrics} \end{pmatrix} + \begin{pmatrix} \text{Safety} \\ \text{Metrics} \end{pmatrix} \qquad (6.1)$$

where each measure can be described by various submeasures that are independent from each other.

For example:

$$\begin{pmatrix} \text{Mission} \\ \text{Metrics} \end{pmatrix} = \frac{\alpha \sum \left[p_{\text{lifetime}} \times \begin{pmatrix} \text{Lifetime} \\ \text{length} \end{pmatrix} \right]}{\begin{pmatrix} \text{Development} \\ \text{Cost} \end{pmatrix} + \begin{pmatrix} \text{Operation} \\ \text{Cost} \end{pmatrix} + \begin{pmatrix} \text{Flight} \\ \text{Cost} \end{pmatrix} + \begin{pmatrix} \text{Hardware} \\ \text{Cost} \end{pmatrix} + \begin{pmatrix} \text{Software} \\ \text{Cost} \end{pmatrix}} \tag{6.2}$$

where $p_{\text{(lifetime)}}$ is the probability of the robot only surviving until the named lifetime, and α is a constant.

$$\begin{pmatrix} \text{GNC} \\ \text{Metrics} \end{pmatrix} = \sum^{\text{biomes}} \left[\frac{\alpha(\text{Velocity}) + \beta(\text{Safety}) + \gamma \begin{pmatrix} \text{Localization} \\ \text{Accuracy} \end{pmatrix}}{\delta(\text{Energy per meter})} \times \begin{pmatrix} \text{Expected percentage} \\ \text{of time in region} \end{pmatrix} \right] + \begin{pmatrix} \text{Mapping} \\ \text{Ability} \end{pmatrix} \tag{6.3}$$

where α, β, γ, and δ are constants. This equation does not take into account different operation modes. However, this can be easily extended by adding every mode multiplied with the probability of being in that mode. The mapping ability can be essentially granularity and information in each region contained by the computational resources.

$$\begin{pmatrix} \text{Payload} \\ \text{Metrics} \end{pmatrix} = \sum \frac{p_{\text{usable results}} \times \begin{pmatrix} \text{Scientific} \\ \text{Importance} \end{pmatrix}}{\alpha(\text{Weight}) + \beta \begin{pmatrix} \text{Energy} \\ \text{Usage} \end{pmatrix}} \tag{6.4}$$

where $p_{\text{usable results}}$ is the probability of usable results and can be a function of the GNC, $\begin{pmatrix} \text{Scientific} \\ \text{Importance} \end{pmatrix}$ is a value that indicates the scientific importance of the results and can be determined by the scientific team based on their desired goals, experiments and findings, and α and β are constants.

$$\frac{1}{(\text{Safety Metrics})} = \sum \left[p_{\text{fault}} \times \begin{pmatrix} \text{Functionality} \\ \text{after Fault} \end{pmatrix} \right] + \begin{pmatrix} \text{Percentage} \\ \text{Un-V\&V Software} \end{pmatrix} \tag{6.5}$$

where V&V software include simulation based V&V, and the $\begin{pmatrix} \text{Functionality} \\ \text{after Fault} \end{pmatrix}$ is a percentage of the original payload activities and GNC abilities after the fault.

The performance metrics can be designed based on different needs and priorities by including different measures and considering the mission as a whole analytically. Such approach allows users to determine important results and issues with the mission operation. The constants in different metrics are tunable based on mission objectives and users. Scientific and industrial users can prefer different constant values, for example, mission lifetime is likely more important to scientists than engineers compared with mission cost.

6.4 Planning and Scheduling (P&S)

6.4.1 P&S Software Design Considerations

As indicated in Section 6.3, planning and scheduling (P&S) are key functions within the operation software to achieve high LoA, such as for onboard goal-oriented autonomy or for ground segment planning and management. Design considerations of a planner and scheduler for planetary robotic missions are summarized as follows:

- *Computational Efficiency*: This is mainly a consideration for the onboard segment that the planner and scheduler should be computationally efficient.
- *GUI*: A usable human interface is important.
- *Human Readable*: It is preferable that both the plan and the reason(s) for choosing the plan are human readable allowing human operators have faith in the planner and scheduler.
- *Multiple Plan Outputs*: It is preferable that more than one plan are generated so that either a rational software agent or the integrated robot planning team can choose the best one.
- *Multidimensional Numerical Properties*: There are many numerical properties mission planning needs to consider such as spatial dimensions, temperatures, power, and energy hence a planner needs to deal with multidimensional numerical problems efficiently.
- *Optimization*: Optimization is a main goal of planning based on number or order of actions used, and resources required, and so on.
- *Plan Reuse*: Many subplans can apply to different scenarios hence reusable. It is efficient if these subplans could be stored in database for reactivation.
- *Language*: The language in which the planning problem and domain are defined should allow the planner to be modular, reusable, human-readable, concise, and memory efficient.
- *Temporal Planning*: Planning against timeline is helpful to mission operations also given the possibility of temporally overlapping actions undertaking concurrently.
- *Uncertainties*: Planning needs to cope with uncertainties of the real-world environments, such as applying fuzziness to deal and plan around probabilistic, conditional events.

6.4.2 Basic Principles & Techniques

Given descriptions of the possible *initial states* of the world or domain, the desired *goals*, and a set of possible *actions*, the planning problem is to find a plan that is guaranteed (from any of the initial states) to generate a sequence of actions that leads to one of the goal states. Effectively, a P&S system takes models of the *domains* and *problems* as inputs and solves planning and scheduling to produce a plan. There are many methods that can generate plans. These methods make different assumptions and compromises to generate a plan as efficiently as possible. The remainder of the section presents various P&S principles and techniques in relation to planetary robotics. The taxonomy presented in Ref. [30] to classify different planning techniques is widely adopted and, hence, used in this section.

6.4.2.1 Classical Methods

Classical planning methods are restricted to a state transition system and serve a baseline for studying P&S. They plan based on a search space, represented by a tree or graph of nodes connected by arcs between the nodes. Commonly known classical techniques are **state-space** and **plan-space** planning. In the state-space planning, the search space containing all possible states of the problem where each node represents a state and the arc represents transition between states. An initial state can lead to a number of other states based on all applicable actions. One of these states will become the current state according to the search that continues until the goal state is achieved. The search can be forward or backward chaining. Forward chaining starts with the available data and uses inference rules to extract more data (e.g., from an end user) until the goal is reached. Backward chaining is the reverse of forward chaining, which uses an inference method working from the goal state. There are many ways to design the inference or search algorithm within the forward or backward chaining, among which the simplest is *heuristic*. A heuristic technique is often used to resolve a planning problem quickly when non-heuristic methods are too slow, or to find an approximated solution when other methods fail to find any exact solution. This is achieved by trading optimality, completeness, accuracy, or precision for speed [31]. In the plan-space planning (PSP), nodes of the search space are partially defined plans and arcs are plan refinements intended to complete a partial plan. Planning starts from an initial node corresponding to an empty plan and searches for a solution by either choosing an action (as in state-space planning) or ordering the actions. Using this planning method, a plan is defined as a set of planning operators together with ordering constraints.

One mandatory input to any planning algorithm is a description of the problem to be solved. For classical methods, the planning domain of any problem is prohibitively large for explicit enumeration of all states and transitions. Three representations can be chosen to describe a classical planning problem as shown in Table 6.4. There are various ways to extend on these basic representations. Examples include the use of logical axioms to infer things about states of the world and the use of more general logical formulas to describe the preconditions and effects of an action. As a result, the extended planning problem representations can be used to describe or used by more sophisticated planning methods.

6.4.2.2 Neoclassical Methods

Another major category of planning is neoclassical techniques, which are also restricted to a state transition system. The main difference to the classical methods

Table 6.4 Three representations of a classical planning problem.

Representation	Description
Set-theoretic	Each state of the world is a set of propositions, and each action is a syntactic expression specifying which propositions belong to the state in order for the action to be applicable and which propositions the action will add or remove in order to make a new state of the world
Classical	The states and actions are such as the ones described for set-theoretic representations except that first-order literals and logical connectives are used instead of propositions. This is the most popular choice for restricted state-transition systems, and is used for many generic planning languages such as PDDL (refer to Section 6.4.2.7 on planning languages)
State-variable	Each state is represented by a tuple of values of n state variables $X_1, \cdots, X_n$, and each action is represented by a partial function that maps this tuple into some other tuple of values of then state variables. This approach is especially useful for representing domains in which a state is a set of attributes that range over finite domains and whose values change over time

is that every node of the search space in the neoclassical methods is a set of several partial plans. This category contains three common techniques:

- The **planning-graph** techniques are based on a reachability structure for a planning problem that is used to efficiently organize and constrain the search space. These techniques output a sequence of sets of actions, for example, $\langle a_1, a_2, a_3, a_4, a_5 \rangle$, which represent all sequences starting a_1 in any order, followed by a_3, a_4, and a_5 in any order. The planning-graph approach considers that actions are fully instantiated and at specific steps, relying on reachability analysis and disjunctive refinement. The reachability analysis addresses the issue of whether a state is reachable from a given state. The disjunctive refinement consists of addressing one of several flaws using a disjunctive resolver.
- The **propositional satisfiability** techniques [32] encode a planning problem into a Boolean Satisfiability (SAT) Problem and then rely on efficient SAT procedures for finding a solution. In other words, "planning as satisfiability" is to map a planning problem to a well-known problem for which there is a well-known solution and effective solving algorithm. The approach follows the below-mentioned formula:
  1. A planning problem is encoded as a propositional formula.
  2. A satisfiability decision procedure determines whether the formula is satisfiable by assigning truth values to the propositional values.
  3. A plan is extracted from the assignments determined by the satisfiability decision procedure.
- The **constraint satisfaction** techniques encode the planning problem to a constraint satisfaction problem (CSP) [33]. For a set of given variables, their

domains and a set of constraints on the values, the general formulation of a CSP is to find a value for each variable that meets the constraints.

6.4.2.3 Solving Strategies

All of the planning methods described previously can be solved with *abstract-search* procedures. The objectives of an abstract search is to find at least one solution, without enumerating the entire planning space. The procedure non-deterministically searches a space in which each node u represents a set of solution plans Π_u, namely the set of all goal states reachable from u. A node u is a structured set of possible actions and constraints. Different planning methods employ the search space as follows:

- In **state-space planning**, u is a sequence of actions. Every solution from u is reachable with a prefix or suffix of actions depending on whether forward or backward search is used, respectively.
- In **plan-space planning** (PSP), u is a set of actions, causal links, ordering constraints, and binding constraints. Every solution reachable from u contains all actions in u and meets all the constraints.
- In **planning-graph algorithms**, u is a subplanning-graph of the entire, meaning a subset of actions and constraints on precondition, effects, and mutual exclusions. Solutions reachable from u contain the actions corresponding to at least one from each level of the subplanning-graph.
- In **SAT-based planning**, u is a set of assigned literals and remaining clauses, each of which is a disjunction of literals that describes actions and states. The solutions reachable from u are the setting of truth values for unassigned literals so that all remaining clauses are satisfied.
- In **CSP planning**, u is a set of CSP variables and constraints, where some values are already assigned to an initial state. Each solution from u includes assignments to all CSP values so as to satisfy the constraints.

The abstract-search procedure includes four main steps:

1) *Refinement* modifies the collection of actions and constraints. For example, the constraint can be removed and the action can be made explicit if only one action meets a constraint.
2) *Branching* generates children v of u. These nodes will be candidates for the next node to visit. For example, in a forward state-space search, each child node corresponds to appending a different action to the end of a partial plan. Not all children are generated.
3) *Pruning* removes nodes that appear to be unpromising for the search. This step is generally domain specific.
4) *Termination* stops the procedure.

The first three steps can be in different orders for different planning methods, for example, a procedure might do branching, then pruning, then refinement. To address the complexity of a planning problem, the abstract-search procedure can be extended using *heuristics* to select the children v, which offers the highest likelihood of reaching a solution. Heuristics can either be domain independent or domain specific. Domain-independent heuristics are techniques intended for use in many different planning domains similar to those heuristic search algorithms for node selection, for example, breadth-first, depth-first, best-first, hill climbing, and A*. Domain-specific heuristics are tailored for a particular kind of domain, for example, using temporal logic to write node-pruning rules that focus the search. Although it takes a considerable amount of effort to tune a heuristic for a problem, the result can lead to much improved planning performance.

Another method to increase planning efficiency is HTN. The HTN method is similar to the classical methods but in which the dependency among actions can be given in the form of task networks. Planning problems are specified by a set of tasks, which can be either the primitive tasks that roughly correspond to actions, or the compound tasks that can be seen as a set of simpler tasks, or the goal tasks that roughly correspond to goals.

6.4.2.4 Temporal Planning

All the planning methods described here are based on the state transition system with the assumption of implicit time. In these models of the planning problem, the dynamics are represented as the sequence of actions where actions and events have instantaneous state changes. But in reality, actions, state changes, and constraints happen over time spans, which may also overlap. Actions take time and as a consequence there is a notion of concurrency. When actions are taken during the same interval of time, they may result in combined effects different from the sum of the effects of the single actions. Goals are often constrained by temporal considerations and more complex than simple state achievements, involving conditions that might have to be achieved not only at the end of the plan or that have to maintained across the plan. An explicit representation of time can significantly extend the expressiveness of a planner, which requires an extension to the representation of the planning domain theory, problem, and solving algorithms. To achieve that, extension to the state transition systems can be pursued in one of following ways:

1) Extending the global state of the world to include time explicitly in the representation of the state transition. The model then views the world as discrete snapshots of the world. This requires the use of logical atoms that extend the usual planning operators to include temporal conditions on those atoms.
2) By replacing the states with a set of functions of time describing parallel evolution. In this model, the dynamics are a collection of partial functions of time describing the local changes of state variables.

Temporality in the planner can be expressed using calculi based on first-order temporal logic or other hybrids. For example, the *point algebra* [34], a symbolic calculus that relates in time a set of instants with qualitative constraints without necessarily ordering them, or the *interval algebra* [35], a symbolic calculus that relates in time a set of intervals with qualitative constraints. More recent methods have added quantitative temporal constraints based on simple temporal problems representations [36].

Temporal planning defines its problem as a set of constraints (conjunctive or disjunctive) over a set of time points or time intervals assigned to actions. The planner typically uses *temporal operators* or *chronicles*, and can be solved by applying generic or specific CSP techniques. The planning solution is represented in a sequence of states containing actions/behaviors of the plan.

Planning with temporal operators extends classical and neoclassical methods by qualifying every proposition within the time period for which it remains true. This can be achieved using temporal operators or *temporally qualified expressions* (*tqe*) in a *temporal database* [30]. A *tqe* is defined as $e(x_1, \cdots, x_k)@[t_s, t_e)$, where e is a temporally flexible relation, x_is are constants or object variables, t_s and t_e are temporal variables such that $t_s < t_e$. The *tqe* asserts that $e(x_1, \cdots, x_k)$ holds true at time t, $\forall t \in [t_s, t_e)$. A temporal database is defined as a set of *tqe*s and a finite set temporal and object constraints, similar to that in a CSP problem. A temporal planning operator is a tuple, $o = \langle \text{name}(o), \text{precon}(o), \text{effect}(o), \text{const}(o) \rangle$, where

- name(*o*) is an expression of the form $o(x_1, \cdots, x_k, t_s, t_e)$ such that *o* is an operator symbol, and $x_1, \cdots, x_k$ are all object variables.
- precon(*o*) and effect(*o*) are *tqe*s.
- const(*o*) is a set of temporal and object constraints.

In an example of planning traverse of a planetary rover (modified from examples presented in Ref. [30]), a few *tqe*s can be specified as $move(r, l, l')$ for rover in motion where r denotes the rover concerned, l and l' denote start and end location of the rover, respectively, $at(r, l)$ denotes rover r in still at location l, and $free(l)$ denotes a location l free from any rover. The temporal operator used by planning can be specified as follows:

name	$move(r, l, l')@[t_s, t_e)$
precond:	$at(r, l)@[t_1, t_s)$
	$free(l')@[t_2, t_3)$
effects:	$at(r, routes)@[t_s, t_e)$
	$at(r, l')@[t_e, t_4)$
	$free(l)@[t_5, t_6)$
consts:	$t_s \leq t_5 < t_3 \leq t_e$
	$adjacent(l, l')$

The temporal database of this example can be captured graphically in Figure 6.12.

Planning with chronicles differs from temporal operators mainly by the use of state-variable representation, which can be more expressive and concise. A simplistic example limited to point algebra is given as follows. A *temporal assertion* on a state variable x is either an event or a persistent condition. A *chronicle* is then defined as a set of state variable $x_1, \cdots, x_j$ in a tuple $\Phi = (F, C)$, where F is a set of temporal assertions about x_i, and C is a set of temporal and object constraints. A *timeline* is a chronicle for a single state variable x. For the planetary rover

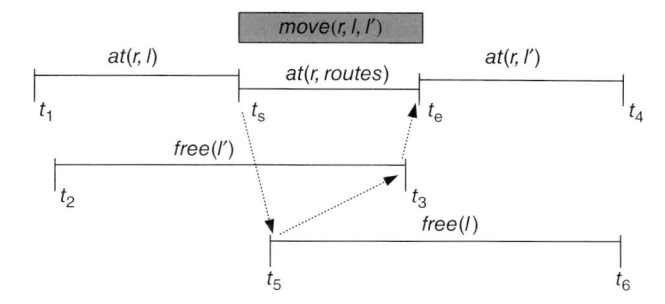

Figure 6.12 An illustration of a temporal database.

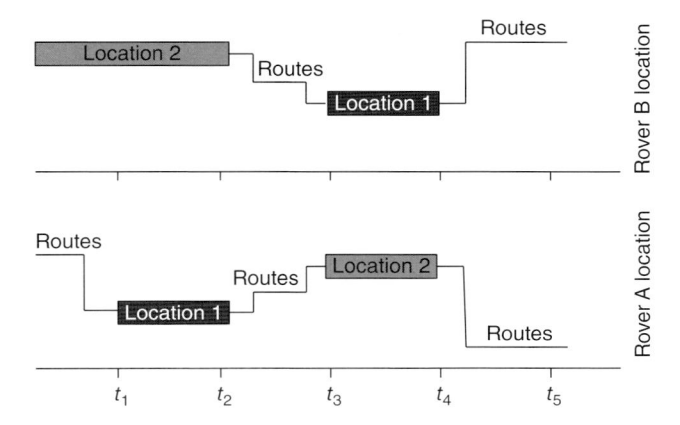

Figure 6.13 An illustration of two timelines for planning with a chronicle.

scenario, each rover would have a timeline, and each location would have a timeline [37] as shown in Figure 6.13. Chronicles can then be formulated as planning operators. A chronicle planning operator o on a set of state variables $X = x_1, \cdots, x_n$ is $o = (name(o), (F(o), C(o)))$, where:

- $name(o)$ is an expression of the form $o(t_s, t_e, t_1, \cdots, v_1, v_2, \cdots)$, where o is an operator, and $t_s, \cdots, v_1, \cdots$ are all the temporal and object variables in o.
- $(F(o), C(o))$ is a chronicle on the state variable X.

This approach is often referred as the constraint-based interval planning or timeline-based planning, and follows the basic idea of using state variables and concurrent threads of time-tagged system properties (i.e., the timelines). It has the major advantage of allowing a seamless integration of P&S techniques, given the fact timelines can represent system states as well as resource profiles.

As seen in the later Section 6.4.2.7, modern planning languages have extended from the classical planning to accommodate temporal planning. For example, the PDDL (from version 2.1) includes durative actions, temporal preconditions, and continuous effects where the domain theories in timeline-based modeling are mostly represented by means of temporal and logical synchronizations among the values taken by the timelines; languages for timeline-based planning have constructs as the *synchronization* in DDL.3 or the *compatibility* in New Domain Definition Language (NDDL) [38] to represent the interaction among the different timelines that model the domain. Conceptually these constructs define valid schema of time intervals and values allowed on timelines. Despite the syntactic differences, they allow the definition of Allen's like quantitative temporal relations mentioned before as well as constraints on the parameters of the related values.

Latest work or development in temporal planning has introduced the concept of HTN and resulted in Hierarchical TimeLine Network (HTLN) that improves the expressiveness of the planner [39].

6.4.2.5 Scheduling

A scheduling problem is how to perform a given set of actions using a limited number of resources in a given time window [30]. A resource is a constraint that can be occupied (e.g., camera in use) or applied (e.g., power). Planning and scheduling are intertwined where the planning focuses on the causal reasoning to determine the correct set of actions to complete a goal and the scheduling focuses on time and resource allocation for the planned set of actions.

A scheduling problem concerns about resources and their temporal availability, actions that need to performed and their resource requirements, constraints on actions and resources, and a cost function. A solution to the scheduling problem, namely a schedule, is an allocation of resources and starting time to the actions that meet all requirements and constraints. An optimal schedule is one that optimizes the cost function.

Figure 6.14 shows a simplistic view of relations between the P&S, which are assumed separated activities. In modern P&S, it is feasible and beneficial to integrate the two, for example, in temporal planning with either temporal operators or chronicles. In both cases, the flaw repair scheme from CSP is a method for planning where resource flaws are detected and resolved.

6.4.2.6 Handling Uncertainties

The planning methods aforementioned rely on three axioms:

- *Determinism:* All actions have deterministic effects.
- *Full Observability:* The planner has full knowledge of the problem domain.
- *Reachability Goals:* Goals can be represented as discrete states.

In the practical world of space missions, none of the above-mentioned assumptions can really be true. For example, *nondeterminism* is a common assumption

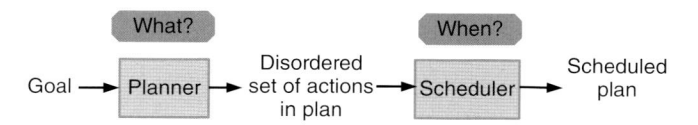

Figure 6.14 Traditional relations between planning and scheduling.

that an action can result in a number of effect, possibly with different likelihoods. Another alternate assumption is *partial observability* in which state variables are assumed to be either unobservable or observable with a degree of uncertainties. In addition, reachability goals can be extended to *extended goals* where goals need to specify requirements of different strengths taking into account nondeterminism [30].

One method to deal with the uncertainties is using **Markov decision processes** (MDPs). The MDP provides a mathematical framework for modeling decision making in situations where outcomes are partly random and partly under the control of a planner. MDPs are stochastic systems where the goals are *utility functions* and solutions are *policies* specifying the action to complete at each state. As a result, the planning is migrated to an optimization problem.

Another method of dealing with uncertainties is planning with **Model Checking**. Model checking is also a nondeterministic state-transition system, where goals are represented by temporal logic and solutions are iterative and conditional plans. Planning uses symbolic model checking algorithms.

Furthermore, neoclassical methods can be extended to cope with uncertainties. A simple approach is to extend actions to include preconditions, effects, and nondeterministic effects. The nondeterministic effect can be solved using extended graph-planning, CSP, or SAT methods.

Automated P&S has been applied in space missions or systems for a few decades, and during this time the techniques have become more complex. What follows is the discussion of some of the more prominent historical planners and schedulers, followed by a discussion into the currently used planning and techniques and frameworks.

6.4.2.7 Planning Languages

The planning language can be seen as a common tool or formalism to define domains and problems concerning a P&S system. In general, such a common formalism can allow greater reuse of research work and more direct comparison of different planning methods hence potentially support faster progress in the field. It often offers a compromise between expressive power and the progress of basic research [40].

The earliest planning language **STRIPS** (Stanford Research Institute Problem Solver) [41] takes the form of an *action language*. An action language is a language for specifying state transition systems, and is commonly used to create formal models based on the effects of actions in the world. A STRIPS instance, $\langle P, O, I, G \rangle$ is composed of

- P: a set of conditions (i.e., propositional variables);
- O: a set of actions where each action is a quadruple $\langle \alpha, \beta, \gamma, \delta \rangle$ with each element being a set of conditions. These four sets of conditions specify "which conditions must be true for the action to be executable," "which conditions must be false," "which conditions are made true by the action," and "which conditions are made false," respectively;

- I: the initial state, given as the set of conditions that are initially true with all others being assumed false (known as the closed world assumption);
- G: specifications of the goal state, given as a pair $\langle N, M \rangle$ that specifies which conditions are true and false in order for a state to be considered a goal state.

Assuming a planning problem where a rover is at location A on Mars, it needs to sample a rock at location B using a robotic manipulator onboard the rover. The problem can be defined in STRIPS as follows:

```
Initial state:
    At(A), ArmLevel(high), RockAt(B), ArmAt(A)
Goal state:
    Sample(Rock)
Actions:
    // move from X to Y
    _Move(X, Y)_
      Preconditions:  At(X)
      Postconditions: not At(X), At(Y)

    // deploy robotic manipulator
    _DeployArm(Location)_
      Preconditions:  At(Location), ArmAt(Location), ArmLevel(high)
      Postconditions: ArmLevel(low), not ArmLevel(high)

    // sample the rock
    _SampleRock(Location)_
      Preconditions:  At(Location), RockAt(Location), ArmLevel(low)
      Postconditions: Sample(Rock)
```

PDDL (Planning Domain Definition Language) is a major modern planning language inspired by precursor languages such as the STRIPS. It is also an attempt to standardize artificial intelligence (AI) planning languages, which made the International Planning Competition (IPC) possible since 1998 and has evolved to multiple official versions and unofficial extensions. For greater modularity, PDDL separates modeling of a planning problem into the *domain* and *problem* description [40]:

- The domain description refers to a domain-name definition, including definitions of: (i) requirements to declare those model-elements to the planner that the PDDL model is using; (ii) object-type hierarchy, similar to class-hierarchy in Object-Oriented Programming; (iii) constant objects that are present in every problem in the domain; (iv) predicates that are templates for logical facts; and (v) possible actions that have parameters (i.e., variables that may be instantiated with objects), preconditions, and effects. Newer versions of PDDL improve on the domain description by allowing real-world planning problems to be modeled more effectively. For example, nonmodal resources such as energy, durative or continuous actions, timed initial literals, and preferences can all be represented.
- The problem description refers to a problem-name definition, including definitions of (i) the related domain-name; (ii) all the possible objects; (iii) initial conditions; (iv) goal states; and (v) plan-metrics.

Assuming the same planning problem where a planetary rover at location A is tasked to sample a rock at location B, the PDDL syntax to model the scenario can be

```
(define (problem sampler-prob)
    (:domain sampler-dom)
    (:objects locationa locationb free rock)
    (:init (location locationa)
           (location locationb)
           (sampler free)
           (at rover locationa)
           (at sampler locationa)
           (at rock locationb))
    (:goal (sampler rock)))
```

PDDL has also become inspiration to many other modern planning languages (such as O-Plan language, SHOP2 language and Opt/PDDL+) among which PDL/DDL and NDDL are commonly used in space applications.

NDDL (New Domain Definition Language, pronounced "noodle") is a domain description language used to model hybrid systems and the context in which they operate. It is NASA's response to the PDDL since 2002 and used in NASA's EUROPA P&S software (previously mentioned in Section 6.3.2 and detailed in Section 6.4.3). NDDL has different representation to PDDL in several aspects: (i) it uses a variable, value representation rather than a propositional, first-order logic and (ii) there is no concept of states or actions, only of intervals (or activities) and constraints between those activities; hence, models in NDDL are schemas for the planning problem unlike the PDDL models. As a result, planning and execution of plans using NDDL can be more robust, which is particularly useful to space missions. However, NDDL's representations of the planning problem can be less intuitive than using PDDL. The NDDL syntax for the aforementioned example on sampling Martian rock can be described as follows [42]:

```
class Instrument{
    RoverName rover;
    InstrumentLocation location;
    InstrumentState state;

    Instrument(Rover r){
            rover = r;
            location = new InstrumentLocation();
            state = new InstrumentState();}

    action TakeSample{
        Location rock;
        eq(10, duration); //duration of TakeSample is 10 time units}

    action Place{
        Location rock;
        eq(3, duration); // duration of Place is 3 time units}

    action Stow{
        eq(2, duration); // duration of Stow is 2 time units}
```

```
    action Unstow{
        eq(2, duration); // duration of Unstow is 2 time units}
}
Instrument :: TakeSample{
    met_by(condition object.state.Placed on);
    eq(on.rock, rock);

    contained_by(condition object.location.Unstowed);

    equals(effect object.state.Sampling sample);
    eq(sample.rock, rock);

    starts(effect object.rover.mainBattery.consume tx);
    eq(tx.quantity, 120); // consume battery power
}
```

DDL (Domain Definition Language) specifies both the components and the relevant physical constraints that influence their possible temporal evolution such as possible state transitions over time of a component, synchronization/coordination constraints among different components, and maximum capacity of resources. Similarly **PDL** (Problem Definition Language) specifies P&S problems. DDL is used in ESA's APSI planning framework (see Section 6.4.4 on APSI). An example of APSI's DDL.3 syntax is provided, showing a state variable type component whose allowed values represent the possible science operations that can be performed [43]. The DDL.3 specifies: (i) in the first line the path of the Java class that implements the component; (ii) the allowed transitions between values; and (iii) the minimum and maximum duration of each value, in this case $[1, +\infty]$.

```
COMP_TYPE GROUND_STATE_VARIABLE SC_SCIENCE
    VALUES
    {
            Nadir_Science()   [1,+INF];
            Radio_Science()   [1,+INF];
            Inertial_Science() [1,+INF];
            <DEFAULT> No_Science() [1,+INF];
    }
    TRANSITIONS
    {
            No_Science() TO { Nadir_Science(); Radio_Science();
                    Inertial_Science();}
            Nadir_Science() TO { No_Science(); }
            Radio_Science() TO { No_Science(); }
            Inertial_Science() TO { No_Science(); }
    }
```

6.4.3
P&S Software Systems

A P&S system developed for real-world space missions can be used within the ground operation software (i.e., ground segment) and/or onboard operation software (i.e., onboard segment). In principle, all the basic P&S techniques described in Section 6.4.2 are applicable to resolve a space problem. Given practical considerations aforementioned in Section 6.4.1, most space P&S systems are based on temporal planning paradigm. Some representative P&S software systems developed by major space agencies are summarized as follows:

- **HSTS** (Heuristic Scheduling Testbed System) [38] is a representation and problem-solving framework that aims to unify planning and scheduling. In advance to classical scheduling, HSTS decomposes a domain into state variables evolving over continuous time. The planner uses multilevel heuristic techniques to manage time and resource constraints imposed by the action scheduler. This allows the description and manipulation of resources far more complex than it is possible in classical scheduling. The inclusion of time and resource capacity into the description of causal justifications allows a fine-grain integration of planning and scheduling and a better adaptation to problem and domain structure. HSTS puts special emphasis on leaving as much temporal flexibility as possible during the planning/scheduling process to generate better plan/schedules with less computation effort.
  Since: 1992.
  Funding Space Agency: NASA.
  Served Missions: Hubble Space Telescope, Extreme Ultraviolet Explorer, Cassini (ground segment).
- **IxTeT** (Indexed Time Table) [25, 44] is a temporal planner that uses time-point and restricted interval algebra. It focuses on the representation and control issues to achieve compromise between the expressiveness of the planner and the efficiency of the search results. Hierarchical planning operators offer an expressive description, with parallelism, durations, effects, and conditions at various moments of the action. It uses partially ordered causal links planning process with CSP for producing flexible and parallel plans.
  Since: 1994.
  Funding Space Agency: CNES.
  Served Mission: N/A.
- **ASPEN** (Automated Scheduling and Planning ENvironment) [45] is a modular, reconfigurable application framework based on AI techniques, which is capable of supporting a variety of P&S space applications and originally designed to work within the ground operation software. It consists of: (i) a PDDL-like modeling language called ADDL to describe the domain model; (ii) a management system for representing and maintaining resource constraints and activities; (iii) a temporal reasoning system for representing and maintaining temporal constraints; (iv) a set of search algorithms for planning, that is, classical planning; and (v) a graphical interface for visualizations. ASPEN provides a variety of core search algorithms, such as DFS, BFS, hill-climbing, and A*, to meet different planning requirements. It also has an iterative repair search algorithm allowing the user to interact with the schedule in order to replan efficiently. Its optimization algorithm can optimize plans for specific purposes such as maximizing science data or minimizing power consumption.

Since: 1997.
Funding Space Agency: NASA.
Served Missions: DATA CHASER, Citizen Explorer (ground segment).
- **CASPER** (Continuous Activity Scheduling Planning Execution and Replanning) [27] is the real-time version of ASPEN and, hence, is capable of continuous replanning for onboard applications. It uses iterative repair to support continuous modification and updating of a current working plan in light of changing operating context.
Since: 1999.
Funding Space Agency: NASA.
Served Missions: Earth Observer 1 (onboard segment).
- **EUROPA** (Extensible Universal Remote Operations Planning Architecture) [26, 42] is the successor of HSTS and principally based on temporal and CSP planning techniques (or constraint-based temporal planning paradigm). It uses NDDL (described in Section 6.4.2.7), a high-level declarative modeling language to describe the planning domains and problems, and is extensible and can accommodate diverse and highly specialized P&S techniques within a common design and development framework around its technology core. EUROPA consists of three major components: (i) a *plan database* is the technology cornerstone of EUROPA for the storage and manipulation of plans as they are initialized and refined, which integrates a rich representation of actions, states, objects, and constraints with powerful algorithms for automated reasoning, propagation, querying, and manipulation; (ii) a *core solver* to find, and automatically fix, flaws in the plan database, which can be configured to plan, schedule, or both, and is easily customized to integrate specialized heuristics and resolution operations; (iii) a *debugger* for the instrumentation and visualization of applications.
Since: 1998.
Funding Space Agency: NASA.
Served Missions: Deep Space 1: RAX (space segment); MER (ground segment).
- **APSI** (Automated Planning and Scheduling Initiative) [46] is a software framework for developing AI-based P&S technologies of space missions. APSI is composed of a set of plug-ins developed in JAVA based on temporal planning theory. A problem is described using components (e.g., a camera payload) and each component describes its valid state transitions. Each component has an associated timeline to represent the temporal state change of the component along the time, limited by a time horizon. Planning decisions over the components are represented as choices for a determined period of time over the set of values that the component can take (i.e., state variables) or consumption/production activities in a period of time (i.e., resources).
Since: 2008.
Funding Space Agency: ESA.
Served Mission: Mars Express, Alphasat and INTEGRAL (ground segment).

6.4.4
P&S Software Development Frameworks

Traditional, integrated P&S software for space or planetary missions can have rather limited shelf-life based on the state-of-the-art progressing capabilities. A modern trend is to develop software frameworks that enable reusability, interoperability, and integration of upcoming P&S techniques. Such frameworks also allow different algorithms or methods get compared in a fair manner and ease the process of new development. The basic components of a P&S software framework include

- problem and solution language, which allow domain consolidation;
- knowledge and reasoning, which allow for domain-based reasoning and inferencing to optimize planning;
- planning algorithms;
- scheduling algorithms, which can be deeply intertwined with the planning algorithms;
- heuristics, which are applicable to both P&S algorithms.

The P&S software systems previously described in Section 6.4.3 can be or have been evolved into generic software frameworks so as to aid further development of generic P&S techniques in the long run, such as NASA's EUROPA framework illustrated in Figure 6.15. ESA's APSI as a project was initiated in 2007 with such a long-term goal in mind hence is further described in this section. The APSI project aimed to bridge the gap between advanced artificial intelligent P&S technologies and the world of space mission planning. Its goal was to design and implement

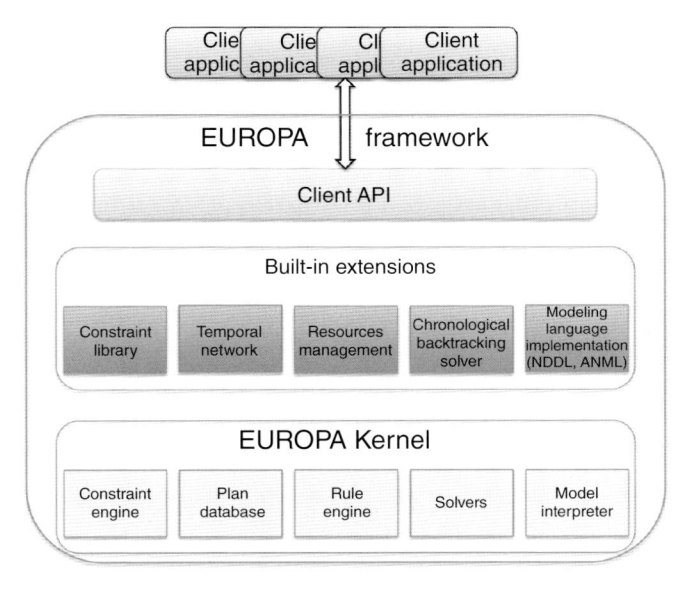

Figure 6.15 EUROPA framework architecture [42].

an experimental platform to improve the cost-effectiveness and flexibility of tool development for mission planning systems. As a result, the APSI framework offers primitives and languages to capture the specifications of an application domain and a given problem, as well as solving algorithms to foster rapid and fast prototyping of P&S applications. Furthermore, plan execution and monitoring services have been added to APSI in a subsequent ESA development of the onboard autonomous software GOAC (mentioned previously in Section 6.3.3 and further detailed in Section 6.7).

The APSI framework is grounded on a few modeling primitives such as timelines, state variables, numerical resources, temporal/value synchronizations among different timelines, problems, and solutions represented as constraint networks on events occurring in time instants or over time intervals. The basic assumption is that the world can be modeled as a set of entities (such as one or more physical subsystems) whose properties vary in time according to some internal logic or as a consequence of external inputs. The intrinsic properties of each entity are that they evolve over time concurrently, and that their behaviors can be affected by external inputs as well. These properties are represented using *timelines*, which are finite sets of ordered, flexible transition points with associated values. The values associated to a transition of the timeline allow the computation of the actual value of the timeline between two consecutive transition points. Each transition is specified a lower and upper bound of occurrence as well as a minimal and maximal duration for the transition. In this context, problem solving consists of controlling the evolution of the components by means of external inputs in order to achieve a desired behavior. Hence, different types of problem (e.g., planning, scheduling, execution, or more specific tasks) can be modeled by identifying a set of inputs and relations among them, which together with the model of the components and a given initial set of possible temporal evolutions, will lead to a set of final behaviors, which satisfy the requested properties such as feasible sequences of states or resource consumption, and so on.

The APSI framework uses modeling primitives based on two classes of components, namely the state variables and resources. These components and their possible evolutions are then "connected" by means of temporal and logical synchronizations. The state variables represent components that can take sequences of symbolic states subject to various (possibly temporal) transition constraints. This primitive permits definition of the *timed automata* as the one illustrated in Figure 6.16a. Here, the automaton represents the constraints that specify the logical and temporal allowed transitions of a timeline. A timeline for a state variable is valid if it represents a *timed word* accepted by the automaton.

The timed automaton (i.e., state variable) is a powerful modeling primitive that is widely studied at the theoretical level [47] as well as implemented in planner designs to find valid timelines. The automaton models the following: (i) values that the timeline can take, possibly as function of numeric or enumerated parameters; (ii) transition constraints on these values, possibly with additional constraints that restrict the transition to a subset of the possible values that the parameters

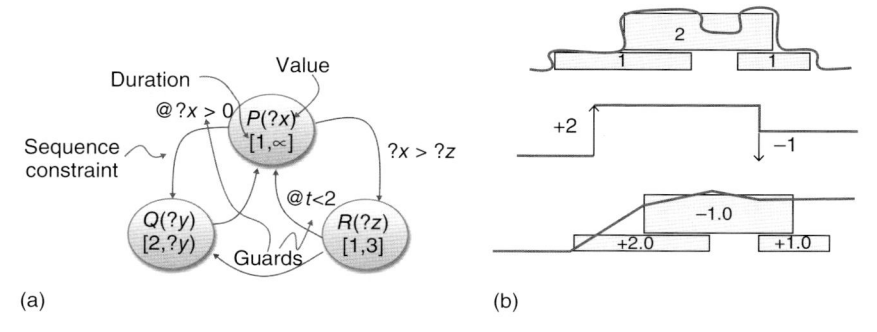

Figure 6.16 APSI modeling components. (a) State variable, (b) resources.

can take, for example, in Figure 6.16a the transition from P(?x) to R(?z) imposes that ?x>?z; (iii) temporal constraints that state the minimal and maximal temporal duration of a value; and 4) guards that restrict the applicability of a transition, either based on the value of a parameter, for example, in Figure 6.16a the transition from P(?x) to Q(?y) is allowed only if ?x>0), or on the relative timing of the transition, for example, in Figure 6.16a the transition from R(?z) to P(?x) is allowed only if R(?z) has been maintained for less than 2 time units.

A resource is any physical or virtual entity of limited availability such that its timeline (or profile) represents its availability over time. A decision on the resource models a quantitative use/production/consumption of the resource over a time interval or in a time instant. Three types of resources are currently available in the APSI: (i) the *reusable resources* that abstract any physical subsystem with a limited capacity where an activity uses a quantity of such resource during a limited interval and then releases it at the end (as seen in Figure 6.16b, top), for example, an electric generator with a maximal available power; (ii) the *consumable resources* that abstract any subsystem with a minimum capacity and a maximum capacity, where consumptions and productions consume and restore a quantity of the resource in specific time instants (as seen in Figure 6.16b, middle), for example, a battery has a minimum amount of charge that has to be guaranteed (>0 for operational or security reasons) and a maximum capacity. Operations can either consume (e.g., by using payloads) or recharge (e.g., by using solar arrays) the battery; (iii) the *linear reservoir resource* that does not have a step-wise constant profile of consumption such as reusable or consumable ones but the activities specify the amount of production/consumption per time (i.e., slope) resulting in a profile of resource that is linear in time (as seen in Figure 6.16b, bottom). As a consequence, the amount of resource available at each transition of the timeline depends on the duration of the time intervals over which this production or consumption has been performed, conversely with the other types of resource where the profile of the resource availability at each transition depends only on when and how much is produced/consumed and not on the duration of the production/consumption.

In timeline-based modeling, the physical and technical constraints that influence the interaction of the subsystems modeled either as state variables or resources, that is, the *Domain Theory*, are represented by means of temporal and logical synchronizations among the values taken by the automata and/or resource allocations in the timelines. Planning languages have constructs to represent the interaction among the different timelines that model the domain, for example, the synchronizations in DDL or the compatibilities in the NDDL (previously introduced in Section 6.4.2). Conceptually, these constructs define valid schema of values allowed in timelines while linking the values of the timelines with the resource allocations. They allow the definition of Allen's like quantitative temporal relations [35] among time points and time intervals as well as constraints on the parameters of the related values (previously described in Section 6.4.2 on temporal planning).

The APSI framework provides services and algorithms for temporal planning with basic building blocks illustrated in Figure 6.17. The main architecture includes databases of information, groups of implemented functionalities, and application program interfaces (APIs), where the databases maintain the information, the functionalities provide services to access and manipulate the databases, and the APIs provide access to the databases and the services. The framework also provides support for implementing solvers and applications. An application is a collection of solvers for the components. Components are the lower level primitives for modeling problems that are typically state variables and reusable resources. Solvers are implemented algorithms for solving conflicts between components' timelines, for example, multicapacity resource schedulers or planners for state variables. An application is built by connecting solvers with the APSI platform services.

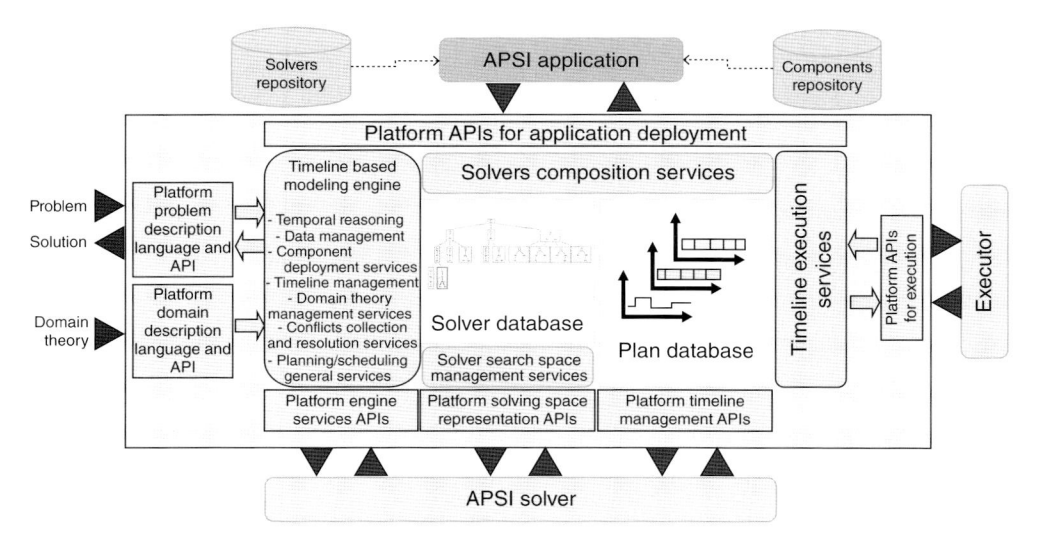

Figure 6.17 APSI framework architecture.

The core of the architecture is constituted by the *plan database* and the *solver database*. The plan database contains all the information on time, data, timelines, and events, which describe the current solution. The solver database contains all the information to properly represent the search space of a solver for the framework and to implement search strategies. The search space of an APSI solver is made of timelines, events, and constraints on time and data. Functionalities include temporal reasoning, data management, timeline extraction and management, conflicts collection and resolution, planning/scheduling general services, search space management services, and timeline execution services.

An example on the rover domain is used here to further demonstrate the general concept and modeling language of the APSI framework. Consider a rover that has to perform experiments in a given environment, whereby it is equipped with a battery that is consumed when the rover is moving or performing experiments or charged when the rover reaches a specific locations during a specific time interval (a map with the distribution on the charging locations can be given), the objective is to allow the rover to autonomously navigate, perform experiments, and monitor its battery level to satisfy the required activities. To model the rover domain, the following subsystems are considered: a mobility system MS, a battery BAT, and one (or more) payload(s) PLD:

- The mobility system MS can be modeled as a state variable taking the following values: AT(?x,?y) when the rover is standing in location $\langle x, y \rangle$ and GOTO(?x,?y) when the rover is moving toward $\langle x, y \rangle$. A transition GOTO(?x,?y) → AT(?x,?y) denotes a successful move to $\langle x, y \rangle$ and a transition AT(?x,?y) → GOTO(?x',?y') denotes the rover starting to move from a point $\langle x, y \rangle$ to a point $\langle x', y' \rangle$.
- The rover payloads PLD can perform experiments in the current position of the rover or can stay idle. A state variable is used to model the payload that admits the following values: IDLE() when the payload is not performing experiments, or RUNNING(? exp) when the unit is performing an experiment exp.
- The battery BAT is modeled as a reservoir resource with a minimum charging value (assuming 0 in this model) and a maximum charging capability max. When the rover is performing an experiment, it consumes the battery depending on the type of experiment being performed. This is modeled via an external function f_{cons}(? exp) that provides the actual consumption of the experiment. When the rover moves, the battery is consumed with a fixed rate hence battery consumption is proportional to the time taken to navigate from one point to another.

As a result, a goal RUNEXPERIMENT(?x,?y,? exp) can be achieved by the rover payload taking a status PLD.RUNNING(? exp), with the mobility system in a status MS.AT(?x,?y). The battery consumption is modeled by synchronizing the status PLD.RUNNING(? exp) of the payload with an activity BAT.ACTIVITY(f_{cons}(? exp)). Besides that, it is necessary to synchronize the activities that refill the battery with the locations and the time intervals in which the rover can recharge its battery. This input can be provided as a timeline with the geographical coordinates of the location and the available solar flux in various time intervals. Such a timeline can have intervals specified with the value CHARGE(?x,?y,?solarflux) when the flux ?solarflux is available in ⟨x,y⟩. Given that, synchronization is required for the charging activities with the rover standing in one of these zones. The actual charging capability generated by a given solar flux is modeled with an external function f_{prod}(?solarflux).

A problem for this rover domain specifies: the initial values for the timelines, the starting and final position of the rover and the experiments to be performed, as well as the initial charging value of the batteries and the visibility of the charging zones. The mix of causal and temporal relations among the operations can be stated using the DDL synchronizations as shown in the following:

```
SYNCHRONIZE MissionTimeline {
  value runExperiment(?x,?y,?exp){
    op1 pld.running(?exp);
    op2 ms.at(?x,?y);
    ref contains op1;
    op1 during op2; }}




SYNCHRONIZE pld {
  value running(?exp) {
    op1 bat.activity(f_cons(?exp))
    ref equals op1; }}
```

```
SYNCHRONIZE ms {
  value goTo(?x,?y) {
    op1 bat.activity(-1.0);
    ref equals op1; }}

SYNCHRONIZE bat {
  value activity(?prod) {
    [?prod > 0.0 ];
    op1 ms.at(?x,?y);
    op2 zone.charge(?x,?y,?solar_
        flux);
    ref during op1;
    op1 during op2;
    ?prod :=
    f_prod(?solar_flux); }}
```

6.5
Reconfigurable Autonomy

6.5.1
Rationale

As previously shown in Table 6.1, goal-oriented autonomy or high LoA increase the capability of planetary robotic systems and missions. The highest LoA or E4 systems are generally designed with respect to specific conditions, environments, software/hardware component abilities and their optimal functioning. If one of these considerations changes, the high LoA functions cannot be achieved. Typical planetary robotic systems have no ability to reconfigure itself based on introspection, some system faults and errors have to be rectified or addressed by human operators remotely. For example, if a rover's navigation camera starts performing suboptimally, its GNC performance will get degraded or the rover will have to revert to a lower LoA. It is, therefore, advantageous to have solutions that autonomously solve errors, optimize, reconfigure at run time, make initial configuration simpler, and also allow "experts to be experts" [48].

In the field of computer science, a system that can deal with faults, changing environments, requirements, software, and hardware can be defined as an *autonomic computing system*. More specifically, the autonomic computing refers to the self-managing characteristics of distributed computing resources, adapting to unpredictable changes to the environment, software, or hardware. The main self-managing characteristics are known as the self-CHOP [49], where CHOP stands for

C self-configuring, which is the automatic configuration of components;
H self-healing, which is the automatic discovery, and correction of faults;
O self-optimization, which is the automatic monitoring and control of resources to ensure the optimal functioning with respect to the defined requirements;
P self-protection, which is the proactive identification and protection from arbitrary attacks, where an attacks are faults coming from an external source.

Extending self-CHOP to include the automatic reconfiguration of components at run-time can then be called *self-reconfiguring*. This is similar to the *self-managed software architecture* [50] in which components automatically configure their interaction in a way that is compatible with an overall architectural specification and achieves the goals of the system. Another important consideration for a reconfigurable system is that it cannot be too computationally expansive, or introduce too many errors relative to the base system. Also it preferably would not be too restrictive for developers, that is, the reconfiguration system should not be focused on one component model and can also deal with black box components. It is clear that self-CHOP and self-reconfiguring are useful to planetary robotic systems in achieving high LoA reliably. The remainder of the section presents major techniques in reconfiguration system in relation to achieve reconfigurable autonomy for planetary robotic systems.

6.5.2
State-of-the-Art Methods

In modern robotics, biologically inspired low-level robot behavior adaptations such as using genetic algorithms

The *self-managed software architectures* is currently a vibrant field that covers a number of distinct solutions such as service-based [52–54], aspect-oriented component-based [55], generic component-based [56, 57], model-based [58–60], self-organization

[61, 62], and ontology-based [63–65]. Each of the methods approaches the problem in a different manner. Some are described here to show the diversity and extensiveness of relevant work.

The **service-based** methods offer service-oriented architecture where services are *loosely coupled* software components that encapsulate functionalities and are available to be remotely accessed by applications over a network or Internet. A "loosely coupled" system is one in which each of its components has, or makes use of, little or no knowledge of the definitions of other separate components [66]. For example, the Service-Oriented Robotic Architecture (SORA) in Ref. [52]

demonstrates that service-oriented approach offers potentially a scalable, flexible, and reliable framework for space robotics. Nevertheless, taking service-based approach is a major paradigm shift for robotics. These services can be automatically composed and configured toward a user goal, which is generally done in one of the two ways: *orchestration* or *choreography*. In service orchestration, a software agent arranges, coordinates, and manages services toward a goal, but allows services to run with self-determination [53, 54]. In service choreography, a software agent controls and chooses all activities that any service performs, and in doing so can reconfigure a system toward a goal at run time. One such system is KnowRob [67], which uses symbol grounding [68] of the problem and the robot's hardware capabilities to search the RoboEarth database [69] for a proposed solution to the robotics problem. KnowRob demonstrates the choreographic method's ability to reconfigure even at the lowest level.

Another example of choreography was demonstrated with Livingstone-2 [60] where a "generic **model-based** autonomy plant engine" operates the system at the choreographic base-level control when given a "Livingstone model" of the system. Following its implementation on the EO-1 mission, it was concluded that the choreographic method introduced more errors and was not as efficient as a traditionally engineered system [70]. For the sake of modern robotics, choreography is impractical since the domain of control is massive and complex. Consequently, no general agent could choreograph all possible actions necessary, as effectively and efficiently as an engineered planner agent.

The **self-organization** methods include distributed reconfiguration system such as the Autonomy Generic Architecture, Tests and Application (AGATA) [61].

AGATA is an autonomous architecture focused on the issues of maintaining a high-level autonomy while attempting to incorporate genericity and modularity, with applications to avionics and spacecraft systems. Its modules are based on a common pattern and connected together to form a global architecture, and each module is in charge of controlling a part of the system and handling the data associated with this part. It takes into account requests and information coming from other modules and can send requests to or ask information of other modules. To avoid potential decision conflicts, it cannot have direct access to the part of the system controlled by any other module. Each module is built on a sense/plan/act pattern and each maintains its own knowledge of the state of the system part it controls on the basis of an internal Unified Modeling Language (UML) model. The architecture is overseen by the generic control module and itself organized into generic subcomponents. The core of this module is a set of four components dedicated to closed-loop control: received request tracking, system state tracking, decision making, and emitted request tracking components. The decision-making component decides whether a control requests is emitted to another modules, physical systems, or information processing services. The AGATA architecture allows for online reactive and deliberative reconfiguration from a bottom-up reconfiguration system perspective. Alternatively, a top-down reconfiguration system can be achieved such as the Requirements-Driven Architecture (RDA) [71], in which a centralized agent monitors and models the system using UML and disjunctive Datalog, then analyzes, plans, and reconfigures based on a requirement engineering goal. Comparing AGATA and RDA leads to the usual comparison of the distributed versus monolithic, such as distribution improving stability while sacrificing utility and efficiency.

The more recently developed system reconfiguration concept is **ontology-based** [63-65],. which mainly consists of an application layer representing the ordinary system with user-defined modules and a network layer representing components integral to the reconfiguration. The purpose of the separation is to minimize the number of possible errors that the reconfiguration can inject into the usual system, reduce potential computation tax that the reconfiguration can inflict, and make the reconfiguration removable if no longer required. The network layer has three major components, namely the *Inspectors*, the *Ontology*, and the *Rational Agent*, which together represent the Monitor, Analyze, Plan, Execution—Knowledge or *MAPE-K* loop. The Inspectors represent the "monitor" within the MAPE-K loop that can update "knowledge" and are generally lightweight in computation. The Ontology represents the "knowledge" within the MAPE-K loop and a model of the system in its entirety using description logic. The main domains of knowledge that require modeling are basic software operation, software–software interaction, external environments, software–environment interaction, and goal requirements. Modeling requirements can include numerics such as confidences, timings, hierarchies of information, and their attributes (e.g., types of maps for rover GNC). The Rational Agent represents the "analysis," "planning," and "execution" of the MAPE-K loop. It reads Ontology and can compute an optimal configuration based upon a metric of utility (defined in the goal). It also manages the Inspectors in order to maximize the coverage while minimizing the computational overheads.

6.5.3
Taxonomy

As exemplified earlier, there exists a large selection of methods for reconfigurable software system. These methods can be characterized by a simple taxonomy of techniques as illustrated in Figure 6.18. The primary attribute in the taxonomy is the method for system *coordination*, which is either orchestration or choreography, representing the level of control the reconfiguration system takes to achieve autonomicity. For low-level control, that is, choreography all activities are chosen and performed by the reconfiguration system (e.g., "turn left"). In high-level control, that is, orchestration modules are instructed to perform certain tasks but have a level of autonomy in the execution thereof (e.g., orchestrate a path planner, which decides to "turn left"). The next taxonomic feature is the *workflow* in which all information flow is either centralized or decentralized. The final taxonomic feature is the *source* of reconfiguration, that is, from a single centralized agent to multiple centralized agent controlling interacting subsystems (i.e., localized), or entirely

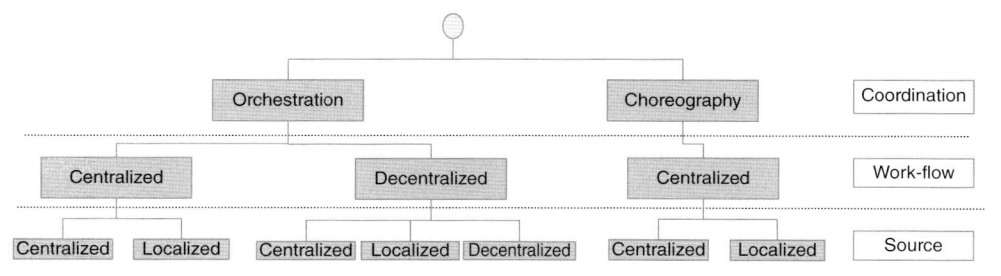

Figure 6.18 A simple taxonomy of reconfiguration systems.

decentralized control. This taxonomy can be illustrated through anatomical analogies: the salivary glands give an excellent example of an orchestrated centralized workflow system, the heart demonstrates an orchestrated decentralized workflow system, and the motile muscles provide a good example of a choreographed centralized workflow system.

Detailed attributes to distinguish different reconfiguration methods are subdivided within the basic properties, the self-CHOP properties and the self-monitoring properties as follows:

Within the basic properties:

- **Interaction level**—*External*, subcomponents are reconfigured via external API and communication redirection. Or *Internal*, subcomponents are altered internally to cause reconfiguration, that is, subcomponents demand a particular design pattern.
- **Model type**—*Symbolic-based* or *Model-based*.
- **Dynamism**—*Static*, all configuration and reconfiguration plans set a build time. Or *Dynamic*, all plans made at run-time.
- **Context-aware**—Whether the system is aware of the external environment and computing hardware environment.
- **Temporal**—Reconfiguration plans contain knowledge of time.
- **Uncertainty**—Probabilities or confidences are considered in plan making.
- **Transparency**—The reconfiguration system and its decisions are human-readable.
- **Extendible**—Both the model and system are extendible.
- **Portable**—Both the system and its controlled subsystems are portable.

Within the self-CHOP properties:

- **Self-installing**—The reconfiguration system autonomously installs new software.
- **Self-configuring**—The reconfiguration system creates a configuration plan at build time.
- **PnP-like**—Plug and play like, that is, new subsystems can be introduced and utilized at run-time.
- **Self-reconfiguring**—The reconfiguration system creates a configuration plan at run-time.

- **Failure prediction**—The system predicts failures and prepares plans to mitigate them.
- **Check pointing**—The system resolves failure states by substituting broken subcomponents.
- **Utility**—The system is optimized by a utility function.
- **Reputation**—The system is optimized via the trustworthiness of subcomponents.
- **Load-balancing**—The system is optimized based on computational and temporal considerations.
- **Safe reconfiguration**—The reconfiguration plans have a safe method for migration.
- **Failure state tracking**—Diagnose fault, via known fault states in the model.

Within the self-monitoring properties:

- **Active or Passive**—*Active*, subsystems are internal states and variable are monitored. Or *Passive*, external information from subsystems are monitored.
- **Performance and Trust**—Monitors in real-time the trustworthiness of subsystems.
- **State or parameter**—*State*, state tracking in real-time. *Parameter*, tracks parameters not states.

There are numerous documented techniques on reconfiguration systems that cover varying range of a subset of the concerned attributes. Table 6.5 presents a collection of the existing designs that offer relatively comprehensive coverage of those attributes and the entirety of the taxonomy. This helps demonstrate the assortment of reconfiguration systems against different measures and classifications. For example, the more recently developed ontology-based method [63] has demonstrated a complete coverage of relevant attributes for system reconfiguration hence is described in detail in the next section using a design example.

6.5.4
Design Examples: Reconfigurable Rover GNC

As aforementioned in Section 6.3, GNC is one of the major subsystems to planetary robots that high LoA is desired. In this section, a self-reconfiguring GNC system for planetary rovers is presented as a design example to demonstrate the state-of-the-art ontology-based reconfigurable autonomy [63, 64] and its application. This technique exhibits a complete coverage of required attributes for reconfiguration systems as shown in Table 6.5 and can work with generic GNC techniques presented previously in Chapters 3 and 4.

A self-reconfiguring GNC system of the planetary rover is an onboard operation software that can alter its high-level goals, midlevel goals, software architecture, component options and attributes, and low-level control options. It can autonomously overcome system errors and faults, unexpected environmental changes, and unexpected capability changes. The resulting reconfigurable autonomous GNC illustrated in Figure 6.19 can be broken into three layers, namely

Table 6.5 Comparison of various reconfiguration systems against 23 design attributes.

Taxonomy	Coordination	Orchestration									Choreography		
	Workflow	Centralized				Decentralized					Centralized		
	Source	Centralized	Centralized	Localized	Localized	Centralized	Centralized	Centralized	Localized	Localized	Decentralized	Centralized	Centralized	Localized
System	Citation	[72]	[73]	[74]	[75]	[76]	[77]	[56]	[57]	[63]	[55]	[61]	[62]	[78]	[59]	[79]	[58]	[80]
Basic properties	Interaction	I	E	I	E	E	E	E	E	E	I	I	I	I	I	I	I
	Model type	S	M	M	M	S	M	M	M	SM	M	M	M	M	M	M	M
	Dynamism	S	D	D	D	D	D	D	D	D	S	D	D	D	S	D	D	D
	Context-aware				✓					✓			✓	✓				✓
	Temporal	✓		✓			✓		✓	✓		✓			✓			
	Uncertainty			✓			✓	✓		✓					✓			
	Transparency	✓	✓		✓		✓	✓	✓	✓		✓		✓		✓	✓	✓
	Extendible	✓	✓		✓	✓	✓	✓	✓	✓	✓	✓		✓		✓	✓	✓
	Portable	✓	✓	✓	✓	✓	✓	✓	✓	✓	✓	✓	✓	✓		✓	✓	✓
Self-configuring	Installation						✓		✓	✓	✓	✓	✓	✓		✓	✓	✓
	Configuration	✓	✓	✓	✓	✓	✓	✓	✓	✓	✓	✓	✓	✓	✓	✓	✓	✓
	PnP-like						✓	✓	✓	✓	✓	✓		✓		✓	✓	✓
	Reconfiguration	✓	✓	✓	✓	✓	✓	✓	✓	✓	✓	✓	✓	✓	✓	✓	✓	✓
Self-healing	Failure state		✓		✓		✓			✓			✓	✓	✓	✓	✓	
	Check-pointing	✓	✓	✓	✓		✓		✓	✓					✓		✓	
Self-optimizing	Utility	✓	✓	✓														
	Reputation	✓	✓															
	Load-balancing			✓						✓								
Self-protecting	Failure prediction													✓	✓	✓	✓	✓
	Safe reconfiguration									✓								✓
Self-monitoring	Active or passive	—	P	P	P	P	P	P	P	P	A	A	P	P	A	P	A	A
	Performance and trust		✓						✓	✓		✓		✓				
	State or parameter	—	S	P	P	P	S	P	P	P	P	S	P	P	S	S	S	S

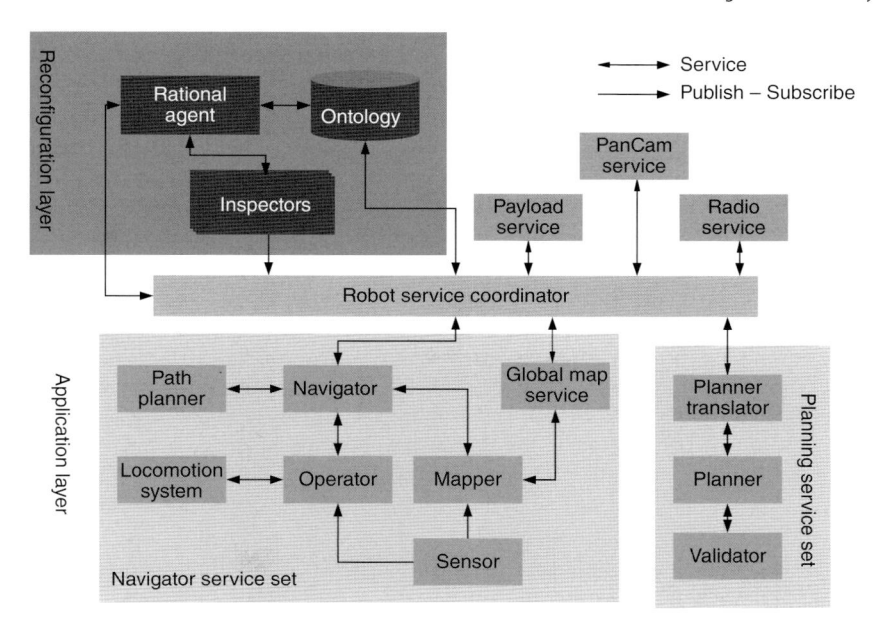

Figure 6.19 Block diagram of a self-reconfiguring rover GNC system.

the application, reconfiguration, and housekeeping layer. The **application layer** can be viewed as the conventional GNC. The **reconfiguration layer** contains the components that perform the reconfiguration. The separation of these two layers exists to minimize the computation cost of the reconfiguration layer as well as the faults that can be injected into the application layer by the reconfiguration layer. The final **housekeeping layer** contains low-level system safety and housekeeping components.

6.5.4.1 Application Layer

The application layer contains a number of reconfigurable components representing major GNC functions as well as one centralized *Robot Service Coordinator* (as shown in Figure 6.19. This layer acts similar to a service-based GNC. Services are unassociated, loosely coupled units of functionality that are self-contained, where a loosely coupled component has or makes use of little or no knowledge of other separate components. Furthermore, each service implements at least one action and a standard interface. This allows high level of reconfigurability and modularity of aforementioned services. The robot service coordinator acts similarly to a service coordinator, which is a generic software agent that organizes and coordinates external services to create an overall system. The Robot Service Coordinator can be tailored to the scenario, for example, the Robot Service Coordinator will always attempt to create a connection with some sort of radio service, and therefore this can be tuned for. The Robot Service Coordinator can reconfigure which service it connects to and its operation flow based upon a reconfiguration plan generated by the ontology-based Rational Agent. During nominal operations, the Robot Service Coordinator operates with a fixed architecture and operations order.

One major service set is the *Planner Service Set*. The planner service, when given a request including a high-level mission goal and domain level knowledge, responds with a plan and schedule for completing the high-level goals. The planner service is designed to be generic hence more reconfigurable. The request format is in PDDL planning language (previously introduced in Section 6.4.2.7) with a language usage marker, indicating language features to ease conversion. This can then be converted to or from another format in order to accommodate the best available P&S software (previously exampled in Section 6.4.3) for individual problems. Another possible associated service is a validator engine, which validates the plan based on some set criteria. Once a plan and schedule has been generated, the Robot Service Coordinator can use it to execute the other services at the scheduled times in order to complete the high-level goals determined.

Another major service set is the *Navigation Service Set*, which is concerned with the GNC functionalities of the rover platform. The services in the set are subdivided for greater reconfigurability. The *Navigator* that determines the location of the rover against a global coordinate can use a variety of different algorithms and sensors such as mono, stereo camera, or LIDAR. The *Path Planner* can use many different algorithms (e.g., A*, Dijkstra) for rover way-point determination. The *Mapper* can cover multiple levels from local traversability mapper to more abstract global mapper that spans the entire mission. The *Operator* covers direct locomotion and steering with the hardware interface, and can also include low-level obstacle avoidance. It should have the ability to tune control strategies via the parameters API. The *Sensor* service provides an interface to the onboard sensors. The *Locomotion System* provides an interface with the rover locomotion hardware. It can receive locomotion commands and outputs odometry and other system information in a publish-subscribe model.

Other important services can include the *Radio Service* that allows the Robot Service Coordinator to perform the downlink to the ground station with operations abstracted behind the basic service interface, or the *Payload Service* (e.g., PanCam) that has a high-level abstract interface.

These services and their connective architecture can be evaluated and optimized based on the environment and faulty hardware. For example in case of different navigation techniques, some are more tolerant to sensor noise while being less accurate in localization, and some techniques are more tolerant to wheel slippage. For smooth transition during reconfiguration, the services would follow a safety procedure for stopping and starting. For example, when reconfiguring a local mapper service from an existing method to another, the global mapper will get updated with the appropriate information before the existing method is removed and the new method is initialized.

6.5.4.2 Reconfiguration Layer

During nominal operations, the reconfiguration layer should use minimal resources and not increase the total number of unrecoverable errors in the system.

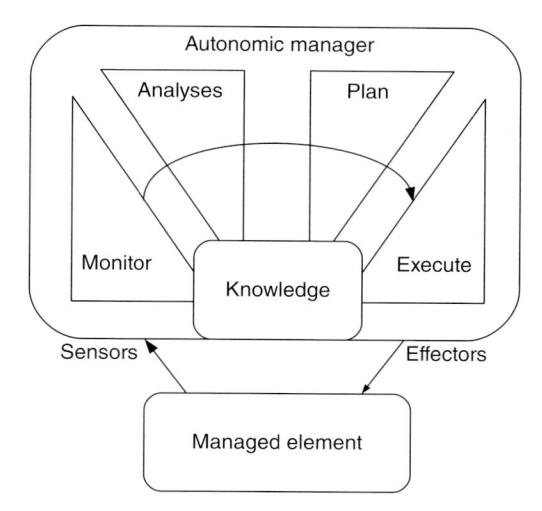

Figure 6.20 The MAPE-K loop.

This layer acts in a MAPE-K loop [49], which is an adaptation control loop as illustrated in Figure 6.20. In essence, the layer can be split into three components, namely the *Inspectors*, *Rational Agent*, and *Ontology*. These components can be mapped to the MAPE-K loop: the Inspectors monitor the system; the Rational Agent analyzes, plans, and executes reconfiguration operations; the Ontology represents the knowledge in the MAPE-K loop.

The primary element of the reconfiguration layer is the Ontology. It contains all the knowledge required for reconfiguration in the system. The Ontology is an extremely extensible knowledge representation, for example, a representation of entities, ideas, and events along with their properties and relations, according to a system of categories. It allows first-order logic (or a description logic) to be applied to the knowledge in the system automatically, allows automatic inferences and knowledge expansion, allows the knowledge of the system to be checked for consistency automatically, and allows plans in the system to be verified automatically through ontology-based model checking.

The next set of elements is the Inspectors that monitor all elements of the system. Their aim is to monitor and update the ontology on the current state of the world, while using as few resources as possible without disrupting the monitored operations. They also passively monitor the application layer. The Inspectors can be of generic or specific type. The generic type performs the basic network checks, resource checks, and state checks (i.e., the service is reporting a world change) of any individual services. The specific type is tailored for a particular service, for example, an Inspector that checks camera performance. Inspectors are reconfigurable to allow the system to decide what level of monitoring is required for an individual subsystem, in order to allow an optimal compromise between computational resource use and level of self-protection to be reached.

The last element is the Rational Agent. For given goals, it uses the information available in the ontology (which can be seen as the world knowledge) to configure the system to make those goals achievable. Furthermore, the Rational Agent will attempt to optimize the system toward the goals based upon some goal set utility function. Moreover, if the state of the world changes (i.e., the state of the Ontology changes) the Rational Agent will reevaluate whether the goals are still achievable. The Rational Agent will also deploy the Inspectors to monitor and protect the system, and to maximize safety while minimizing computational resources based on some goal-set function. The Rational Agent utilizes traditional planning algorithms to find a solution to a single high-level control loop also determines how to optimize the planning problem, how to optimally deploy Inspectors, and which plans to verify.

6.5.4.3 Housekeeping Layer

The housekeeping layer is separate to the first two layers as it involves safety checks and operation modes that correspond to low LoA activities (up to ECSS LoA E2, as described previously in Section 6.2.3). Safety checks and modes are low-level checks for faults and errors, which can put the rover into safety mode (meaning a full halt state that requires further instruction from the ground station). The purpose of keeping this layer separate from self-reconfiguration is to ensure basic continuity and safety of the robot. Furthermore, the housekeeping components are generally low-level software/hardware scripted processes, which are not reconfigurable.

6.5.4.4 Ontology Design

An ontology is best described as a description logic built on top of a description logic, where a description logic is a decidable fragment of a first-order logic. This allows for generic inference engines (and other tools) to be used on tailored description logics. The nature of ontology places it between model-based and symbolic-based descriptions, allowing easy use of a modeling language with the extensibility and logic techniques of a symbolic-based language.

Here, the ontology design involves designing an Ontology and an Ontology Manager. The Ontology holds information necessary for reconfiguration a planetary rover and is able to describe the entire world domain. Such domain knowledge includes the rover hardware, software, environment, maps, processes, temporality, and potential fuzziness. To combat the inherent complexity, the Ontology is modular and reusable with an upper ontology as shown in Figure 6.21. The upper ontology module is the common ontology shared by all modules and can be seen as the base syntax. For greater modularity and readability, the upper ontology is subdivided into modules describing base logic, numerics, temporality, fuzziness, confidences, processes, and block diagrams, and so on. Other ontology modules cover functions of the software, hardware and environment all of which are extensible.

The Ontology allows for subsumption and logical rules of concepts to be applied to knowledge held within. Not only does it allow for an efficient, extensible

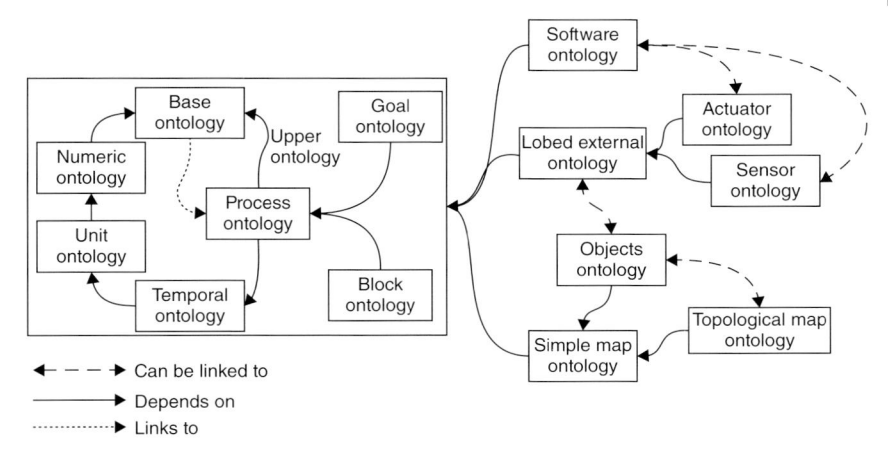

Figure 6.21 Modules of the Ontology.

method of containing information, it allows for new knowledge to be discovered. Since complex elements of knowledge domain are contained in respective modular ontologies, they can be used or omitted in a efficient way. For example, a module that contains the logic for the Navigation Service can introduce the "$subRegionOf$" property. This $subRegionOf$ property is transitive and asymmetric, that is, $subRegionOf(A, B)$ and $subRegionOf(B, C)$ implies $subRegionOf(A, C)$ but not $subRegionOf(C, A)$. If this module is not used, this level of complexity of transitivity and asymmetry may be omitted from the reasoner. Not having to reason about transitive properties could reduce the complexity of the Ontology from $\mathcal{ROIQ}$ to $\mathcal{ALCOIQ}$, or from NEXPTIME-complete to NEXPTIME-hard Decidable. This reduction in complexity is crucial for efficient reasoning. First-order predicate logic allows complex inferences to be made automatically about the domain of the world. This enables complex validation and knowledge creation to occur automatically and gives readable explanations of the logical steps.

The domain knowledge verification and plan validation are important considerations for the rover GNC because they increase the overall reliability and safety of the system. Another use for the Ontology is to verify domain knowledge and validate plan in an automatic and efficient manner. For example, consider the reconfiguration plan proposed by the Rational Agent in which the domain can be defined as the tuple $\langle I, G, R, W, S, P \rangle$, where I is the initial state of the world, G is the goal state of the world, R is the finite resources of the system, W is the world rules of the system, S is the safety criteria, and P is the plan for reconfiguration. The plan P is composed of connected services, where services have the form of Input, Output, Precondition, and Effect (IOPE). If two services are connected via a publish-subscribe model, the two output and input criteria of both services are individuals in the Ontology and thus have two sets of logical restrictions applied to them. When these exemplar services are connected, a logical rule is created to say that they are the same individual, and thus they are restricted to both sets of

374 | *6 Mission Operations and Autonomy*

logical restrictions. If there exists no contradiction in this union of logical rules, then the connection is valid and viable. In a similar fashion, the criteria for plan validity is whether the plan is consistent with the initial and goal state of the world, and whether the plan has no inconsistency with the finite resources, world rules, and safety criteria. This verification also works for inferred domain information, for example, the consequence of two services interacting in a novel fashion.

The Ontology Manager manages access to the Ontology and runs the inference tools. This allows multiple components to concurrently access and alter the Ontology safely. The Ontology Manager also prepares downlink reports about changes of the Ontology's state for the ground station.

6.5.4.5 Rational Agent Design

The Rational Agent is where the reconfiguration originates. It has both reactive and deliberative components, and is initiated by changes in the Ontology. A block diagram on its work flow is shown in Figure 6.22. When the Rational Agent is informed by the Ontology Manager that the Ontology has changed, it goes through a few steps to attempt rectify or optimize the system. To guarantee speed, these steps are reactive. The reactive component has precalculated criteria when triggered, react with a corresponding precalculated plan. The precalculated plans and criteria are prepared by a deliberative fault (or world-change) injection analysis to calculate the most likely changes to the system. In other words, the Rational Agent can create these reactive plans by analyzing the high-risk factors in the

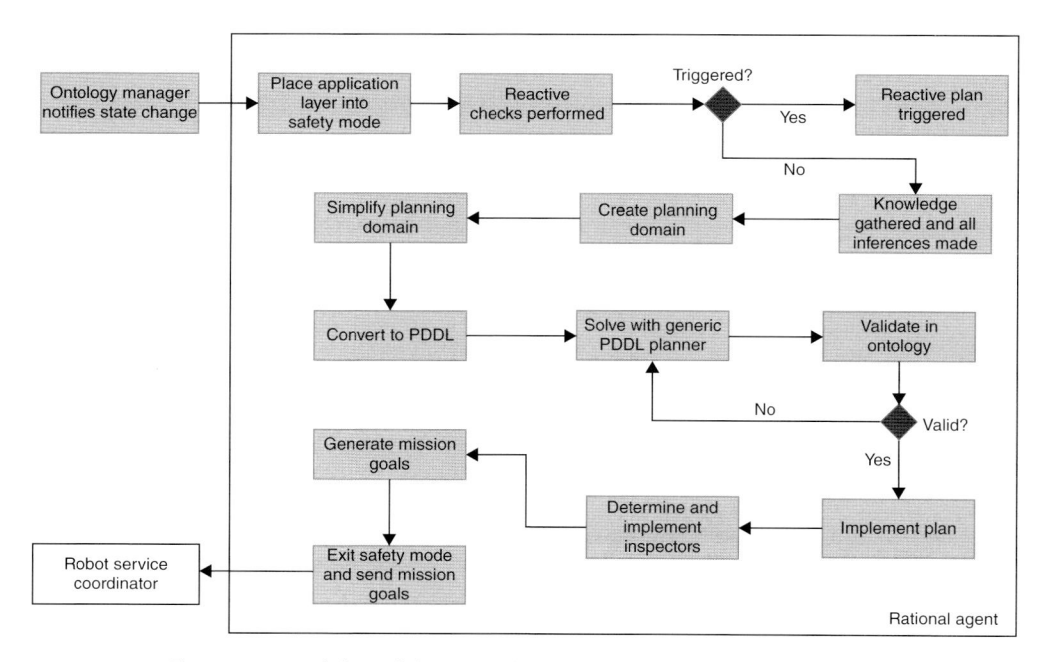

Figure 6.22 Work flow of the Rational Agent for system reconfiguration.

current plan's service configurations. If none of the reactive criteria are fulfilled, the Rational Agent will move to the deliberative component that uses more complex methods hence can deal with more complex or even novel scenarios. The deliberative component gathers information relevant to the reconfiguration process and convert it to generate a reconfiguration plan. This can be achieved using first-order logic in the Ontology's inference engine, after which planning algorithms and heuristics can be selected or used to resolve the new configuration. Here, the planning problem and domain are modeled using PDDL, and a generic planner is used to find the reconfiguration plan. The Rational Agent implements the plan once it is validated by the Ontology, otherwise it iterate the process. Once the plan has been implemented but before the application layer has been brought out of the safety mode, the Rational Agent needs to select suitable Inspectors based on risk metrics defined in the Ontology. This can be achieved by every service configuration (including external knowledge) in the Ontology to have a value indicating its potential risk of needing reconfiguration. Similarly, every Inspector can have a threshold value whom if the risk metrics surpass the Inspector needs to be selected and implemented. This gives users flexibility to tune or adjust the Inspector's coverage. Finally, the Rational Agent sends the chosen goals to the Robot Service Coordinator within the application layer for execution, following which it returns to the default reactive mode.

The key design challenges and solutions within the Rational Agent's work flow are further described as follows:

- The planning problem within the Rational Agent is most crucial since it establishes the new configuration concerning the application layer on the rover GNC. It has an initial state that includes the initial world state and the primary communication of the quasi-uplink from the Rational Agent. Its goal state includes the final world state and a connection to the quasi-downlink to the Rational Agent. The possible actions in the planning domain are the services available in the application layer in the form of IOPE. The input and output refer to their communication requirements (e.g., publishing rate), and the precondition and effect refer to the system's world state. The precondition and effect may include energy requirements, context requirements, world state changes, and service requirements, for example, whether a camera has previously been connected to the Robot Service Coordinator. The solution to the planning problem is a set of services, their parameter configurations and connections.
- The planning domain formulated in PDDL needs to be efficient and flexible allowing new configuration to be produced quickly and optimized. The first-order logic rules are used offering extensibility as new rules can be introduced easily. To minimize the planning domain sent to the planner, the domain can be viewed as a graph of linked services hence be potentially pruned for unreachable services using techniques such as the planning-graph and heuristic search algorithms.
- The planner of the Rational Agent is made generic which allows consideration of different planning algorithms and heuristics. Here, the PDDL planner is used implementing heuristic algorithms such as the A^*.

- The high-level goals are attributed to respective payload services. The Rational Agent selects the goal based on optimization of a cost function or priority function.

6.5.4.6 Impact on Mission Operations

Standard mission operations for a planetary rover have been previously described in Section 6.2.2. Now by introducing reconfigurable autonomy, operations representing the highest LoA can be further secured onboard the rover hence maximize the utility of the space segment. In this case, the initial mission uplink process of a reconfigurable rover GNC system can update its ontology, which is merged and validated by the Ontology Manager autonomously. This validation checks for knowledge inconsistencies, which adds an extra level of safety to the system. The updates to the Ontology can include a list of goals to be achieved in the next loop with a function that priorities these goals if not all goals are achievable. In addition, the updates can include alterations, for example, map changes, utility function changes, or software function changes. Once the ontology has been updated, the Rational Agent will reconfigure the application layer in order to complete the planned goals, and choose the achievable goals in the time allowed based on the priority function. Then, the Rational Agent passes the chosen goals to the Robot Service Coordinator to complete in a quasi-uplink. While the Robot Service Coordinator is completing the goals, the Inspectors watch for changes in the world state. If changes occur, the Rational Agent reevaluates whether the goals or configurations need adapting, and reconfigures the application layer if they do. Once the downlink is due, the payload information and changes to the Ontology are transmitted to the ground station. Separate to this process the housekeeping layer performs its standard process.

An example to demonstrate the reconfiguration of mission goals is given as follows:

- **Scenario**: In the first uplink, the rover is given the following high-level goals in order of priority: to perform a science objective at way-point A, to take a PanCam image at way-point B, and to take a PanCam image at way-point C. Furthermore, there are unmissable goals including to stay stationary during one communication window and to return to the start point way-point D at the end of a standard cycle operation. The navigation techniques or the Navigator services available are based on a stereo-camera and a LIDAR where the stereo camera has higher localization errors but lower energy usage. The daily power limit only allows for the science objective at way-point A to be achieved with the stereo camera if the traverse is direct. The PanCam goals require minimal power, hence the LIDAR can be used. The cause for reconfiguration is an unexpected power drain due to a navigation choice leading to a dead end. This in turn causes the science goal to be no longer possible. The system aims to maximize the goal priority function in between the uplink and next downlink communication windows.
- **Results**: Figure 6.23 highlights the activities of the reconfiguration layer during the initial uplink. Upon receiving the uplink, the Ontology Manager validates

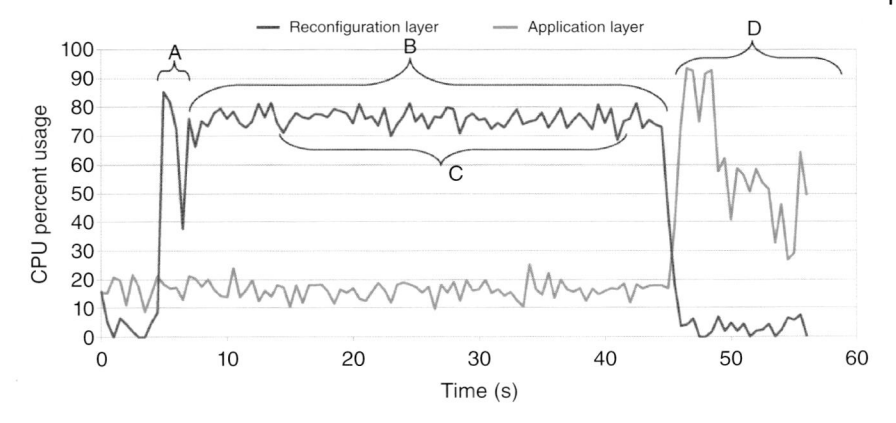

Figure 6.23 CPU percentage usage during the initial uplink reconfiguration of the self-reconfiguring GNC system: (A) the Ontology Manager is receiving, validating, and merging with the uplink instructions; (B) the Rational Agent plans, validates, and executes reconfiguration; (C) the Rational Agent uses the generic PDDL planner; (D) the application layer is initialized.

and merges the uplink message into the Ontology in 2.4 s. Since the Ontology is altered, the Rational Agent develops and validates a plan in 36.8 s where the generic planner uses 27.4 s of the overall time. The selected goal or configuration plan is to only perform the science objective at way-point A as this is given a weight of importance more than the combined weight of the other goals. The application layer is then configured as illustrated in Figure 6.25a and subsequently initialized. While traversing toward way-point A, the rover gets stuck in a gulley and needs to back-track. The Inspector in charge of monitoring the Navigator service reports to the Ontology of the world state change. The Ontology Manager then informs the Rational Agent who places the rover into a safety mode within 0.5 s (see Figure 6.24. The Rational Agent then calculates a plan for completing the Ontology's goals. The resulted plan is to reroute to the two PanCam goals, as the science objective is deemed no longer achievable. As a result, a new configuration of the rover GNC system's application layer is selected as shown in Figure 6.25b. The rover now does not require energy saving for the science payload, hence can use LIDAR as the primary Navigator option for better accuracy and system optimization. The rover exits safety mode within 8.5 s. The Robot Service Coordinator then calculates a schedule that takes 10.5 s. The application layer completes the schedule including one radio stop and returns to way-point D in 21 min 3 s.

6.6 Validation and Verification

V&V are independent procedures that are used together for checking that a product, service, or system meets requirements and specifications, hence fulfills its

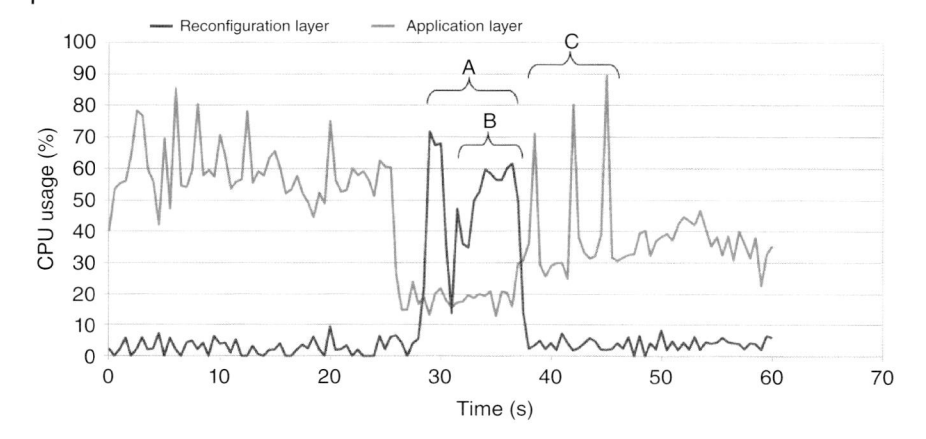

Figure 6.24 CPU percentage usage during the mission goal reconfiguration of the self-reconfiguring GNC system: (A) the Rational Agent plans, validates, and executes reconfiguration; (B) the Rational Agent uses the generic PDDL planner; (C) the application layer initiates scheduler.

intended purpose. For space applications in general, V&V is of great importance to the mission operation hardware and software. When comes to operation software, V&V is mandatory and increasingly challenging given the increase in LoA hence sophistication in associated autonomy designs, which often involve creating novel behaviors and making novel decisions. Engineers, designers, and scientists have difficulties to V&V artificially intelligent models and solutions by simple, human inspection. The quality and reliability of any AI-based system can be hard to assess due to the architectural complexity, the heterogeneity of semantics and of the algorithms involved, as well as the multitude of enabled behaviors. Therefore, automated V&V techniques or tools play an important role in mission operation software design and development.

The *validation* checks whether models, knowledge bases, and control accurately represent the expert knowledge and mission objectives; in other words, validation is to do with "building the right system" or asking "are the problem and domain models correct?." The *verification* checks whether the system and its components meet the specified requirements; in other words, it is to do with "building the system right" or asking "is the solution correct?." V&V can be applied at different stages of the knowledge engineering lifecycle including the domain, plan, planner/solver, and plan execution, and so on.

Both dynamic and static methods can be used for V&V of space mission operation software. The dynamic approach finds issues via simulation tools and experimental testings, which may have limited creation of certainties in the V&V result but is less computational. The static approach (such as in terms of formal methods such as model checking) attempts to mathematically verify the autonomous software, and hence is computationally heavy. There are also development efforts mainly driven by space agencies to establish V&V frameworks that aim to integrate different theoretical approaches into an unified test facility. The ESA's Harwell

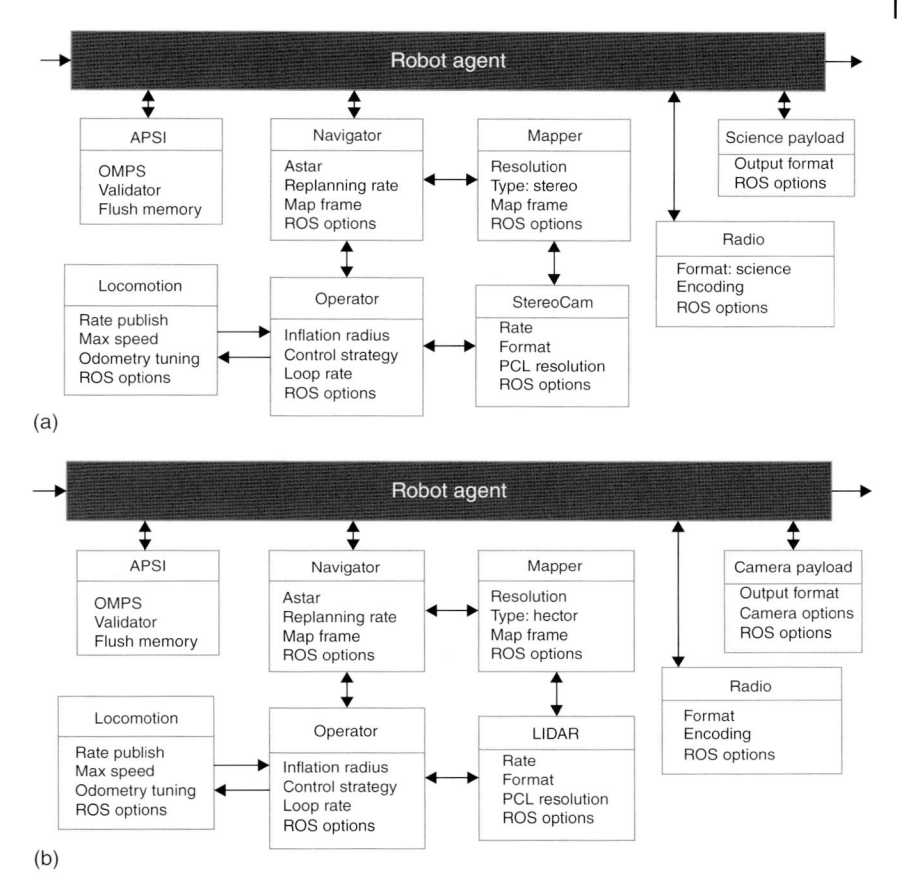

Figure 6.25 Configuration for the application layer of the self-reconfiguring GNC system. (a) First configuration, (b) second configuration.

Robotics and Autonomy Facility (HRAF) [81] is an example of such generic V&V facilities for planetary robotics. Some underpinning techniques for V&V in the space context are described further in the remainder of this section.

6.6.1 Simulation Tools

Two different types of simulators are often considered in relation to V&V of planetary robotic mission operations:

- **Realistic Simulators:** These simulators have exactly the same software as onboard the robot and have a extremely realistic model of the environment. This realism makes the possibility unaffordable to change the time factor, but can reproduce the activities performed onboard with high fidelity. Since the time scale cannot be changed, these testbeds are used offline to check whether

the software to be uploaded to the robot behaves as expected. This type of simulators can be purely software based or having hardware in the loop the V&V. To certain extend, field trials at planetary analog sites on Earth can be considered realistic simulations, which are common for testing specific payload operations, autonomous software packages and end-to-end subsystems.
- **Low-fidelity Simulators:** These simulators are generally created in computer software environment and have lower fidelity than realistic simulators. They can be used online to validate the plan since time factor can be adjusted. For instance, a 10-h activity in real time could run in 1 h using the simulator. This type of simulators is required in the frame of the tactical process when the plan need to be validated and/or verified for the next mission day in hours.

To give some examples, NASA's ROAMS [82] and ESA's 3DROV [83] are two software-based simulators with varying fidelity for planetary rovers. The **ESA's 3DROV** is designed for end-to-end simulation of rover operations. It includes models of the planetary environment, the mechanical, electrical, and thermal subsystems of the rover and of a generic onboard controller. The port-based modeling approach was adopted for the rover's physical subsystems and different levels of fidelity are foreseen for these models, depending on the scope of the simulation. Scientific instrument models are included for simulating science mission scenarios. The simulator also offers the ability to attach onboard algorithms for testing. The following key building blocks are identified within 3DROV [83]:

- The *simulation framework* relies on ESA's SIMSAT [84] and is responsible for the proper execution and scheduling of the simulation run. It provides the necessary mechanisms for the communication of the simulated models, the online 2D visualization of the simulation data, and the simulation control.
- The *control station* serves as the virtual rover's ground control station. It provides the means to setup mission scenarios via a 3D graphical environment and upload them to the rover while also assigned to display the onboard telemetry.
- The *generic controller* assumes the role of onboard flight software (as a SIMSAT component) and controls all the rover operations. Algorithms or software modules can be implemented within the generic controller (such as replacing or adding functionalities) for testing in the virtual environment.
- The *rover s/s component* includes models of rover physical subsystems, sensors, and scientific instruments. Although developed in different modeling environments, they comprise C/C++ code encapsulated in SIMSAT modules. Users can define or customize their own rover models.
- The *environment model* is responsible for the ephemeris and timekeeping of the system, for generating terrain and atmospheric conditions, and for tracking the rover location. Within the framework it interfaces with most of the rover models (such as contact model, thermal model, etc.) to provide them with the necessary environmental information.
- The *visualization environment* acts as the front-end of 3DROV, providing real-time visualization of the simulation run as well as the control station to assist

the preparation of activities. For example, it can be used to visualize 3D simulation independently or work with the rover physical s/s model and the generic controller in order to provide information on the wheel–terrain interaction or synthetic images to feed vision-based navigation algorithms.

The planetary exploration simulation environment created by the 3DROV tool can support V&V in early system design and technical assessment considering the robotic system into context, specific engineering studies (e.g., mobility, autonomy, operations), and early insight to rover operations. It makes relevant environmental information available early on in the system design process, setting up representative terrain, atmospheric conditions, temperature, and illumination, as well as introducing representative environmental components down to the effects of wind-blown dust. The tool simulates how the rover subsystems interact with each other, modeling the rover physical motion across the surface and the performance of the onboard scientific instruments, solar panels and power, electrical and thermal subsystems, down to the required detail level. The simulation also extends beyond the Martian surface to take account of orbital factors, such as calculating when the Sun is rising and times when the rover lines up either directly with Earth or else to a relay orbiter, making ground communications feasible.

6.6.2
Model Checking

It is crucial for high LoA functions such as the P&S to be validated and verified. This can be achieved through **formal methods** such as model checking as used by the Mars Express Science Plan Opportunities Coordination Kit (MrSPOCK), an use case of ESA's APSI framework described previously in Section 6.4.4. The MrSPOCK [85] solves the long-term planning problem for the Mars Express spacecraft, a multiobjective optimization problem that requires satisfaction of a number of temporal and causal constraints. In general, the goal of MrSPOCK is to develop a preplanning optimization tool for the spacecraft operations. It focuses on the generation of a preoptimized skeleton long-term plan, which can be subject to refinement on behalf of the science and engineering team. To improve effectiveness and confidence of using MrSPOCK, V&V of the autonomous software implemented based on model checking. In this effort, model validation and solver V&V have been addressed using two prominent model checking tools, namely NuSMV and UPPAAL.

NuSMV [86] is a model checker for temporal logics. It has a dedicated modeling language, which permits the definition of concurrent finite state systems in an expressive, compact, and modular way. The SMV specification uses variables with finite types, grouped into a hierarchy of module declarations. Each module states its local variables, their initial value, and how they change from one state to the next. The properties are expressed in Computation Tree Logic (CTL). CTL is a branching-time temporal logic, which means that it supports reasoning over both the breadth and the depth of the tree of possible executions.

UPPAAL [87], whose acronym comes from joining the names of UPPsala and AALborg universities that built it, is a tool box for modeling, simulation, and verification of real-time systems. The verifier covers the exhaustive dynamic behavior of the system for proving safety and bounded liveness properties. A UPPAAL model consists of a set of timed automata, a set of clocks, global variables, and synchronizing channels. A node in an automaton may be associated with an invariant, for enforcing transitions out of the node. An arc may be associated with guards, for controlling when this transition can be taken. On any transition, local clocks may get reset and global variables may get reassigned. Channels are used in order to synchronize transitions on different automata. Analogously to NuSMV, verified properties are stated in CTL.

In order to perform validation, MrSPOCK domain and plan are encoded in a new model, while a property assuring model or plan validity is defined. Both the model and the property are provided as inputs to model checking tool as shown in Figure 6.26. The translation from the MrSPOCK planning model to a model checker formal model requires introduction of a well-defined set of state variables and clocks [88]. State variables range on domain states, while clocks are used to represent time progression. For each timeline in the model, a state variable automaton is introduced whose states correspond to possible values of the timeline, while the transitions represent the value changes. In addition, the so-called "observer automaton" is introduced to check the consistency of the temporal constraints defined among different timelines. Once the translated model is available as the input for the model checkers, the MrSPOCK model can be validated with respect to various properties and requirements by setting a CTL formula for the model checker. For instance, it can be verified that whenever a given activity is performed by a subsystem, the results must be processed by another subsystem. Whenever the CTL formula provided in input does not hold, the model checker

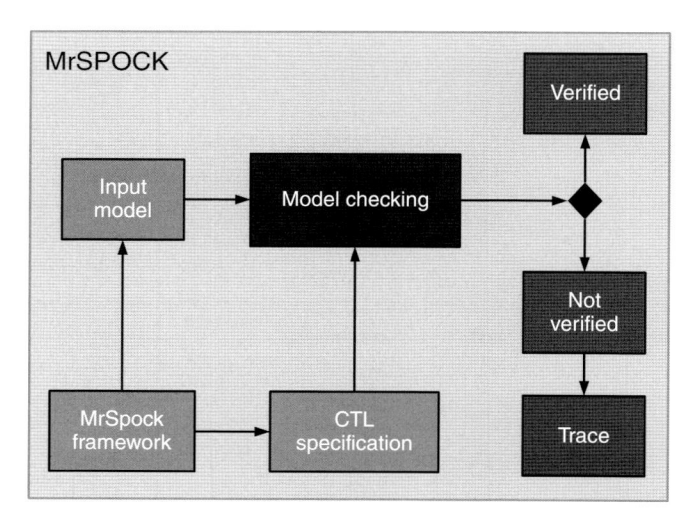

Figure 6.26 Plan validation using model checking within MrSPOCK.

produces an execution trace proving that the system reached an error state. The reported trace can be used to identify the domain inconsistency and to diagnose the conditions it originated from.

Plans generated by MrSPOCK provide a set of decisions/activations over the timelines. For each timeline, a generated plan provides a set of activations at fixed or flexible time points or planned timeline; therefore, a plan describes the sequence of values the timelines have to assume in a given time frame. The observer automaton in the translated model is extended to check not only the domain constraints but also the synchronization between the values defined in the generated plan and the possible values modeled. Therefore, in the observer automaton, there is a transition that triggers whenever a timeline in the model and the value decided by the planner cannot be aligned. Once the input model is completed and forwarded to the model checkers, it is possible to formulate and verify plan properties. In particular, using the observer automaton, the plan validity property can be formulated: for each timeline, a no-error status for the monitor is always requested.

Planner validation is based on a plan verification tool that checks the solution generated by MrSPOCK with respect to the specified properties. Plan verification requires an input model that encodes both the MrSPOCK domain specification and the generated plan. In this case, the model checker can verify whether the generated plan is actually a good controller for the controlled systems, that is, the model checker verifies whether changes to plan executions and state variables can be synchronized or not.

The MrSPOCK work has demonstrated how V&V can have practical impact in the P&S. It is worth noting that the solving system of MrSPOCK is based on a hybrid approach where the solving and optimization are performed by different modules. Not all the domain constraints can be explicitly represented in the plan domain; therefore, the soundness of the generated plan with respect to the domain model does not necessarily ensure the soundness of the produced solution with respect to the real world model. The use of an independent solution verifier has major value, not only for model validation and plan verification but also for testing consistencies of the generated plans with respect to the implicit requirements, for example, those to be enforced by heuristics or optimization processes. In addition, from the end-user perspective the V&V tools offer an independent testing environment, which may enhance users' trust on the complex and sometimes counterintuitive solutions generated by MrSPOCK.

6.6.3
Ontology-based System Models

In Section 6.5, ontology has demonstrated abilities to model knowledge of the planetary robots and the world using formal methods (such as first-order logic), which are then used to govern and validate reconfiguration of robotic systems. Ontology can also model systems using informal methods such as modeling languages and allow knowledge representations to be highly granular, extensible, human usable, and equipped with the tools available to the formal methods for model checking, logical inferencing, and theorem proving, and so on. Hence, ontology-based system models can serve as a natural way for automatic V&V.

One of the most widely used modeling languages is the Systems Modeling Language (SysML), which implements Model-Based System Engineering (MBSE) philosophy hence supports specification, design, analysis, V&V of systems, and systems-of-systems. The models in SysML are split into (i) structural/component models, such as block diagrams and parametric diagrams; (ii) behavioral models, such as state machine models, activity models, sequence models and use case models; and (iii) other engineering analytic models, such as requirement models. With a planetary robotic system modeled in SysML, the V&V can be performed to the mission operation software at any stage of the mission development, starting from the theoretical conceptualization and requirement definition, to system preliminary and critical design. This includes automatic simulation and automatic formal V&V allowing complex issues to be identified early and throughout the development cycle. Moreover, MBSE offers added benefits of reduced risk via improved estimates and continuous analysis, improved communications in large multilayered teams, and use of best system engineering practices [89].

System models developed for a space system in SysML can have an underlying ontology that defines formal relationships between entities. The ontology can be further used to perform model checking [90], automated test case generations [91], or automated theorem proving [92]. This allows both the logic and model to be easily extensible and human readable. The ontology descriptions can be made highly modular including the logic complexity can be modularized to increase efficiency. This extends the capabilities of the ontology to support *composite V&V*, which focuses V&V of the composed of self-contained subsystems. In this case, system V&V problems are modularized into composition of subsystem problems. Thus, run-time V&V of composed system becomes plausible, for example, reconfigurable autonomy in a highly generic and modular system model.

6.7
Case Study: Mars Rovers' Goal-Oriented Autonomous Operation

The general concepts of mission operation software (covering both ground and space segment) have been introduced and explained previously in Section 6.3 with subsequent sections describing various underlining technologies. This section further presents a design example on Mars rovers' goal-oriented autonomous operation (i.e., LoA E4) and demonstrates its design process and results leveraging the state-of-the-art techniques. This case study primarily builds on the results of ESA's GOAC project [28] (previously introduced in Section 6.3.3).

6.7.1 Design Objectives

The GOAC project aimed to tackle some classical challenges of the three-layered autonomous software designs particularly within the executive and deliberative layers, while applying to the Mars rovers to achieve LoA E4 as follows:

- to reach a coherent and consistent manner to handle high-level goal commanding, in a deterministic way,
- to decompose these goals into lower-level commands that comprise a plan to achieve the commanded goal,
- to perform the forwarding of the lower-level commands by an executive to the functional layer of the rover,
- to evaluate the correct execution of the plan, detecting when a goal has been achieved, and in case the plan for that goal fails perform a reevaluation of the plan when needed,
- to maximize the results obtained by the rover via an optimal usage of resources such as power,
- to increase the level of abstraction, developing an architecture that will allow users to deal with classical problems for multiagent planning in a model-based environment.

Keeping the decisions generated by the deliberative layer deterministic is key to the robot in critical environments, as this leads to possibilities for V&V. Being deterministic, it is assured that under the same sequence of inputs from the world, the rover will behave in the same manner. This requirement particularly important to space or planetary applications since it reduces uncertainties while consolidating a test campaign.

6.7.2 Onboard Software Architecture

To achieve goal-oriented autonomy or LoA E4, the GOAC is expected to ensure P&S capabilities onboard the rover. Once the onboard plan is generated, an executive is responsible for its correct execution, and determines in real time whether the plan is matched or possible for replanning dynamically. In the classical three-layer architectures previously introduced in Section 6.3.3, planning/deliberation and execution are separate tasks within autonomous software or "the controller." This means that it takes the controller some time to produce a plan. Once a plan is produced, it is dispatched for execution. If for some reason the plan fails, the executive layer must stop executing and the deliberation layer must produce a new plan. This approach can be inefficient. For GOAC, the deliberation and execution are tightly coupled, allowing a more efficient way of replanning when needed. A detailed illustration of the GOAC software architecture is shown in Figure 6.27.

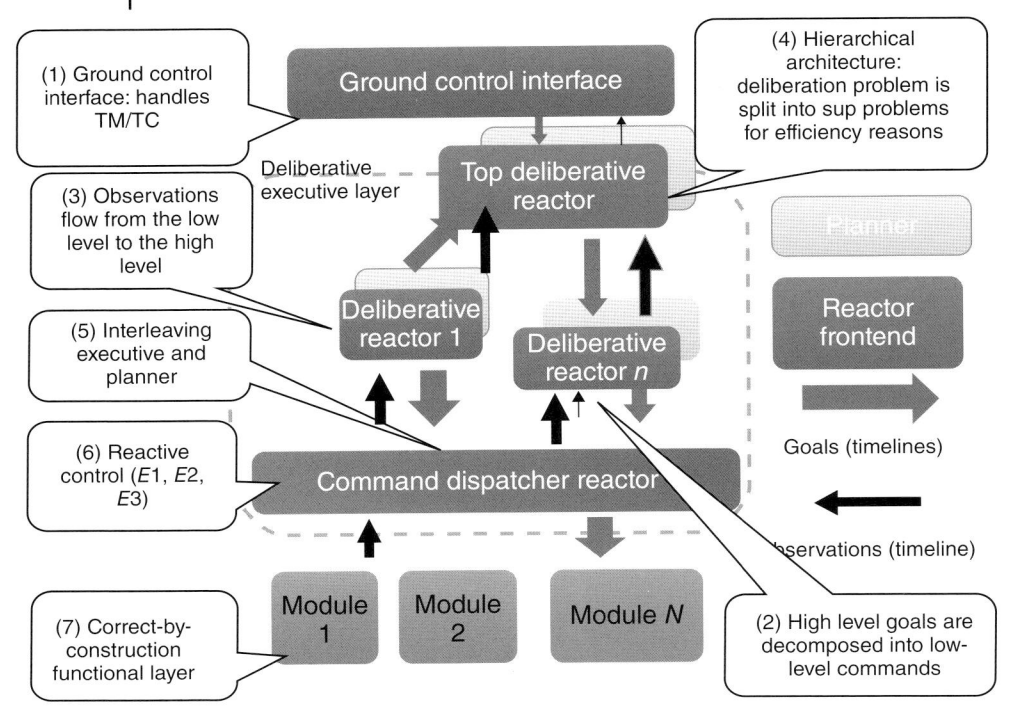

Figure 6.27 GOAC architecture.

The GOAC architecture builds on the T-REX agent [93] that can be viewed as the coordinator of a set of concurrent control loops. Each control loop is embodied in a teleo-reactor (or reactor in short) that encapsulates all details of how to accomplish its control objectives. Reactors are in charge of solving particular problems, such as mission operations, navigation or scientific target detection, by applying the principle of *divide and conquer?* to the problem of real-time deliberation. There is a natural hierarchy of reactors, from the reactor in charge of high-level abstraction such as the mission level with longest deliberation horizon and greatest latency, to the reactor dealing with low-level abstraction such as controlling the hardware beneath, forwarding commands or collect sensory observations in relation to the functional layer. Two different types of reactors exist in GOAC: *deliberative reactors* (that have an associated planning instance) and *reactive reactors* (that do not require deliberation hence have no associated planning instance). Each reactor has a different deliberation horizon (i.e., the maximum time that it can deliberate) and a different latency (i.e., time required to provide a response). It uses a different instance of the planner who takes into account the corresponding deliberation horizon and latency.

Although the number and type of reactors are configurable, the GOAC architecture only has two reactive reactors: a **Ground Control Interface Reactor** that is in charge of handling incoming telecommands from ground and providing the required telemetry, and a **Command Dispatcher Reactor** that is in charge of handling lower levels of autonomy (e.g., E1–E3) and sending the proper commands to the functional layer. Furthermore, the command dispatcher is the lowest-level reactor with the lowest deliberation horizon and the minimum latency because it does not plan nor uses any planning instance. Effectively the command dispatcher is able to handle LoA E1–E3 without any other reactor, except the ground control interface reactor. Partitioning the controller in different reactors increases robustness since any failure in the controller can be localized to a reactor, allowing graceful system degradation.

Each deliberative reactor uses a planning instance or timeline-based planner, where the APSI (closely described in Section 6.4.4) is the default planner in GOAC. However, it is possible to change the planner without altering the architecture as long as the planner uses timelines to deliberate. The DDL.3 language (previously introduced in Section 6.4.2.7) is used in GOAC to provide the domain description of the rover in its operational environment. By relying on domain models, the planner remains a generic software component of the controller. In addition, it is possible to embed the definition of a mission in the models. Since models are interpreted by the planner when the controller software starts, it is easy to change the possible behaviors of the rover by just uploading a new model.

The communication between reactors is exclusively based on **timelines**. Timelines are sequences of procedures that encapsulate state evolution, which in turn are composed of tokens. Each token describes a procedure invocation, the state variables on which it can occur, the parameter values of the procedure, and the time values defining the interval of execution of the procedure. Each timeline is owned by a single reactor (i.e., only that reactor can modify it) but can be observed by many different reactors who subscribe to the timeline owned by the given reactor.

Observations capture the current value of a timeline and are asserted by the owner of a timeline, for example, an observation could be the rover is positioned at the location (X, Y). The observations flow from the functional layer to the lowest command dispatcher reactor, enforcing changes in its timelines that are forwarded to the upper reactors. This in turn cause changes in the timelines of the upper reactors or new observations.

Commands are sent to the reactors as **goals** in the form of tokens, that is, predicates with start and end time bounds defining the temporal scope over which they hold. These goals are decomposed by a reactor into sets of subgoals that are assigned to lower-level reactors. Therefore, commands flow from higher-level reactors to lower-level reactors, and in the end trigger actions at the functional layer requested by the command dispatcher reactor. Command propagation for all the reactors is handled by executing a single process that is invoked on each tick. The expansion of commands is performed from the top to the bottom. An example for the "goal" command can be "take a picture at point X, Y with pitch angle Z and yaw angles W today between 10:00 and 14:00." The predicate "take a picture" is possible status (token) of a state variable that can be "operation being performed" (timeline).

The low-level, functional layer controls the hardware, which performs V&V as well. This is the most important layer with respect to online verification techniques to mitigate dangerous situations derived from deadlocks and violation of constraints. Use of the real-time verification (i.e., BIP) guarantees that the functional layer will not commit an action that will harm the rover. Unlike the formal methods in V&V such as model checking that has the state explosion problem, describing the functional layer using the BIP formal language, can perform offline analysis of the modules and express constraints at run time. The correct-by-construction techniques help to enforce a set of restrictions, for example, the drill cannot be deployed while the rover is moving, to increase the robustness of the system. In addition, the GenoM framework [22] is used to implement the modules of the functional layer. It handles all aspects related to the communication between modules hence allows developers to focus on the module definition (including interface between offered services and exported data) and algorithms.

As a result, this architecture offers the following attributes:

1) **Scalability**: The number of elements in the architecture can be changed adapting to the complexity of the system, without affecting the functionality of the system.
2) **Selectable level of autonomy**: System can be commanded to work in LoA E1, E2, E3, or E4. For instance, reducing LoA from E4 to E3 can be achieved by disabling all the deliberative reactors and canceling their plans.
3) **Model-based**: For the deliberative reactors, problems are modeled using a planning domain language. In the functional layer, both GenoM and BIP allow generation of code based on stubs. Therefore, the architecture can be built on models, which ease the instantiation of the controller for a particular robotic platform.
4) **Collaborative scenario**: Multiple elements of the architecture collaborate to achieve common goals.
5) **Planning at different levels**: A high-level planner can coordinate and execute a global plan. The low-level planning can resolve planning problems in short time scales. This architecture can offer several different layers of deliberation with different deliberation horizon and latency. It is, therefore, possible to set in the temporal space the level of abstraction of the domain.
6) **Goal commanding**: The architecture offers capability to handle high-level goals at LoA E4.
7) **Correct-by-construction**: The design involves techniques to avoid violation of constraints during the development of the system as well as when the system is running, which addresses the V&V.

6.7.3
Implementation and Validation

The GOAC architecture has been implemented and validated using two test cases as illustrated in Figure 6.28.

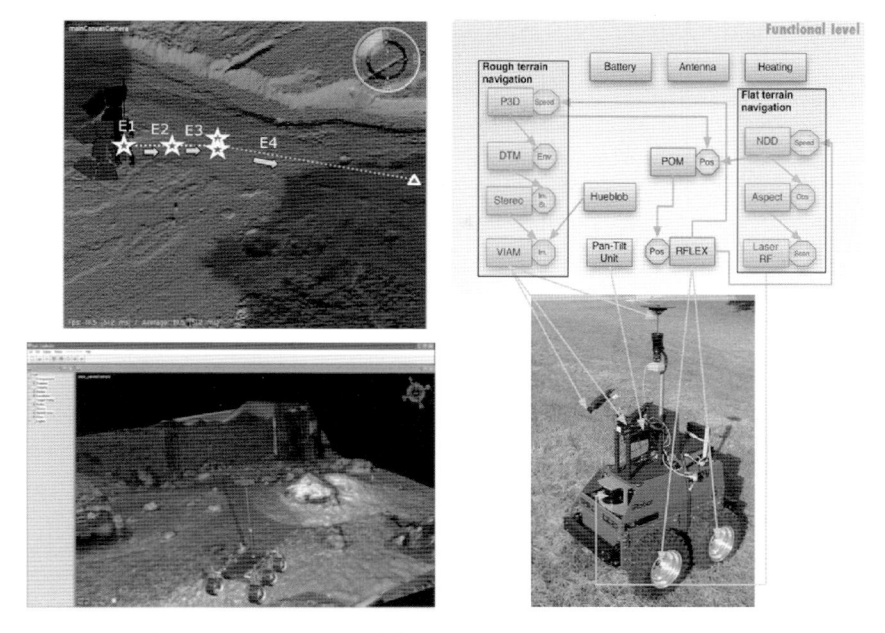

Figure 6.28 The 3DROV environment (left column, Courtesy TRASYS) and the DALA rover (right column, Courtesy LAAS-CNRS).

First, the system was tested using the **simulation tool 3DROV** (previously explained in Section 6.6.1), where computer simulation models of a Mars rover and terrain are used in the 3DROV simulation environment. Several mission scenarios have been applied including opportunistic science, replanning due to navigation failure, replanning due to experiment failure and testing the different (E1 – E4) autonomy levels. The simulations were carried out using the Linux Instance running inside a virtual machine, with a base memory of 768 MB and 32 GB of virtual disk, running in an Intel(R) Core(TM)2 Duo CPU E8400 at 3.0 GHz with 2 GB of RAM. Figure 6.29a and 6.29b illustrates the CPU and memory usage, respectively, during the execution of GOAC in 3DROV. The main test results show the following:

- The base use of CPU when GOAC is not running is 3%.
- The average use of CPU by GOAC is 35% (38% − 3%).
- The use of CPU is relatively stable. The peaks correspond to the time when the GOAC is planning.
- The deliberative reactor and embedded planner use more memory than the other processes. This is mainly due to the Java Virtual Machine used where the planner APSI runs in a Java instance.
- The peaks of CPU usage correspond to the time when APSI is actively planning for the initial goal or replanning. Other times the use of CPU by the planner is low.
- The CPU usage by the 3DROV simulator, T-REX executive, and interprocess communication is negligible.

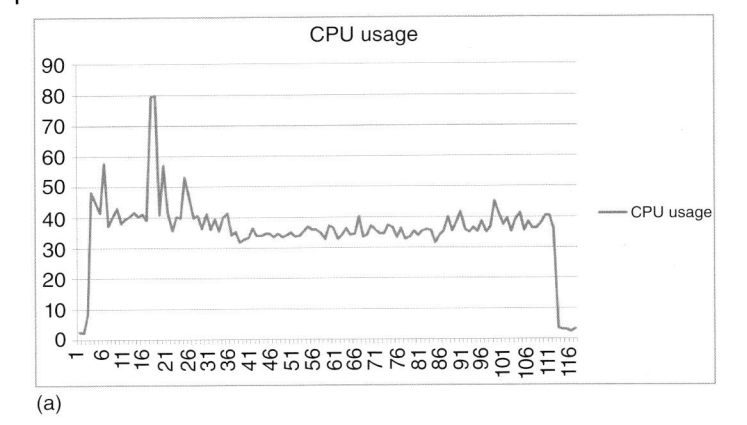

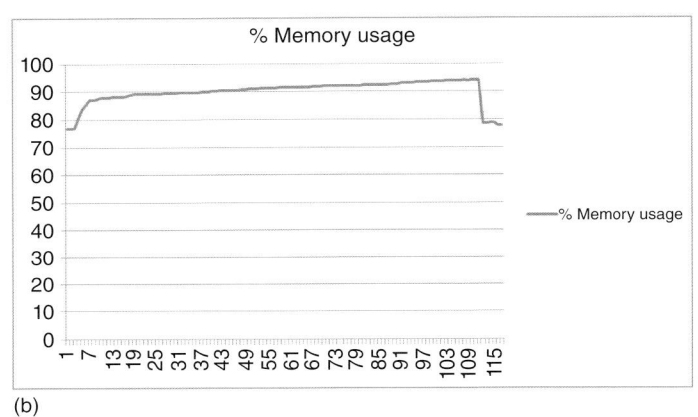

Figure 6.29 GOAC test results using 3DROV.

- The memory usage by the 3DROV simulator and interprocess communication is negligible.

Reliability and goal achievement in the 3DROV tests: The main experiments worked fine without any deviation in the successive tests. However, the 3DROV simulator can get frozen after running for a period of time (days) due to memory leak caused by the complexity of its environment model. The problem was isolated and addressed every time the system got restarted periodically. Though changes were done to the simulation to increase speed by a multiplying factor, it was not allowed to run in real time. Furthermore, the possibilities have been demonstrated to use two deliberative reactors (i.e., two planning instances with different domains), using different latency and deliberation horizons. The look-ahead and latency of the deliberative reactors play important roles in the performance of the onboard controller. As a general rule of thumb, the more complex the domain model is, the higher LoA the GOAC can achieve. In addition, the synchronization rules have significant impact on how long it takes the planner to produce a plan and to be fine-tuned. The 3DROV simulator uses a generic controller hence cannot consider a full representative model for the functional layer (including GenoM and BIP). To do so, a second test case is carried out as discussed later.

Second, the system was tested using a **physical rover called DALA** from LAAS. DALA is an iRobot ATRV platform that provides a large number of sensors and effectors. The mission scenarios to test a combination of the following capabilities: (i) navigation under uncertainties due to the characteristics of the terrain; (ii) navigation to take different sets of pictures at given locations; (iii) communication with the orbiter during visibility windows; (iv) opportunistic science; and (v) monitoring of the resources. Effectively, the rover had a number of concurrent tasks to achieve:

- to navigate safely in an a priori unknown environment using two navigation modes based on a rough map of the environment (flat terrain vs rough terrain);
- to take high-resolution images of an user-defined list of locations (simulating science targets);
- to communicate with orbiters or a lander during visibility windows;
- to continuously monitor the environment for opportunistic science and take appropriate actions when something interesting is detected;
- to monitor and control the thermal status of the platform and the payload;
- to monitor and control the power usage and energy consumption.

In this test case, the DALA rover simulated some typical Martian day (sol) operation. Its functional layer ran in a single processor (a Linux machine with two Intel Pentium 4, 3.06 GHz CPUs with 512 KB cache each, and 1 GB of total memory). The T-REX agent (with the reactors) and the APSI planner ran in a separate machine. The functional model was tested using two different configurations: (i) configuration including the GenoM functional components of the robot and (ii) configuration including both GenoM and BIP. As commented before, the BIP allows a set of restrictions at the functional layer to be defined while the extra safety feature causes extra CPU usage. The run-time performance of BIP and GenoM for various modules within the functional layer is summarized in Table 6.6. The results show the CPU execution time taken by each module within the functional layer (in 0.01 s). The average value over five runs is used.

It is evident that the execution time with BIP enabled took approximately 10 times more CPU time than the GenoM-only scenario driven by the BIP engine. Such CPU demand can be reduced with improved real-time BIP by limiting number of interactions and/or restrictions evaluated at each loop in the BIP modules. However, the time duration of a complete experiment was approximately 4.3 min in average for both executions (GenoM or GenoM+BIP), where the second configuration takes only about 4 s longer (considered negligible). This is partly because when using the GenoM+BIP configuration CPUs are used at higher capacity (52% usage) than the GenoM-only case (6.3% usage). In addition, executing actions in the real world take longer time (in seconds) than the time difference caused by BIP, for example, moving DALA from (x, y) coordinates $(0, 0)$ to $(4, 0)$ takes approximately 30 s, moving the PTU toward the left or right front wheels of the

Table 6.6 GOAC test results using DALA: CPU time taken by different modules within the functional layer.

Module	GenoM (0.01 s)	GenoM+BIP (0.01 s)
LaserRF	120	1 947
Aspect	192	2 362
NDD	43	2 009
RFLEX	168	10 763
Antenna	56	1 102
Battery	69	1 219
Heating	92	1 029
PTU	126	2 394
Hueblob	690	1 850
VIAM	1046	3 225

rover takes approximately 5 s, and transmitting a single picture to the orbiter takes approximately 5 s.

Reliability and goal achievement in DALA tests: The high-level goals were successfully achieved by the DALA rover during majority of the outdoor test runs. Failures in most cases were due to hardware problems such as the camera, network bandwidth or serial port. The aim to demonstrate high-level planning/execution framework and the ability to handle undesirable/unsafe situations via the BIP framework was achieved.

6.7.4
Integration with Ground Operation

The GOAC software design onboard the rover is the driver for achieving LoA E4 in a relevant planetary mission. Nevertheless, this design allows varying LoA (from E1 to E4) to be selected depending on its operational situation. For example, the SOWG and RIPT previously mentioned in Section 6.2.2 sometimes may want to take direct control over the rover and not rely on the rover to identify science targets or perform certain engineering activities. As a result, the LoA becomes a state variable of the rover that can be changed either autonomously by the robot (e.g., when a failure is detected) or by the ground station.

Integrating GOAC with the ground operation for different LoA needs to establish the following issues:

- *Type of activity execution control*: For LoA E1, the ground station uplinks commands for immediate execution. This is not practical when there is no direct communication links with the rover unless results of the commands can be evaluated in the long term. The LoA E2 relies on time-tagged commands, which are not reliable to cope with uncertainties and unknowns of the environment. The

LoA E3 can be adaptive or allow event-driven procedure execution as demonstrated by the MER mission rovers. The LoA E4 allows goal-oriented operations where the ground control is more concerned about "what to do" rather than "how to do it."

- *Level of abstraction for the planners handled on board and on ground*: At E1 and E2, planning on ground is captured in all levels of abstraction either by human operators or machines, which produces low-level commands for uplink. At E4, planning can be performed entirely onboard based on goals and generated low-level commands. Intermediate levels of abstraction in planning such as represented by Onboard Control Procedures (OBCPs) can also be commanded at E3 onboard.
- *End-to-end control architecture that distributes responsibilities between the ground and space segment*: This determines where/how the deliberation takes place, including the explicit consideration of alternative courses of actions, generating alternatives and choosing one of the possible alternatives. After goal commanding is issued, the LoA E4 requires deliberation to take place onboard, while E1 and E2 restrict the deliberation on ground. Modern mission operations often adopt a mixed initiative that combines *automated machine initiative* and *human initiative* in the deliberation or decision-making process similar to that during the MER mission. In the mixed-initiative operations, human operators are in control, guidance, and monitoring of the automated machine reasoning methods, and overriding them if deemed necessary. This requires a continuous dialogue between the operators and the planner who share decision-making responsibilities. The distribution of responsibilities can range from E1 to E4.

From the mission operation perspective, a planning problem has a set of goals, which still have not been justified while a planning solution represents a plan where all the goals have been justified. A plan is the structure represented as a temporal network containing a set of activities used by the deliberation or the planner. The plan structure is ideally the same between the ground and onboard operations at different LoA, meaning all the elements (such as goals and commands) within the temporal network share a common representation. The GOAC has attempted to provide a generic planning structure to allow smooth interaction between the ground and onboard autonomy as well as smooth transition between different LoA, as described in the following:

- *Goal commanding for E4*: LoA E4 autonomy requires the plan to be produced onboard the rover based on uplinked goals from the ground. The plan, however, may be incomplete for various reasons, such as the goal is not fully supported due to unjustified preconditions or a replanning activity or being uplinked incorrectly. In these cases, new supporting goals should be added. In addition, there can be uncertainties when executing a plan due to change of resources, which may lead to invalid plans at the execution time. In this situation, the planner onboard must fix the flaws. As part of the downlink process, the onboard operation software should update its ground counterpart in terms of plan change so that the overall plan structure is synchronized.

- *Event-action for E3*: At LoA E3, granularity of the produced plan needs to be time-tagged and event-based sequences of commands. Therefore, the temporal network is constrained to fixed time boundaries as soon as a final plan is selected for uplink (unlike E4). The sequence of commands comprises of a mission timeline containing a set of time-tag commands, and a table of events and actions. These can be seen as sets of fully scheduled procedures (i.e., OBCPs and actions) that will be triggered when certain events arise. For example, a procedure can represent a action of "take a picture of a rock" that is only triggered or executed when the rover reaches certain position and location. Therefore, the procedure adapts to the final position of the rover (to orientate the camera) and the local time (to configure the camera with respect to the light conditions), and finally gets scheduled to allocate all the actions within the execution time intervals.
- *Time-tagged commanding for E2*: The LoA E2 executes preplanned time-based commands, which are uplinked from the ground in the form of batch files. These batch files are common to both nominal and nonnominal situations, such as FDIR functions to guarantee the rover safety. The on-ground planner in this LoA must generate a fully defined plan based on commands fixed on time. This can be achieved by means of constraints describing the time bounds for the execution of each command. The onboard software must understand these commands for execution.

6.7.5 Design Remarks

A few extra remarks worth mentioning within the GOAC design:

- *Use of legacy software*: GOAC builds various existing software legacies, each of which offers specific solutions, including the T-REX for interleaved planning and execution, the APSI for timeline-based planning, the GenoM for concrete component-based functional layer, and the BIP for real-time verification.
- *Design life cycle*: GOAC employs the spiral or iterative design life cycle rather than the waterfall. The two implementation and test campaigns (one using computer simulators 3DROV and one using realistic testbeds DALA) overlap in schedule and complement each other. Test results or lessons learned from both have benefited one or the other, as well as the overall generic architecture design.

6.8 Future Trends

Future planetary exploration missions using robots are envisaged to become more ambitious and complex. Highest LoA for the robotic systems, subsystems and components will be required, reached, and surpassed. Avenues for further advancement involve topics such as autonomicity and multiagent systems, where many open issues still exist.

6.8.1
Autonomic Robotics

Planetary robotic systems have been moving from automation toward autonomy. The next step is autonomicity [94], which is an extension to autonomy by adding self-management characteristics to the system such as self-CHOP [49]. Autonomicity can be achieved from top-down, as a fourth layer in the three-layer control architecture. This new layer can configure, heal, optimize, protect, and manages the three lower levels of autonomous components and their interactions. Alternatively, autonomicity can be achieved from a bottom-up, self-organizing system in which individual components act as agents attempting to complete independent goals via reasoned interactions. Some initial attempts on autonomicity for space applications have been presented previously in Section 6.5, including:

- An example of a bottom-up autonomic system is the AGATA [61]. All AGATA components have a generic schema, which carry a model of itself and a self- generating model of the modules it interacts with, a state tracking sub component, and decision-making subcomponent. This allows the system to self-organize; however, this distribution of reasoning and distrust of components can lead to nonoptimal results.
- An example of a top-bottom can be seen in the ontology-based reconfigurable autonomy [63] where a central Rational Agent self-reconfigures the lower level autonomy modules. This method is demonstrated in achieving a self-reconfiguring GNC system and reconfiguring the GNC software and software architecture for optimized and increased robustness at all levels of control (i.e., from high-level goal-oriented decision making to low-level hardware control). This work demonstrates the potential of improved robustness and self-optimization that autonomicity can offer.

6.8.2
Common Robot Operating System

Given the success of Robot Operating System (ROS) within the terrestrial robotics community, there are initiatives particularly in Europe to establish a similar generic operating system for space or planetary robotic systems. The main motivation is to allow a common, potentially open source framework to better facilitate and coordinate R&D effort in robotics for space from academia and industry [95]. This will be a low-level software system (opposite to autonomicity mentioned earlier) or mid-ware that interfaces the robot hardware and onboard operation software. The main functions are envisaged to include hardware abstraction, low-level control drivers, file/data system operation and management, real-time tasks (scheduling, communication and synchronization), FDIR, networking, command processing, telemetry/telecommand monitoring, and API for the above-mentioned, and so on.

6.8.3
MultiAgent Systems

Another route to increase the capability and robustness of a planetary robotic mission is to increase the number of robots. Multiple robots have the benefit of parallelism and added redundancy. Tasks can be divided and conquered among a team of robots, which can potentially complete more complex tasks than a single comprehensive robot. Existing R&D efforts have been made in promoting a multiple-tiered robotic system that combines the planetary orbiter, aerobot, and surface rover [96, 97], or the cooperative rovers with high LoA working together toward common mission goals [98], or the fractional robots that can physically combine to gain increased capabilities being together [99].

These multiagent systems can be achieved through complex bottom-up algorithms such as self-organizing systems [100] but optimal solutions are not guaranteed without an overarching agent. Alternatively, the system could have top-down control such as the tiered design in Ref. [96] where the high-level decisions are made by higher level robots or a planner that oversees the multirobot system such as in Ref. [98]; however, this requires more complex autonomous agents in the higher tiers, managing more components.

References

1. Washington, R., Golden, K., Bresina, J., Smith, D.E., Anderson, C., and Smith, T. (1999) Autonomous rovers for Mars exploration. *Proceedings of IEEE Aerospace Conference*, vol. 1, pp. 237–251.
2. Pedersen, L., Smith, D.E., Deans, M., Sargent, R., Kunz, C., Lees, D., and Rajagopalan, S. (2005) Mission planning and target tracking for autonomous instrument placement. *Proceedings of IEEE Aerospace Conference*, pp. 34–51.
3. ESA Requirements and Standards Division (2008) ECSS: Ground Systems and Operations - Monitoring and Control Data Definition, ecss-e-st-70-31c edition.
4. ESA Requirements and Standards Division (2008) ECSS: Space Segment Operability, ecss-e-st-70-11c edition.
5. ESA Publications Division (2003) ECSS: Ground Systems and Operations - Telemetry and Telecommand Packet Utilization, ecss-e-70-41a edition.
6. CCSDS official website http://public.ccsds.org/publications/default.aspx (accessed 08 April 2015).
7. Trucco, L.J., Franceschetti, P., Martino, M., and Trichilo, M. (2008) ExoMars rover operation control center design concept and simulations. *Proceedings of Advanced Space Technologies in Robotics and Automation (ASTRA)*.
8. Erickson, J. and Grotzinger, J. (2014) MSL Extended Mission Plan, http://mars.nasa.gov/files/msl/2014-MSL-extended-mission-plan.pdf (accessed 08 April 2015).
9. Truszkowski, W., Hallock, H., Rouff, C., Karlin, J., Rash, J., Hinchey, M., and Sterritt, R. (2009) *Autonomous and*

Autonomic Systems: With Applications to NASA Intelligent Spacecraft Operations and Exploration Systems, Springer Science & Business Media.

10. Eggleston, J., Haddow, C., and Affaitati, F. (2009) EGOS Core components. *Proceedings of SpaceOps*.
11. Nerri, M. (2009) Smart software for complex space mission data systems at the European space agency. *Proceedings of AIAA SPACE*.
12. Reggestad, V., Livanos, N.A.I., Antoniou, P., and Zois, E. (2012) Introducing parallelization & performance optimization in SIMULUS based operational simulators. *Proceedings of International ICST Conference on Simulation Tools and Techniques*, pp. 220–222.
13. Garcıa-Matamoros, M.A., Kuijper, D., and Righetti, P.L. (2003) NAPEOS: ESA/ESOC navigation package for earth observation satellites. *Proceedings of the European Workshop on Flight Dynamics Facilities, Darmstadt, Germany*.
14. Del Rey, I., Navarro, V., and Pe nataro, J.R. (2010) DABYS: EGOS generic database system. *Proceedings of SpaceOps*.
15. Aghevli, A., Bachmann, A., Bresina, J., Greene, K., Kanefsky, B., Kurien, J., McCurdy, M., Morris, P., Pyrzak, G., Ratterman, C. et al. (2006) Planning applications for three Mars missions with Ensemble. International Workshop on Planning and Scheduling for Space.
16. Norris, J.S., Powell, M.W., Vona, M.A., Backes, P.G., and Wick, J.V. (2005) Mars exploration rover operations with the science activity planner. *Proceedings of IEEE International Conference on Robotics and Automation*, pp. 4618–4623.
17. Reeves, G.E. and Snyder, J.F. (2005) An overview of the Mars exploration rovers' flight software. *Proceedings of IEEE International Conference on Systems, Man and Cybernetics*, vol. 1, pp. 1–7.
18. Biesiadecki, J.J. and Maimone, M.W. (2006) The Mars exploration rover surface mobility flight software driving ambition. *Proceedings of IEEE Aerospace Conference*, p. 15.
19. Woods, M., Shaw, A., Barnes, D., Price, D., Long, D., and Pullan, D. (2009) Autonomous science for an ExoMars Rover–like mission. *Journal of Field Robotics*, **26** (4), 358–390.
20. Helmick, D., McCloskey, S., Okon, A., Carsten, J., Kim, W., and Leger, C. (2013) Mars Science Laboratory algorithms and flight software for autonomously drilling rocks. *Journal of Field Robotics*, **30** (6), 847–874.
21. Ingrand, F., Lacroix, S., Lemai-Chenevier, S., and Py, F. (2007) Decisional autonomy of planetary rovers. *Journal of Field Robotics*, **24** (7), 559–580.
22. Ceballos, A., De Silva, L., Herrb, M., Ingrand, F., Mallet, A., Medina, A., and Prieto, M. (2011) GenoM as a robotics framework for planetary rover surface operations. *Proceedings of Advanced Space Technologies in Robotics and Automation (ASTRA)*, pp. 12–14.
23. Ingrand, F. and Py, F. (2002) An execution control system for autonomous robots. Proceedings of IEEE International Conference on Robotics and Automation, vol. 2, pp. 1333–1338.
24. Ingrand, F.F., Chatila, R., Alami, R., and Robert, F. (1996) PRS: a high level supervision and control language for autonomous mobile robots. Proceedings of IEEE International Conference on Robotics and Automation, vol. 1, pp. 43–49.
25. Ghallab, M. and Laruelle, H. (1994) Representation and control in IxTeT, a temporal planner. *Proceedings of AIPS*, pp. 61–67.
26. Muscettola, N., Nayak, P.P., Pell, B., and Williams, B.C. (1998) Remote agent: to boldly go where no AI system has gone before. *Artificial Intelligence*, **103** (1), 5–47.
27. Chien, S., Sherwood, R., Tran, D., Cichy, B., Rabideau, G., Castano, R., Davis, A., and Boyer, D. (2005) Using autonomy flight software to improve science return on earth observing one. *Journal of Aerospace Computing,*

Information and Communication, **2**, 196–216.

28. Ceballos, A., Bensalem, S., Cesta, A., De Silva, L., Fratini, S., Ingrand, F., Ocon, J., Orlandini, A., Py, F., Rajan, K. et al. (2011) A goal-oriented autonomous controller for space exploration. *Proceedings of ESA Symposium on Advanced Space Technologies in Robotics and Automation (ASTRA)*, vol. 11.

29. Basu, A., Bensalem, S., Bozga, M., Combaz, J., Jaber, M., Nguyen, T.-H., and Sifakis, J. (2011) Rigorous component-based system design using the BIP framework. *IEEE Software*, **28**, 41–48.

30. Ghallab, M., Nau, D., and Traverso, P. (2004) *Automated Planning: Theory & Practice*, Elsevier.

31. Wilkins, D.E. (2014) *Practical Planning: Extending the Classical AI Planning Paradigm*, Morgan Kaufmann.

32. Biere, A., Heule, M., and van Maaren, H. (2009) *Handbook of Satisfiability*, vol. **185**, IOS Press.

33. Ghédira, K. and Dubuisson, B. (2013) *Foundations of CSP*, John Wiley & Sons, Inc., pp. 1–28.

34. Vilain, M., Kautz, H., and Beek, P. (1986) Constraint propagation algorithms for temporal reasoning, in *Readings in Qualitative Reasoning About Physical Systems*, Morgan Kaufmann, pp. 377–382.

35. Allen, J.F. (1983) Maintaining knowledge about temporal intervals. *Communications of the ACM*, **26** (11), 832–843.

36. Dechter, R., Meiri, I., and Pearl, J. (1991) Temporal constraint networks. *Artificial Intelligence*, **49** (1), 61–95.

37. Frank, J. (2013) What is a timeline? *Proceedings of Knowledge Engineering for Planning and Scheduling Workshop of International Conference on Automated Planning and Scheduling*, pp. 1–8.

38. Muscettola, N. (1994) *HSTS: Integrating Planning and Scheduling*, Morgan Kaufmann, pp. 169–210.

39. Victoria, J. M. D., Fratini, S., Policella, N., von Stryk, O., Gao, Y., and Donati, A. (2014) Planning mars rovers with hierarchical timeline networks. *Acta Futura*, **9** (6), 21–29, DOI: 10.2420/AF09.2014.21

40. Fox, M. and Long, D. (2003) PDDL2.1: an extension to PDDL for expressing temporal planning domains. *Journal of Artificial Intelligence Research*, **20**, 61–124.

41. Bylander, T. (1994) The computational complexity of propositional strips planning. *Artificial Intelligence*, **69** (1), 165–204.

42. Barreiro, J., Boyce, M., Do, M., Frank, J., Iatauro, M., Kichkaylo, T., Morris, P., Ong, J., Remolina, E., Smith, T. et al. (2012) EUROPA: a platform for AI planning, scheduling, constraint programming, and optimization. *Proceedings of 22nd International Conference on Automated Planning & Scheduling (ICAPS)*.

43. Cesta, A., Fratini, S., Oddi, A., and Pecora, F. (2008) APSI Case#1: pre-planning science operations in Mars Express. *Proceedings of International Symposium on Artificial Intelligence, Robotics and Automation in Space (i-SAIRAS)*.

44. Laborie, P. and Ghallab, M. (1995) Ix-TeT: an integrated approach for plan generation and scheduling. *Proceedings of INRIA/IEEE Symposium on Emerging Technologies and Factory Automation*, vol. 1, pp. 485–495.

45. Chien, S., Rabideau, G., Knight, R., Sherwood, R., Engelhardt, B., Mutz, D., Estlin, T., Smith, B., Fisher, F., Barrett, T. et al. (2000) ASPEN: automated planning and scheduling for space mission operations. *Proceedings of SpaceOps*, pp. 1–10.

46. Fratini, S. and Cesta, A. (2012) The APSI framework: a platform for timeline synthesis. *Proceedings of AAAI Workshop on Planning and Scheduling with Timelines*, pp. 8–15.

47. Alur, R. and Dill, D.L. (1994) A theory of timed automata. *Theoretical Computer Science*, **126** (2), 183–235.

48. Dennis, L.A., Fisher, M., Aitken, J.M., Veres, S.M., Gao, Y., Shaukat, A., and Burroughes, G. (2014) *Reconfigurable Autonomy*, KI-Künstliche Intelligenz, pp. 1–9.

49. Kephart, J.O. and Chess, D.M. (2003) The vision of autonomic computing. *Computer*, **36** (1), 41–50.
50. Kramer, J. and Magee, J. (2007) Self-managed systems: an architectural challenge. Future of Software Engineering, pp. 259–268.
51. Nikdel, P., Hosseinpour, M., Badamchizadeh, M.A., and Akbari, M.A. (2014) Improved Takagi-Sugeno fuzzy model-based control of flexible joint robot via Hybrid-Taguchi genetic algorithm. *Engineering Applications of Artificial Intelligence*, **33**, 12–20.
52. Flueckiger, L., To, V., and Utz, H. (2008) Service-oriented robotic architecture supporting a lunar analog test. *Proceedings of International Symposium on Artificial Intelligence, Robotics, and Automation in Space (iSAIRAS*, Citeseer.
53. Leite, L.A.F., Oliva, G.A., Nogueira, G.M., Gerosa, M.A., Kon, F., and Milojicic, D.S. (2013) A systematic literature review of service choreography adaptation. *Service Oriented Computing and Applications*, **7** (3), 199–216.
54. Yeung, W.L. (2011) A formal and visual modeling approach to choreography based web services composition and conformance verification. *Expert Systems with Applications*, **38** (10), 12772–12785.
55. Costa-Soria, C., Pérez, J., and Carsí, J.A. (2011) An aspect-oriented approach for supporting autonomic reconfiguration of software architectures. *Informatica: An International Journal of Computing and Informatics*, **35** (1), 15–27.
56. Dhouib, S., Kchir, S., Stinckwich, S., Ziadi, T., and Ziane, M. (2012) RobotML, a domain-specific language to design, simulate and deploy robotic applications, in *Simulation, Modeling, and Programming for Autonomous Robots*, Springer-Verlag, pp. 149–160.
57. Dormoy, J., Kouchnarenko, O., and Lanoix, A. (2012) Using temporal logic for dynamic reconfigurations of components, in *Formal Aspects of Component Software*, Springer-Verlag, pp. 200–217.
58. Delaval, G. and Rutten, E. (2010) Reactive model-based control of reconfiguration in the fractal component-based model, in *Component-Based Software Engineering*, Springer, pp. 93–112.
59. Williams, B.C. and Nayak, P.P. (1996) A model-based approach to reactive self-configuring systems. *Proceedings of the National Conference on Artificial Intelligence*, pp. 971–978.
60. Hayden, S., Sweet, A., and Christa, S. (2004) Livingstone model-based diagnosis of earth observing one. *Proceedings of the AIAA 1st Intelligent Systems Conference*.
61. Charmeau, M.-C. and Bensana, E. (2005) AGATA: a lab bench project for spacecraft autonomy. International Symposium on Artificial Intelligence Robotics and Automation in Space (iSAIRAS).
62. Liu, L., Thanheiser, S., and Schmeck, H. (2008) A reference architecture for self-organizing service-oriented computing, in *Architecture of Computing Systems*, Springer-Verlag, pp. 205–219.
63. Burroughes, G. and Gao, Y. (2016) Ontology-based self-reconfiguring guidance, navigation and control for planetary rovers. *Journal of Aerospace Information Systems*, DOI:10.2514/1.I010378.
64. Shaukat, A., Burroughes, G., and Gao, Y. (2015) Self-reconfigurable robotics architecture utilising fuzzy and deliberative reasoning. SAI Intelligent Systems Conference 2015, pp. 258–266.
65. Shaukat, A., Bajpai, A., and Gao, Y. (2015) Reconfigurable SLAM utilising fuzzy reasoning. 13th ESA Symposium on Advanced Space Technologies in Robotics and Automation (ASTRA).
66. Bertoli, P., Pistore, M., and Traverso, P. (2010) Automated composition of web services via planning in asynchronous domains. *Artificial Intelligence*, **174** (3-4), 316–361.
67. Tenorth, M. and Beetz, M. (2013) KnowRob: a knowledge processing infrastructure for cognition-enabled robots. *International Journal of Robotics Research*, **32** (5), 566–590.

68. Chrisley, R. (2003) Embodied artificial intelligence. *Artificial Intelligence*, **149** (1), 131–150.
69. Tenorth, M., Perzylo, A.C., Lafrenz, R., and Beetz, M. (2012) The RoboEarth language: representing and exchanging knowledge about actions, objects, and environments. IEEE International Conference on Robotics and Automation, pp. 1284–1289.
70. Hayden, S.C., Sweet, A.J., and Shulman, S. (2004) Lessons learned in the Livingstone 2 on earth observing one flight experiment. Proceedings of the AIAA 1st Intelligent Systems Technical Conference, pp. 1–15.
71. Dalpiaz, F., Giorgini, P., and Mylopoulos, J. (2009) An architecture for requirements-driven self-reconfiguration, in *Advanced Information Systems Engineering*, Springer-Verlag, pp. 246–260.
72. Valls, M.G., Lopez, I.R., and Villar, L.F. (2013) iLAND: an enhanced middleware for real-time reconfiguration of service oriented distributed real-time systems. *IEEE Transactions on Industrial Informatics*, **9** (1), 228–236.
73. Cardellini, V., Casalicchio, E., Grassi, V., Iannucci, S., Lo Presti, F., and Mirandola, R. (2012) MOSES: a framework for QoS driven runtime adaptation of service-oriented systems. *IEEE Transactions on Software Engineering*, **38** (5), 1138–1159.
74. Calinescu, R., Grunske, L., Kwiatkowska, M., Mirandola, R., and Tamburrelli, G. (2011) Dynamic QoS management and optimization in service-based systems. *IEEE Transactions on Software Engineering*, **37** (3), 387–409.
75. Hallsteinsen, S., Geihs, K., Paspallis, N., Eliassen, F., Horn, G., Lorenzo, J., Mamelli, A., and Papadopoulos, G.A. (2012) A development framework and methodology for self-adapting applications in ubiquitous computing environments. *Journal of Systems and Software*, **85** (12), 2840–2859.
76. Esfahani, F.S., Azrifah, M., Murad, A., Nasir, Md., Sulaiman, B., and Udzir, N.I. (2011) Adaptable decentralized service oriented architecture. *Journal of Systems and Software*, **84** (10), 1591–1617.
77. Deussen, P.H., Höfig, E., Baumgarten, M., Mulvenna, M., Manzalini, A., and Moiso, C. (2010) Component-ware for autonomic supervision services. *International Journal On Advances in Intelligent Systems*, **3** (1 and 2), 87–105.
78. Dragone, M., Abdel-Naby, S., Swords, D., MP O'Hare, G., and Broxvall, M. (2013) A programming framework for multi-agent coordination of robotic ecologies, in *Programming Multi-Agent Systems*, Springer, pp. 72–89.
79. Vizcarrondo, J., Aguilar, J., Exposito, E., and Subias, A. (2012) ARMISCOM: autonomic reflective middleware for management service composition. Global Information Infrastructure and Networking Symposium (GIIS), pp. 1–8.
80. Yoon, Y., Ye, C., and Jacobsen, H.-A. (2011) A distributed framework for reliable and efficient service choreographies. *Proceedings of the 20th International Conference on World Wide Web*, pp. 785–794.
81. Allouis, E., Blake, R., Gunes-Lasnet, S., Jorden, T., Maddison, B., Schroeven-Deceuninck, H., Stuttard, M., Truss, P., Ward, K., Ward, R., and Woods, M. (2013) A facility for the verification & validation of robotics & autonomy for planetary exploration. *Proceedings of Advanced Space Technologies in Robotics and Automation (ASTRA)*.
82. Jain, A., Balaram, J., Cameron, J., Guineau, J., Lim, C., Pomerantz, M., and Sohl, G. (2004) Recent developments in the ROAMS planetary rover simulation environment. Aerospace Conference, 2004. Proceedings. 2004 IEEE, vol. 2, IEEE, pp. 861–876.
83. Poulakis, P., Joudrier, L., Wailliez, S., and Kapellos, K. (2008) 3DROV: a planetary rover system design, simulation and verification tool. *Proceedings of International Symposium on Artificial Intelligence, Robotics and Automation in Space (i-SAIRAS)*.
84. Eggleston, J., Boyer, H., van der Zee, D., Pidgeon, A., de Nisio, N., Burro, F.,

and Lindman, N. (2005) *Proceedings of 6th International Symposium on Reducing the Costs of Spacecraft Ground Systems and Operations (RCSGSO), ESA SP-601*, Darmstadt, Germany, European Space Agency.

85. Cesta, A., Cortellessa, G., Fratini, S., and Oddi, A. (2011) MrSPOCK - steps in developing an end-to-end space application. *Computational Intelligence*, **27** (1), 83–102.

86. Cimatti, A., Clarke, E., Giunchiglia, E., Giunchiglia, F., Pistore, M., Roveri, M., Sebastiani, R., and Tacchella, A. (2002) NuSMV 2: an opensource tool for symbolic model checking. *Proceedings of International Conference on Computer Aided Verification*, pp. 359–364.

87. Behrmann, G., David, A., Larsen, K.G., Hakansson, J., Petterson, P., Yi, W., and Hendriks, M. (2006) Uppaal 4.0. Proceedings of International Conference on Quantitative Evaluation of Systems, pp. 125–126.

88. Cesta, A., Finzi, A., Fratini, S., Orlandini, A., and Tronci, E. (2010) Validation and verification issues in a timeline-based planning system. *Knowledge Engineering Review*, **25** (03), 299–318.

89. INCOSE (2007) Systems Engineering Vision 2020, incose-tp-2004-004-02 edition.

90. Huang, H., Tsai, W.-T., and Paul, R. (2005) Automated model checking and testing for composite web services. Object-Oriented Real-Time Distributed Computing, 2005. ISORC 2005. 8th IEEE International Symposium on, IEEE, pp. 300–307.

91. Bai, X., Dong, W., Tsai, W.-T., and Chen, Y. (2005) WSDL-based automatic test case generation for web services testing. Service-Oriented System Engineering, 2005. SOSE 2005. IEEE International Workshop, IEEE, pp. 207–212.

92. Schneider, M. and Sutcliffe, G. (2011) Reasoning in the OWL 2 Full ontology language using first-order automated theorem proving, in *Automated Deduction–CADE-23*, Springer-Verlag, pp. 461–475.

93. McGann, C., Py, F., Rajan, K., Thomas, H., Henthorn, R., and McEwen, R. (2007) T-REX: a model-based architecture for AUV control. *Proceedings of 3rd Workshop on Planning and Plan Execution for Real-World Systems*.

94. Truszkowski, W.F., Hinchey, M.G., Rash, J.L., and Rouff, C.A. (2006) Autonomous and autonomic systems: a paradigm for future space exploration missions. *IEEE Transactions on Systems, Man, and Cybernetics Part C: Applications and Reviews*, **36** (3), 279–291.

95. Gao, Y., Samperio, R., Shala, K., and Cheng, Y. (2012) Modular design for planetary rover autonomous navigation software using ros. *Acta Futura*, **5** (1), 9–16.

96. Fink, W., Dohm, J.M., Tarbell, M.A., Hare, T.M., and Baker, V.R. (2005) Next-generation robotic planetary reconnaissance missions: a paradigm shift. *Planetary and Space Science*, **53** (14), 1419–1426.

97. Isarabhakdee, P. and Gao, Y. (2009) Cooperative control of a multi-tier multi-agent robotic system for planetary exploration. *Proceedings of International Joint Conference on Artificial Intelligence (IJCAI) - Workshop on AI in Space*.

98. Victoria, J.M.D., Yeomans, B., Gao, Y., and von Stryk, O. (2015) Autonomous mission planning and execution for two collaborative Mars rovers. *13th ESA Symposium on Advanced Space Technologies in Robotics and Automation (ASTRA)*.

99. Toglia, C., Kennedy, F., and Dubowsky, S. (2011) Cooperative control of modular space robots. *Autonomous Robots*, **31** (2-3), 209–221.

100. Mushet, G.S., Mingotti, G., Colombo, C., and McInnes, C.R. (2014) Self-organizing satellite constellation in geostationary Earth Orbit. *IEEE Transactions on Aerospace and Electronic Systems.*, **51** (2), 910–923.

Index

a

active compliance control methods   273
admittance control   278
Advanced Stirling Radioisotope Generator (ASRG)   82
air-bearing tables (ABT)   300
alpha particle X-ray spectrometer (APXS)   38
Analytical Laboratory Drawer (ALD)   44
angular random walk (ARW)   204
Apollo Lunar Roving Vehicle (LRV) navigation system   184
application program interface (API)   366
artificial intelligence   169
Automated Planning and Scheduling Initiative (APSI)   356, 358
– architecture   360
– consumable resources   359
– domain theory   360
– linear reservoir resource   359
– reusable resources   359
– rover domain   361
– state variables   358
– timelines   358
Automated Scheduling and Planning ENvironment (ASPEN)   355
automatic direction finding (ADF) technique   204
autonomic robotics   395
Autonomous Sciencecraft Experiment (ASE)   339
autonomous-navigation   141
autonomy
– communication latency   321
– definition   321
– environment uncertainty   321
– mission operation *see* mission operation
– operational cost   322
– reconfigurable *see* reconfigurable autonomy
Autonomy Generic Architecture, Tests and Application (AGATA)   364

b

Beagle 2 arm   260
Beagle 2 lander   28
bilateral control   291

c

calibration   106
– error-influence   125
– geometric-calibration   121
– radiometric-calibration   124
Canadian Space Agency (CSA)   200, 240
CASPER software   340
Chang'e 3 mission   42
Chemistry and Mineralogy (CheMin)   41
CHOP   363
classical planning methods   344
Close-Up Imager (CLUPI)   44
collision avoidance   288
command dispatcher reactor   387
Command Packet File (CPF)   335
commercial off-the-shelf (COTS)   195
Common Allocation Scheduling Tool (CAST)   335
compliance control   278
computed torque control 281
constraint satisfaction problem (CSP)   345
Consultative Committee for Space Data Systems (CCSDS)   325
Continuous Activity Scheduling Planning Execution and Re-planning (CASPER)   356
coordinate measurement machines (CMM)   298
Curiosity rover   39, 266
Curiosity rover navigation system   188

d

DALA rover   391
DARPA Robotics Challenge (DRC)   310
Deep Space 1 (DS1)   155
Deep Space Network (DSN) communication system   200
design specifications
– autonomous capabilities   266
– dexterous workspace   264
– mass   265
– number of degrees of freedom   265
– on/off-board processing   266
– payload   265
– power   265
– reachable workspace   265
– redundancy   266
– size   265
– software control algorithms   266
design trade-offs
– arm kinematics   268
– sensors   269
– structure and material   269
Designing LoA   294
dexterous workspace   265
digital elevation model (DEM)   129, 135, 186, 204, 211, 220
digital signal processor (DSP)   198
direct force control methods   274
direct-to-Earth (DTE) link   38
Domain Definition Language (DDL)   354
Domain-Specific Views   325
Doppler effect   195
dual-arm manipulation   304
dust deposition   55
Dust Removal Tool (DRT)   41
dust storms   268
dynamic control   278
– electrical motor model   281
– feedforward plus feedback control   281
– mechanical motor model   280
– torques   279
dynamic model-based control methods   274

e

Earth's Deep Space Network   182
electric heaters   92
Electrical Power Subsystem (EPS) design   68
electro-dynamic screen (EDS)   81
ESA's Ground Operation System (EGOS)   332
European Cooperation for Space Standardization (ECSS)   5, 289
– level of autonomy (LoA)   329
– onboard and ground segment   323
– SSM and the Domain-Specific View   325
– telemetry and telecommand packet utilization   325
European Space Agency (ESA)   199
European Space Operations Centre (ESOC)   332
ExoMars mission   27, 111, 119, 202, 207
ExoMars rover   43
Extended Information Filter (EIF)   215
Extended Kalman Filter (EKF)   215
Extensible Universal Remote Operations Planning Architecture (EUROPA)   356
extra-vehicular activities (EVAs)   184
extrinsic calibration   121

f

feature integration theory (FIT)   159
feedforward plus feedback control   281
field of view (FOV) sensors   202
field programmable gate array (FPGA)   111
force control
– active compliance control methods   273
– admittance control   278
– compliance control   278
– direct   274
– dynamic model   274
– hybrid force/motion control   275
– impedance control   275
– indirect   274
– parallel force/motion control   275
– robust and adaptive robot   273
– static model   274
– stiffness control   277
force torque sensors   270
fuel cells (FCs)   88

g

geographic information system (GIS) methods   209
Geographical Image Database Server (GIDS)   154
geometric-calibration   121
Goal-Oriented Autonomous Controller (GOAC)   384, 385, 340
– 3DROV environment   389
– DALA   391
– DDL.3 language   387
– deliberative reactors   386
– design   394
– LoA   392
– reactive reactors   386
– 3-layered autonomous software designs   385
– T-REX agent   386
– timelines   387

graphical processing unit (GPU)   198
graphical user interface (GUI)   331
gravity   49
ground control interface reactor   386
ground vision processing   128
– compression-decompression   129
– offline-localization   133
– 3D mapping   129
– visualization-simulation   135
ground-based vision processing   7
ground-penetrating radar (GPR)   44
Guidance, Navigation and Control (GNC)   332
– application layer   369
– housekeeping layer   372
– mission operations impact   376
– ontology design   372
– planetary rovers   367
– Rational Agent design   374

h
H.E.A.R.D scale   69
HazCams   190
Heuristic Scheduling Testbed System (HSTS)   355
Hierarchical TimeLine Network (HTLN)   350
High Resolution Imaging Science Experiment (HiRISE) camera   183
high-level robotic arm control strategies   270
– LoA   289
– path planning   289
– tele-manipulation *see* tele-manipulation
– Shared control/E3 LoA   293
– Supervisory control/E2 LoA   293
– Traded control/E4 LoA   294
HiRISE imagery   209
Huygens lander   25

i
Imager for Mars Pathfinder (IMP) camera   144
impedance control   275
incremental bundle adjustment (IBA) algorithm   133, 185
Indexed Time Table (IxTeT)   355
indirect force control methods   274
Inertial Measurement Units (IMUs)   192, 195, 203
infrared (IR) cameras   245
"Infrared Spectrometer for ExoMars" (ISEM)   44
instrument deployment device (IDD)   38, 259

International Planning Competition (IPC)   352
iterated closest point (ICP) algorithm   211

j
Johnson Space Center (JSC)   311

k
kinematic calibration   298
kinematic control   271

l
laser-induced breakdown spectrometer (LIBS)   41
level of autonomy (LoA)   289, 329
light detection and ranging (LiDAR)   118, 195
localization technologies and systems
– autonomous navigation   219
– celestial localization   214
– control strategy   227
– example systems   218
– global path planning   225
– inertial measurement units   203
– local path-planning and obstacle avoidance   224
– mapping   220
– orbiting asset/earth-based localization   212
– orientation estimation   201
– particle filters   216
– path planning   223
– rover-to-orbiter digital elevation model matching   211
– rover-to-orbiter horizon matching   211
– rover-to-orbiter imagery matching   209
– simultaneous localization and mapping   216
– speed sensing   208
– star trackers   203
– Sun finding   202
– terrain assessment   222
– 3D visual odometry   208
– vision techniques   204
– visual odometry   205
– wheel odometry   205
loop heat pipes   95
Low Complexity Lossless Compression (LOCO) software   129
low intensity low temperature (LILT) silicon solar cells   30
low-level robotic arm control strategies
– dynamic control *see* dynamic control
– force control *see* force control
– position control   271
– trajectory interpolation   285

Index

low-level robotic arm control strategies (*contd.*)
– visual servoing 283
low power high torque actuators (LPHTAs) 263
Lunar Reconnaissance Orbiter (LRO) 143, 183
lunar-vision 142
Lunokhod 1, 31
Lunokhod II 184
Lunokhod missions 77

m

manipulability 269
mapping process 220
Markov decision processes (MDPs) 351
Mars Ascent Vehicle (MAV) 63
Mars exploration missions 162
Mars Exploration Rovers (MERs) 36, 259, 290, 322
– curiosity rover navigation system 188, 189
– digital elevation model (DEM) 186
– in-mission flight software 188
– offline processing 185
– path-planning 186
– terrain assessment module 187
– visual odometry 185, 188
Mars Local Geodesic (MLG) frame 229
Mars Multispectral Imager for Subsurface Studies (Ma-MISS) 44
Mars Orbiter Laser Altimeter (MOLA) data 186
Mars Organic Molecule Analyzer (MOMA) 44
Mars Pathfinder (MPF) 34
Mars Polar Lander (MPL) 23
Mars rovers GOAC *see* Goal-Oriented Autonomous Controller (GOAC)
Mars Sample Return (MSR) campaign 63
Mars Science Laboratory robotic arm 261
Mars Surveyor 98 robotic arm 256
Mars Surveyor 2001 robotic arm 256
Mars Surveyor Program (MSP) 23
Martian regolith 258
Martian soil 165
MastCam 39
mechanical motor model 280
Micro electro-mechanical systems (MEMS) 204
Miniature Integrated Camera And Spectrometer (MICAS) 155
Mission Control System (MCS) software 334
mission operation 322
– CCSDS standards 325
– design considerations software 331
– ECSS 324
– ground operation software 332
– ground segment 323
– Mars rovers *see* Goal-Oriented Autonomous Controller (GOAC)
– onboard operation software 336
– onboard segment 323, 329
– P and S *see* planning and scheduling (P&S) 343
– performance measures 341
– procedures 326
– V and V *see* Validation and Verification (V&V)
Mission Specific Data Processing (MSDP) 153
Mixed nitiative Activity Plan GENerator (MAPGEN) 334
mobile manipulation
– autonomous 309
– DRC 310
– JSC 311
– research platforms 309
Monitor, Analyse, Plan, Execution - Knowledge (MAPE-K) loop 365
Mössbauer Spectrometer (MB) 38
multi-agent systems 396
Multimission Image Processing Lab (MIPL) 145
Multi-Mission Radioisotope Thermoelectric Generator (MMRTG) 82

n

NASA 199
NASA Discovery Program 25
NASA MRO Context Camera (CTX) 6m imagery 210
NASA Resource Prospector mission 241
Navigation Cameras (NavCam) 41, 119, 190, 229
Navigation Service Set 370
Near Earth Asteroid Rendezvous (NEAR) 156
neoclassical methods 344
New Domain Definition Language (NDDL) 353
Noise Equivalent Spectral Radiance (NESR) 125

o

onboard-vision-processing
– autonomous-navigation 141
– compression-modes 138
– stereo-perception-software-chain 139

– visual-odometry   140
on-line trajectory generation   286
Open Procedural Reasoning System (OpenPRS)   339
optical depth (OD)   55

p

"Pan-Tilt Units"(PTU)   130
Path Planning module   232
path planning   289
phase spectrum of quaternion Fourier transform (PQFT)   158
Phoenix Mars Lander robotic arm   258
Phoenix mission   24
photovoltaic/solar cells   80
plan-space planning (PSP)   346
planetary landers   105
planetary robotic arms
– Beagle 2 arm   260
– Mars Surveyor '98/'01   256
– MSL   261
– Phoenix   258
planetary robotic system design
– atmosphere and vacuum   51
– autonomous functions   76
– Beagle 2 lander   28
– Chang'e 3 rover   42
– cold case   92
– curiosity rover   39
– drivers identification   17
– dust   55
– ExoMars rover   43
– functional analysis   14
– functional requirements   15
– gravity   49
– heat management   94
– heat provision   92
– lander mission   19
– launch environment   59
– length of local day   53
– liquid   56
– locomotion subsystem options   17
– long-term mission concepts   47
– Luna sample return landers   20
– Lunokhod   1, 31
– Lunokhod   2, 32
– Mars 2020 rover   46
– Mars Sample Return (MSR) campaign   63
– Mars surveyor lander family and successors   23
– mass   59
– medium-term mission concepts   47
– mission concept   19
– mission definition   14
– mission-driven system design drivers   58
– operation sequence   72
– orbital characteristics   52
– Philae lander   29
– planetary bodies and moons   56
– power generation   80
– power storage   84
– power subsytem   79
– Prop-M rovers   33
– rock   54
– scoring system   17
– SFR design evaluation   68
– SFR subsystem design drivers   65
– SFR system design drivers   64
– Sojourner rover   34
– spirit and opportunity rovers   36
– Sun distance   53
– surface conditions   54
– surface deployment   60
– surface operation   63
– tade-off options   96
– target environment   59
– technical requirements   15
– teleoperation *vs*. onboard autonomy   77
– temperature   49
– thermal subsystem   90
– Viking landers   21
– warm case   92
planetary robotics
– performance characteristics   4
– robotic agents   5
– robotic explorers   6
– robotic manipulators   8
– robotics assistants   6
Planetary Robotics Vision Scout (PRoViScout) field trials   241
planetary robots   106
planetary surface navigation   7, 181, 228
– Canadian Space Agency (CSA)   240
– cooperative robotics and mobility concepts   242
– ExoMars rover   229
– future navigation technologies   235
– future rover missions   234
– Mars 2020 rover   233
– Mars precision lander   234
– NASA and CSA   236
– orbital imagery   245
– processing capabilities   244
– Resource Prospector mission   234
– SEEKER field trial   239
– sensors   244
– SLAM Systems   242

PLanetary Undersurface TOol (PLUTO)   28
Planner Service Set   370
planning and scheduling (P&S)
– APSI framework *see* Automated Planning and Scheduling Initiative (APSI)
– ASPEN   355
– CASPER   356
– classical planning methods   344
– EUROPA   356
– handling uncertainties   350
– HSTS   355
– IxTeT   355
– neoclassical methods   344
– planning languages *see* planning languages
– scheduling problem   350
– solving strategies   346
– temporal planning   347
Planning Domain Definition Language (PDDL)   352
planning languages
– DDL   354
– NDDL   353
– PDDL   352
– PDL   354
– STRIPS   351
planning-graph algorithms   346
position adjustable workbench (PAW)   28, 260
position and velocity sensors   270
Powder Acquisition Drilling System (PADS)   41
Predicted Events File (PEF)   335
printed circuit board (PCB)   50
Problem Definition Language (PDL)   354
Prop-M rovers   33

r

radiators   94
radio service   370
radiometric-calibration   124
Rational Agent design   374
reachable workspace   265
reconfigurable autonomy
– AGATA   364
– GNC *see* Guidance, Navigation and Control (GNC)
– model-based autonomy plant engine   364
– ontology-based cite method   365
– rationale   362
– self-managed software architectures   363
– service-based methods   363
– taxonomy   365
Re-engineered Space Flight Operations Schedule (RSFOS)   335
Reliable Autonomous Surface Mobility (RASM) system   221
Remote Agent Experiment (RAX)   339
Remote Micro Imager (RMI)   41
Resource Prospector mission   234
Robot Operating System (ROS)   395
Robot Sequencing and Visualization Program (RSVP)   136
Robot Service Coordinator   369
robotic arm control
– collision avoidance   288
– high-level control strategies *see* high-level robotic arm control strategies   289
– low-level control strategies *see* low-level robotic arm control strategies
– trajectory generation   270
robotic arm system design
– design specifications   264
– design trade-offs *see* design trade-offs
– environmental design considerations   267
– performance requirements   264
robotics vision   105
Rock Abrasion Tool (RAT)   38
Rover Environmental Monitoring Station (REMS)   41
Rover Integrated Planning Team (RIPT)   327
rover navigation system
– Apollo Lunar Roving Vehicle (LRV) navigation system   184
– communication system   182
– computational resources   198
– design decisions   194
– "direct-to-Earth" communication system   183
– Earth   182
– environmental requirements   192
– functional components   194
– gravity differences   183
– interface requirements   193
– localization technologies and systems *see* localization technologies and systems
– Lunokhod I and II   184
– Mars   182
– Mars Exploration Rovers (MERs)   185
– performance requirements   191
– planetary surface navigation *see* planetary surface navigation
– resource requirements   193
– rover control strategy   199
– sensor   194
– software   196
– Sojourner Microrover   184
– Sun-sensing techniques   183

Rover Operation Control Centre (ROCC) 334
Rover Sequencing and Visualization Program (RSVP) 335

s
Sample Analysis at Mars (SAM) 41
sample-fetching rover (SFR) 234
Satellite Control and Operation System 2000 (SCOS-2000) 334
Science Activity Planner (SAP) 335
Science Operations Working Group (SOWG) 327
self-collision avoidance 288
sensor
– absolute-localization 120
– active vision sensing strategies 117
– calibration 120
– dedicated-navigation-vision-sensors 118
– imaging sensor 116
– navigation-perception-stereo-vision 119
– passive optical vision sensors 116
– vision sensor 114
– visual-localization 119
sensors 194, 244
Sequence Generator (SEQGEN) 335
"Sequence Review Tool (SEQREVIEW)" 335
Service Module (SVM) 45
Service-Oriented Robotic Architecture (SORA) 363
Shackleton crater 267
signal-to-noise ratio (SNR) 116
Simultaneous Localization and Mapping (SLAM) 154, 215, 216
software control algorithms 266
soil analysis 164
Sojourner Microrover 184
Sojourner rover 34
Soviet Lunokhod 183
space robotics 157
space segment operability 324
Space System Model (SSM) 325
Spacecraft Language Interpreter and Collector (SLINC) 335
Spacecraft Sequence File (SSF) 335
speed sensors 195
Speeded Up Robust Features (SURF) 217
Stanford Research Institute Problem (STRIPS) 351
state-space planning 346
static model-based control methods 274
stereo correlation 146
stiffness control 277
Stirling cycle generator 71

Stirling process 83
Stored Chemical Energy Power Systems (SCEPS) 84
Sun 202
Sun-sensing techniques 183
super-resolution restored (SRR) images 133
supercritical fluids (SCF) 93

t
Technology Readiness Level (TRL) 69
tele-manipulation
– bilateral control 291
– performance parameters 291
– unilateral control 290
teleoperation 199
temporal planning 347
testing
– coefficient of thermal expansion (CTE) 299
– kinematic calibration 298
– purpose of 297
– strategy 295
thermal control system (TCS) 90
thermal protection system (TPS) 60
3D digital elevation maps (DEM) 45
3D visual odometry 208
three-layer architecture 339

u
unilateral control 290
USSR's Luna program 20

v
Validation and Verification (V&V)
– 3DROV 380
– dynamic and static methods 378
– low-fidelity simulators 380
– ontology-based system models 383
– realistic simulators tool 379
Venera landers 20
vision and image processing
– artificial intelligence and cybernetics vision 169
– ExoMars PanCam ground data processing chain 152
– ExoMars-control-chaincontrol 150
– ExoMars-onboard-vision-testing-and-verification 151
– ground vision ground processing see ground vision processing
– lunar-vision 142
– mapping-by-vision-sensors 112
– MER and MSL missions 109, 145
– onboard-requirements 110
– onboard-vision-processing 138

vision and image processing (*contd.*)
– pathfinder   144
– physical environment   113
– planetary saliency models   157
– science-autonomy   166
– sensor-fusion   166
– sensor   114
– Viking Lander camera system   143
vision sensors   7, 114, 194

visual odometry (VO)   133, 185, 188, 193, 197, 205, 214, 230
visual servoing   283

w
wheel odometry   205
whole-body motion control   306
WISDOM data   168
Wrist and Turret Actuators (WATERs)   263